T0139921

Lecture Notes in Physics

Volume 968

The Lecture Notes in Physics

The series Lecture Notes in Physics (LNP), founded in 1969, reports new developments in physics research and teaching-quickly and informally, but with a high quality and the explicit aim to summarize and communicate current knowledge in an accessible way. Books published in this series are conceived as bridging material between advanced graduate textbooks and the forefront of research and to serve three purposes:

- to be a compact and modern up-to-date source of reference on a well-defined topic
- to serve as an accessible introduction to the field to postgraduate students and nonspecialist researchers from related areas
- to be a source of advanced teaching material for specialized seminars, courses and schools

Both monographs and multi-author volumes will be considered for publication. Edited volumes should, however, consist of a very limited number of contributions only. Proceedings will not be considered for LNP.

Volumes published in LNP are disseminated both in print and in electronic formats, the electronic archive being available at springerlink.com. The series content is indexed, abstracted and referenced by many abstracting and information services, bibliographic networks, subscription agencies, library networks, and consortia.

Proposals should be sent to a member of the Editorial Board, or directly to the managing editor at Springer:

Dr Lisa Scalone
Springer Nature
Physics Editorial Department
Tiergartenstraße 17
69121 Heidelberg, Germany
lisa.scalone@springernature.com

More information about this series at http://www.springer.com/series/5304

Kristof T. Schütt • Stefan Chmiela •
O. Anatole von Lilienfeld •
Alexandre Tkatchenko • Koji Tsuda •
Klaus-Robert Müller
Editors

Machine Learning Meets Quantum Physics

Editors
Kristof T. Schütt
Machine Learning
Technical University of Berlin
Berlin, Germany

Stefan Chmiela
Machine Learning Group
Technical University of Berlin
Berlin, Germany

O. Anatole von Lilienfeld
Institute of Physical Chemistry
and MARVEL
University of Basel
Basel, Switzerland

Alexandre Tkatchenko
Department of Physics and Materials
Science
University of Luxembourg
Luxembourg, Luxembourg

Koji Tsuda
Graduate School of Frontier Sciences
University of Tokyo
Kashiwa, Japan

Klaus-Robert Müller
Computer Science
Technical University of Berlin
Berlin, Germany

ISSN 0075-8450 ISSN 1616-6361 (electronic)
Lecture Notes in Physics
ISBN 978-3-030-40244-0 ISBN 978-3-030-40245-7 (eBook)
https://doi.org/10.1007/978-3-030-40245-7

This Springer imprint is published by the registered company Springer Nature Switzerland AG.
The registered company address is: Gewerbestrasse 11, 6330 Cham, Switzerland

To Sandro!
7.3.1965–2.10.2018

Contents

Introduction

1

Kristof T. Schütt, Stefan Chmiela, O. Anatole von Lilienfeld, Alexandre Tkatchenko, Koji Tsuda, and Klaus-Robert Müller

Abstract

Rational design of molecules and materials with desired properties requires both the ability to calculate accurate microscopic properties, such as energies, forces, and electrostatic multipoles of specific configurations, and efficient sampling of potential energy surfaces to obtain corresponding macroscopic properties. The tools that provide this are accurate first-principles calculations rooted in quantum mechanics, and statistical mechanics, respectively. Both of these come with a high computational cost that prohibits calculations for large systems or sampling-intensive applications, like long-timescale molecular dynamics simulations, thus presenting a severe bottleneck for searching the vast chemical compound space. To overcome this challenge, there have been increased efforts to accelerate quantum calculations with machine learning (ML).

K. T. Schütt · S. Chmiela
Machine Learning, Technical University of Berlin, Berlin, Germany

O. A. von Lilienfeld
Institute of Physical Chemistry and MARVEL, University of Basel, Basel, Switzerland
e-mail: anatole.vonlilienfeld@unibas.ch

A. Tkatchenko
Department of Physics and Materials Science, University of Luxembourg, Luxembourg, Luxembourg

K. Tsuda
Graduate School of Frontier Sciences, University of Tokyo, Kashiwa, Japan
e-mail: tsuda@k.u-tokyo.ac.jp

K.-R. Müller (✉)
Computer Science, Technical University of Berlin, Berlin, Germany

MPII Saarbrücken, Germany and Korea University, Seoul, South Korea
e-mail: klaus-robert.mueller@tu-berlin.de

K. T. Schütt et al. (eds.), *Machine Learning Meets Quantum Physics*, Lecture Notes in Physics 968, https://doi.org/10.1007/978-3-030-40245-7_1

The success of ML methods, and artificial intelligence (AI) in general, rests on their ability to uncover and exploit regularities within data, without an explicit concept of the underlying principles that cause them (e.g., [1, 2]). This allows the description of phenomena that are not yet fully understood or computationally too expensive to calculate exactly. Recently, AI was able to beat the champion in Go, one of the most challenging classical games [3]. Traditional rule-based algorithms have failed to perform at an even remotely comparable level in the face of the enormous search space that needs to be traversed. The proposed AI method, on the other hand, was able to develop more efficient strategies in a reinforcement learning scheme. Similarly, ML and AI have been used to enable medical decision making (e.g., [4–7]) or contribute to analyzing data in the neurosciences (e.g., [8–11]). While a careful application of ML can aid in revealing insights like that, it is crucial to ascertain their robustness beyond any doubt. The field of explainable AI addresses this long standing challenge, by highlighting the reliability of problem-solving behaviors of learning machines to which standard performance evaluation metrics are oblivious (e.g., [12, 13]).

For the field of quantum chemistry, training nonlinear regression models, such as Gaussian processes or neural networks, on accurate reference calculations allows to shortcut quantum mechanics and directly predict chemical properties [14–16]. ML predictions of atomic forces can drive molecular dynamics simulations, while uncertainty estimates and active learning guide the acquisition of more reference calculations to further improve the previously trained models [17, 18]. Having immediate access to such accurate predictions of chemical properties allows for extensive, large-scale studies that would have been infeasible with conventional quantum chemistry simulations. Beyond that, chemical insights can be obtained by developing interpretable models and employing techniques from explainable AI [19]. This book has emerged from the 2016 IPAM Long Program *Understanding Many-Particle Systems with Machine Learning* as well as the NeurIPS Workshops 2017 and 2018 on *Machine Learning for Molecules and Materials*. It aims to provide researchers that are new to this quickly developing field both with the fundamental concepts and a cross-section through the various lines of research. While the focus lies on physics-based machine learning that models electronic and atomistic properties, the book also touches on related fields of chemo- and materials-informatics. In the following, we give an overview of the topics covered in the book. As the subject of this book is inherently interdisciplinary, its first part starts off with an introduction of the fundamental concepts and techniques. This includes a chapter on the basics of quantum and statistical mechanics for modeling molecules and materials as well as chapters on two of the most popular machine learning techniques in this field—kernel methods and neural networks.

Incorporating prior physical knowledge into the machine learning models is crucial to obtain accurate predictions given as few computationally expensive reference calculations as possible. Part II is concerned with various approaches to achieve this. In particular, it covers the design of n-body force-fields based on physical principles such as smoothness and invariances, modeling response properties as well as choosing the kernel complexity for optimal generalization.

While manual design of kernels and features may be informed by prior knowledge, some aspects need to be chosen using heuristics. In contrast, deep neural networks aim to *learn* a representation of the data. Part III presents deep learning approaches to model molecules and materials and shows how to extract insights from the learned representation.

The second half of the book focuses on applications, starting with ML-enhanced atomistic simulations in Part IV. The chapters cover the application of neural networks and kernel methods to drive molecular dynamics simulations, computing uncertainty estimates as well as dealing with long time-scales. Part V addresses the discovery of novel molecules and materials and the exploration of chemical compound space. This includes ML predictions across chemical compound space as well as approaches from chemo- and materials-informatics such as recommender systems and graph generative models.

We expect that combining approaches from all these directions of research—ranging from prediction of chemical properties over the sampling of potential energy surfaces to recommender systems and generative models for molecular graphs—will enable researchers to build powerful multi-step frameworks to facilitate targeted, rational design.

Acknowledgments All editors gratefully acknowledge support by the Institute of Pure and Applied Mathematics (IPAM) at the University of California Los Angeles during the long program on *Understanding Many-Particle Systems with Machine Learning*.

References

1. C.M. Bishop, *Pattern Recognition and Machine Learning* (Springer, Berlin, 2006)
2. Y. LeCun, Y. Bengio, G. Hinton, Nature **521**(7553), 436 (2015)
3. D. Silver, A. Huang, C.J. Maddison, A. Guez, L. Sifre, G. Van Den Driessche, J. Schrittwieser, I. Antonoglou, V. Panneershelvam, M. Lanctot et al., Nature **529**(7587), 484 (2016)
4. D. Capper, D.T. Jones, M. Sill, V. Hovestadt, D. Schrimpf, D. Sturm, C. Koelsche, F. Sahm, L. Chavez, D.E. Reuss et al., Nature **555**(7697), 469 (2018)
5. A. Meyer, D. Zverinski, B. Pfahringer, J. Kempfert, T. Kuehne, S.H. Sündermann, C. Stamm, T. Hofmann, V. Falk, C. Eickhoff, Lancet Respir. Med. **6**(12), 905 (2018)
6. P. Jurmeister, M. Bockmayr, P. Seegerer, T. Bockmayr, D. Treue, G. Montavon, C. Vollbrecht, A. Arnold, D. Teichmann, K. Bressem et al., Sci. Transl. Med. **11**(509), eaaw8513 (2019)
7. D. Ardila, A.P. Kiraly, S. Bharadwaj, B. Choi, J.J. Reicher, L. Peng, D. Tse, M. Etemadi, W. Ye, G. Corrado, et al., Nat. Med. **25**(6), 954 (2019)
8. J. Gemignani, E. Middell, R.L. Barbour, H.L. Graber, B. Blankertz, J. Neural Eng. **15**(4), 045001 (2018)
9. T. Nierhaus, C. Vidaurre, C. Sannelli, K.R. Müller, A. Villringer, J. Physiol. (2019). https://doi.org/10.1113/JP278118
10. J.D. Haynes, G. Rees, Nat. Rev. Neurol. **7**(7), 523 (2006)
11. K.N. Kay, T. Naselaris, R.J. Prenger, J.L. Gallant, Nature **452**(7185), 352 (2008)
12. S. Lapuschkin, S. Wäldchen, A. Binder, G. Montavon, W. Samek, K.R. Müller, Nat. Commun. **10**(1), 1096 (2019)
13. W. Samek, *Explainable AI: Interpreting, Explaining and Visualizing Deep Learning* (Springer, Berlin, 2019)
14. J. Behler, M. Parrinello, Phys. Rev. Lett. **98**(14), 146401 (2007)

15. A.P. Bartók, M.C. Payne, R. Kondor, G. Csányi, Phys. Rev. Lett. **104**(13), 136403 (2010)
16. M. Rupp, A. Tkatchenko, K.R. Müller, O.A. Von Lilienfeld, Phys. Rev. Lett. **108**(5), 058301 (2012)
17. Z. Li, J.R. Kermode, A. De Vita, Phys. Rev. Lett. **114**(9), 096405 (2015)
18. J. Behler, Int. J. Quantum Chem. **115**(16), 1032 (2015)
19. W. Samek, G. Montavon, A. Vedaldi, L.K. Hansen, K.R. Müller, *Explainable AI: Interpreting, Explaining and Visualizing Deep Learning*, vol. 11700 (Springer, Berlin, 2019)

Part I
Fundamentals

Preface

In recent years, applying machine learning techniques to model atomistic systems has led to promising advances towards accelerated sampling of potential energy surfaces and inverse chemical design. This quickly developing field is inherently interdisciplinary and brings together fundamental techniques from material modeling and quantum chemistry as well as machine learning.

The first part of the book aims to introduce important concepts from these domains. Chapter 2 [1] briefly covers material modeling, including the necessary fundamentals of statistical and quantum mechanics. In the following, Chaps. 3 [2] and 4 [3] introduce kernel methods and neural networks, respectively. Those are among the most common machine learning techniques in the field and applied in many chapters throughout the book.

Berlin, Germany	Kristof T. Schütt
Berlin, Germany	Stefan Chmiela
Basel, Switzerland	O. Anatole von Lilienfeld
Luxembourg, Luxembourg	Alexandre Tkatchenko
Kashiwa, Japan	Koji Tsuda
Berlin, Germany	Klaus-Robert Müller
September 2019	

References

1. J. Hermann, in *Machine Learning for Quantum Simulations of Molecules and Materials*, ed. by K.T. Schütt, S. Chmiela, A. von Lilienfeld, A. Tkatchenko, K. Tsuda, K.-R. Müller. Lecture Notes in Physics (Springer, Berlin, 2019)

2. W. Pronobis, K.-R. Müller, in *Machine Learning for Quantum Simulations of Molecules and Materials*, ed. by K.T. Schütt, S. Chmiela, A. von Lilienfeld, A. Tkatchenko, K. Tsuda, K.-R. Müller. Lecture Notes in Physics (Springer, Berlin, 2019)
3. G. Montavon, in *Machine Learning for Quantum Simulations of Molecules and Materials*, ed. by K.T. Schütt, S. Chmiela, A. von Lilienfeld, A. Tkatchenko, K. Tsuda, K.-R. Müller. Lecture Notes in Physics (Springer, Berlin, 2019)

2 Introduction to Material Modeling

Jan Hermann

Abstract

This introductory chapter presents material modeling, or computational materials science, as a discipline that attempts to computationally predict properties of materials and design materials with given properties, and discusses how machine learning can help in this task. At the center of material modeling is the relationship between the atomic structure and material properties. This relationship is introduced together with the quantum and statistical mechanics as the two main tools for material modeling from first principles, which should also guide the development of new machine learning approaches to computational materials science. Material modeling operates with an abundance of technical terms, and some of these are explained in a glossary at the end of the chapter.

2.1 Introduction

Material modeling, or computational materials science, refers to the problem of computational prediction of material properties, and also to the inverse problem of computational discovery of materials with given properties [1–4]. Traditionally, modeling of materials is based on the underlying fundamental physical laws—quantum and statistical mechanics—expressed in the form of mathematical equations, which provide an in principle exact framework for building the right models. This bottom-up approach is limited by the complexity of the equations, and by the difficulty in recognizing a priori which of the degrees of freedom in the

J. Hermann (✉)
TU Berlin, Machine Learning Group, Berlin, Germany

FU Berlin, Department of Mathematics and Computer Science, Berlin, Germany
e-mail: jan.hermann@fu-berlin.de

K. T. Schütt et al. (eds.), *Machine Learning Meets Quantum Physics*, Lecture Notes in Physics 968, https://doi.org/10.1007/978-3-030-40245-7_2

description of a system are relevant for the solution of the equations, and which can be omitted to reduce the complexity. Material modeling sometimes resorts to the opposite top-down approach as well, in which known materials and their properties are used to parametrize or simplify the models, but such techniques are often used in an ad-hoc fashion without any systematic theoretical guidelines. Here, machine learning as a discipline can fill this void and provide precisely such guidelines in the form of a solid mathematical framework for inferring unknown models from known data, extracting the relevant degrees of freedom along the way. Delivering on such a promise, this book presents a series of works that pave the way towards incorporating machine learning into material modeling. The challenge in such a program is to avoid creating an entirely orthogonal branch of research to the established techniques. Instead, the goal is to either incorporate machine learning into existing methods, thus enhancing them, or use the existing methods to bootstrap the machine learning approaches. For this reason, this introductory chapter gives a brief introduction into the traditional bottom-up approach of material modeling in the form of fundamental physical laws.

One possible way to characterize learning is that it involves inferring probability distributions from data that are assumed to be sampled from those distributions. In the common meaning of the word learning, the "data" may be observed instances of behavior and the probability distributions would be the "rules" according to which one behaves, both in terms of what behavior is possible (marginal distributions) and which behavior is appropriate for a given situation (conditional distributions). In some types of learning problems, with seemingly or evidently deterministic rules, the formulation in terms of probabilities may seem superfluous, but even in such cases it enables a rigorous handling of the fact that one always works with limited amount of data, and as such there can never be a certainty about the underlying inferred rules. In traditional machine learning tasks, e.g., image recognition, natural language processing, or optimal control, it is in general assumed that very little is known about the underlying distributions besides some general and often obvious properties such as locality or certain invariances. Either the data being modeled are too complex to derive anything meaningful about the distributions from theory, or, more commonly, there is no theory available.

When applying machine learning to molecules and materials, on the other hand, one has the whole toolbox of quantum and statistical mechanics to derive all sorts of general approximations or exact constraints to the distributions that underlie the data. The approximations can serve as a basis which a machine learning model improves, or as motivation for constructing specialized features and architectures. The constraints, such as symmetries or asymptotic behavior, can improve data efficiency and numerical stability of the learning algorithms, and in the case of unsupervised learning may even be a necessity to avoid learning unphysical solutions. In many applications, the methods of quantum and statistical mechanics do not enter the machine-learning techniques directly, but they are used to generate the data that are used to train the machine-learning models. In such cases, the question of feasibility and accuracy of the generated data becomes an integral part of the design of the machine learning approach.

This chapter aims to give a brief introduction to the problem of modeling of molecules and materials intended for a reader with only the most rudimentary knowledge of physics and chemistry. It attempts to provide a context for the technical parts of the book, and place the subsequent chapters in the broader map of materials science. The topic and scope of this chapter preclude any chances at being comprehensive, but it should give an explanation of some of the core principles and ideas, and point to topics of possible further study.

Section 2.2 presents the traditional problems encountered in material modeling in the framework of the relationship between material structure and material properties. Sections 2.3 and 2.4 present quantum mechanics and statistical mechanics—the two main disciplines of physics and theoretical chemistry that provide the theoretical framework for modeling molecules and materials. The chapter is concluded in Sect. 2.4 by a glossary of common terms that are frequently used in the subsequent technical chapters and where they are assumed to be understood.

2.2 Structure–Property Relationship

The ultimate goal of material modeling is to replace real-world lab experiments using materials and measuring instruments with (cheaper) computer simulations in the task of predicting the physical and chemical properties of a material [5]. This is the so-called *forward* problem in materials science. The corresponding *inverse* problem is to produce a (usually novel) material that has a set of desired properties, using computational techniques—this branch of materials science is called materials design or discovery [6]. The inverse problem is usually considered to be much harder than the forward problem, at least with existing techniques.

The reason for calling the material property prediction a forward problem is that, in principle, it has an exactly known unique solution. The mathematical solutions of the Schrödinger equation of quantum mechanics, applied to materials, can give answers to many questions about the properties of a material [7]. The rest of the questions can be, again in principle, answered by the tools of statistical mechanics [8, 9]. Unfortunately, the Schrödinger equation can be solved analytically only for the simplest "molecule"—a single hydrogen atom—and the numerical techniques that give sufficiently accurate solutions are computationally too expensive for many materials of interest. A large portion of the efforts in material modeling then deals with a careful analysis of these numerical models and their errors, with attempts to have these errors under control, and ultimately with devising more accurate, more general, and more computationally efficient models. The first technical part of the book, Representations, covers attempts to entirely avoid having to deal with the Schrödinger equation by learning the properties obtained from its solutions directly from data. Similarly, statistical mechanics can provide analytical answers only for the most rudimentary models, such as noninteracting harmonic oscillators, hard spheres, or two-dimensional lattices, and relies on statistical sampling from various types of thermodynamic distributions (*ensembles*) for more realistic systems. The biggest (unsolved) problem is then to generate representative samples from these distributions. The second part of the book, Atomistic Simulations, deals with applications of machine learning to this sampling problem. The inverse problem of materials science, as every inverse problem, is likewise solvable in principle, by enumerating all possible materials and solving the forward problem for all of them. But this is unfeasible in most cases due to the sheer combinatorial size of the problem space in question. The third part of the book, Discovery, covers the use of machine learning to accelerate materials design by reducing the effective search space.

2.2.1 Atomic Structure

An important part of the discussion of material properties is the question of unique specification of a material or a molecule. All materials are composed of atoms, and specifying a material is usually understood as specifying its *atomic structure*. Disregarding quantum mechanics for a moment, the state of a piece of material is fully given by listing the positions and velocities of all its atoms, and to which element each of them belongs—this is called a *microstate*. However, since one gram of a typical solid or liquid contains on the order of 10^{21}–10^{23} atoms, this is practically impossible and in fact not necessary. The vast majority of the degrees of freedom involved in such a specification are constantly changing in any given piece of material, and as such do not pertain to any kind of permanent structure.

Which of the degrees of freedom constitute the atomic structure depends largely on the type of material and sometimes on the material property being investigated in the structure–property relationship. Although the term "material" is usually reserved for solid objects, consider water vapor for a moment as an example of a simple material. Water vapor consists of individual water molecules, each of which consists of one oxygen (O) and two hydrogen (H) atoms and whose geometry can be specified by the two O–H distances and the angle between them. At any given moment, the distances and angles in all the molecules in a drop of water are narrowly distributed around some mean values. (The actual width of these distributions increases with temperature and is a subject matter of statistical mechanics.) The same distributions would be found if one repeatedly measured the geometry of a single molecule over a period of time. The individual molecules move fast along straight lines between relatively rare collisions, spinning as they do, so the relative positions of the molecules change rapidly. It then follows that the only structurally persistent motif is the mean geometry of a water molecule—two O–H distances and one H–O–H angle. In the case of water vapor, this is all that can be said about its atomic structure, and many properties of water vapor can in fact be determined from studying just a single molecule as a representative of all the molecules.

Going to liquid water, the geometry of the individual molecules is unchanged, and they still constantly vibrate, move around, and spin, but they are now *condensed* and essentially touching each other. Many interesting water properties do not follow from the properties of a single molecule like in vapor anymore, but rather are a result of the statistical characteristics of the relative positions and movement of the water molecules, which can be very complex. However, these characteristics do uniquely follow from the interactions between individual water molecules. In principle, there could be different "kinds" of water, for which the statistical characteristics and hence material properties would be different, but this is not the case, and neither is for vast majority of liquids. As a result, all it takes to specify liquid water as a material is the geometry of a water molecule.

Moving on to solid water—ice—the water molecules themselves are again unchanged, they are condensed similarly to liquid water, and they still vibrate around the most likely geometry, but now their relative positions and rotations are frozen. As in the case of the *intra*molecular degrees of freedom (O–H distances, H–O–H angles), the *inter*molecular degrees of freedom are not sharp values, but narrowly distributed around some mean values, nevertheless they are not entirely free as in the liquid or vapor. Fortunately, it is not necessary to specify all the relative positions of all the molecules in a piece of ice. Common ice is a *crystal*, which means that its water molecules are periodically arranged with a relatively short period. Thus one needs to specify only the positions of several water molecules and the periodic pattern to express the atomic structure of ice. Unlike in the case of liquid water, the water molecules in ice can be arranged in different structural arrangements, and as a result there are different types (*phases*) of ice, each with different material properties. (Almost all ice found naturally on Earth is of one form, called I_h.) In general, a crystal is any material with periodic atomic structure. This relative structural simplicity enables them to be specified and characterized precisely, because a microscopic region of a crystal at most several nanometers across uniquely determines the atomic structure of the whole macroscopic piece of a material. Many common solid materials are crystalline, including metals, rocks, or ceramics. Usually a macroscopic piece of a material is not a single crystal, but a large array of very small crystals that stick together in some way. This structure on top of the microscopic atomic structure of a single crystal can sometimes have an effect on some properties of a material, but often is inconsequential and not relevant for material modeling.

Other solid materials, including glasses, plastics, and virtually all biological matter, are not crystalline but amorphous—their atomic structure has no periodic order. Still, specifying the structure of such materials does not require specifying all the atomic positions. Consider a protein as an instance of complex biological matter. Proteins are very large molecules consisting of hundreds to hundreds of thousands atoms, with typical sizes in thousands of atoms, that play a central role in all life on Earth. All proteins are constructed as linear chains, where each chain link is one of 22 amino acids, which are small organic molecules. Like in a common chain, the overall three-dimensional shape of a protein is relatively easy to change, whereas the linear sequence of the links (amino acids) is fixed. Unlike in a common chain, the amino acids themselves have internal degrees of freedom, with some of them hard as the distances and angles in the water molecules, and some of them relatively soft. When a protein molecule is constructed in a cell from the amino acids, its three-dimensional shape is essentially undetermined, and all that can be said about its atomic structure in that moment is the linear sequence of the amino acids—this is called the *primary* structure. Most proteins in living organisms are not found having random shapes, but very specific *native* forms. As a result of the complex quantum-mechanical interactions within the protein and of the protein with its environment, the initially random shape *folds* into the native shape, referred to as the secondary and tertiary structure, at which point the protein can fulfill its biological function. This process does not only fix the overall three-dimensional shape of the protein, but also some of the previously free internal degrees of freedom within the amino acids. In the native state, the atomic structure of a protein can be characterized much more precisely. Finally, many proteins can be extracted from the cells into solutions and then crystallized, which freezes all the degrees of freedom, just like when ice is formed from water. At that point the atomic structure can be specified most precisely, with all atoms of the protein vibrating only slightly around their mean positions.

The previous examples demonstrate that what is considered to be the atomic structure of a material depends on a given context. The myriads of degrees of freedom in the atomic structure of a piece of a material can be under given circumstances (temperature, pressure, time scale) divided into hard ones and soft ones, with the caveat that there is no sharp boundary between the two. The soft degrees of freedom are described by distributions with large variance, sometimes multi-modal, and often with strong correlations between them. These distributions follow the laws of statistical mechanics, and in many cases are detrimental to the molecular and material properties, but are not considered to be part of the atomic structure per se. The hard degrees of freedom can be described by sharp probability distributions with small variance, and their mean values constitute the atomic structure of a molecule or a material. The chapters of the first part of the book deal with how best to represent the atomic structure so that machine learning approaches can then efficiently learn the relationship between the structure and the properties. These constitute both properties that follow directly from the structure and are independent of the soft degrees of freedom, as well as those that follow from the statistics of the soft degrees of freedom, which are in turn constrained by the hard degrees of freedom. The following section discusses the material properties in more detail.

2.2.2 Molecular and Material Properties

Having clarified what is meant by the structure of a molecule or a material, the two problems of materials science can be stated as being able to maximize the following two conditional probability distributions:

$$\begin{aligned} \text{forward:} &\qquad P(\text{property}|\text{structure}) \\ \text{inverse (materials design):} &\qquad P(\text{structure}|\text{property}) \end{aligned}$$

The atomic structure was introduced above as a specification of a material that uniquely determines its properties, which would suggest that the forward problem should be formulated in terms of a function, rather than a distribution. Nevertheless, it is useful to replace the deterministic function with a probability distribution for two reasons. First, the forward problem is strictly deterministic only under idealized conditions (e.g., isolated molecule, zero temperature) and when the problem is set up in such a way that the atomic structure is indeed specified fully. For instance, temperature effects or random defects in crystals smear many sharp properties into Gaussian distributions. Furthermore, a given material modeling problem may specify atomic structure only partially, such as when one is given only the structure of a molecule, of which a molecular crystal is composed, but not the complete structure—the periodic arrangement of the molecules. Second, when the forward problem is to be learned from data, rather than derived from first principles, the uncertainty, and hence probability, arises naturally from the finite amount of available data. In contrast to the forward problem, its inverse is inherently probabilistic, because molecular and material properties are never guaranteed to uniquely specify a molecule or a material.

Listing all the material properties that scientists have ever measured would cover many books on its own, so this introduction will keep to classifying them into general categories. The molecular and material properties can be tentatively divided into two categories: electronic properties on the one hand and thermodynamic properties on the other.

Atoms themselves are not indivisible, but consist of heavy nuclei (which can be almost always considered point-like in materials science) and thousand-times lighter electrons, which form electronic "clouds" around the nuclei. The behavior of the electrons deviates strongly from classical mechanics, and is best described by quantum mechanics. The closest (but still bad) classical analogy for the electrons would be perhaps that of a liquid floating around the nuclei, rather than particles moving around them. In any case, because most of the mass of an atom is concentrated in its nucleus, its position is likewise associated with the position of the nucleus. Under almost all circumstances, the atoms moving around can be described as nuclei moving around and being instantly followed by their respective clouds of electrons. The mathematical formulation of this idea is called the Born–Oppenheimer approximation, which underlies most of material modeling. The hard degrees of freedom that constitute the atomic structure are then associated with fixed electronic clouds, and the material properties that can be explained by quantum mechanics from these stationary electronic clouds are considered to be electronic properties. In contrast, the thermodynamic properties can be explained by statistical mechanics from the often complex statistics of the soft degrees of freedom in the motion of the atoms (atomic nuclei).

Electronic properties of molecules and materials are essentially properties of the collection of their electrons (electronic cloud) under the influence of the nuclei located at positions given by the atomic structure. As will be discussed in a bit more detail in Sect. 2.3, the electronic cloud can be in different discrete quantum states, and under temperatures found on Earth, most matter is found in the lowest-energy state, called the *ground* state. The higher-energy states, called *excited* states, can be reached, for instance, by exposing a material to visible light or UV radiation. In general, one can then distinguish between the ground-state electronic properties and the excited-state properties, the latter usually referring to the *transitions* between the ground and the excited states. Some of the ground-state properties include the atomization energy (energy released when a molecule is formed from its constituent atoms), the dipole moment and polarizability (shape and responsiveness of the electronic cloud), and the vibrational spectra (the hardness of the hard degrees of freedom). (See the Glossary for more details.) It is mostly the ground-state electronic properties that are targeted by the approaches in the first part of this book. The excited-state properties include UV and optical absorption spectra (which determine the color and general photosensitivity of a material), or the ability to conduct electrical current.

Thermodynamic properties stem from the motion of the atoms along the soft degrees of freedom in a material. Virtually all the soft degrees of freedom are not entirely free, but are associated with some barriers. For instance, in liquid water the molecules can rotate and move around, but not equally easily in all directions, depending on a particular configuration of the neighboring molecules at any given moment. The ability to overcome these barriers is directly related to the average speed of the atoms, which in turn is expressed by the temperature of the material. At absolute zero, 0 K, the atoms cease almost all motion, only slightly vibrating around their mean positions as dictated by quantum mechanics. At that point, all the degrees of freedom in the material can be considered hard, and virtually all material properties that can be observed and defined at absolute zero can be considered electronic properties. Calculations of material properties that neglect the motion of atoms essentially model the materials as if they were at zero temperature. As the temperature is increased, the atoms move faster and fluctuate farther from their mean positions, and the lowest barriers of the least hard degrees of freedom can be overcome, which turns them into soft degrees of freedom. The statistical characteristics of the resulting atomic motion, and material properties that follow from it, can be then obtained with the tools of statistical mechanics. (Statistical mechanics is valid and applicable at all temperatures, but only at higher temperatures does the atomic motion become appreciably complex.)

The study of thermodynamic properties—thermodynamics—can be divided to two main branches, equilibrium and nonequilibrium thermodynamics. The equilibrium refers to a state of a material in which the hard degrees of freedom and the probability distributions of the soft degrees of freedom remain unchanged over time. Traditionally, thermodynamic properties refer to those properties related to the motion of atoms that can be defined and studied at equilibrium. This includes the melting and boiling temperatures, the dependence of density on pressure and temperature, the ability to conduct and absorb heat, or the ability of liquids to dissolve solid materials. It is mostly properties of this kind that motivate the development of the approaches presented in the second part of this book. On the other hand, nonequilibrium thermodynamics deals with *processes* that occur in materials when they are out of equilibrium, that is, when the distributions of the degrees of freedom change over time. Examples of such processes are chemical reactions and their rates, transport of matter through membranes, or the already mentioned folding of a protein to its native state.

2.3 Quantum Mechanics

Quantum mechanics is the set of fundamental laws according to which all objects move [10]. In the limit of macroscopic objects, with which we interact in everyday life, the laws of quantum mechanics reduce to classical mechanics, which was first established by Newton, and for which all of us build good intuition when jumping from a tree, riding on a bus, or playing billiard. On the microscopic scale of atoms, however, the laws of quantum mechanics result in a behavior very different from that of the classical mechanics.

One of the fundamental differences between the two is that in quantum mechanics, an object can be simultaneously in multiple states. In classical mechanics, an object can be exclusively either at point $\mathbf{r}_1$ or at point $\mathbf{r}_2 \neq \mathbf{r}_2$, but not in both. In quantum mechanics, an object, say an electron, can be in any particular combination of the two states. Mathematically, this is conveniently expressed by considering the two position states, denoted $|\mathbf{r}_1\rangle$, $|\mathbf{r}_2\rangle$, to be basis vectors in a vector space (more precisely a Hilbert space), and allowing the object to be in a state formed as a linear combination of the two basis vectors,

$$|\psi\rangle := c_1|\mathbf{r}_1\rangle + c_2|\mathbf{r}_1\rangle$$

Note that $c_1|\mathbf{r}_1\rangle + c_2|\mathbf{r}_1\rangle \neq |c_1\mathbf{r}_1 + c_2\mathbf{r}_2\rangle$, that is, the object is not simply in a position obtained by adding the two position vectors together, but rather it is somewhat at $\mathbf{r}_1$ and somewhat at $\mathbf{r}_2$. Generalizing the two position vectors to the infinite number of positions in a three-dimensional space, the general state of an object can be expressed as

$$|\psi\rangle := \int \mathrm{d}\mathbf{r}\,\psi(\mathbf{r})\,|\mathbf{r}\rangle$$

where $\psi(\mathbf{r})$, called a *wave function* plays the role of the linear coefficients c_1, c_2 above. It is for this reason that electrons in molecules are better to be thought of as electronic clouds (corresponding to $\psi(\mathbf{r})$) rather than point-like particles.

Another fundamental law of quantum mechanics is that each in principle experimentally observable physical quantity ("observable" for short) is associated with a linear operator, $\hat{L}$, acting on the Hilbert space of the object states, and the values of the quantity that can be actually measured are given by the eigenvalues, λ_i, of the operator,

$$\hat{L}|\psi_i\rangle = \lambda_i|\psi_i\rangle$$

One of the most important operators is the energy operator, called Hamiltonian, which determines which energies of an object can be measured,

$$\hat{H}|\psi_i\rangle = E_i|\psi_i\rangle$$

This particular eigenvalue equation is called a Schrödinger equation, and since energy plays a key role in all physical phenomena, its solution (the eigenvalues and eigenstates) enables determination of many electronic properties of both the ground state ($|\psi_0\rangle$) and the excited states ($|\psi_n\rangle, n > 0$). The abstract eigenvalue equation can be transformed into a differential equation by expressing the state vectors using the wave function,

$$\hat{H}\psi_i(\mathbf{r}) = E_i\psi_i(\mathbf{r})$$

where $\hat{H}$ becomes a differential operator. For example, consider a hydrogen atom, consisting of a single electron described by position $\mathbf{r}$ and moving around a nucleus fixed at position $\mathbf{R}$. Such a system can be considered the simplest "molecule." In this case, the Schrödinger equation for the wave function of the electron (in atomic units) has the form

$$\left(-\frac{1}{2}\nabla^2 - \frac{1}{|\mathbf{r}-\mathbf{R}|}\right)\psi_i(\mathbf{r}) = E_i\psi_i(\mathbf{r})$$

where ∇^2 is the Laplace operator. In more complex molecules with multiple nuclei and electrons, the Hamiltonian contains more terms, and the wave function is a function of the coordinates of all the electrons, but the general structure of the problem remains the same.

The Schrödinger equation for the hydrogen atom can be solved analytically $\left(E_0 = -\frac{1}{2}, \psi_0(\mathbf{r}) \propto \mathrm{e}^{-|\mathbf{r}-\mathbf{R}|}\right)$, but this is not possible for more complex molecules. Direct methods for numerical solution of differential equations are inapplicable because of the dimensionality of the wave function ($3N$ dimensions for N electrons). Many methods of quantum chemistry, such as the Hartree–Fock or the coupled-cluster method, attempt to solve this issue by cleverly selecting a subspace in the full Hilbert space of the electrons spanned by a finite basis, and finding the best possible approximate eigenstates within that subspace. This turns the differential Schrödinger equation into an algebraic problem, which can be solved numerically at a feasible computational cost. Another class of methods, such as the density functional theory, also change the original Hamiltonian such that the eigenvalues (and in some cases also the eigenstates) are as close to the true eigenvalues as possible. In all the approximate quantum-mechanical methods, however, the computational cost grows asymptotically at least as $O(N^3)$ (or much faster for the more accurate methods), and their use becomes unfeasible from a certain system size. These include the density functional theory ($O(N^3)$), the Hartree–Fock (HF) method ($O(N^3)$), the Møller–Plesset perturbation theory to second order (MP2, $O(N^3)$), the coupled-cluster method with single, double, and perturbative triple excitations (CCSD(T), $O(N^7)$), and the full configuration interaction (FCI, $O(\exp(N))$).

Once the Schrödinger equation is solved, evaluating the electronic properties from the known eigenvalues and eigenstates of the Hamiltonian is often straightforward. In particular, many properties such as the dipole moment, polarizability, or atomic forces can be calculated as integrals of the corresponding operators over a given eigenstate, or as derivatives of such integrals, which can be transformed to integrals over derivatives of the operators by the Hellmann–Feynman theorem [11]. Besides that, the solution to the Schrödinger equation also provides a direct link between the quantum mechanics of the electrons and the statistical mechanics of the atoms. The electronic Hamiltonian has terms depending on the positions of the nuclei, $\mathbf{R}_i$, and as a result, the energies of the eigenstates likewise depend on the positions of the nuclei. The energy of a particular eigenstate as a function of the nuclear positions, $V(\mathbf{R}_1, \ldots)$, is called a *potential energy surface*, and in principle completely determines the dynamics of the motion of the atoms.

2.4 Statistical Mechanics

In the context of material modeling, statistical mechanics deals with the motion in the soft degrees of freedom of the atoms in a material, and its central idea is that most of the detail in that motion ($>10^{21}$ variables) can be safely omitted, while the physically relevant characteristics can be expressed in a smaller number of collective degrees of freedom. These collective variables can range from microscopic, in the form of a coarse-grained description of an atomistic modeling, to macroscopic, such as temperature or pressure. In all cases, the remaining degrees of freedom beyond the collective ones are treated in a statistical fashion, rather than explicitly.

The fundamental concept in statistical mechanics is that of a microstate, $\mathbf{s}$, which comprises the positions, $\mathbf{R}_i$, and velocities $\mathbf{v}_i$, of all the atoms in a material, $\mathbf{s} \equiv (\mathbf{R}_1, \mathbf{v}_1, \mathbf{R}_2, \ldots) \equiv (\mathbf{v}, \mathbf{R})$. The total energy, H, of a given microstate consists of the kinetic part, arising from the velocities, and the potential part, which is determined by the potential energy surface,

$$E(\mathbf{s}) = \sum_i \frac{1}{2} m_i v_i^2 + V(\mathbf{R}_1, \ldots)$$

One of the central results of statistical mechanics is that given that a material is kept at a constant temperature T (the so-called canonical ensemble), the probability density of finding it at any particular microstate is proportional to the so-called Boltzmann factor (in atomic units),

$$P(\mathbf{s}) \propto \mathrm{e}^{-\frac{E(\mathbf{s})}{T}} \quad \Rightarrow \quad P(\mathbf{R}) \propto \mathrm{e}^{-\frac{V(\mathbf{R})}{T}}$$

The latter proportionality follows from the fact that the kinetic and potential parts of the total energy are independent. Close to absolute zero temperature, the Boltzmann factor is very small (in relative sense) for all but the lowest-energy microstates, which correspond to small atomic velocities and atomic positions close to the minimum of the potential energy surface. This coincides with the picture of all the degrees of freedom in the atomic motion being hard close to absolute zero. As the temperature rises, the microstates with higher energy become more likely, corresponding to higher velocities and atomic positions further from the energy minimum. To see how this simple principle can be used to calculate thermodynamic properties of materials, assume that one can enumerate all the possible microstates, calculate the sum of all the Boltzmann factors—the *partition function* [12]—and thus normalize the probabilities above,

$$Z(T) = \int \mathrm{d}\mathbf{s}\,\mathrm{e}^{\frac{-E(\mathbf{s})}{T}}, \qquad P(\mathbf{s}) = \frac{1}{Z(T)}\mathrm{e}^{-\frac{E(\mathbf{s})}{T}}$$

The mean total energy, for instance, can be then calculated directly from the partition function,

$$\langle E\rangle = \int \mathrm{d}\mathbf{s}\, P(\mathbf{s})E(\mathbf{s}) = T^2\frac{\partial \ln Z}{\partial T}$$

A quantity closely related to the partition function is the *free energy*,

$$F(T) = T \ln Z(T)$$

Whereas the total partition function of two combined systems is a product of their respective partition functions, the logarithm in the definition of the free energy makes it an additive quantity. The multiplication by the temperature allows the free energy to have a physical interpretation—the change in the free energy between two states of a system is the maximum amount of work that can be extracted from a process (is available—free) that takes the system from one state to the other.

One of the reasons that makes the free energy (and partition function) a powerful tool is that it can be calculated not only for the set of all possible microstates, but also for physically meaningful subsets. For instance, consider the melting temperature of a solid as an example. To use the free energy, one can characterize the melting temperature as the point of inversion of the probabilities of the atoms of a material appearing solid on the one hand or liquid on the other (inversion of free energies of a solid and of a liquid),

$$p_\text{solid}(T) \propto \int_\text{solid} \mathrm{d}\mathbf{s}\,\mathrm{e}^{-\frac{E(\mathbf{s})}{T}} = Z_\text{solid}(T) = \mathrm{e}^{-\frac{F_\text{solid}(T)}{T}}, \qquad p_\text{liquid}(T) \propto \mathrm{e}^{-\frac{F_\text{liquid}(T)}{T}}$$

where the integrals run over all microstates that correspond to the solid or liquid forms of matter. The melting temperature can then be calculated as the temperature at which the free energies of the solid and liquid forms of a given material are equal (the probabilities of finding the two forms are identical). In the case of melting, the average potential energy of the solid microstates is lower (Boltzmann factors larger) than that of the liquid microstates, but there is many more liquid microstates than solid microstates, so the free energies (total probabilities) end up being equal.

The computational difficulty in statistical mechanics lies in the evaluation of the integrals over microstates such as those above. Even if one modeled only several hundreds atoms of a material at a time, the number of degrees of freedom involved in a microstate prohibits analytical evaluation of the integrals, and even direct numerical integration is unfeasible. Fortunately, Monte Carlo techniques present a general approach to evaluate such high-dimensional integrals by replacing them with sums over representative samples of the microstates. The task is then to generate statistically significant and diverse microstates, that is, microstates with large Boltzmann factors that completely span the physically relevant microstate subspaces. This can be achieved by many different techniques, the most common one being *molecular dynamics*. In this approach, the atoms are let to move according to the laws of classical mechanics along trajectories influenced by the potential energy surface, and the microstate samples are obtained by taking periodic snapshots of this dynamical system. This approach is justified by the so-called *ergodic principle*, which states that integral averages over the space of microstates (also called the *phase space*) are equal to time averages over sufficiently long times when a system is let to evolve under appropriately chosen dynamical conditions. This technique is used in most chapters in the second part of the book.

Glossary

Atomic units The standard units of measurement (SI units), including the meter, kilogram, or joule, are convenient for macroscopic settings, but result in very small values of quantities in the microscopic world. The atomic units have been designed to alleviate this inconvenience, and also to simplify physical equations by making the numerical values of common physical constants equal to one (in atomic units). For instance, the atomic unit of energy is called Hartree, and the ground-state electronic energy of a hydrogen atom is $-\frac{1}{2}$ Hartree. In SI units, this would be equal to approximately $-2.18 \cdot 10^{-18}$ J.

Boltzmann distribution A piece of material in equilibrium at temperature T can be found in a microstate $\mathbf{s}$ (positions and velocities of all atoms) with a probability that is proportional to $e^{-H(\mathbf{s})/k_B T}$. This probability distribution of the microstates is called a Boltzmann distribution. Since the energy of a microstate consists of two independent parts—the kinetic and potential energy—the distribution of the positions of the atoms also follows a Boltzmann distribution. The statistical ensemble of microstates following the Boltzmann distribution is called a canonical ensemble.

Chemical bond All chemistry is a consequence of the motion of electrons in molecules, which is in general complicated. Comparing the electronic motion across different molecules reveals common patterns, and chemical bonding is one of the most widely recognized of such patterns. For instance, when two carbon atoms get close to each other, between one to three pairs of electrons tend to concentrate between the two atoms, depending on other atoms in the neighborhood. This in turn attracts the two atoms together. This effect is an example of a chemical bond and is what holds the atoms in diamond together.

Computational cost Usually called computational complexity in computer science, the computational cost of approximate methods for solving the fundamental physical equation of quantum mechanics and statistical mechanics involved in material modeling is one of their three key properties besides accuracy and universality. The main determining factor of a given factor is usually the size of the system being modeled, that is, the number of atoms.

Configuration vs. conformation The degrees of freedom in the positions of atoms in a material can be under given circumstances, such as temperature, divided into hard and soft. The hard degrees are essentially fixed for the purpose of a given modeling task and determine the *configuration*. The soft degrees of freedom can change through the simulation and their particular arrangement is called a *conformation*. For instance, the sequence of amino acids in a protein is a configuration, whereas any particular three-dimensional shape of the amino acid chain is a conformation.

Crystal The atomic structure of many solids is characterized by a small pattern of atoms (usually units to hundreds) periodically repeated throughout the three-dimensional space. Such a material is called a crystal. Most crystalline materials do not consist of single large crystals, but of many small crystals randomly stitched together, or forming a powder. Although crystals are usually modeled as being perfectly periodic, real-world crystals have various defects that make the perfect crystal only an approximation.

Density functional theory (DFT) Tracking all the interactions and correlations in the motion of electrons in molecules and materials becomes quickly unfeasible as the system size grows. DFT attempts to alleviate this problem by reformulating the electronic problem such that the electrons do not interact explicitly one with each other, but rather only via the total density of electrons, which makes the problem mathematically tractable. DFT is in principle an exact theory, but its practical realizations (the different functionals of the electron density) achieve only approximate description of the electronic motion.

Dipole moment When an electrical charge is distributed continuously in space, such as in the electronic cloud of molecules, its dipole moment is simply the mathematical first moment of the density of the charge, $\mathbf{p} = \int \mathrm{d}\mathbf{r}\,\mathbf{r}n(\mathbf{r})$. Its importance lies in the fact that when two molecules are sufficiently far apart, their electric interaction can be Taylor expanded around the infinite separation, and the leading term of this expansion depends on the dipole moments of the molecules.

Electronic property Material property that can be explained by the electronic structure of a material without regard for the statistics of the motion of the atoms.

Electronic structure The electronic structure of a molecule or a material refers to the particular arrangement of electrons in it and their collective properties. The electronic structure can be obtained by solving the Schrödinger equation of quantum mechanics, and used to predict and explain electronic properties.

Excitation energy The electrons in a molecule or a crystal can be collectively in different states with different electronic properties. Most matter on Earth is in the lowest-energy state, called the ground state, but electrons can be *excited* to higher-energy states using light or chemical reactions, which supply the necessary excitation energy. The excited states usually do not persist for long and fall back to the ground state, with the excitation energy being released back in various forms.

First principles (ab initio) Approximate methods of material modeling that are considered to be based on first principles can be straightforwardly derived from fundamental physical laws without introducing much room for tuning. They stand in opposition to empirical approaches which use flexible models that can be optimized to reproduce available data. The first principles and empirical methods are not two binary categories, but opposite extremes on a spectrum.

Force field The electronic structure methods that are able to calculate the true electronic energy for a given position of nuclei are usually too costly for them to be used for molecular dynamics simulations, which model often large systems and where the energy must be evaluated many times. Fortunately, the absolute value of the electronic energy is irrelevant for the molecular dynamics, only the forces exerted on the atoms matter. Force fields are usually relatively simple sets of functions that take the positions of atoms as an input, and map them to the forces acting on the atoms. Force fields are usually highly empirical and their parametrization from data is a tedious task.

Free energy Free energy is always associated with some subset of microstates and is a direct measure of a probability, p, to find a system in that subset, $F = -T \ln p$. The subsets are usually either coarse-grained degrees of freedom (e.g., all microstates of a protein with a given overall three-dimensional shape fixed, regardless of the internal degrees of freedom of individual amino acids) or macroscopic states (e.g., all microstates with a given total volume of the system). Only differences between the free energies of different microstate subsets are physically relevant.

HOMO–LUMO gap The difference between the energies of the highest-occupied molecular orbital (HOMO) and the lowest-unoccupied molecular orbital (LUMO). This quantity can only be defined within approximate models of the electrons in molecules and is not physically observable, but it is an approximation of the lowest excitation energy.

Hamiltonian The Hamiltonian is a physicist's way of uniquely specifying a given physical system, by relating the energy of a system to its internal degrees of freedom. Given a Hamiltonian, the behavior of a system can be in principle calculated using the laws of quantum mechanics, or approximately by classical mechanics. In quantum mechanics, the Hamiltonian is mathematically expressed as an operator acting on the Hilbert space of potential states of the system. In classical mechanics, the Hamiltonian is just a function mapping from the internal degrees to energy.

Intermolecular interactions Molecules are aggregations of atoms that stick together via strong *intramolecular* interactions that can be broken apart only in chemical reactions (a single water molecule). *Intramolecular* interactions are relatively weaker forces between different molecules that determine the relative motion of molecules around each other (molecules in liquid water). Both intra- and intermolecular interactions are the end result of the single underlying Coulomb interaction between electrons and nuclei in molecules.

Many-body interactions Effective models of systems composed of multiple interacting bodies (electrons, atoms, molecules) often describe collective property or behavior as resulting from a simple aggregate effect of the property or behavior of individual bodies and of pairs of bodies (pairwise interactions). In many cases, such an effective description captures a large part of the collective behavior. Collective behavior that cannot be expressed in terms of individual bodies and pairwise interactions is said to results from many-body interactions.

Materials design One of two main branches of materials science that deals with discovery of novel materials with desired material properties. Sometimes also referred to as the inverse problem of material modeling. Computational materials design usually attempts to predict the atomic structure of the material with the desired properties, which can then be in principle prepared by synthetic experimental techniques.

Material modeling One of two main branches of materials science that deals with prediction of properties of a given material. The general approach is to approximate the real materials with simplified model systems (with fewer degrees of freedom or simpler interactions), whose properties can be calculated using the laws of quantum and statistical mechanics.

Metastable state The electrons in materials can be in different quantum states. Likewise the atoms in materials can be in different "states," which is a shorthand term for subsets of microstates that share some relevant physical feature. In both cases, if the system is allowed to interact with an environment that can "disturb" it, it can transition between the different states at any given moment with some probabilities. Some of the states are stable, which means that the probability of transitioning to any other state is low enough that it most likely does not happen on the relevant time scale. Unstable states are so short-lived, that the system never exhibits behavior that could be associated with any particular unstable state. Metastable states are between stable and unstable states in the sense that the system in a metastable state most likely transitions to other states during the relevant time scale, but it stays long enough in it that it exhibits behavior characteristic of that state.

Molecular dynamics The atoms of a material constantly transition from one microstate to another, and the statistical distribution of the microstates determines many material properties. Molecular dynamics generates samples from this distribution by evolving the positions of atoms along classical trajectories. The particular trajectories (sequences of microstates) are inconsequential, but the overall generated statistics can be used to calculate various thermodynamic properties.

Molecular geometry Specification of the charges and positions of atomic nuclei in a molecule. Specifying the charges is equivalent to specifying the chemical identities of the atoms. The fixed nuclei are surrounded by the electronic cloud which determines the electronic properties of a molecule. The atoms of a molecule (the atomic nuclei) are in constant motion and a fixed molecular geometry corresponds to molecule frozen in time or a molecule at absolute zero temperature, when the atomic motion is greatly reduced.

Molecular symmetry The geometry of many common molecules is symmetrical with respect to rotations, inversions, and reflections around a point. As a result of this symmetry, any observable function of the molecular geometry has to by invariant or equivariant with respect to these symmetry operations.

Observable Physical laws operate with quantities. Some of these quantities can be in principle measured, and those are called physical observables, regardless of whether it is feasible to actually perform the measurement. Other quantities are only auxiliary intermediates used to formulate the physical laws, but cannot be measured directly. Examples of observables are distance, mass, square of a wave function, or an energy difference. Examples or non-observables are a wave function or absolute energy.

Periodic boundary conditions One gram of a typical material contains on the order of 10^{21}–10^{23} atoms. To make their modeling tractable, a common approximation is to consider a sufficiently large box containing the atoms, which is then periodically repeated throughout the space. How large the box should be depends on the material and property in question. The errors caused by using a sufficiently large box are called finite size effects. The atomic structure of crystals is in fact periodic, so in the case of crystals periodic boundary conditions are not really an approximation.

Potential energy surface The dependence of the energy of a molecule or a material on the positions of the atoms. The potential energy for the atoms is a result of the electronic motion and can be calculated by solving the Schrödinger equation of quantum mechanics. Each electronic state (ground state and excited states) has its own potential energy surface, which can cross. Such effects are important when studying the dynamics of excited states and electronic mechanisms of chemical reactions.

Quantum chemistry Chemistry is the study of chemical reactions, and much of chemical knowledge preceded the discovery of quantum physics. Every since that discovery, quantum chemistry attempts to explain chemical properties of molecules from first principles using the tools of quantum mechanics. Quantum chemistry relies heavily on numerical calculations and the computational power of modern computers.

Schrödinger equation The central eigenvalue equation of quantum mechanics that, given a specification of a system in the form of a Hamiltonian operator, determines the possible quantum states in which the system can be found and the energies of those states. Depending on the basis in which the abstract operator equation is expressed, the Schrödinger equation can be either a differential equation or an algebraic matrix equation. Except for the simplest quantum-mechanical systems, the Schrödinger equation cannot be solved exactly, necessitating various approximations and numerical techniques.

References

1. J.G. Lee, *Computational Materials Science: An Introduction*, 2nd edn. (CRC Press, Boca Raton, 2017)
2. R. LeSar, *Introduction to Computational Materials Science: Fundamentals to Applications* (Cambridge University Press, Cambridge, 2013)
3. K. Ohno, K. Esfarjani, Y. Kawazoe, *Computational Materials Science: From Ab Initio to Monte Carlo Methods*, 2nd edn. (Springer, Berlin, 2018)
4. A.R. Leach, *Molecular Modelling: Principles and Applications*, 2nd edn. (Pearson Education, Edinburgh, 2001)
5. S. Yip, A. Wanda (eds.), *Handbook of Materials Modeling* (Springer, Berlin, 2020)
6. A.R. Oganov, G. Saleh, A.G. Kvashnin (eds.), *Computational Materials Discovery* (Royal Society of Chemistry, Cambridge, 2018)
7. L. Piela, *Ideas of Quantum Chemistry*, 2nd edn. (Elsevier, Amsterdam, 2014)
8. R.K. Pathria, P.D. Beale, *Statistical Mechanics*, 3rd edn. (Elsevier, Amsterdam, 2011)
9. L.D. Landau, E.M. Lifschitz, *Statistical Physics*, vol. 5, 3rd edn. (Pergamon Press, Oxford, 1980)
10. J.J. Sakurai, J. Napolitano, *Modern Quantum Mechanics*, 2nd edn. (Pearson Education, San Francisco, 2011)
11. R.P. Feynman, Phys. Rev. **56**(4), 340 (1939). https://doi.org/10.1103/PhysRev.56.340
12. D. Yoshioka, in *Statistical Physics* (Springer, Berlin, 2007), pp. 35–44

Kernel Methods for Quantum Chemistry

3

Wiktor Pronobis and Klaus-Robert Müller

Abstract

Kernel ridge regression (KRR) is one of the most popular methods of non-linear regression analysis in quantum chemistry. One of the main ingredients of KRR is the representation of the underlying physical system which mainly determines the performance of predicting quantum-mechanical properties based on KRR. Several such representations have been developed for both, solids and molecules; all of them with different advantages and limitations. These descriptors correspond to a similarity measure between two chemical compounds which is represented by the kernel. As recent approaches define the kernel *directly* from the underlying physical system, it is important to understand the properties of kernels and how these kernel properties can be used to improve the performance of machine learning models for quantum chemistry. After reviewing key representations of molecules, we provide an intuition on how the choice of the kernel affects the model. This is followed by a more practical guide of two complementary kernel methods, one for supervised and one for unsupervised learning, respectively. Finally, we present a way to gain an understanding about the model *complexity* by estimating the effective dimensionality induced by the data, the representation, and the kernel.

W. Pronobis
Technische Universität Berlin, Berlin, Germany

K.-R. Müller (✉)
Technische Universität Berlin, Berlin, Germany

Max Planck Institute for Informatics, Saarbrücken, Germany

Department of Brain and Cognitive Engineering, Korea University, Seoul, South Korea
e-mail: klaus-robert.mueller@tu-berlin.de

K. T. Schütt et al. (eds.), *Machine Learning Meets Quantum Physics*, Lecture Notes in Physics 968, https://doi.org/10.1007/978-3-030-40245-7_3

3.1 Introduction

Kernel-based learning methods [1–4] allow an efficient convex solution of highly non-linear optimization problems often encountered in quantum chemistry. A common task for the practitioner is to find a (kernel) representation of the problem at hand which encodes the distribution of the data in a complete, unique, and efficient way [5], favorably taking into account the inherent symmetries of the system such as rotational, translational, and atomic indexing invariance. As typical settings for a chemist or physicist include a low number of data points paired with a highly non-linear learning problem, kernel-based formulations are considered as suitable and powerful methods of choice. In view of these considerations, it is important to understand the kernel properties relevant for an efficient solution in a possibly much higher-dimensional, sometimes even unknown feature space induced by the kernel [6]. This is especially true, as it is non-trivial how a kernelized formulation can circumvent the so-called *curse of dimensionality* [7–9]. We provide insights on how the choice of the kernel helps solving these problems by introducing a function class of limited complexity from which the final model is chosen.

This chapter is structured as follows. Section 3.2 introduces a set of molecular descriptors and shows how they incorporate prior physical knowledge. This is followed by a theoretical background on kernels in Sect. 3.3, where we discuss ways to tailor the kernel to the problem at hand. Section 3.4 provides two examples of kernel methods often used in the quantum chemistry domain, namely the supervised KRR and the unsupervised kernel principal component analysis. Finally, in Sect. 3.5 we present an estimation procedure for the effective dimensionality of the underlying learning problem using kernels. The conclusion Sect. 14.7 completes the chapter.

3.2 Representations of Physical Systems

A crucial ingredient of kernel-based methods is the representation of the physical system at hand. Such representations have been developed for a variety of physical systems [10–13]. Note that neural networks in contrast to kernel methods learn their representation from data [14–16]. This representation, together with the kernel and the data, defines the function class from which the model is chosen, e.g., to predict a given quantum-mechanical property. In this section, we discuss three exemplary molecular descriptors and show how chemical knowledge can be included to improve the predictive performance of KRR.

A physical system will be defined by a set of 4-dimensional points $\{(Z_i, \boldsymbol{r}_i)\}_{i=1}^{N}$, where Z_i is the atomic number and $\boldsymbol{r}_i$ is the position of the atom i in three-dimensional space, respectively. While the system size N is well defined for molecules (by the total number of atoms of the molecule), one workaround for solids is to use atomic environment descriptors together with a cutoff distance to limit the number of neighboring atoms used to compute the atomic representation.

Alternatively, any molecular descriptor can also be combined with a modified distance metric to account for the periodic boundary conditions [17]. A raw encoding of the physical system by the atomic positions is unsuited for use in combination with machine learning methods as it neglects invariance with respect to basic symmetry operations. Instead, a representation is defined

$$R : \{(Z_i, \boldsymbol{x}_i)\}_{i=1}^{N} \to \mathbb{R}^{N_F} \tag{3.1}$$

with the number of features N_F. Such a mapping should encode the underlying chemical system in a complete, unique, and efficient way, including as much problem symmetries as possible. One way to incorporate translational and rotational invariance is to use pairwise atomic distances to construct the representation R. For molecules, a pioneering work which utilizes this observation is the Coulomb matrix (CM) [18] which is defined as

$$C_{ij} = \begin{cases} 0.5Z_i^{2.4}, & i = j \\ \frac{Z_i Z_j}{\|\boldsymbol{r}_i - \boldsymbol{r}_j\|}, & i \neq j \end{cases} \tag{3.2}$$

Being composed based on inverse pairwise distances, the off-diagonal elements of the CM account well for Coulomb interaction terms of the atomization energy. The diagonal elements of the CM correspond to a polynomial fit of atomic energies to nuclear charge [18]. From the set of all pairwise distances, a given molecule can be uniquely reconstructed, which is not the case for the following representations of this section. For equilibrium molecules, a variant of the CM has been proposed which sorts the row (or equivalently column) norms, and which better suits the feature comparison needed for applying kernel methods [19]. The CM is a *global* descriptor in the sense that it lacks a direct encoding of local atomic environment features. Due to its simplicity and predictive power, the CM provides the basis for various following molecular descriptors. Being composed of two-body terms, the three-body interactions of a given molecule are implicitly learned by the intrinsic feature mapping of the kernel (see Sect. 3.3). Although sorting of the rows solves some of its problems, one possible flaw of the CM is the comparison of different kinds of atom combinations within the distance metric which brings us to the next descriptor.

The bag-of-bonds (BOB) molecular representation is a development of the CM which rearranges the elements of the CM into bags defined by a given bond type [20]. Within each feature group, the elements of BOB are sorted, thereby ensuring atomic permutation invariance. Due to this grouping, chemically more similar elements are compared with each other as compared to the CM. In addition, three-body interactions in molecules can possibly be better implicitly learned by the kernel. Similarly to the bag-of-words descriptor used in natural language processing and information retrieval applications, BOB encodes the frequencies of bonds present in a given molecule. As such, the BOB descriptor is inspired by interatomic potentials, which model a quantum-mechanical property as a sum over such potentials. In fact, a Taylor series expansion in combination with KRR yields a low-order approximation of the BOB model by a sum over bonds and pairwise potentials [20]. This important finding indicates that the BOB model is better able to *learn* optimal pairwise potentials as compared to the CM, which is beneficial for some extensive properties like the atomization energy and the polarizability, respectively [21]. For BOB, the Laplace kernel performs better than the Gaussian kernel, indicating that the Laplace kernel is better able to utilize non-local information in chemical compound space [20].

Building on the promising feature rearrangement of BOB, a set of pairwise and three-body descriptors have been proposed to tackle some of the sorting problems encountered in the CM and BOB [22]. The invariant pairwise interaction descriptors F_{2B} are composed of the 2-body terms

$$F_{2B}(Z_1, \boldsymbol{r}_1, Z_2, \boldsymbol{r}_2) := \{\|\boldsymbol{r}_1 - \boldsymbol{r}_2\|^{-m}\}_{m=1,\ldots,M} \tag{3.3}$$

where the final set of F_{2B,Z_1,Z_2} descriptors are defined by the sum over all pairs of atoms with the given atomic numbers Z_1, Z_2, respectively. The grouping according to pairs of atom types is extended for the three-body descriptors F_{3B} which are composed of the explicit three-body terms

$$F_{3B}(Z_1, \boldsymbol{r}_1, Z_2, \boldsymbol{r}_2, Z_3, \boldsymbol{r}_3) := \{\|\boldsymbol{r}_{12}\|^{-m_1}\|\boldsymbol{r}_{13}\|^{-m_2}\|\boldsymbol{r}_{23}\|^{-m_3}\}_{m_1,m_2,m_3=1,\ldots,P}. \tag{3.4}$$

where $\|\boldsymbol{r}_{ij}\| := \|\boldsymbol{r}_i - \boldsymbol{r}_j\|$ for $i, j = 1, 2, 3$. According to the pairwise terms, the F_{3B,Z_1,Z_2,Z_3} descriptors are defined by the sum over all triples of atoms with the given ordered atomic numbers Z_1, Z_2, Z_3, respectively. A local variant of the F_{3B,Z_1,Z_2,Z_3} descriptors is composed by summing over all triples which are formed by two sets of bonded atoms which have a common atom. A diverse set of possible two- and three-body molecular interaction terms are thereby explicitly encoded into the $F_{2B} + F_{3B}$ representation. For the concatenated features $F_{2B} + F_{3B}$, a Gaussian kernel is better suited than a Laplace kernel [22]. In addition to leveraging the sorting problems of CM and BOB, one possible advantage of $F_{2B} + F_{3B}$ can be its fixed descriptor size, making them readily applicable in combination with artificial neural network architectures. A feature importance analysis has been performed for the $F_{2B} + F_{3B}$ descriptors [22] to gain insights into the two- and three-body interactions in equilibrium molecules. Figure 3.1 schematically shows the discussed representations of this section for the example of a water molecule.

The features $F_{2B} + F_{3B}$ explicitly encode local chemical information into the representation. In the next section, we will see how to incorporate prior knowledge *directly* into the kernel.

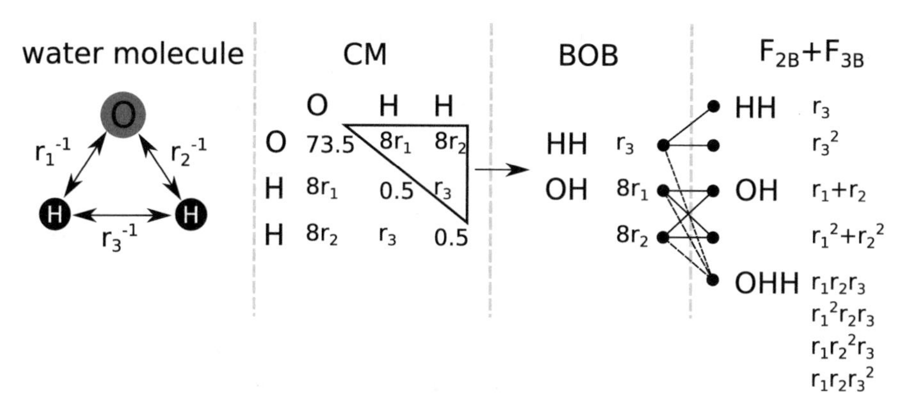

Fig. 3.1 Molecular representations of the water molecule (left) defined by a set of three pairwise distances. From the Coulomb matrix (CM), the off-diagonal elements are reordered by the bag-of-bonds (BOB) descriptor. These two-body terms are then combined to atomic index invariant two-body and three-body features F_{2B} and F_{3B}, respectively

3.3 Implicit Feature Mapping: The Kernel Trick

The second important ingredient of KRR (the first being the representation) is the kernel. But what is a kernel in general and how can it be useful? With a kernel, the data can be nonlinearly mapped onto a feature space, where the learning may become easier and where optimal generalization can be guaranteed. A key concept here is that this mapping can be done *implicitly* by the choice of the kernel. This implicit feature mapping to a possibly much higher-dimensional space is very flexible. More intuitively, the kernel encodes a real valued similarity measure between two chemical compounds. This similarity measure is primarily encoded by the representation of the physical system which is then used in combination with standard non-linear kernel functions like the Gaussian or Laplace kernel. Alternatively, the similarity measure can be encoded *directly* into the kernel, leading to a variety of kernels in the chemistry domain, e.g., for predicting the atomization energy with KRR, local kernels have been developed which compare atomic environments across molecules with each other.

One way to better understand the role of the kernel is to apply existing learning methods in a projected space $\phi : \mathbb{R}^{n_i} \rightarrow \mathbb{R}^{n_o}$ with the input and feature dimension n_i and n_o, respectively. Specifically, it is required that a given algorithm (together with predictions based on this algorithm) works solely on scalar products of type $\boldsymbol{x}^\top \boldsymbol{y}$ which can then be translated into scalar products in feature space $\phi(\boldsymbol{x})^\top \phi(\boldsymbol{y})$. Then, it turns out that such scalar products in feature space can be done *implicitly*, replacing them with an evaluation of the kernel function $k(\boldsymbol{x}, \boldsymbol{y}) := \phi(\boldsymbol{x})^\top \phi(\boldsymbol{y})$ [23]. This is known as the kernel trick [24] and interestingly enough, many algorithms can be kernelized this way [3]. Using the kernel trick, one never has to explicitly perform the potentially computationally expensive transformation $\phi(\cdot)$.

Table 3.1 List of Mercer kernels often used in the quantum chemistry domain

Name	Kernel $k(\boldsymbol{x}, \boldsymbol{y})$
Gaussian	$\exp(-\Vert\boldsymbol{x} - \boldsymbol{y}\Vert_2^2/(2\sigma^2))$
Laplace	$\exp(-\Vert\boldsymbol{x} - \boldsymbol{y}\Vert_1/\sigma)$
Polynomial	$(\boldsymbol{x}^\top \cdot \boldsymbol{y} + c)^d$
Matérn	$\frac{2^{1-\nu}}{\Gamma(\nu)}\left(\frac{\sqrt{2\nu}}{l}\Vert\boldsymbol{x} - \boldsymbol{y}\Vert\right)^\nu K_\nu\left(\frac{\sqrt{2\nu}}{l}\Vert\boldsymbol{x} - \boldsymbol{y}\Vert\right)$

For the Matérn kernel, Γ denotes the Gamma function and K_ν is the modified Bessel function of the second kind, respectively

The kernel function $k(\cdot)$ thus allows to reduce some of the intrinsic difficulties of the non-linear mapping $\phi(\cdot)$. The question remains, which kernel functions allow for such implicit feature mappings. Mercer's theorem [25] guarantees that such a mapping exists, if for all elements f of the Hilbert space L^2 defined on a compact set $C \subset \mathbb{R}^{n_i}$

$$\int_C f(\boldsymbol{x})k(\boldsymbol{x}, \boldsymbol{y})f(\boldsymbol{y})d\boldsymbol{x}d\boldsymbol{y} > 0. \tag{3.5}$$

From the kernel and a set of input samples $\{\boldsymbol{x}_i\}_{i=1}^N$, we can construct a discrete version of Mercer's theorem by composing the matrix

$$\mathbf{K} := \begin{pmatrix} k(\boldsymbol{x}_1, \boldsymbol{x}_1) & \cdots & k(\boldsymbol{x}_1, \boldsymbol{x}_N) \\ & \vdots & \\ k(\boldsymbol{x}_N, \boldsymbol{x}_1) & \cdots & k(\boldsymbol{x}_N, \boldsymbol{x}_N) \end{pmatrix}. \tag{3.6}$$

Mercer's theorem now implies that the matrix $\mathbf{K}$ is a Gram matrix, i.e., positive-semidefinite for any set of inputs $\{\boldsymbol{x}_i\}_{i=1}^N$. Thus, practically if the matrix $\mathbf{K}$ would have negative eigenvalues, then it will not fulfill Mercer's theorem. Examples of popular kernels in the quantum chemistry domain are shown in Table 3.1.

For some kernels like the Gaussian kernel, the feature map $\phi(\cdot)$ can be infinite dimensional. Due to the *curse of dimensionality*, it is then a question whether such a feature mapping to a much higher-dimensional space is a good idea at all, especially as the training set size increases (which corresponds to the dimension of the linear span of the projected input samples in feature space). As it turns out, one can still leverage the feature mapping if the learning algorithm is kept *simple* [3]. The intuitive *complexity* of the learning problem induced by the kernel, the data, and the learning algorithm is a measure of how well a kernel matches the data. We will review and investigate one such complexity measure in Sect. 3.5. For now, note that translation invariant kernels have natural regularization properties which help reducing the complexity of a learning algorithm. The Gaussian kernel, for example, is smooth in all its derivatives [26].

While there is a wide variety of representations of physical systems, it is less obvious how to encode prior knowledge into the kernel (see Zien et al. [27] for the first kernels engineered to reflect prior knowledge). In the quantum chemistry domain, this is typically done by limiting the similarity measure to *local* information [13, 28]. The definition of such locality depends on the chemical system as it limits correlations *between* such localized kernels and emphasized local correlations. Due to the scalar product properties of the mapping $\phi(\cdot)$ in feature space, these local kernels can be combined by a sum to yield a new kernel function.

To conclude this section, we will describe a method for choosing a good kernel among a set of candidate kernels for a given learning problem, a procedure that is commonly called model selection [19]. Typically, a class of kernels is defined by a set of hyperparameters which, e.g., control the scaling of the kernel with respect to the data in the chosen distance metric. These hyperparameters have to be determined (i.e., a kernel is selected from a given class) in order to minimize the generalization error, a measure of how good unseen data can be predicted [19]. Note that minimizing a given criterion on the training data alone with respect to the hyperparameters usually results in poor generalization due to *overfitting*. The most common procedure to estimate the generalization error is cross-validation. In cross-validation, the data set is divided into k subsets of equal sample size. Then, the model is trained on the remaining $k - 1$ subsets and evaluated on the k-th subset, called the *validation set*. The average of the error over the k validation sets is a good estimate of the generalization error. After heuristically choosing a set of hyperparameters (kernels), this cross-validation scheme yields the best hyperparameters among the set which are then evaluated on an unseen test set. Repeating this procedure for different test splits is called *nested cross-validation*. Both cross-validation and nested cross-validation are schematically shown in Fig. 3.2.

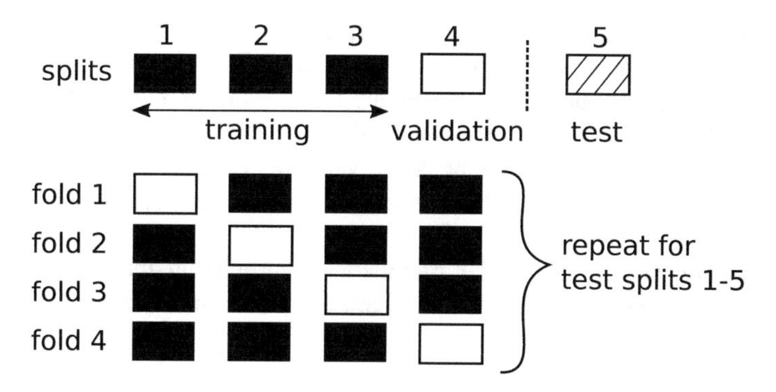

Fig. 3.2 Schematic fourfold cross-validation (inner loop) together with fivefold nested cross-validation (outer loop), respectively

3.4 Kernel Methods

After reviewing key concepts of kernels, we present two practical applications which have been extensively used in the quantum chemistry domain, one for supervised and one for unsupervised learning, respectively.

3.4.1 Kernel Ridge Regression

A typical setting in machine learning problems includes the prediction of response variables $\{y_i\}_{i=1}^N$ for a set of samples $\{\boldsymbol{x}_i\}_{i=1}^N$. The kernel trick introduced in the previous section can be applied to the linear ridge regression model. In ridge regression, a cost function typically given by

$$C(\boldsymbol{w}) := \frac{1}{N}\sum_{i=1}^{N}(y_i - \boldsymbol{w}^\top \boldsymbol{x}_i)^2 + \lambda \cdot \|\boldsymbol{w}\|^2 \tag{3.7}$$

is minimized with respect to the weight coefficients $\boldsymbol{w}$, where λ is a regularization parameter of the model which penalizes the norm of the weights. Given the regularization parameter λ, the weights which minimize Eq. (3.7) are given by

$$\boldsymbol{w}_{\text{ridge}} = (\lambda \cdot \mathbf{I} + \mathbf{X}^\top \mathbf{X})^{-1}\mathbf{X}^\top \boldsymbol{y}, \tag{3.8}$$

with the design matrix $\boldsymbol{X}$ which rows are composed of the inputs $\{\boldsymbol{x}_i\}_{i=1}^N$ and the identity matrix $\mathbf{I}$, respectively. Increasing the complexity regularizer λ results in smoother functions, thereby avoiding purely interpolating the training data and reducing overfitting (see [19]). Due to its form, the ridge regression model often exhibits good stability in terms of generalization error. However, for most real-world problems the linear model is not powerful enough to accurately predict quantum-mechanical properties as it is difficult to find features of the underlying system which linearly correlate with the response variables $\{y_i\}_{i=1}^N$. As the simplicity of the linear ridge regression model turns out to be the main limitation, there is a need for a non-linear variant.

This non-linear variant can be provided by kernelizing the ridge regression model. In kernel ridge regression, the parameters of the model $\boldsymbol{\alpha} := (\alpha_1, \ldots, \alpha_N)$ are calculated by

$$(\lambda \cdot \mathbf{I} + \mathbf{K}) \cdot \boldsymbol{\alpha} = \boldsymbol{y}, \tag{3.9}$$

with the already introduced Gram matrix $\mathbf{K}$ and $\boldsymbol{y} := (y_1, \ldots, y_N)$, respectively. From the parameters $\boldsymbol{\alpha}$, a new prediction for a sample $\boldsymbol{x}$ is given by

$$y_{\text{est}} = \sum_{i=1}^{N} \alpha_i \cdot k(\boldsymbol{x}, \boldsymbol{x}_i) \tag{3.10}$$

Due to its nice practical and theoretical properties, kernel ridge regression has been extensively used in the quantum chemistry domain [29–33]. Note that the formally same solution of Eq. (3.9) is also obtained when training Gaussian processes and starting from the framework of Bayesian statistics [34].

3.4.2 Kernel Principal Component Analysis

Kernel principal component analysis (kernel PCA) [35–37] is a kernelized extension to one of the most popular data dimensionality reduction techniques, namely principal component analysis (PCA). To recall, PCA is an unsupervised method that uses an orthogonal transformation to project the high-dimensional data onto a linearly uncorrelated set of low-dimensional variables called the *principal components.* These principal components are defined in a compact way in the sense that a given component accounts for the highest variance under the constraint of being orthogonal to the preceding ones, the first principal component having the largest possible variance.

PCA can be kernelized by virtue of the kernel trick: the evaluation of the data on the m-th principal component equals the m-th eigenvector of the kernel matrix [38]. As PCA requires the data to be centered which is not guaranteed in feature space, one common preliminary step is to centralize the kernel beforehand by

$$\mathbf{K}' := \mathbf{K} - \mathbf{1}_N \cdot \mathbf{K} - \mathbf{K} \cdot \mathbf{1}_N + \mathbf{1}_N \cdot \mathbf{K} \cdot \mathbf{1}_N, \tag{3.11}$$

where $\mathbf{1}_N$ is the $N \times N$-matrix with entries $1/N$. From the normalized eigenvectors $\{\boldsymbol{u}_i\}_{i=1}^N$ of the centralized kernel $\mathbf{K}'$, we compute the m-th principal component of a new sample $\boldsymbol{x}$ by

$$p_m(\boldsymbol{x}) = \sum_{i=1}^{N} \boldsymbol{u}_{m,i} \cdot k(\boldsymbol{x}, \boldsymbol{x}_i), \tag{3.12}$$

where $\boldsymbol{u}_{m,i}$ is the i-th element of the eigenvector $\boldsymbol{u}_m$. Kernel PCA is often used in the quantum chemistry domain to display the data in its first *two* principal components [39–42], along with the label information if present. Such a projection separates the structure of the data as induced by the kernel from the response variable, possibly learning something about the difficulties to predict a given response variable.

Kernel PCA can be used in a supervised fashion by projecting the label vector $\boldsymbol{y}$ on the normalized eigenvectors of the centralized kernel matrix

$$z_i := \boldsymbol{u}_i^\top \boldsymbol{y} \qquad i = 1, \ldots, N \tag{3.13}$$

where we call the $\{z_i\}_{i=1}^N$ the *kernel PCA coefficients.* Analyzing the kernel PCA coefficients allows to gain additional information about the complexity of the learning problem at hand [43], as we show in the next section.

3.5 Relevant Dimension Estimation

In the previous sections, we indicated that one of the governing factors for the efficiency of the model is the complexity of the learning problem. Due to the potentially infinite dimensional implicit feature space induced by the kernel, it is a priori unclear how to assess this complexity in a feasible way.

One way to tackle this problem is to analyze where the relevant information of the learning problem is contained [43]. Specifically, the relevant information can be bounded by a number of leading kernel PCA-components already encountered in the previous section, under the mild assumption that the kernel asymptotically represents the learning problem and is sufficiently smooth (see [43] for a detailed definition of these aspects). This leads to an estimation of a relevant dimension d_{RD} representing the complexity of the learning problem at hand. Modeling the kernel PCA coefficients $\{z_i\}_{i=1}^N$ by two zero-mean Gaussians yields the following maximum likelihood fit for the relevant dimension

$$d_{\mathrm{RD}} := \operatorname*{argmin}_{1 \leq d \leq N} \left(\frac{d}{N} \log \sigma_1^2 + \frac{N-d}{N} \log \sigma_2^2 \right) \tag{3.14}$$

with

$$\sigma_1^2 = \frac{1}{d} \sum_{i=1}^{d} z_i^2 \tag{3.15}$$

$$\sigma_2^2 = \frac{1}{N-d} \sum_{i=d+1}^{N} z_i^2 . \tag{3.16}$$

Alternatively, d_{RD} can be estimated by a leave-one-out cross-validation approach [43]. In addition to the useful side-product in form of the relevant dimension in the kernel feature space, we can gain important insights into the interplay between the kernel and data, respectively. Specifically, the boundedness of the relevant information in a possibly small amount of kernel PCA-components shows that ideally, the kernel induces a feature space in such a way that makes optimal use of the potentially high number of feature space dimensions. This dimensionality reduction is paired with an intrinsic regularization by the kernel, yielding an efficient model which generalizes well. One additional application is to use the relevant dimension to estimate the noise of a given regression task, which allows to assess whether the difficulty of the learning problem lies in its intrinsic high-dimensionality or alternatively in present noise.

3.6 Conclusion

In this chapter, we examined how the choice of the representation of a given chemical system and the kernel affects the model. We reviewed three recent molecular descriptors, the Coulomb matrix, the bag-of-bonds, and the $F_{2B} + F_{3B}$ representations, for their ability to encode prior physical knowledge. In view of highly non-linear learning problems often encountered in quantum chemistry, we discussed important kernel properties to better understand machine learning methods based on kernels. Specifically, we examined an efficient way to choose an appropriate kernel by a common model selection scheme. This was followed by a practical guide of two popular kernel methods, the supervised kernel ridge regression and the unsupervised kernel principal component analysis. As more descriptors are continuously developed, we assert the importance to better understand their interplay with the kernel and data, respectively. In the final part of this chapter, we provided an analysis tool which captures this connection in form of the relevant dimension estimation, a way how to measure the effective complexity of the learning problem at hand when using a specific kernel.

References

1. C. Cortes, V. Vapnik, Mach. Learn. **20**(3), 273 (1995)
2. V. Vapnik, S.E. Golowich, A.J. Smola, in *Advances in Neural Information Processing Systems* (1997), pp. 281–287
3. K.-R. Müller, S. Mika, G. Rätsch, K. Tsuda, B. Schölkopf, IEEE Trans. Neural Netw. **12**(2), 181 (2001). https://doi.org/10.1109/72.914517
4. B. Schölkopf, A.J. Smola, *Learning with Kernels: Support Vector Machines, Regularization, Optimization, and Beyond* (MIT Press, Cambridge, 2002)
5. B. Huang, O.A. von Lilienfeld, J. Chem. Phys. **145**(16), 161102 (2016). https://doi.org/10.1063/1.4964627
6. B. Schölkopf, S. Mika, C.J. Burges, P. Knirsch, K.-R. Müller, G. Rätsch, A.J. Smola, IEEE Trans. Neural Netw. **10**(5), 1000 (1999)
7. P. Indyk, R. Motwani, in *Proceedings of the Thirtieth Annual ACM Symposium on Theory of Computing* (ACM, New York, 1998), pp. 604–613
8. J.H. Friedman, Data Min. Knowl. Disc. **1**(1), 55 (1997)
9. J. Rust, J. Econ. Soc. **1997**, 487–516 (1997)
10. K.T. Schütt, H. Glawe, F. Brockherde, A. Sanna, K.R. Müller, E.K.U. Gross, Phys. Rev. B **89**, 205118 (2014). https://doi.org/10.1103/PhysRevB.89.205118
11. S. Chmiela, H.E. Sauceda, I. Poltavsky, K.-R. Müller, A. Tkatchenko, Comput. Phys. Commun. **240**, 38 (2019). https://doi.org/10.1016/j.cpc.2019.02.007
12. F.A. Faber, L. Hutchison, B. Huang, J. Gilmer, S.S. Schoenholz, G.E. Dahl, O. Vinyals, S. Kearnes, P.F. Riley, O.A. von Lilienfeld, J. Chem. Theory Comput. **13**(11), 5255 (2017). https://doi.org/10.1021/acs.jctc.7b00577
13. F.A. Faber, A.S. Christensen, B. Huang, O.A. von Lilienfeld, J. Chem. Phys. **148**(24), 241717 (2018). https://doi.org/10.1063/1.5020710
14. K.T. Schütt, F. Arbabzadah, S. Chmiela, K.R. Müller, A. Tkatchenko, Nat. Commun. **8**, 13890 (2017)
15. K.T. Schütt, P.-J. Kindermans, H.E.S. Felix, S. Chmiela, A. Tkatchenko, K.-R. Müller, in *Advances in Neural Information Processing Systems* (2017), pp. 991–1001

16. G. Montavon, W. Samek, K.-R. Müller, Digit. Signal Process. **73**, 1 (2018)
17. S. Chmiela, Towards exact molecular dynamics simulations with invariant machine-learned models. Dissertation, Technische Universität Berlin (2019). https://doi.org/10.14279/depositonce-8635
18. M. Rupp, A. Tkatchenko, K.-R. Müller, O.A. von Lilienfeld, Phys. Rev. Lett. **108**, 058301 (2012). https://doi.org/10.1103/PhysRevLett.108.058301
19. K. Hansen, G. Montavon, F. Biegler, S. Fazli, M. Rupp, M. Scheffler, O.A. von Lilienfeld, A. Tkatchenko, K.-R. Müller, J. Chem. Theory Comput. **9**(8), 3404 (2013). https://doi.org/10.1021/ct400195d
20. K. Hansen, F. Biegler, R. Ramakrishnan, W. Pronobis, O.A. von Lilienfeld, K.-R. Müller, A. Tkatchenko, J. Phys. Chem. Lett. **6**(12), 2326 (2015). https://doi.org/10.1021/acs.jpclett.5b00831
21. W. Pronobis, K.T. Schütt, A. Tkatchenko, K.-R. Müller, Eur. Phys. J. B **91**(8), 178 (2018). https://doi.org/10.1140/epjb/e2018-90148-y
22. W. Pronobis, A. Tkatchenko, K.-R. Müller, J. Chem. Theory Comput. **14**(6), 2991 (2018). https://doi.org/10.1021/acs.jctc.8b00110
23. B.E. Boser, I.M. Guyon, V.N. Vapnik, in *Proceedings of the Fifth Annual Workshop on Computational Learning Theory* (ACM, New York, 1992), pp. 144–152
24. K.-R. Müller, A. Smola, G. Rätsch, B. Schölkopf, J. Kohlmorgen, V. Vapnik, in *Advances in Kernel Methods—Support Vector Learning*, pp. 243–254 (1999)
25. M. James, F.A. Russell, Philos. Trans. R. Soc. Lond. A **209**(441–458), 415 (1909). https://doi.org/10.1098/rsta.1909.0016
26. A.J. Smola, B. Schölkopf, K.-R. Müller, Neural Netw. **11**(4), 637 (1998). https://doi.org/10.1016/S0893-6080(98)00032-X
27. A. Zien, G. Rätsch, S. Mika, B. Schölkopf, T. Lengauer, K.-R. Müller, Bioinformatics **16**(9), 799 (2000)
28. A.P. Bartók, R. Kondor, G. Csányi, Phys. Rev. B **87**, 184115 (2013). https://doi.org/10.1103/PhysRevB.87.184115
29. G. Montavon, K. Hansen, S. Fazli, M. Rupp, F. Biegler, A. Ziehe, A. Tkatchenko, A.V. Lilienfeld, K.-R. Müller, in *Advances in Neural Information Processing Systems* (2012), pp. 440–448
30. R. Ramakrishnan, O.A. von Lilienfeld, CHIMIA Int. J. Chem. **69**(4), 182 (2015)
31. G. Ferré, T. Haut, K. Barros, J. Chem. Phys. **146**(11), 114107 (2017)
32. S. Chmiela, A. Tkatchenko, H.E. Sauceda, I. Poltavsky, K.T. Schütt, K.-R. Müller, Sci. Adv. **3**(5), e1603015 (2017)
33. D. Hu, Y. Xie, X. Li, L. Li, Z. Lan, J. Phys. Chem. Lett. **9**(11), 2725 (2018). https://doi.org/10.1021/acs.jpclett.8b00684
34. C.K. Williams, C.E. Rasmussen, *Gaussian Processes for Machine Learning*, vol. 2 (MIT Press, Cambridge, 2006)
35. B. Schölkopf, A. Smola, K.-R. Müller, in *International Conference on Artificial Neural Networks* (Springer, Berlin, 1997), pp. 583–588
36. Z. Liu, D. Chen, H. Bensmail, Biomed Res. Int. **2005**(2), 155 (2005)
37. D. Antoniou, S.D. Schwartz, J. Phys. Chem. B **115**(10), 2465 (2011)
38. B. Schölkopf, A. Smola, K. Müller, Neural Comput. **10**(5), 1299 (1998). https://doi.org/10.1162/089976698300017467
39. Y.M. Koyama, T.J. Kobayashi, S. Tomoda, H.R. Ueda, Phys. Rev. E **78**(4), 046702 (2008)
40. X. Han, IEEE/ACM Trans. Comput. Biol. Bioinform. **7**(3), 537 (2010)
41. A. Varnek, I.I. Baskin, Mol. Inf. **30**(1), 20 (2011)
42. X. Deng, X. Tian, S. Chen, Chemom. Intell. Lab. Syst. **127**, 195 (2013)
43. M.L. Braun, J.M. Buhmann, K.-R. Müller, J. Mach. Learn. Res. **9**, 1875 (2008)

Introduction to Neural Networks

4

Grégoire Montavon

Abstract

Machine learning has become an essential tool for extracting regularities in the data and for making inferences. Neural networks, in particular, provide the scalability and flexibility that is needed to convert complex datasets into structured and well-generalizing models. Pretrained models have strongly facilitated the application of neural networks to images and text data. Application to other types of data, e.g., in physics, remains more challenging and often requires ad-hoc approaches. In this chapter, we give an introduction to neural networks with a focus on the latter applications. We present practical steps that ease training of neural networks, and then review simple approaches to introduce prior knowledge into the model. The discussion is supported by theoretical arguments as well as examples showing how well-performing neural networks can be implemented easily in modern neural network frameworks.

4.1 Introduction

Neural networks [1–4] are a machine learning paradigm where a large number of simple computational units called neurons are interconnected to form complex predictions [5, 6]. Neurons are typically organized in layers, where each layer compresses certain components of the input data to expand other components that are more task-relevant [7, 8]. Unlike kernels [9–11], for which the problem representation is fixed and high-dimensional, neural networks are able to learn their own representation and keep it confined to finitely many neurons, thereby providing

G. Montavon (✉)
Electrical Engineering and Computer Science, Technische Universität Berlin, Berlin, Germany
e-mail: gregoire.montavon@tu-berlin.de

K. T. Schütt et al. (eds.), *Machine Learning Meets Quantum Physics*,
Lecture Notes in Physics 968, https://doi.org/10.1007/978-3-030-40245-7_4

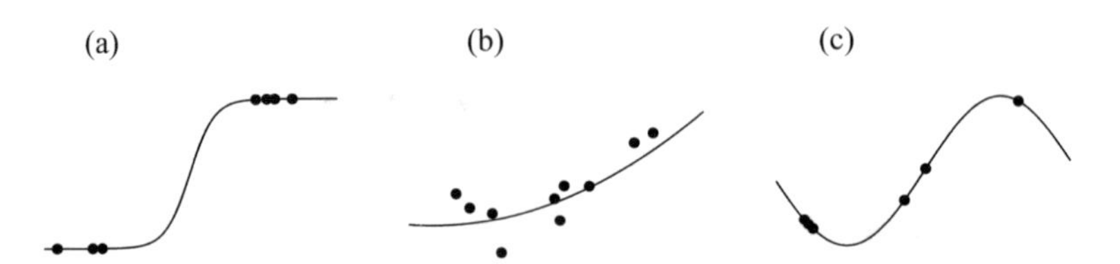

Fig. 4.1 Three types of machine learning problems. (**a**) Classification. (**b**) Regression. (**c**) Interpolation. Curves represent the ground truth, and points represent the available training data. The horizontal axis corresponds to the input space and the vertical axis the output space. This chapter focuses on the problem of interpolation

scalability. Neural networks have reached widespread successes in applications such as image recognition [12–14] and text understanding [15, 16]. These successes have motivated the use of deep neural networks beyond image and text classification, for example, in physics [17–19].

Physics applications are however characterized by a strong heterogeneity in problems and datasets, preventing an outright application of the models and techniques developed for images and text. For example, learning the energy potential of a physical system requires to produce real-valued predictions that match the energy targets exactly. This is clearly not a classification task, nor a typical regression. It is better described as an interpolation problem (cf. Fig. 4.1).

In practice, the type of task being considered has significant implications on the design of the neural network and its learning procedure. For example, noise injection techniques [20] are very useful for classification by forcing the function to be low-varying near the data. However, it would likely be detrimental in the context of regression or interpolation, where the ground truth function is clearly not flat in these regions. The interpolation task also distinguishes itself from regression by an absence of target noise, in other words, the training error must reach zero. This requires to carefully address the question of optimization. Furthermore, learning in physical systems significantly differs from other applications in terms of the representations and neural network structures that are needed for the prediction task. Overall, these multiple differences with the well-established image/text recognition pipelines have so far prevented a straightforward and systematic application of neural networks in physics, requiring instead the development of ad-hoc designs [18, 21–24].

This chapter provides a bottom-up introduction to neural networks, by first presenting the basics and then demonstrating on illustrative toy examples how simple models can be adapted to more complex scenarios. Section 4.2 presents the computation of the forward pass, backpropagation, and the simple gradient descent procedure for learning the model. Section 4.3 presents best practices to quickly bring the training error of a model to zero. These practices are analyzed from the perspective of the Hessian matrix that characterizes locally the function to optimize. Section 4.4 tackles the question of generalization, by illustrating how invariance can be added in the input representation, and how the model can be structured to incorporate prior knowledge about the task. Using concrete code examples, we show that simple neural networks and their well-generalizing extensions can be implemented easily and concisely in modern neural network frameworks such as PyTorch.[1] Finally, we give in Sect. 4.5 an overview of the question of model selection and validation with a focus on aspects that are specific to neural networks. In particular, we discuss recent techniques to explain and visualize the predictions of a neural network, and comment on the relation between the prediction accuracy of a model and its physical plausibility.

4.2 Neural Network Basics

At an abstract level, a neural network can be seen as a function

$$f : \mathbb{R}^d \times \Theta \to \mathbb{R}.$$

It takes as input a data point $\boldsymbol{x} \in \mathbb{R}^d$ and a vector of parameters $\theta \in \Theta$ that must be learned from the data. The output $f(\boldsymbol{x}, \theta) \in \mathbb{R}$ is a real-valued prediction for the input $\boldsymbol{x}$. The function f is structured as an interconnection of many simple neurons, usually organized in a layered structure, where neurons at a given layer receive as input the output of the neurons from the previous layer. A simple neural network is illustrated in Fig. 4.2.

Consider the neuron with index k. Its output a_k is computed in two steps:

$$z_k = \sum_j a_j w_{jk} + b_k$$
$$a_k = \rho(z_k)$$

The first step performs a weighted sum over all neurons j that neuron k receives as input. The weights $(w_{jk})_j$ and bias b_k are the neuron parameters that are learned from the data. The second step applies a predefined nonlinear function $\rho \colon \mathbb{R} \to \mathbb{R}$

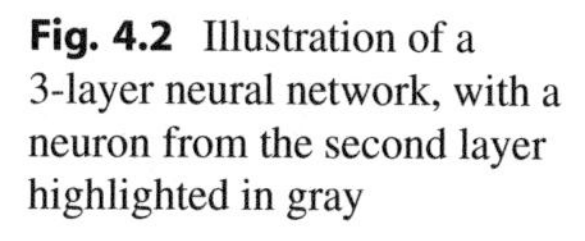
Fig. 4.2 Illustration of a 3-layer neural network, with a neuron from the second layer highlighted in gray

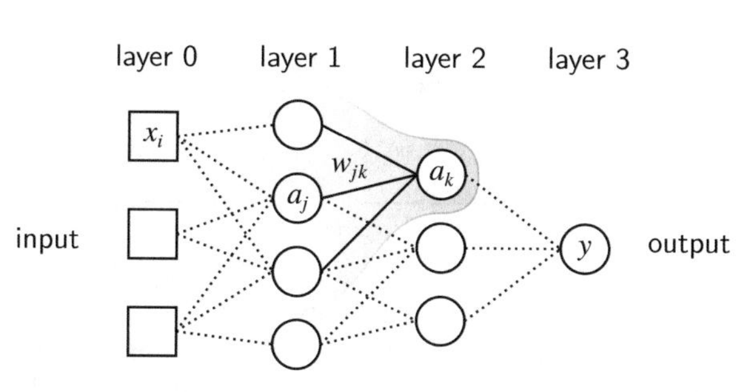

[1] https://pytorch.org/.

called activation function. Specific choices of activation function are discussed in Sect. 4.3.3.

4.2.1 The Forward Pass

Consider the layered network of Fig. 4.2. The whole processing from input $\boldsymbol{x} = (x_i)_i$ to output y can be written as the sequence of computations:

$$\begin{aligned}
\forall_j: \quad & z_j = \textstyle\sum_i x_i w_{ij} + b_j & a_j = \rho(z_j) \quad & \text{(layer 1)}\\
\forall_k: \quad & z_k = \textstyle\sum_j a_j w_{jk} + b_k & a_k = \rho(z_k) \quad & \text{(layer 2)}\\
& y = \textstyle\sum_k a_k v_k + c & & \text{(layer 3)}
\end{aligned}$$

The letters i, j, and k are indices for the neurons at each layer. Weights and bias parameters of each neuron constitute the elements of the parameter vector θ. In practice, when neurons in consecutive layers are densely connected, it is convenient to replace neuron-wise computations by layer-wise computations. In particular, we can rewrite the neural network above as:

$$\begin{aligned}
\boldsymbol{z}^{(1)} &= W^{(1)}\boldsymbol{x} + \boldsymbol{b}^{(1)} & \boldsymbol{a}^{(1)} = \rho(\boldsymbol{z}^{(1)}) \quad & \text{(layer 1)}\\
\boldsymbol{z}^{(2)} &= W^{(2)}\boldsymbol{a}^{(1)} + \boldsymbol{b}^{(2)} & \boldsymbol{a}^{(2)} = \rho(\boldsymbol{z}^{(2)}) \quad & \text{(layer 2)}\\
y &= \boldsymbol{v}^\top \boldsymbol{a}^{(2)} + c & & \text{(layer 3)}
\end{aligned}$$

where $[W^{(1)}]_{ji} = w_{ij}$, $\boldsymbol{b}^{(1)} = (b_j)_j$, $\boldsymbol{a}^{(1)} = (a_j)_j$, etc., and where the function ρ applies element-wise. Matrix operations occurring in this formulation considerably simplify the implementation and improve the computational efficiency. This layer-wise formulation can be further generalized to let the forward pass apply to N data points in parallel, by using matrix–matrix multiplications in place of matrix–vector operations. For example, in PyTorch, the forward pass can be implemented as:

```
def forward(X):
    Z1 = X.matmul(W1)  + b1;  A1 = rho(Z1)
    Z2 = A1.matmul(W2) + b2;  A2 = rho(Z2)
    Y  = A2.matmul(v)  + c
    return Y
```

This function takes as input a `torch` array of shape `N × d` containing the data. The function `matmul` performs matrix–matrix multiplications, and the function `rho` implements the nonlinear activation function. Note that weight matrices `W1` and `W2` are transposed versions of matrices $W^{(1)}$ and $W^{(2)}$. Neural network frameworks such as PyTorch or Keras[2] also come with a collection of predefined layers, including linear layers, convolutions, and a variety of nonlinear activation functions.

4.2.2 The Backward Pass

A crucial aspect of neural networks is the existence of an algorithm called error backpropagation [25, 26] to efficiently propagate the error feedback to the multiple layers of the network where the weights and biases need to be adjusted. Error backpropagation is an iterative application of the chain rule for derivatives to neural network graphs. Let $\delta_{\text{out}} = \partial(\cdot)/\partial y$ be the derivative of some quantity $(\cdot)$ that depends on the input and parameters through the output y. We start from the output and apply the chain rule to get the derivatives in the previous layer:

$$\forall_k : \quad \frac{\partial(\cdot)}{\partial z_k} = \frac{\partial a_k}{\partial z_k}\frac{\partial y}{\partial a_k}\frac{\partial(\cdot)}{\partial y} = \rho'(z_k)\, v_k \delta_{\text{out}}$$

We store the result of this computation in some variable δ_k (i.e., $\delta_k \leftarrow \partial(\cdot)/\partial z_k$), so that it does not need to be recomputed. Application of the (multivariate) chain rule is then used to propagate the derivatives one layer below:

$$\forall_j : \quad \frac{\partial(\cdot)}{\partial z_j} = \frac{\partial a_j}{\partial z_j} \cdot \Big(\sum_k \frac{\partial z_k}{\partial a_j}\frac{\partial(\cdot)}{\partial z_k}\Big) = \rho'(z_j) \textstyle\sum_k w_{jk}\delta_k$$

We store the result in δ_j (i.e., $\delta_j \leftarrow \partial(\cdot)/\partial z_j$). A further step of backpropagation similar to the one above gives $\delta_i \leftarrow \partial(\cdot)/\partial x_i$. The whole backpropagation procedure can be written as:

$$\begin{array}{lllll} \forall_k : & \partial(\cdot)/\partial v_k = a_k\delta_{\text{out}} & \partial(\cdot)/\partial c = \delta_{\text{out}} & \delta_k = \rho'(z_k)\, v_k \delta_{\text{out}} & \text{(layer 3)} \\ \forall_j : & \partial(\cdot)/\partial w_{jk} = a_j\delta_k & \partial(\cdot)/\partial b_k = \delta_k & \delta_j = \rho'(z_j) \sum_k w_{jk}\delta_k & \text{(layer 2)} \\ \forall_i : & \partial(\cdot)/\partial w_{ij} = x_i\delta_j & \partial(\cdot)/\partial b_j = \delta_j & \delta_i = \sum_j w_{ij}\delta_j & \text{(layer 1)} \end{array}$$

Computations are applied from top to bottom and from left to right. The first two columns compute derivatives with the model weights and biases, and make use of the quantities δ_j, δ_k discussed before. These quantities are obtained iteratively by computations in the last column. This last column forms the backbone of the backpropagation procedure. Derivatives with respect to weights and biases constitute the elements of the gradient that we will use in Sect. 4.2.3 to train the neural network. Overall, the backward pass runs in linear time with the forward

[2]https://keras.io/.

pass. Like for the forward pass, the backward pass can also be rewritten in terms of matrix multiplications.

In practice, however, modern neural network frameworks such as PyTorch or Keras come with an *automatic differentiation* mechanism, where the backward computations are generated automatically from the forward pass. Consequently, the backward pass does not have to be implemented manually. For example, to get the gradient of the nth data point's output w.r.t. the first layer weights, we write in PyTorch:

```
W1.requires_grad_(True)
Y = forward(X)
Y[n].backward()
print(W1.grad)
```

The first line declares that the gradient w.r.t. `W1` needs to be collected. The second line applies the forward pass. The third line invokes for the desired scalar quantity the automatic differentiation mechanism which stores the gradient in the variable `W1.grad`.

4.2.3 Optimizing Neural Networks

Having described how to compute activations at each layer of a neural network and the gradient w.r.t. the parameters, we now focus on the problem of learning these parameters from the data. Assume a dataset of N examples $\mathcal{D} = ((\boldsymbol{x}_1, t_1), (\boldsymbol{x}_2, t_2), \ldots, (\boldsymbol{x}_N, t_N))$ where each example consists of an input $\boldsymbol{x}_n \in \mathbb{R}^d$ and an associated target $t_n \in \mathbb{R}$ representing what needs to be predicted for this example. The problem can be formalized by defining an error function $\mathcal{E}(\theta)$ that measures the average divergence on the training data between the predictions $f(\boldsymbol{x}_n, \theta)$ and the targets t_n:

$$\mathcal{E}(\theta) = \frac{1}{N} \sum_{n=1}^{N} \left[f(\boldsymbol{x}_n, \theta) - t_n \right]^2 \tag{4.1}$$

The best fitting neural network is then obtained as a solution to the optimization problem $\min_{\theta \in \Theta} \mathcal{E}(\theta)$. Because the function $\mathcal{E}(\theta)$ has potentially many (local) minima, we usually start at some given position in the parameter space and iterate the map

$$\theta \leftarrow \theta - \gamma \cdot \nabla_\theta \mathcal{E}(\theta)$$

until it converges to some local minimum $\theta^\star$. The hyperparameter γ is the learning rate. The error gradient is obtained with automatic differentiation (cf. Sect. 4.2.2). Assuming our dataset is stored in `torch` arrays `X` and `T` of size `N` × `d` and `N` respectively, the gradient descent procedure can be implemented compactly as:

```
for i in range(n_iter):
    Y = forward(X)
    E = ((Y-T)**2).mean()
    E.backward()

    with torch.no_grad():
        for p in [W1,b1,W2,b2,v,c]:
            p -= gamma * p.grad
            p.grad = None
```

The statement `torch.no_grad()` temporarily deactivates the automatic differentiation mechanism so that the parameters of the model can be updated. Importantly, because the call to `backward()` increments parameter gradients rather than replacing them by their new value, we need to manually set these gradients to `None` after each iteration.

4.3 Efficient Training of Neural Networks

The optimization procedure above assumes that the error function $\mathcal{E}(\theta)$ is such that a step along the gradient direction will result in a new parameter θ' at which the error $\mathcal{E}(\theta')$ is significantly lower. In practice, this is not necessarily the case. The error function might instead exhibit pathological curvature, which makes the gradient direction of little use. Section 4.3.1 looks at the Hessian of the error function at the local minimum $\theta^\star$, from which we can infer how difficult it is to optimize the function locally. Building on this Hessian-based analysis, Sects. 4.3.2–4.3.5 present common practices to make neural network training more efficient.

4.3.1 Hessian-Based Analysis of the Error Function

To assess how quickly gradient descent converges, we need to inspect the shape of the error function $\mathcal{E}(\theta)$ locally. A Taylor expansion at $\theta^\star$ gives:

$$\mathcal{E}(\theta) = \mathcal{E}(\theta^\star) + \frac{1}{2}(\theta - \theta^\star)^\top H\big|_{\theta^\star}(\theta - \theta^\star) + \cdots$$

where $H = \partial^2\mathcal{E}/\partial\theta^2$ is the Hessian matrix containing all second-order derivatives. Its eigenvalues $\lambda_1 \geq \lambda_2 \geq \cdots \geq \lambda_{\dim(\theta)} \geq 0$ give information about the local curvature of the error function and how well gradient descent will perform. When all the eigenvalues are strictly positive, the difficulty to converge to $\theta^\star$ is given by the condition number $\lambda_1/\lambda_{\dim(\theta)}$. An example of error function annotated with its eigenvalues is given in Fig. 4.3.

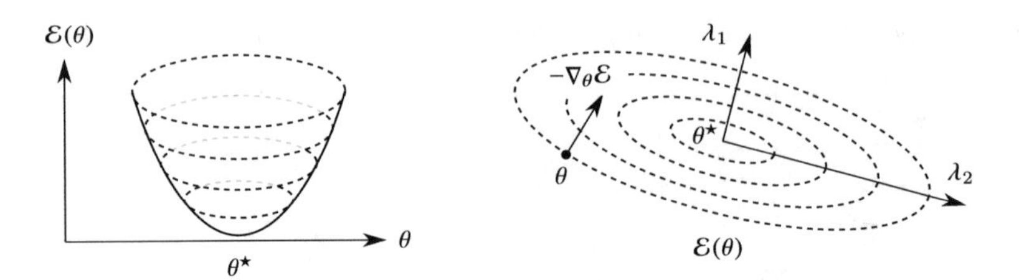

Fig. 4.3 Error function viewed in three dimensions (left) and from the top (right) with $\theta \in \mathbb{R}^2$. The highest eigenvalues correspond to the direction of highest curvature. A large ratio λ_1/λ_2 makes the function ill-conditioned and harder to optimize with gradient descent

A general way of computing the Hessian in a neural network makes use of the chain rule for second derivatives [27]:

$$H = \frac{\partial Y}{\partial \theta}^\top \frac{\partial^2 \mathcal{E}}{\partial Y^2} \frac{\partial Y}{\partial \theta} + \frac{\partial \mathcal{E}}{\partial Y} \frac{\partial^2 Y}{\partial \theta^2}$$

The vector Y represents the neural network outputs, here, $Y = [f(\boldsymbol{x}_1, \theta), \ldots, f(\boldsymbol{x}_N, \theta)]$. While the Hessian is high-dimensional and hard to analyze, certain fragments of it have a simpler analytical form [28, 29]. Let $\langle \cdot \rangle$ denote the mean over the training data, and $\delta_k = \partial y/\partial z_k$. The matrix $H_k = (\partial^2 \mathcal{E}/\partial w_{jk} \partial w_{j'k})_{jj'}$ focuses on the parameters of neuron k and takes the simple form:

$$[H_k]_{jj'} = 2\big\langle a_j\, a_{j'}\, \delta_k^2 \big\rangle + 2\Big\langle a_j \cdot \frac{\partial \delta_k}{\partial w_{j'k}} \cdot (y - t) \Big\rangle \tag{4.2}$$

The first term is a product of incoming activations $a_j\, a_{j'}$ modulated by δ_k^2 measuring the sensitivity to the output. The second term disappears when the error at the minimum is zero.

Overall, this Hessian-based analysis will serve to theoretically motivate a number of heuristics and best practices that are commonly used to prepare the data, build the neural network, and train it efficiently. These practices are presented in the next sections.

4.3.2 Normalizing the Input Data

A recommendation by LeCun et al. [27] is to apply a transformation to the data before giving it as input to the neural network. The transformation applies:

$$x_i \leftarrow \big(x_i - m_i\big)/s_i$$

to each input feature $i = 1 \ldots d$, where m_i and s_i are the mean and standard deviation of feature i computed on the training data. This transformation is known as *standardization*. It is easy to compute and has been shown empirically to ease the task of optimization. Standardization can be motivated from the Hessian-based analysis. Consider the special case where x_i, $x_{i'}$, and δ_j are *independent*. The Hessian matrix $H_j = 2\langle \boldsymbol{x}\boldsymbol{x}^\top \delta_j^2 \rangle$ computed on the original data has in that case its condition number upper-bounded as $\lambda_1/\lambda_d \leq (\max_i s_i^2 + \|\boldsymbol{m}\|^2)/\min_i s_i^2$. A mean that is distant from the origin or a large spread in standard deviations is likely to raise the condition number significantly and make optimization harder. The standardization procedure thus clearly contributes in this simple scenario to lower the condition number. When x_i and $x_{i'}$ are instead strongly correlated, the condition number would remain high, and application of an additional transform, e.g., data whitening, becomes needed to ease optimization.

During training, the variables δ_j and $(x_i, x_{i'})$ are likely to become dependent. The neuron j (and the associated sensitivity term δ_j^2) might indeed specialize to very specific input patterns. If these patterns are distant from the origin, e.g., in the tails of the input distribution, this will lead to a poorly conditioned Hessian matrix, even after standardization or whitening. A technique to improve optimization in this case is *thermometer coding* [30, 31]. It applies to each standardized feature the nonlinear expansion

$$\forall_{i=1}^{d} : x_i \leftarrow \Big(\dots , \sigma(x_i + 2) , \sigma(x_i + 1) , \sigma(x_i + 0) , \sigma(x_i - 1) , \sigma(x_i - 2) , \dots \Big)$$

where σ is a sigmoid function, for example, the hyperbolic tangent. Each term of the expansion "zooms-in" on a particular range of values of the input. Thermometer coding is then followed by another step of standardization. The effect of thermometer coding is illustrated for a simple one-dimensional distribution in Fig. 4.4. We highlight in blue a specific region of the input distribution. In certain dimensions, it collapses to a constant value while in other dimensions it becomes well-expanded, making it easier to learn the local variations of the target function.

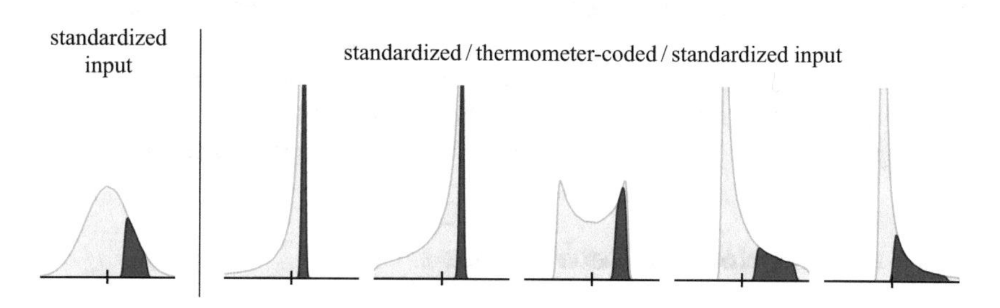

Fig. 4.4 On the left, standardized input distribution with a selection of the data shown in blue. On the right, the same distribution after thermometer coding and another standardization step

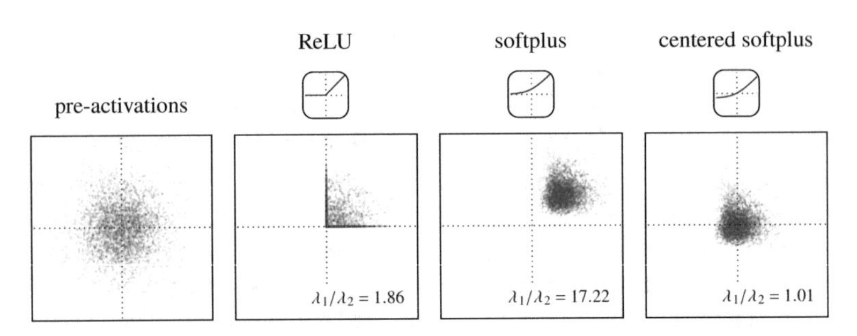

Fig. 4.5 Distribution of activations resulting from application of different nonlinearities, and the measured condition number. A low condition number makes optimization easier

4.3.3 Choosing the Activation Function

Another important parameter of the neural network is the choice of the activation function ρ. The ReLU activation given by

$$\rho(z_j) = \max(0, z_j)$$

was recommended by Glorot et al. [32]. It has become the "default" nonlinearity and is commonly used in state-of-the-art models [12,16,33]. Similarly to Sect. 4.3.2, this choice of nonlinearity can be motivated from the Hessian-based analysis. Let $\boldsymbol{a} = (a_j)_j$ be the vector of activations received by neuron k, and $H_k = 2\langle \boldsymbol{a}\boldsymbol{a}^\top \delta_k^2 \rangle$ its Hessian matrix. Figure 4.5 shows the distribution of input activations $(a_j)_j$ obtained from various nonlinearities, as well as the condition number of a neuron k built on top of these activations. (Here, we assume constant sensitivity $\delta_k^2 = 1$.)

Application of the ReLU function results in activations for which the condition number is low. This hints at a good optimization behavior. For certain problems (cf. Sect. 4.4.4), a further desirable property of an activation function is *smoothness*. The softplus activation $\rho(z_j) = \beta^{-1} \cdot \log(1 + \exp(\beta \cdot z_j))$ is a smooth variant of the ReLU. However, it tends to produce activations that are not well centered, causing the condition number to be high. The centered softplus activation $\rho(z_j) = \beta^{-1} \cdot \big[\log(1 + \exp(\beta \cdot z_j)) - \log(2)\big]$ corrects for this behavior by ensuring $\rho(0) = 0$.

4.3.4 Initialization and Network Size

A common practice for initializing a neural network is to draw the weights randomly from the Gaussian distribution $w_{jk} \sim \mathcal{N}(0, \sigma_{jk}^2)$ where σ_{jk}^2 is a variance parameter. A motivation for randomness is to break symmetries in the parameter space. This avoids scenarios where two neurons with same biases, incoming weights, and outgoing weights, would always receive the same updates and never become different. For neural networks with ReLU activations, it was recommended in [34] to set the variance to

$$\sigma_{jk}^2 = 2 \cdot (\sum_j 1_{jk})^{-1},$$

i.e., inversely proportional to the number of incoming connections, as a way to keep activations at each layer on the same scale. Similarly to Sects. 4.3.2 and 4.3.3, this scaling can be motivated from a Hessian-based analysis (see also [28]). A block-diagonal approximation of the whole Hessian is first constructed from the Hessian matrices of each neuron at each layer: $H = \mathrm{diag}\{H_j, H_{j'}, H_{j''}, \ldots, H_k, H_{k'}, H_{k''}, \ldots, H_{\mathrm{out}}\}$. Eigenvalues of H are then given by the collection of eigenvalues of the different blocks. Reducing the condition number can be achieved by requiring each block to have eigenvalues on a similar scale. In particular, remembering that $[H_k]_{jj'} \approx 2\langle a_j a_{j'} \delta_k^2 \rangle$ we can impose the stronger requirement that (1) activations at each layer are on the same scale, and (2) sensitivities at each layer are also on the same scale. The weight scaling heuristic above implements the first requirement. Fulfilling the second requirement would require an alternate scaling of the weights, with variance inversely proportional to the number of *outgoing* connections. These seemingly contradicting requirements can be reconciled by using the same number of neurons in each layer.

One last parameter of the neural network that influences training time is its size. On the one hand, a small network will be fast to evaluate; however, it may not be expressive enough to reach zero error. On the other hand, a large neural network will be able to represent a broader class of functions and reach zero training error more easily; however, each training iteration will take longer. Therefore, the number of neurons in the network should be large enough to easily and accurately represent the target values, but not larger than necessary.

4.3.5 Learning Rate, Momentum, and Mini-Batches

Once we have built and initialized the neural network, the next step is to train it. The most important training hyperparameter is the learning rate γ. A simple way to choose it is to start with a fairly large learning rate, e.g., $\gamma = 1.0$, and if learning diverges, repeatedly slash the learning rate by a factor 10.

Because techniques outlined in Sects. 4.3.2–4.3.4 are heuristics based only on fragments of the true Hessian matrix, the conditioning of the optimization problem is likely to remain suboptimal. This remaining ill-conditioning can be handled at training time by incorporating *momentum* in the gradient descent:

$$g \leftarrow \eta \cdot g + \nabla_\theta \mathcal{E}(\theta)$$

$$\theta \leftarrow \theta - \gamma \cdot g$$

where $0 \leq \eta < 1$ is a hyperparameter. Using a strong momentum speeds up convergence along the directions of the parameter space associated with low Hessian eigenvalues. Typical choices of momentum are $\eta = 0.9$ or $\eta = 0.99$. Other popular choices of optimization algorithms include the Adam optimizer [35] which further improves convergence by incorporating a second-order moment. This optimizer is available in neural network frameworks such as PyTorch or Keras.

When the dataset is composed of more than a few hundreds training examples, evaluation of the error function and its gradient becomes very expensive, both computationally and in terms of memory. In that case, it is common to use *stochastic gradient descent* [36, 37], a technique that substitutes at each iteration the error gradient $\nabla_\theta \mathcal{E}(\theta)$ by the gradient of an estimate of the error built from a single example:

$$\widehat{\nabla}_\theta \mathcal{E}(\theta) = \nabla_\theta \big[f(\boldsymbol{x}_n, \theta) - t_n\big]^2 \qquad n \sim \text{random}\{1, \ldots, N\}.$$

The index n is selected randomly at each iteration. In practice, it is common to not consider a single example, but to estimate the gradient at each iteration from a *mini-batch* containing somewhere between 10 and 100 randomly selected examples.

4.4 Improving Neural Network Generalization

Application of the practical steps of Sect. 4.3 will help the neural network to reach quickly a low training error $\mathcal{E}(\theta)$. Training error however does not tell us how well the model will generalize to new data points. For this, we need to impose a preference among all models that correctly fit the training data, for those that are likely to generalize well to new data points.

4.4.1 Model Regularization

Regularization consists of applying a preference for simple models [38]. For example, among all models that fit the training data, we might retain the one that has the fewest variations or that is the smoothest. In the context of neural networks, a common regularization technique is *weight decay* [39]. The neural network is trained to match the target values while at the same time pulling its weights to smaller values. This procedure can be implemented by adding the squared weights to the objective function:

$$\mathcal{E}_{\text{reg}}(\theta) = \mathcal{E}(\theta) + \lambda \cdot \Big[\textstyle\sum_{ij} w_{ij}^2 + \sum_{jk} w_{jk}^2 + \sum_k v_k^2\Big]$$

The larger the hyperparameter λ, the smaller the weights will get. In practice, on interpolation tasks, λ should be set high enough for the weights to decay quickly enough, but not too high in order to still be able to fit the training data perfectly. In

PyTorch, weight decay can be implemented by adding the squared weights to the original error function:

```
E += lambd * sum((p**2).sum() for p in [W1,W2,v])
```

We then call `backward()` on this new quantity to get the gradients.

Other popular regularization techniques that can be applied in this context include pruning [40] or bagging [41]. Regularization techniques are generally applicable and make relatively few assumptions about the task. Next sections will show that generalization can be further improved by introducing problem-specific *prior knowledge* into the model and the input representation.

4.4.2 Invariant Input Representations

The problem of building invariant input representations has been extensively studied in the context of predicting molecular electronic properties [42–46]. We discuss below two approaches to introduce invariance: feature selection and data extension.

Feature selection [47] aims to map the data to a new representation that retains task-relevant features while discarding information to which the model should be invariant. Consider the toy problem of Fig. 4.6 where we would like to learn an energy function by integrating prior knowledge such as translation and rotation invariance. One representation that readily incorporates this invariance is the matrix

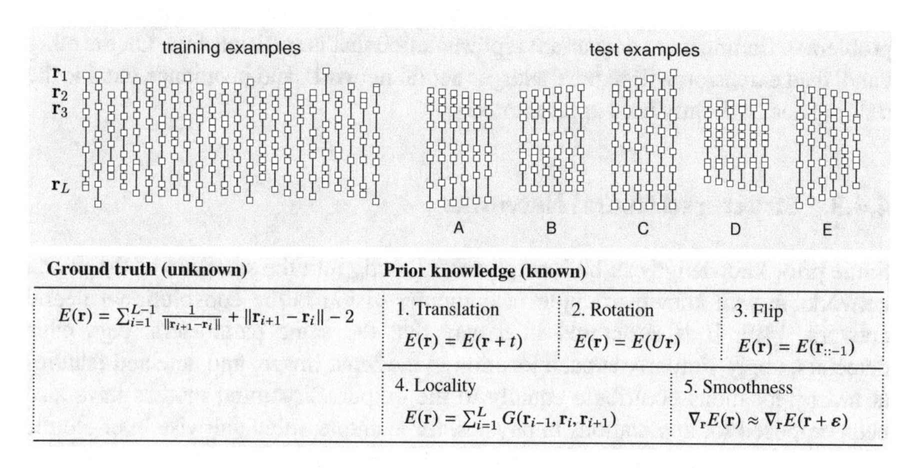

Fig. 4.6 Toy dataset representing various configurations of a one-dimensional system of eight elements. The system is governed by some ground truth energy function assumed to be unknown, but for which we have some prior knowledge that we would like to incorporate in our model

of pairwise distances:

$$\Phi(\mathbf{r}) = \begin{pmatrix} \|\mathbf{r}_1 - \mathbf{r}_2\|, & \|\mathbf{r}_1 - \mathbf{r}_3\|, & \dots & \|\mathbf{r}_1 - \mathbf{r}_L\| \\ & \|\mathbf{r}_2 - \mathbf{r}_3\|, & & \|\mathbf{r}_2 - \mathbf{r}_L\| \\ & & \ddots & \vdots \\ & & & \|\mathbf{r}_{L-1} - \mathbf{r}_L\| \end{pmatrix} \tag{4.3}$$

Training a neural network on such representation ensures that the predictions of the model also have the desired invariance. An important property to verify when building a representation is the absence of collision in feature space between systems with different energies, i.e., $E(\mathbf{r}) \neq E(\mathbf{r}') \Rightarrow \Phi(\mathbf{r}) \neq \Phi(\mathbf{r}')$. If two systems collide, it is indeed no longer possible for the neural network to produce different predictions.

Another approach to implementing invariance is *data augmentation* [31, 48]. Here, instead of building a better representation, we augment the data with all transformations that leave the quantity to predict invariant. In the toy example of Fig. 4.6, the prior knowledge we have about flip invariance can be induced by building the extended dataset:

$$\mathcal{D}_{\text{new}} = \bigcup_{(\mathbf{r},E)\in\mathcal{D}} \{(\mathbf{r}, E), (\mathbf{r}_{::-1}, E)\} \tag{4.4}$$

Data extension delegates to the neural network the task of learning the required invariance. Neural networks work well with data extension, as they can scale to potentially very large datasets. The main advantage of data extension is to avoid the problem of finding a good invariant representation that is collision-free. On the other hand, data extension will require a larger neural network, and invariance outside the training data will only hold approximately.

4.4.3 Structured Neural Networks

Some prior knowledge can be incorporated directly into the structure of the neural network. A well-known example in computer vision is the convolutional neural network [49]. It is structured in a way that the same parameters, e.g., edge detectors, apply similarly at each location in the input image, and detected features at nearby locations contribute equally to the output. Structured models have also been proposed for applications in physics, for example, fitted pairwise inter-atomic potentials [21], fitted local environments [22], or more recently, message passing neural networks [18, 50, 51]. Consider again the toy example of Fig. 4.6, where we now focus on implementing the prior on the locality of the energy function.

Figure 4.7 depicts two possible neural network architectures for predicting energies. The first one is a plain network that takes as input a flattened version of the representation $\Phi(\mathbf{r})$, normalizes it, and maps it to the output through a sequence

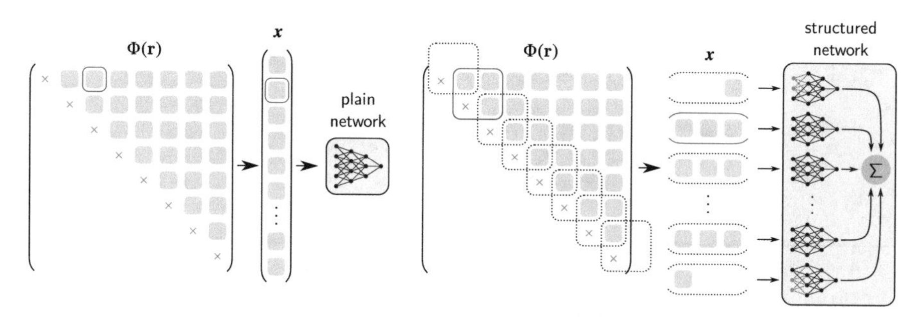

Fig. 4.7 Plain and structured neural networks that predict the energy of systems from their feature representation

of fully connected layers. The second one is a structured network that applies the same subnetwork multiple times to distinct groups of variables, and sums the corresponding outputs into a global energy prediction. Starting from the plain neural network of Sect. 4.2.1, we get the following sequence of layers:

$$
\begin{aligned}
&\forall_{m=1}^{M}, \forall_j : && z_{mj} = \textstyle\sum_i x_{mi} w_{ij} + b_j && a_{mj} = \rho(z_{mj}) && \text{(layer 1)}\\
&\forall_{m=1}^{M}, \forall_k : && z_{mk} = \textstyle\sum_j a_{mj} w_{jk} + b_k && a_{mk} = \rho(z_{mk}) && \text{(layer 2)}\\
&\forall_{m=1}^{M} : && y_m = \textstyle\sum_k a_{mk} v_k + c && && \text{(layer 3)}\\
& && y = \textstyle\sum_{m=1}^{M} y_m && && \text{(layer 4)}
\end{aligned}
$$

While this neural network architecture appears more complicated, it can actually be implemented very easily from a standard neural network. For this, a `torch` tensor `X` of dimensions `M × N × 3` is prepared, containing the datasets of size `N × 3` received by the different subnetworks. Then, we take the readily implemented forward function from Sect. 4.2.1, and get the output of our structured neural network by simply applying:

```
Y = forward(X).sum(dim=0)
```

What happens is that the `matmul` operation used in the `forward` function views the added dimension of `X` as a batch and thus propagates the added dimension from layer to layer. The operation `.sum(dim=0)` applies the top-layer pooling along the dimension of the batch.

To demonstrate the practical benefit of the locality prior implemented by the structured network, a plain and a structured network are trained on the toy problem of Fig. 4.6. For the plain network, we choose layer sizes (28 × 5)-25-25-1. For the structured model, layer sizes for each subnetwork are (3 × 5)-25-25-1. The factor 5 for the input layer corresponds to the thermometer coding expansion. Weight decay

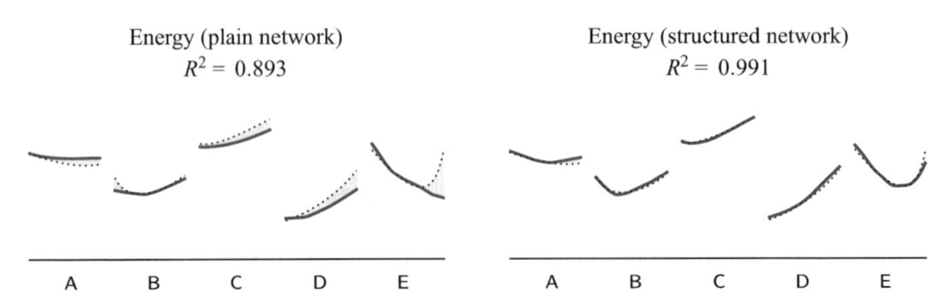

Fig. 4.8 Energy predictions along the test trajectories of the toy dataset. Results are shown for the plain and structured neural networks illustrated in Fig. 4.7. Shaded area represents the discrepancy between the prediction (solid blue line) and the ground truth (dotted line) and the R^2 score is the coefficient of determination quantifying the overall model accuracy

with $\lambda = 0.001$ is applied to both networks. Predictions on the test trajectories are shown in Fig. 4.8. We observe that the structured neural network predicts energies much more accurately than the plain network.

While using a structured network will be the preferred choice most of the time, there are however two potential limitations to keep in mind: First, the newly imposed restriction on the structure of the model makes the error function potentially harder to optimize. Use of a strong momentum (cf. Sect. 4.3.5) is becoming especially needed here. Second, the toy example of Fig. 4.6 is artificially structured to only contain short-range dependencies. This is not representative of all real-world systems, as some of them may exhibit long-range dependencies. For such systems, the structured neural network considered here will likely see its performance saturate as more data is being observed, and in the limit case be outperformed by a plain neural network.

4.4.4 Smoothness of the Prediction Function

The last prior knowledge we would like to incorporate for the toy example of Fig. 4.6 is smoothness. Here, the predicted energy function $E(\mathbf{r})$ is considered to be smooth if its gradient $\nabla_{\mathbf{r}} E(\mathbf{r})$ is continuous. The sequence of computations that leads to the gradient includes (1) the mapping from the system's coordinates to the neural network input, (2) the forward pass, (3) the backward pass, and (4) the gradient of the input mapping:

$$\boldsymbol{x} = \text{normalize}(\Phi(\mathbf{r})) \qquad \text{(neural network input)}$$

$$\vdots$$

$$\forall_{m=1}^{M}, \forall_k : \qquad a_{mk} = \rho\left(\textstyle\sum_j a_{mj} w_{jk} + b_k\right) \qquad \text{(forward pass)}$$

$$\vdots$$

$$\forall_{m=1}^{M}, \forall_j : \qquad \delta_{mj} = \rho'(z_{mj}) \textstyle\sum_k w_{jk} \delta_{mk} \qquad \text{(backward pass)}$$

$$\vdots$$

$$\nabla_{\mathbf{r}} E(\mathbf{r}) = \sum_{m=1}^{M} \sum_i \frac{\partial x_{mi}}{\partial \mathbf{r}} \cdot \delta_{mi} \qquad \text{(gradient of input)}$$

The mapping to the neural network input is composed of distance functions, standardization, and thermometer coding, all of them are continuous. Weighted sums occurring in the forward pass and backward pass are continuous. The gradient $\partial x_{mi}/\partial \mathbf{r}$ is continuous. The activation function ρ is also continuous. However, when ρ is the standard ReLU activation, the derivative ρ' used in the backward pass is *not* continuous. This discontinuity causes the overall gradient $\nabla_{\mathbf{r}} E(\mathbf{r})$ to be discontinuous, and the energy prediction $E(\mathbf{r})$ to be non-smooth. To enable smoothness, we can replace the ReLU activation by the centered softplus activation presented in Sect. 4.3.3 whose derivative is continuous.

To verify the effect of the activation function on the gradient continuity and the resulting accuracy of the energy prediction, we consider the structured model of Sect. 4.4.3 applied to our toy example. The neural network is trained once using the original ReLU activations, and once with the centered softplus activations. The quantity $\|\nabla_{\mathbf{r}} \mathbf{E}(\mathbf{r})\|$ is plotted in Fig. 4.9 (top) for both models along the test trajectories. The discontinuous behavior for the neural network with ReLU activations is clearly visible. Figure 4.9 (bottom) plots predicted energies for the

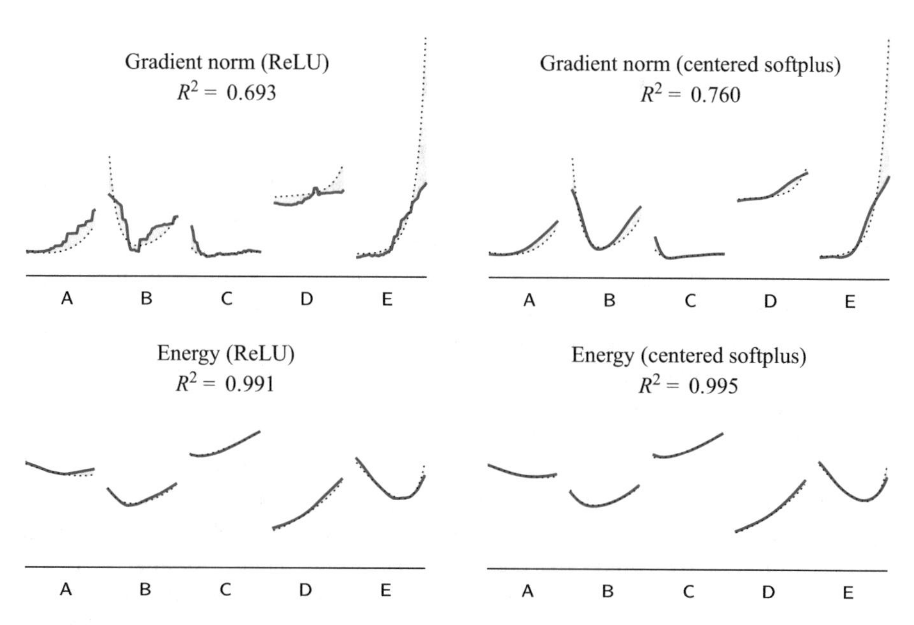

Fig. 4.9 Top: Gradient norm $\|\nabla_{\mathbf{r}} E(\mathbf{r})\|$ as derived from the energy predictions, and computed along the test trajectories of the toy dataset. Bottom: Energy prediction $E(\mathbf{r})$. Results are shown for structured neural networks with ReLU or centered softplus activations. Shaded area represents the discrepancy between the prediction (solid blue line) and the ground truth (dotted line), and R^2 is the coefficient of determination

same models and trajectories. We observe that the neural network with the centered softplus activation outperforms the non-smooth ReLU variant, reaching near-perfect predictions. Overall, this experiment demonstrates the benefit in terms of prediction accuracy of implementing the smoothness prior.

Generally, the smoothness implemented by the centered softplus function makes it particularly suitable in the context of modeling physical systems. However, the smoothness of the prediction is likely to be useful only as long as the true function to approximate is smooth. Interpolating non-smooth functions with a smooth model may lead to poor results.

4.5 Model Selection, Evaluation, and Understanding

So far, we have introduced techniques for learning a neural network efficiently and in a way that it generalizes well. However, which regularization technique to use, how to set the hyperparameters, and which prior knowledge to incorporate will typically be problem-dependent. Section 4.5.1 explains how to perform model selection quantitatively and automatically, and what are the implicit assumptions that are made about the task. Section 4.5.2 describes another set of techniques that are not subject to the same restrictions and that aim to present the neural network decision structure to a human.

4.5.1 Model Selection and Evaluation

A common procedure for selecting the best model and hyperparameters consists of randomly splitting the dataset in three parts. The first part is used to train a set of candidate models with different hyperparameters. The second part is used to evaluate each trained model and select the one with highest prediction accuracy. The last part of the data is used to produce an unbiased estimate of prediction accuracy for the selected model. The procedure is illustrated in Fig. 4.10. A more advanced version of this selection procedure, k-fold cross-validation, averages the evaluation on multiple data splits, so that a larger fraction of the data can be allocated to training. (See, e.g., [52] for further discussion.)

A first question to consider is which candidate models to select from. Given the flexibility of neural networks, taking all possible combinations of structures and hyperparameters would be computationally infeasible. Randomized or predictive approaches [53, 54] search only over a limited and sufficient collection of models, allowing to strongly reduce the computational requirements. Another common practice is to set most hyperparameters to best guesses, and exhaustively search for only two or three hyperparameters. In addition, for a more robust selection procedure, it is also recommended to include in the list of candidates, simple non-neural network models such as linear regression or Gaussian kernels.

The second ingredient that enters in the model selection procedure is the metric that we use to evaluate the model performance. Independently of the exact

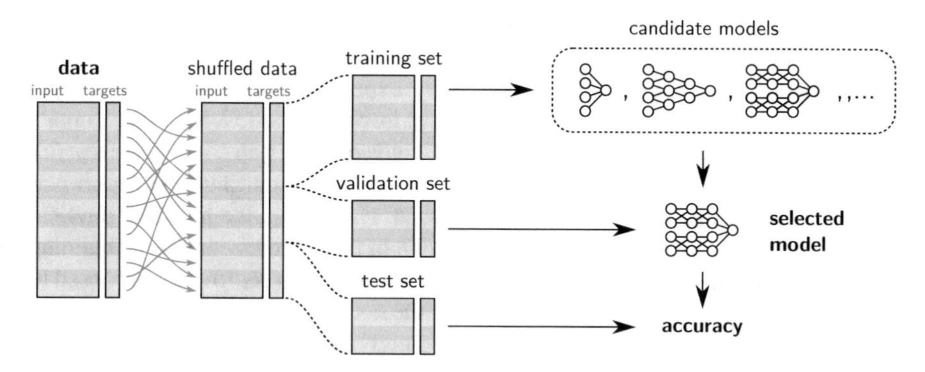

Fig. 4.10 Illustration of a typical procedure for model selection and evaluation. The data is shuffled and partitioned in three sets, with each set having a particular role

objective used to train the model, the evaluation metric should evaluate the *true* model performance, reflecting what the end-user would consider to be a good prediction. Examples of common evaluation metrics are the mean square error, the R^2 coefficient of determination, or the mean absolute error. We may also more specifically measure, e.g., whether the prediction correctly ranks systems from lowest to highest energy, or whether quantities derived from the prediction such as forces are accurate.

A limitation of this simple model validation technique, and more generally of any technique based on cross-validation, is the assumption that we can truly evaluate and quantify the usefulness of the model. Such assessment may be impossible for various reasons. For example, we do not always know in advance on which input distribution the end-user will apply the model. The model may perform well on the current data but fail to extrapolate to new scenarios. Furthermore, the selection procedure assumes that the user is able to quantify what he wants from a model. In practice, the user may not only be interested in getting accurate predictions, but also to extract insight from the model. The insights to be extracted from the model are in essence not quantifiable by the user.

4.5.2 Understanding Neural Network Predictions

It is often desirable for a user to be able to understand what the neural network model has learned [55, 56]. The structures learned by a well-functioning neural network could, for example, allow the user to refine his own problem's intuition. Furthermore, a neural network that closely aligns with the user's own physics knowledge is likely to extrapolate better outside the available data. Although prior knowledge about the task can be induced in the model through various mechanisms, e.g., input representation, neural network structure, and smooth nonlinearities, we still would like to verify that the trained model has used these mechanisms in the way they were intended.

Interpretability is commonly achieved by exposing to the user, which input features were relevant for the model to arrive at its prediction [57,58]. For example, if the input of our neural network is the collection of pairwise distances (cf. Eq. (4.3)), the produced explanation would highlight the contribution of each pair $(\mathbf{r}_i, \mathbf{r}_j)$ to the neural network output, that is, in which pairs the modeled physical system stores its energy. A first approach to obtain these scores consists of building interpretability structures directly into the model. For example, [59,60] proposed to incorporate a top-level summing structure in the network so that summands readily give the contributions of their associated features. The summing structure is also present in recent neural networks for predicting molecular electronic properties [61,62], as well as in the simple structured neural network of Sect. 4.4.3. For this last example, we note that summands are not bound to a single input feature, but depend on three input features. Thus, one still needs to determine which of these three features are actually being used. Various methods have been proposed to determine the relevance of individual input features [58, 63–65]. In the following, we focus on layer-wise relevance propagation [58], which identifies the contribution of input features via a reverse propagation pass in the neural network.

4.5.3 Layer-Wise Relevance Propagation

Layer-wise relevance propagation (LRP) [58] is a technique to systematically attribute the prediction of a neural network to its input features. It is applicable to general neural network structures including CNNs [58,66] and LSTMs [67]. LRP is based on a conservative reverse propagation procedure in the neural network that is illustrated in Fig. 4.11.

The notation $[y]_j$ indicates the share of the output that flows backward through neuron j. In the context of deep networks with ReLU nonlinearities, a practical propagation rule is LRP-γ [66]:

$$[y]_j = \sum_k \frac{a_j g(w_{jk})}{g(b_k) + \sum_j a_j g(w_{jk})} [y]_k$$

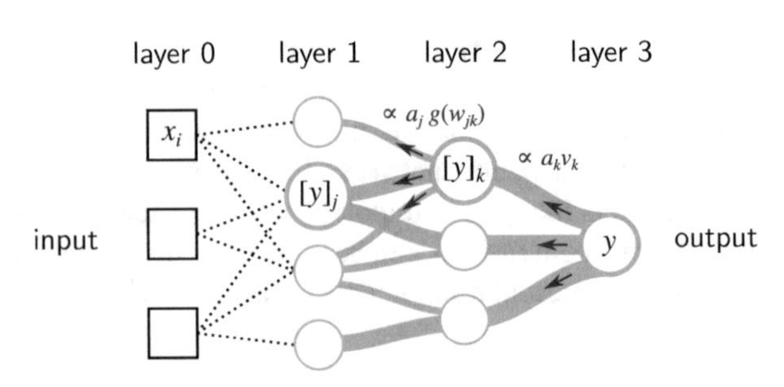

Fig. 4.11 Illustration of the LRP procedure where the prediction is propagated from the output towards the input features, by means of propagation rules

where $g(w) = w + \gamma \max(0, w)$, with a hyperparameter $\gamma \geq 0$. A large value for γ prevents the emergence of strong positive and negative scores in redistribution process but it also makes the resulting explanation less sensitive to local variations of the prediction. Typical values for γ are between 0 and 1. When neurons are linear (e.g., in the top layer), we must choose $\gamma = 0$. The LRP-γ propagation rule is applicable to all layers that receive positive activations as input. For the first layer where inputs $(x_i)_i$ are signed, we can replace the contribution terms $a_j g(w_{jk})$ by the symmetrized variant $g(x_i)g(w_{ij}) + g(-x_i)g(-w_{ij})$ where contributions from positive and negative input values are treated equally.

We now come to the question of implementation. A direct approach for implementing LRP is to manually iterate from the top layer to the bottom layer. However, the implementation can be made more concise by leveraging automatic differentiation. We first observe that, when expressing the propagated scores as $[y]_j = a_j c_j$ and $[y]_k = a_k c_k$, the standard LRP-γ propagation procedure can be rewritten as:

$$c_j = \sum_k g(w_{jk}) \frac{\rho(\sum_j a_j w_{jk} + b_k)}{g(b_k) + \sum_j a_j g(w_{jk})} c_k$$

This equation is similar to the one used for gradient propagation (cf. Sect. 4.2.2), except for the ReLU derivative which is replaced by a more complex term, and for the weight w_{jk} that becomes $g(w_{jk})$. Our strategy will be to redefine neurons of the forward pass in a way that their output remains the same, but results in the LRP computation when calling the automatic differentiation mechanism. This can be achieved by constructing the neuron:

$$a_k = \big(g(b_k) + \textstyle\sum_j a_j g(w_{jk})\big) \cdot \Big[\rho\big(\textstyle\sum_j a_j w_{jk} + b_k\big) \Big/ \big(g(b_k) + \textstyle\sum_j a_j g(w_{jk})\big)\Big]_{\text{const.}}$$

where the right-hand side is treated as constant. In PyTorch, the forward pass of Sect. 4.2.1 can be reimplemented as:

```
def forward(X):
    # layer 1 ...

    # layer 2
    Z2 = A1.matmul(g(W2)) + g(b2)
    A2 = (A1.matmul(W2) + b2).clamp(min=0)
    A2 = Z2 * (A2 / Z2).data

    # layer 3 ...
```

where `().data` detaches the variable from the graph used for automatic differentiation, making it effectively constant. Once the new architecture has been built, the

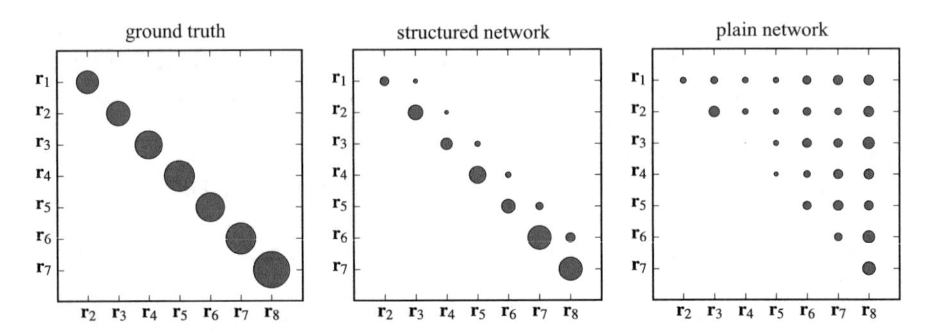

Fig. 4.12 Contribution of input features to the predicted energy, summed over the whole test set. Circle size indicates the magnitude of the contribution

LRP explanation is produced by simply calling:

```
def explain(X):
    X.requires_grad_(True)
    Y = forward(X).sum(); Y.backward()
    return X * X.grad
```

We consider the neural networks trained in Sect. 4.4.3 and use LRP to identify how much of the energy is attributed to each pair $(\mathbf{r}, \mathbf{r}')$. $\text{LRP}_{\gamma=0}$ is applied in the top layer, $\text{LRP}_{\gamma=1}$ in the second layer, and its symmetrized variant in the first layer. Once the explanation in terms of the neural network input is obtained, the scores are sum-pooled over the dimensions of the thermometer coding expansion. Relevance scores are then summed over all trajectories in the test set to get a dataset-wide explanation. Results are shown in Fig. 4.12.

In both cases, the inferred contributions do not sum to the ground truth, indicating that part of the predicted energy is modeled by neurons biases. We also observe that the plain network uses long-range dependencies to support its prediction. The structured network, which we have found to be more accurate on the prediction task, structurally prevents long-range dependencies from being used. As a result, inferred energy contributions are also closer to the ground truth.

4.5.4 What Did the Neural Network Actually Learn?

Clearly, the model selection procedure of Sect. 4.5.1 would have resulted in selecting the structured network over the plain network on the basis of higher predicting accuracy. The relation between high predicting accuracy and the structure of the network gives some clue to the user on the actual structure of the physical system. Yet, the LRP analysis reveals that even when the model reaches high test set accuracy, the decision structure of the learned model may still differ in certain aspects from the ground truth [68]. We note that this observation is not necessarily contradictory: The representation $\Phi(\mathbf{r})$ given as input is indeed an overcomplete description of the system's state $\mathbf{r}$. Thus, nothing a priori prevents the machine learning model to base its prediction, e.g., on features $(\|\mathbf{r}_1 - \mathbf{r}_2\|, \|\mathbf{r}_1 - \mathbf{r}_3\|)$ rather than $(\|\mathbf{r}_1 - \mathbf{r}_2\|, \|\mathbf{r}_2 - \mathbf{r}_3\|)$. Both of them indeed contain all information about the relative positions of $(\mathbf{r}_1, \mathbf{r}_2, \mathbf{r}_3)$. In fact, the model may well choose to use all three distances in order to spread its dependency on multiple correlated features to increase robustness, or because the nonlinear relation to predict is more easily implemented in terms of indirect interactions.

To summarize, rather than converging to a physically plausible problem representation, the exact strategy implemented by the neural network, if not explicitly regularized for physical plausibility, might instead be driven by more technical factors, such as statistical robustness or representation power.

4.6 Conclusion

In this chapter, we have introduced neural networks, with a focus on their use in the context of physical systems modeling. Owing to their very high representation power and scalability, neural networks have been the object of sustained interest for solving a wide range of problems in the sciences, for example, in the field of atomistic simulations.

As neural network models are also intrinsically complex, training them can be a delicate task. We have outlined a number of practical steps that can be taken to facilitate this process. Once a neural network has been trained successfully, i.e., has minimized its training objective, its ability to generalize to new examples must be considered. The latter can be improved by application of common regularization techniques, and by refining its structure and input representation in a way that it incorporates prior knowledge on the modeled physical system. These techniques let the neural network learn complex problems from a fairly limited number of data points.

Furthermore, while a neural network model should be able to predict accurately, it is also important to make the model interpretable, so that the learned structures can be explored and assessed qualitatively by the user.

Acknowledgments This work was supported by the German Ministry for Education and Research as Berlin Center for Machine Learning (01IS18037I). The author is grateful to Klaus-Robert Müller for the valuable feedback.

References

1. C.M. Bishop, *Neural Networks for Pattern Recognition* (Oxford University Press, New York, 1995)
2. G. Montavon, G.B. Orr, K. Müller (eds.), in *Neural Networks: Tricks of the Trade*, 2nd edn. Lecture Notes in Computer Science, vol. 7700 (Springer, Berlin, 2012)
3. J. Schmidhuber, Neural Netw. **61**, 85 (2015)

4. Y. LeCun, Y. Bengio, G. Hinton, Nature **521**(7553), 436 (2015)
5. G. Cybenko, Math. Control Signals Syst. **2**(4), 303 (1989)
6. Z. Lu, H. Pu, F. Wang, Z. Hu, L. Wang, in *Advances in Neural Information Processing Systems*, vol. 30 (2017), pp. 6231–6239
7. K. Fukushima, Biol. Cybern. **36**, 193 (1980)
8. G. Montavon, M.L. Braun, K. Müller, J. Mach. Learn. Res. **12**, 2563 (2011)
9. C. Cortes, V. Vapnik, Mach. Learn. **20**(3), 273 (1995)
10. K. Müller, S. Mika, G. Rätsch, K. Tsuda, B. Schölkopf, IEEE Trans. Neural Netw. **12**(2), 181 (2001)
11. B. Schölkopf, A. J. Smola, in *Learning with Kernels: Support Vector Machines, Regularization, Optimization, and Beyond*. Adaptive Computation and Machine Learning Series (MIT Press, Cambridge, MA, 2002)
12. A. Krizhevsky, I. Sutskever, G. E. Hinton, in *Neural Information Processing Systems* (2012), pp. 1106–1114
13. K. Simonyan, A. Zisserman, in *Third International Conference on Learning Representations* (2015)
14. M. Oquab, L. Bottou, I. Laptev, J. Sivic, in *IEEE Conference on Computer Vision and Pattern Recognition* (2014), pp. 1717–1724
15. R. Collobert, J. Weston, L. Bottou, M. Karlen, K. Kavukcuoglu, P.P. Kuksa, J. Mach. Learn. Res. **12**, 2493 (2011)
16. Y. Kim, in *Proceedings of the Conference on Empirical Methods in Natural Language Processing* (2014), pp. 1746–1751
17. P. Baldi, P. Sadowski, D. Whiteson, Nat. Commun. **5**, 4308 (2014)
18. K. T. Schütt, F. Arbabzadah, S. Chmiela, K.R. Müller, A. Tkatchenko, Nat. Commun. **8**, 13890 (2017)
19. A. Mardt, L. Pasquali, H. Wu, F. Noé, Nat. Commun. **9**(5) (2018)
20. L. Holmström, P. Koistinen, IEEE Trans. Neural Netw. **3**(1), 24 (1992)
21. S. Hobday, R. Smith, J. Belbruno, Model. Simul. Mater. Sci. Eng. **7**(3), 397 (1999)
22. J. Behler, M. Parrinello, Phys. Rev. Lett. **98**(14), 146401 (2007)
23. K. Yao, J.E. Herr, D.W. Toth, R. Mckintyre, J. Parkhill, Chem. Sci. **9**(8), 2261 (2018)
24. B. Nebgen, N. Lubbers, J.S. Smith, A.E. Sifain, A. Lokhov, O. Isayev, A.E. Roitberg, K. Barros, S. Tretiak, J. Chem. Theory Comput. **14**(9), 4687 (2018)
25. D.E. Rumelhart, G.E. Hinton, R.J. Williams, Nature **323**(6088), 533 (1986)
26. P.J. Werbos, in *System Modeling and Optimization* (Springer, Berlin, 1982), pp. 762–770
27. Y. LeCun, L. Bottou, G.B. Orr, K. Müller, in *Neural Networks: Tricks of the Trade*, 2nd edn. Lecture Notes in Computer Science, vol. 7700 (Springer, Berlin, 2012), pp. 9–48
28. J. Lafond, N. Vasilache, L. Bottou (2017). CoRR abs/1705.09319
29. A. Botev, H. Ritter, D. Barber, in *Proceedings of the 34th International Conference on Machine Learning* (2017), pp. 557–565
30. Y. Jeon, C. Choi, in *International Joint Conference Neural Network* (1999), pp. 1685–1690
31. G. Montavon, M. Rupp, V. Gobre, A. Vazquez-Mayagoitia, K. Hansen, A. Tkatchenko, K.-R. Müller, O. A. von Lilienfeld, New J. Phys. **15**(9), 095003 (2013)
32. X. Glorot, A. Bordes, Y. Bengio, in *International Conference on Artificial Intelligence and Statistics* (2011), pp. 315–323
33. M.D. Zeiler, M. Ranzato, R. Monga, M.Z. Mao, K. Yang, Q.V. Le, P. Nguyen, A.W. Senior, V. Vanhoucke, J. Dean, G.E. Hinton, in *IEEE International Conference on Acoustics, Speech and Signal Processing* (2013), pp. 3517–3521
34. K. He, X. Zhang, S. Ren, J. Sun, in *IEEE International Conference on Computer Vision* (2015), pp. 1026–1034
35. D.P. Kingma, J. Ba, in *Third International Conference on Learning Representations* (2015)
36. L. Bottou, in *Proceedings of Neuro-Nîmes*, vol. 91 (EC2, Nimes, 1991)
37. L. Bottou, in *Neural Networks: Tricks of the Trade*, 2nd edn. Lecture Notes in Computer Science, vol. 7700 (Springer, Berlin, 2012), pp. 421–436

38. V.N. Vapnik, *The Nature of Statistical Learning Theory*, 2nd edn. Statistics for Engineering and Information Science (Springer, Berlin, 2000)
39. A. Krogh, J.A. Hertz, in *Advances in Neural Information Processing Systems*, vol. 4 (1991), pp. 950–957
40. R. Reed, IEEE Trans. Neural Netw. **4**(5), 740 (1993)
41. L. Breiman, Mach. Lear. **24**(2), 123 (1996)
42. M. Rupp, A. Tkatchenko, K.-R. Müller, O.A. von Lilienfeld, Phys. Rev. Lett. **108**, 058301 (2012)
43. K. Hansen, F. Biegler, R. Ramakrishnan, W. Pronobis, O.A. von Lilienfeld, K.-R. Müller, A. Tkatchenko, J. Phys. Chem. Lett. **6**(12), 2326 (2015)
44. F.A. Faber, L. Hutchison, B. Huang, J. Gilmer, S.S. Schoenholz, G.E. Dahl, O. Vinyals, S. Kearnes, P.F. Riley, O.A. von Lilienfeld, J. Chem. Theory Comput. **13**(11), 5255 (2017)
45. S. Chmiela, A. Tkatchenko, H.E. Sauceda, I. Poltavsky, K.T. Schütt, K.-R. Müller, Sci. Adv. **3**(5), e1603015 (2017)
46. S. Chmiela, H.E. Sauceda, K.-R. Müller, A. Tkatchenko, Nat. Commun. **9**, 3887 (2018)
47. I. Guyon, A. Elisseeff, in *Feature Extraction—Foundations and Applications*. Studies in Fuzziness and Soft Computing, vol. 207 (Springer, Berlin, 2006), pp. 1–25
48. P.Y. Simard, Y. LeCun, J.S. Denker, B. Victorri, in *Neural Networks: Tricks of the Trade*, 2nd edn. Lecture Notes in Computer Science, vol. 7700 (Springer, Berlin, 2012), pp. 235–269
49. Y. LeCun, P. Haffner, L. Bottou, Y. Bengio, in *Shape, Contour and Grouping in Computer Vision* (Springer, Berlin, 1999), pp. 319–345
50. J. Gilmer, S.S. Schoenholz, P.F. Riley, O. Vinyals, G.E. Dahl, in *Proceedings of the 34th International Conference on Machine Learning* (2017), pp. 1263–1272
51. K.T. Schütt, H.E. Sauceda, P.-J. Kindermans, A. Tkatchenko, K.-R. Müller, J. Chem. Phys. **148**(24), 241722 (2018)
52. K. Hansen, G. Montavon, F. Biegler, S. Fazli, M. Rupp, M. Scheffler, O. A. von Lilienfeld, A. Tkatchenko, K.-R. Müller, J. Chem. Theory Comput. **9**(8), 3404 (2013)
53. J. Bergstra, Y. Bengio, J. Mach. Learn. Res. **13**, 281 (2012)
54. J. Bergstra, R. Bardenet, Y. Bengio, B. Kégl, in *Advances in Neural Information Processing Systems*, vol. 24 (2011), pp. 2546–2554
55. Z.C. Lipton, ACM Queue **16**(3), 30 (2018)
56. W. Samek, G. Montavon, A. Vedaldi, L.K. Hansen, K.-R. Müller (eds.), *Explainable AI: Interpreting, Explaining and Visualizing Deep Learning*. Lecture Notes in Computer Science, vol. 11700 (Springer, Berlin, 2019)
57. D. Baehrens, T. Schroeter, S. Harmeling, M. Kawanabe, K. Hansen, K. Müller, J. Mach. Learn. Res. **11**, 1803 (2010)
58. S. Bach, A. Binder, G. Montavon, F. Klauschen, K.-R. Müller, W. Samek, PLoS One **10**(7), e0130140 (2015)
59. R. Caruana, Y. Lou, J. Gehrke, P. Koch, M. Sturm, N. Elhadad, in *Proceedings of the 21th ACM SIGKDD International Conference on Knowledge Discovery and Data Mining* (2015), pp. 1721–1730
60. B. Zhou, A. Khosla, À. Lapedriza, A. Oliva, A. Torralba, in *IEEE Conference on Computer Vision and Pattern Recognition* (2016), pp. 2921–2929
61. K. Yao, J.E. Herr, S.N. Brown, J. Parkhill, J. Phys. Chem. Lett. **8**(12), 2689 (2017)
62. K.T. Schütt, M. Gastegger, A. Tkatchenko, K.-R. Müller, in *Explainable AI: Interpreting, Explaining and Visualizing Deep Learning*. Lecture Notes in Computer Science, vol. 11700 (Springer, Berlin, 2019)
63. M.T. Ribeiro, S. Singh, C. Guestrin, in *Proceedings of the 22nd ACM SIGKDD International Conference on Knowledge Discovery and Data Mining* (2016), pp. 1135–1144
64. R.C. Fong, A. Vedaldi, In *IEEE International Conference on Computer Vision* (2017), pp. 3449–3457
65. M. Sundararajan, A. Taly, Q. Yan, in *Proceedings of the 34th International Conference on Machine Learning* (2017), pp. 3319–3328

66. G. Montavon, A. Binder, S. Lapuschkin, W. Samek, K.-R. Müller, in *Explainable AI: Interpreting, Explaining and Visualizing Deep Learning*. Lecture Notes in Computer Science, vol. 11700 (Springer, Berlin, 2019)
67. L. Arras, J. Arjona-Medina, M. Widrich, G. Montavon, M. Gillhofer, K.-R. Müller, S. Hochreiter, W. Samek, in *Explainable AI: Interpreting, Explaining and Visualizing Deep Learning*. Lecture Notes in Computer Science, vol. 11700 (Springer, Berlin, 2019)
68. S. Lapuschkin, S. Wäldchen, A. Binder, G. Montavon, W. Samek, K.-R. Müller, Nat. Commun. **10**, 1096 (2019)

Part II

Incorporating Prior Knowledge: Invariances, Symmetries, Conservation Laws

Preface

When attempting to apply machine learning to atomistic simulations or prediction throughout chemical compound space, one will quickly face a crucial challenge—*how to smartly encode the input?* A naïve choice could be to stack the atomic types and positions into a vector and apply a standard non-linear regression method, such as a neural network or a Gaussian process with a radial basis function (RBF) kernel. While it is true that those methods are universal approximators and will converge to the true solution eventually, it may require a huge amount of training data. Generating this data will not be feasible for many of the applications presented in this book, due to the high computational cost of accurate quantum chemical reference calculations.

In order to substantially increase the data efficiency of machine learning approaches, we need to incorporate all available prior knowledge about atomistic systems: We know that our input is a set of categorical atom types and three-dimensional position vectors. We know that the order of the atoms does not change the chemical properties of the system and that many chemical properties are invariant or covariant to rotation and translation. These properties can be achieved by what is commonly called *feature engineering* or *descriptor design*, where the system is encoded in such a way that invariances and symmetries are reflected [1]. Finding suitable descriptors for molecules and materials has been a major aspect of research in recent years [2–7].

This part of the book will introduce several novel aspects when incorporating prior knowledge into the descriptor. Glielmo et al. [8] cover the construction of nonparametric n-body force fields using Gaussian processes (GPs). Several GP kernels are reviewed, in particular with respect to constraining the interaction order n. Chapter 6 [9] was originally published in the *Handbook of Materials Modeling* [10]. We reprint it here in an adapted version as it gives an excellent introduction to using the SOAP (Smooth Overlap of Atomic Positions) representation and incorporating physical principles into the machine learning framework.

Faber et al. [11] demonstrate how models for the accurate prediction of energies—here using the FCHL descriptor—can be generalized to incorporate response properties. In Chap. 7, [12] present the sGDML framework, consisting of vector-valued kernels that are constrained to model energy-conserving force fields only. In addition sGDML encompasses an efficient matching procedure that automatically extracts rigid and fluxional symmetries contained in the reference data.

In Chap. 9, [13] describe the development of machine learning models that generalize across phase transitions—a feature the authors call *physical extrapolation*, i.e., the extrapolation in physical input space, while interpolating in feature space. The authors discuss controlling model complexity using the Bayesian information criterium to avoid overfitting in the unknown domain.

While this part of the book encompasses mainly kernel-based learning methods, where prior domain knowledge allows the incorporation of invariances into representations, we will in part of the book also encounter representation learning using deep neural networks, where incorporating prior knowledge is mainly achieved through the choice of architecture [14–18].

Berlin, Germany — Kristof T. Schütt
Berlin, Germany — Stefan Chmiela
Basel, Switzerland — O. Anatole von Lilienfeld
Luxembourg, Luxembourg — Alexandre Tkatchenko
Kashiwa, Japan — Koji Tsuda
Berlin, Germany — Klaus-Robert Müller
September 2019

References

1. O.A. von Lilienfeld, R. Ramakrishnan, M. Rupp, A. Knoll, Int. J. Quantum Chem. **115**(16), 1084 (2015)
2. B.J. Braams, J.M. Bowman, Int. Rev. Phys. Chem. **28**(4), 577 (2009)
3. J. Behler, J. Chem. Phys. **134**(7), 074106 (2011)
4. A.P. Bartók, R. Kondor, G. Csányi, Phys. Rev. B **87**(18), 184115 (2013)
5. K.T. Schütt, H. Glawe, F. Brockherde, A. Sanna, K.-R. Müller, E. Gross, Phys. Rev. B **89**(20), 205118 (2014)
6. F. Faber, A. Lindmaa, O.A. von Lilienfeld, R. Armiento, Int. J. Quantum Chem. **115**(16), 1094 (2015)
7. F.A. Faber, L. Hutchison, B. Huang, J. Gilmer, S.S. Schoenholz, G.E. Dahl, O. Vinyals, S. Kearnes, P.F. Riley, O.A. von Lilienfeld (2017). arXiv:1702.05532
8. A. Glielmo, C. Zeni, A. Fekete, A. De Vita, in *Machine Learning for Quantum Simulations of Molecules and Materials*, ed. by K.T. Schütt, S. Chmiela, A. von Lilienfeld, A. Tkatchenko, K. Tsuda, K.-R. Müller. Lecture Notes in Physics (Springer, Berlin, 2019)
9. M. Ceriotti, M.J. Willatt, G. Csányi, in *Machine Learning for Quantum Simulations of Molecules and Materials*, ed. by K.T. Schütt, S. Chmiela, A. von Lilienfeld, A. Tkatchenko, K. Tsuda, K.-R. Müller. Lecture Notes in Physics (Springer, Berlin, 2019)
10. M. Ceriotti, M.J. Willatt, G. Csányi, *Handbook of Materials Modeling: Methods: Theory and Modeling* (2018), pp. 1–27

11. F.A. Faber, A.S. Christensen, O.A. von Lilienfeld, in *Machine Learning for Quantum Simulations of Molecules and Materials*, ed. by K.T. Schütt, S. Chmiela, A. von Lilienfeld, A. Tkatchenko, K. Tsuda, K.-R. Müller. Lecture Notes in Physics (Springer, Berlin, 2019)
12. S. Chmiela, H.E. Sauceda, A. Tkatchenko, K.-R. Müller, in *Machine Learning for Quantum Simulations of Molecules and Materials*, ed. by K.T. Schütt, S. Chmiela, A. von Lilienfeld, A. Tkatchenko, K. Tsuda, K.-R. Müller. Lecture Notes in Physics (Springer, Berlin, 2019)
13. R.A. Vargas-Hernández, R.V. Krems, in *Machine Learning for Quantum Simulations of Molecules and Materials*, ed. by K.T. Schütt, S. Chmiela, A. von Lilienfeld, A. Tkatchenko, K. Tsuda, K.-R. Müller. Lecture Notes in Physics (Springer, Berlin, 2019)
14. K.T. Schütt, F. Arbabzadah, S. Chmiela, K.-R. Müller, A. Tkatchenko, Nat. Commun. **8**, 13890 (2017)
15. J. Gilmer, S.S. Schoenholz, P.F. Riley, O. Vinyals, G.E. Dahl, in *Proceedings of the 34th International Conference on Machine Learning* (2017), pp. 1263–1272
16. K.T. Schütt, H.E. Sauceda, P.-J. Kindermans, A. Tkatchenko, K.-R. Müller, J. Chem. Phys. **148**(24), 241722 (2018)
17. N. Lubbers, J.S. Smith, K. Barros, J. Chem. Phys. **148**(24), 241715 (2018)
18. N. Thomas, T. Smidt, S. Kearnes, L. Yang, L. Li, K. Kohlhoff, P. Riley (2018). arXiv:1802.08219

Building Nonparametric n-Body Force Fields Using Gaussian Process Regression

5

Aldo Glielmo, Claudio Zeni, Ádám Fekete, and Alessandro De Vita

Abstract

Constructing a classical potential suited to simulate a given atomic system is a remarkably difficult task. This chapter presents a framework under which this problem can be tackled, based on the Bayesian construction of nonparametric force fields of a given order using Gaussian process (GP) priors. The formalism of GP regression is first reviewed, particularly in relation to its application in learning local atomic energies and forces. For accurate regression, it is fundamental to incorporate prior knowledge into the GP kernel function. To this end, this chapter details how properties of smoothness, invariance and interaction order of a force field can be encoded into corresponding kernel properties. A range of kernels is then proposed, possessing all the required properties and an adjustable parameter n governing the interaction order modelled. The order n best suited to describe a given system can be found automatically within the Bayesian framework by maximisation of the marginal likelihood. The procedure is first tested on a toy model of known interaction and later applied to two real materials described at the DFT level of accuracy. The models automatically selected for the two materials were found to be in agreement with physical intuition. More in general, it was found that lower order (simpler) models should

A. Glielmo (✉) · C. Zeni · Á. Fekete
Department of Physics, King's College London, London, UK
e-mail: aldo.glielmo@kcl.ac.uk; claudio.zeni@kcl.ac.uk

A. De Vita
Department of Physics, King's College London, London, UK

Dipartimento di Ingegneria e Architettura, University of Trieste, Trieste, Italy

K. T. Schütt et al. (eds.), *Machine Learning Meets Quantum Physics*, Lecture Notes in Physics 968, https://doi.org/10.1007/978-3-030-40245-7_5

be chosen when the data are not sufficient to resolve more complex interactions. Low n GPs can be further sped up by orders of magnitude by constructing the corresponding tabulated force field, here named “MFF”.

5.1 Introduction

The no free lunch (NFL) theorems proven by D. H. Wolpert in 1996 state that no learning algorithm can be considered better than any other (and than random guessing) when its performance is averaged uniformly over all possible functions [1]. Although functions appearing in real-world problems are certainly not uniformly distributed, this remarkable result seems to suggest that the search for the “best” machine learning (ML) algorithm able to learn any function in an “agnostic” fashion is groundless, and strongly justifies current efforts within the physics and chemistry communities aimed at the development of ML techniques that are particularly suited to tackle a *given* problem, for which prior knowledge is available and exploitable.

In the context of machine learning force field (ML-FF) generation, this resulted in a proliferation of different approaches based on artificial neural networks (NN) [2–11], Gaussian process (GP) regression [12–17] or linear expansions on properly defined bases [18–20]. Particularly within GP regression (the method predominantly discussed in this chapter), a considerable effort was directed towards the inclusion of the known physical symmetries of the target system (translations, rotations and permutations) in the algorithm as a prior piece of information. Among these, rotation symmetry proved the most cumbersome one to deal with, and received special attention. This typically involved either building explicitly invariant descriptors (as the Li et al. feature-matrix based on internal vectors [13]) or imposing the symmetry via an invariant [21] or covariant [14] integral to learn energies or forces. Clearly, many more detailed recipes than those featuring in the list above would be possible in virtually all situations, making the problem of selecting a single model for a particular task both interesting and unavoidable. In the following, we will argue that a good way of choosing among competing explanations is to follow the long-standing Occam’s razor principle and select the simplest model that is still able to provide a satisfactory explanation [22–24].

This general idea has found rigorous mathematical formulations. Within statistical learning theory, the complexity of a model can be measured by calculating its Vapnik–Chervonenkis (VC) dimension [25, 26]. The *VC dimension* of a model then relates to its *sample complexity* (i.e., the number of points needed to effectively train it) as one can prove that the latter is bounded by a monotonic function of the former [26, 27]. Similar considerations can also be made in a Bayesian context by noting that models with prior distributions concentrated around the true function (i.e., *simpler* models) have a lower sample complexity and will hence learn faster [28]. The above considerations suggest that a principled approach to learn a force field is to incorporate as much prior knowledge as is available on the function to be learned and the particular system at hand. When prior knowledge is not enough to

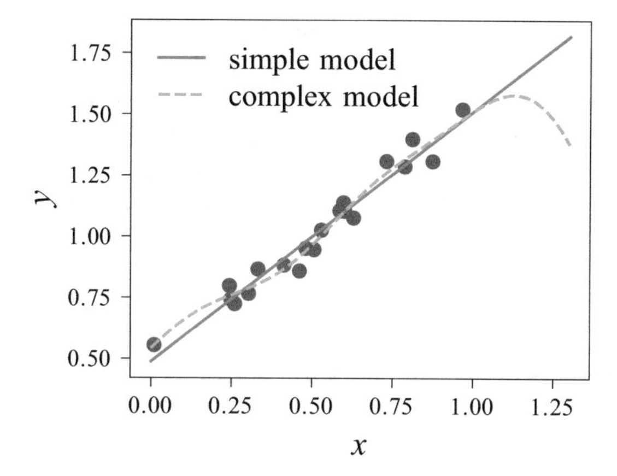

Fig. 5.1 A simple linear model (blue solid line) and a complex GP model (green dashed line) are fitted to some data points. In this situation, if we have prior knowledge that a linear trend underpins the data, we should enforce the blue model a priori; otherwise we should select the blue model by Occam's razor after the data becomes available, since it is the simplest one. The advantages of this choice lie in the greater interpretability and extrapolation power of the simpler model

decide among competing models, these should all be trained and tested, after which the simplest one that is still compatible with the desired target accuracy should be selected. This approach is illustrated in Fig. 5.1, where two competing models are considered for a one dimensional dataset.

In the rest of this chapter, we provide a step-by-step guide to the incorporation of prior knowledge and to model selection in the context of Bayesian regression based on GP priors (Sect. 5.2) and show how these ideas can be applied in practice (Sect. 5.3). Section 5.2 is structured as follows. In Sect. 5.2.1, we give a pedagogical introduction to GP regression, with a focus on the problem of learning a local energy function. In Sect. 5.2.2, we show how a local energy function can be learned in practice when using a database containing solely total energies and/or forces. In Sect. 5.2.3, we then review the ways in which physical prior information can (and should) be incorporated in GP kernel functions, focusing on smoothness (5.2.3.1), symmetries (5.2.3.2) and interaction order (5.2.3.3). In Sect. 5.2.4, we make use of the preceding section's results to define a set of kernels of tunable complexity that incorporate as much prior knowledge as is available on the target physical system. In Sect. 5.2.5, we show how Bayesian model selection provides a principled and "automatic" choice of the simplest model suitable to describe the system. For simplicity, throughout this chapter only systems of a single chemical species are discussed, but in Sect. 5.2.6, we briefly show how the ideas presented can be straightforwardly extended to model multispecies systems.

Section 5.3 focuses on the practical application of the ideas presented. In particular, Sect. 5.3.1 describes an application of the model selection method described in Sect. 5.2.5 to two different Nickel environments, represented as different subsets of a general Nickel database. We then compare the results obtained from this Bayesian model selection technique with those provided by a more heuristic model selection approach and show how the two methods, while being substantially different and optimal in different circumstances, typically yield similar results. The final Sect. 5.3.2 discusses the computational efficiency of GP predictions, and explain how a very simple procedure can increase by several orders of magnitude the evaluation speed of certain classes of GPs when on-the-fly training is not needed. The code used to carry out such a procedure is freely available as part of the "MFF" Python package [29].

5.2 Nonparametric n-body Force Field Construction

The most straightforward well-defined local property accessible to QM calculations is the force on atoms, which can be easily computed by way of the Hellman–Feynman theorem [30]. Atomic forces can be machine learned directly in various ways, and the resulting model can be used to perform molecular dynamics simulations, probe the system's free energy landscape, etc. [13, 14, 16, 31, 32]. We can however also define a *local energy* function $\varepsilon(\rho)$ representing the energy ε of an atom given a representation ρ of the set of positions of all the atoms surrounding it within a cutoff distance. Such a set of positions is typically called an *atomic environment* or an *atomic configuration*, and ρ could simply be a list of the atomic species and positions expressed in Cartesian coordinates, or any suitably chosen representation of these [13, 15, 21, 33].

Although local energies are not well-defined in quantum calculations, in the following section we will be focusing on GP models for learning this somewhat accessory function $\varepsilon(\rho)$, as this makes it easier to understand the key concepts [34]. We will also assume for simplicity that our ML model is trained on a database of local configurations and energies, although in practice $\varepsilon(\rho)$ is machine-learned from the atomic forces and total energies produced by QM codes. The details of how this can be practically done will be discussed in Sect. 5.2.2.

5.2.1 Gaussian Process Regression

In order to learn the local energy function $\varepsilon(\rho)$ yielding the energy of the atomic configuration ρ, we assume to have access to a database of reference calculations $\mathcal{D} = \{(\varepsilon_i^r, \rho_i)\}_{i=1}^N$ composed by N local atomic configurations $\boldsymbol{\rho} = (\rho_1, \ldots, \rho_N)^T$ and their corresponding energies $\boldsymbol{\varepsilon}^r = (\varepsilon_1^r, \ldots, \varepsilon_N^r)^T$. It is assumed that the energies have been obtained as

$$\varepsilon_i^r = \varepsilon(\rho_i) + \xi_i \tag{5.1}$$

where the noise variables ξ_i are independent zero mean Gaussian random variables ($\xi_i \sim \mathcal{N}(0, \sigma_n^2)$). This noise in the data can be imagined to represent the combined uncertainty associated with both training data and model used. For example, an important source of uncertainty is the *locality error* resulting from the assumption of a finite cutoff radius, outside of which atoms are treated as non-interacting. This assumption is necessary in order to define local energy functions but it never holds exactly.

The power of GP regression lies in the fact that $\varepsilon(\rho)$ is not constrained to be a given parametric functional form as in standard fitting approaches, but it is rather assumed to be distributed as a Gaussian stochastic process, typically with zero mean

$$\varepsilon(\rho) \sim \mathcal{GP}\left(0, k\left(\rho, \rho'\right)\right) \tag{5.2}$$

where k is the *kernel function* of the GP (also called *covariance function*). This notation signifies that for any finite set of input configurations $\boldsymbol{\rho}$, the corresponding set of local energies $\boldsymbol{\varepsilon} = (\varepsilon(\rho_1), \ldots, \varepsilon(\rho_N))^T$ will be distributed according to a multivariate Gaussian distribution whose covariance matrix is constructed through the kernel function:

$$\begin{cases} p(\boldsymbol{\varepsilon} \mid \boldsymbol{\rho}) & = \mathcal{N}(\mathbf{0}, \boldsymbol{K}) \\ \boldsymbol{K} & = \begin{pmatrix} k(\rho_1, \rho_1) & \cdots & k(\rho_1, \rho_N) \\ \vdots & \ddots & \vdots \\ k(\rho_N, \rho_1) & \cdots & k(\rho_N, \rho_N) \end{pmatrix} \end{cases}. \tag{5.3}$$

Given that both ξ_i and $\varepsilon(\rho_i)$ are normally distributed, and since the sum of two Gaussian random variables is also a Gaussian variable, one can write down the distribution of the reference energies ε_i^r of Eq. (5.1) as a new normal distribution whose covariant matrix is the sum of the original two:

$$\begin{cases} p\left(\boldsymbol{\varepsilon}^r \mid \boldsymbol{\rho}\right) & = \mathcal{N}(\mathbf{0}, \boldsymbol{C}) \\ \boldsymbol{C} & = \boldsymbol{K} + \mathbf{1}\sigma_n^2. \end{cases} \tag{5.4}$$

Building on this closed form (Gaussian) expression for the probability of the reference data, we can next calculate the *predictive distribution*, i.e., the probability distribution of the local energy value ε^* associated with a new target configuration ρ^*, for the given training dataset $\mathcal{D} = (\boldsymbol{\rho}, \boldsymbol{\varepsilon}^r)$ —the interested reader is referred to the two excellent references [35, 37] for details on the derivation. This is:

$$\begin{cases} p\left(\varepsilon^* \mid \rho^*, \mathcal{D}\right) & = \mathcal{N}\left(\hat{\varepsilon}\left(\rho^*\right), \hat{\sigma}^2\left(\rho^*\right)\right) \\ \hat{\varepsilon}\left(\rho^*\right) & = \boldsymbol{k}^T \boldsymbol{C}^{-1} \boldsymbol{\varepsilon}^r \\ \hat{\sigma}^2\left(\rho^*\right) & = k\left(\rho^*, \rho^*\right) - \boldsymbol{k}^T \boldsymbol{C}^{-1} \boldsymbol{k} \end{cases}, \tag{5.5}$$

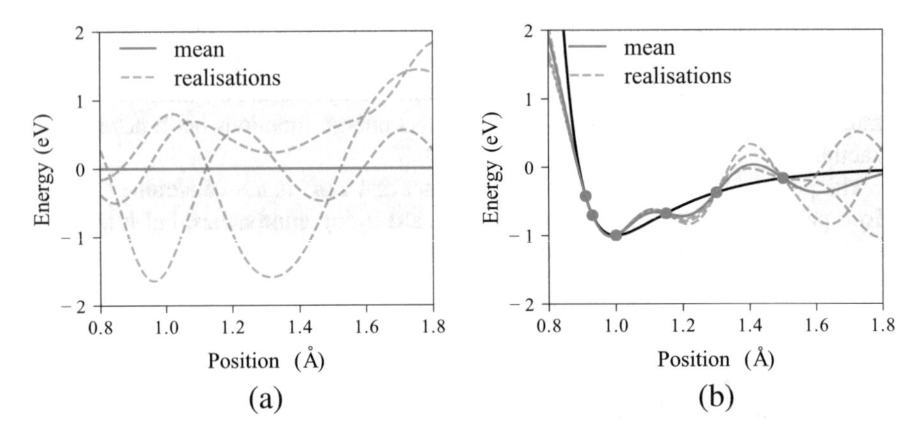

Fig. 5.2 Pictorial view of GP learning of a LJ dimer. Panel (**a**): mean, standard deviation and random realisations of the prior stochastic process, which represents our belief on the dimer interaction before any data is seen. Panel (**b**): posterior process, whose mean passes through the training data and whose variance provides a measure of uncertainty

where we defined the vector $\boldsymbol{k} = (k(\rho^*, \rho_1), \ldots, k(\rho^*, \rho_N))^T$. The mean function $\hat{\varepsilon}(\rho)$ of the predictive distribution is now our "best guess" for the true underlying function as it can be shown that it minimises expected error.[1]

The mean function is often equivalently written down as a linear combination of kernel functions evaluated over all database entries

$$\hat{\varepsilon}(\rho) = \sum_{d=1}^{N} k(\rho, \rho_d)\alpha_d, \tag{5.6}$$

where the coefficients are readily computed as $\alpha_d = (\boldsymbol{C}^{-1}\boldsymbol{\varepsilon})_d$. The posterior variance of ε^* provides a measure of the uncertainty associated with the prediction, normally expressed as the standard deviation $\hat{\sigma}(\rho)$.

The GP learning process can be thought of as an update of the prior distribution Eq. (5.2) into the posterior Eq. (5.5). This update is illustrated in Fig. 5.2, in which GP regression is used to learn a simple Lennard Jones (LJ) profile from a few dimer data. In particular, Fig. 5.2a shows the prior GP (Eq. (5.2) while Fig. 5.2b shows the posterior GP, whose mean and variance are those of the predictive distribution Eq. (5.5). By comparing the two panels, one notices that the mean function (equal

[1]Choosing a squared error function $L = (\bar{\varepsilon}(\rho) - \varepsilon)^2$, the expected error under the posterior distribution reads $\langle L \rangle = \int d\varepsilon\, p(\varepsilon \mid \rho, \mathcal{D})(\bar{\varepsilon}(\rho) - \varepsilon)^2$. Minimising this quantity with respect to the unknown optimal prediction $\bar{\varepsilon}(\rho)$ can be done by equating the functional derivative $\delta\langle L \rangle / \delta\bar{\varepsilon}(\rho)$ to zero, yielding the condition $(\bar{\varepsilon}(\rho) - \langle \varepsilon \rangle) = 0$, proving that the optimal estimate corresponds to the mean $\hat{\varepsilon}(\rho)$ of the predictive distribution in Eq. (5.5). One can show that choosing an absolute error function $L = |\bar{\varepsilon}(\rho) - \varepsilon|$ makes the mode of the predictive distribution the optimal estimate, this however coincides with the mean in the case of Gaussian distributions.

to zero in the prior process) approximates the true function (black solid line) by passing through the reference calculations. Clearly, the posterior standard deviation (uniform in the prior) shrinks to zero at the points where data is available (as we set the intrinsic noise σ_n to zero) to then increase again away from them. Three random function samples are also shown for both prior and posterior process.

5.2.2 Local Energy from Global Energies and Forces

The forces acting on atoms are well-defined local property accessible to QM calculations, easily computed by way of the Hellman–Feynman theorem [30]. As a consequence, GP regression can in principle be used to learn a force field directly on a database of quantum forces, as done, for instance, in Refs. [13, 14, 31]. Local atomic energies on the contrary cannot be computed in QM calculations, which can only provide the *total* energy of the full system. However, the material presented in the previous section, in addition to being of pedagogical importance, is still useful in practice since local energy functions can be learned from observations of total energies and forces only.

Mathematically this is possible since any sum, or derivative, of a Gaussian process is also a Gaussian process [35], and the main ingredients needed for learning are hence the covariances (kernels) between these Gaussian variables. In the following, we will see how kernels for total energies and forces can be obtained starting from a kernel for local energies, and how these derived kernels can be used to learn a local energy function from global energy and force information.

Total Energy Kernels The total energy of a system can be modelled as a sum of the local energies associated with each local atomic environment

$$E(\{\rho_a\}) = \sum_{a=1}^{N_a} \varepsilon(\rho_a) \tag{5.7}$$

and if the local energy functions ε in the above equation are distributed according to a zero mean GP, then also the global energy E will be GP variable with zero mean. To calculate the kernel functions $k^{\varepsilon E}$ and k^{EE} providing the covariance between local and global energies and between two global energies, one simply needs to take the expectation with respect to the GP of the corresponding products

$$\begin{aligned} k^{\varepsilon E}\left(\rho_a, \{\rho_b'\}\right) &= \left\langle \varepsilon(\rho_a) E\left(\{\rho_b'\}\right)\right\rangle \\ &= \sum_{b=1}^{N_a'} \left\langle \varepsilon(\rho_a) \varepsilon\left(\rho_b'\right)\right\rangle \\ &= \sum_{b=1}^{N_a'} k(\rho_a, \rho_b). \end{aligned} \tag{5.8}$$

$$k^{EE}\left(\{\rho_a\},\{\rho'_b\}\right) = \left\langle E(\{\rho_a\})E\left(\{\rho'_b\}\right)\right\rangle = \sum_{a=1}^{N_a}\sum_{b=1}^{N'_a}\left\langle \varepsilon(\rho_a)\varepsilon\left(\rho'_b\right)\right\rangle = \sum_{a=1}^{N_a}\sum_{b=1}^{N'_a} k(\rho_a,\rho_b). \tag{5.9}$$

Note that we have allowed the two systems to have a different number of particles N_a and N'_a and that the final covariance functions can be entirely expressed in terms of local energy kernel functions k.

Force Kernels The force $\mathbf{f}(\{\rho_a\}^p)$ on an atom p at position $\mathbf{r}_p$ is defined as the derivative

$$\mathbf{f}\left(\{\rho_a\}^p\right) = -\frac{\partial E\left(\{\rho_a\}^p\right)}{\partial \mathbf{r}_p}, \tag{5.10}$$

where by virtue of the existence of a finite cutoff radius of interaction, only the set of configurations $\{\rho_a\}^p$ that contain atom p within their cutoff function contribute to the force on p. Being the derivative of a GP-distributed quantiy, the force vector is also distributed according to a GP [35] and the corresponding kernels between forces and between forces and local energies can be easily obtained by differentiation as described in Refs. [35, 36]. They read

$$\mathbf{k}^{\varepsilon\mathbf{f}}\left(\rho_a,\{\rho_b\}^p\right) = -\sum_{\{\rho_b\}^q}\frac{\partial k(\rho_a,\rho_b)}{\partial \mathbf{r}_q^{\mathrm{T}}} \tag{5.11}$$

$$\mathbf{K}^{\mathbf{ff}}\left(\{\rho_a\}^p,\{\rho_b\}^q\right) = \sum_{\{\rho_a\}^p}\sum_{\{\rho_b\}^q}\frac{\partial^2 k(\rho_a,\rho_b)}{\partial \mathbf{r}_p\partial \mathbf{r}_q^{\mathrm{T}}}. \tag{5.12}$$

Total Energy–Force Kernel Learning from both energies and forces simultaneously is also possible. One just needs to calculate the extra kernel $\mathbf{k}^{\mathbf{f}E}$ comparing the two quantities in the database

$$\mathbf{k}^{\mathbf{f}E}\left(\{\rho_a\}^p,\{\rho'_b\}\right) = -\sum_{\{\rho_a\}^p}\sum_{b=1}^{N'}\frac{\partial k(\rho_a,\rho_b)}{\partial \mathbf{r}_p}. \tag{5.13}$$

To clarify how the kernels described above can be used in practice, it is instructive to look at a simple example. Imagine having a database made up of a single snapshot coming from an ab initio molecular dynamics of N atoms, hence containing a single energy calculation and N forces. Learning using these quantities would involve building a $N+1 \times N+1$ block matrix $\mathbb{K}$ containing the covariance between every pair

$$\mathbb{K} = \begin{pmatrix} k^{EE}\left(\{\rho_a\}, \{\rho_b\}\right) & \mathbf{k}^{E\mathbf{f}}\left(\{\rho_a\}, \{\rho_b\}^1\right) & \cdots & \mathbf{k}^{E\mathbf{f}}\left(\{\rho_a\}, \{\rho_b\}^N\right) \\ \mathbf{k}^{\mathbf{f}E}\left(\{\rho_a\}^1, \{\rho_b\}\right) & \mathbf{K}^{\mathbf{ff}}\left(\{\rho_a\}^1, \{\rho_b\}^1\right) & \cdots & \mathbf{K}^{\mathbf{ff}}\left(\{\rho_a\}^1, \{\rho_b\}^N\right) \\ \vdots & \vdots & \ddots & \vdots \\ \mathbf{k}^{\mathbf{f}E}\left(\{\rho_a\}^N, \{\rho_b\}\right) & \mathbf{K}^{\mathbf{ff}}\left(\{\rho_a\}^N, \{\rho_b\}^1\right) & \cdots & \mathbf{K}^{\mathbf{ff}}\left(\{\rho_a\}^N, \{\rho_b\}^N\right) \end{pmatrix}. \tag{5.14}$$

As is clear from the above equation, each block is either a scalar (the energy–energy kernel in the top left), a 3×3 matrix (the force–force kernels) or a vector (the energy–force kernels). The full dimension of $\mathbb{K}$ is hence $(3N+1) \times (3N+1)$.

Once such a matrix is built and the inverse $\mathbb{C}^{-1} = [\mathbb{K} + \mathbb{I}\sigma_n^2]^{-1}$ computed, the predictive distribution for the value of the latent local energy variable can be easily written down. For notational convenience, it is useful to define the vector $\{x_i\}_{i=1}^N$ containing all the quantities in the training database and the vector $\{t_i\}_{i=1}^N$ specifying their type (meaning that t_i is either E or $\mathbf{f}$ depending on the type of data point contained in x_i). With this convention, the predictive distribution for the local energy takes the form

$$\begin{aligned} p(\varepsilon^* \mid \rho^*, \mathcal{D}) &= \mathcal{N}\left(\hat{\varepsilon}(\rho^*), \hat{\sigma}^2(\rho^*)\right) \\ \hat{\varepsilon}(\rho^*) &= \sum_{ij} k^{\varepsilon t_i}(\rho^*, \rho_i)\mathbb{C}_{ij}^{-1} x_j \\ \hat{\sigma}^2(\rho^*) &= k(\rho^*, \rho^*) - \sum_{ij} k^{\varepsilon t_i}(\rho^*, \rho_i)\mathbb{C}_{ij}^{-1} k^{t_j \varepsilon}(\rho_j, \rho^*), \end{aligned} \tag{5.15}$$

where the products between x_j, $\mathbb{C}_{ij}^{-1}$ and $k^{t_j\varepsilon}$ are intended to be between scalars, vectors or matrices depending on the nature of the quantities involved.

5.2.3 Incorporating Prior Information in the Kernel

Choosing a Gaussian stochastic process as prior distribution over the local energies $\varepsilon(\rho)$ rather than a parametrised functional form brings a few key advantages. A much sought advantage is that it allows greater flexibility: one can show that in general a GP corresponds to a model with an infinite number of parameters, and with a suitable kernel choice can act as a "universal approximator": capable of learning any function if provided with sufficient training data [35]. A second one is a greater ease of design: the kernel function must encode all prior information about the local energy function, but typically contains very few free parameters (called *hyperparameters*) which can be tuned, and such tuning is typically straightforward. Third, GPs offer a coherent framework to predict the uncertainty associated with the predicted quantities via the posterior covariance. This is typically not possible for classical parametrised n-body force fields.

All this said, the high flexibility associated with GPs could easily become a drawback when examined from the point of view of computational efficiency. Broadly, it turns out that for maximal efficiency (which takes into account both accuracy and speed of learning and prediction) one should constrain this flexibility in physically motivated ways, essentially by incorporating prior information in the kernel. This will reduce the dimensionality of the problem, e.g., by choosing to learn energy functions of significantly fewer variables than those featuring in the configuration ρ ($3N$ for N atoms).

To effectively incorporate prior knowledge into the GP kernel, it is fundamental to know the relation between important properties of the modelled energy and the corresponding kernel properties. These are presented in the remainder of this section for the case of local energy kernels. Properties of smoothness, invariance to physical symmetries and interaction order are discussed in turn.

5.2.3.1 Function Smoothness

The relation between a given kernel and the smoothness of the random functions described by the corresponding Gaussian stochastic process has been explored in detail [35, 37]. Kernels defining functions of arbitrary differentiability have been developed. For example, on opposite ends we find the so-called *squared exponential* (k_{SE}) and *absolute exponential* (k_{AE}) kernels, defining, respectively, infinitely differentiable and nowhere differentiable functions:

$$k_{SE}(d) = e^{-d^2/2\ell^2} \tag{5.16}$$

$$k_{AE}(d) = e^{-d/\ell}, \tag{5.17}$$

where the letter d represents the distance between two points in the metric space associated with the function to be learned (e.g., a local energy). The *Matérn* kernel [35, 37] is a generalisation of the above-mentioned kernels and allows to impose an arbitrary degree of differentiability depending on a parameter ν:

$$k_{M,\nu}(d) = \frac{2^{1-\nu}}{\Gamma(\nu)} \left(\sqrt{2\nu}\frac{d}{\ell}\right)^{\nu} K_\nu \left(\sqrt{2\nu}\frac{d}{\ell}\right), \tag{5.18}$$

where Γ is the gamma function and K_ν is a modified Bessel function of the second kind.

The relation between kernels and modelled function differentiability is illustrated by Fig. 5.3, showing the three kernels mentioned above (Fig. 5.3a) along with typical samples from the corresponding GP priors (Fig. 5.3b). The absolute exponential kernel has been found useful to learn atomisation energy of molecules [38–40], especially in conjunction with the discontinuous Coulomb matrix descriptor [38]. In the context of modelling useful machine learning force fields, a relatively smooth

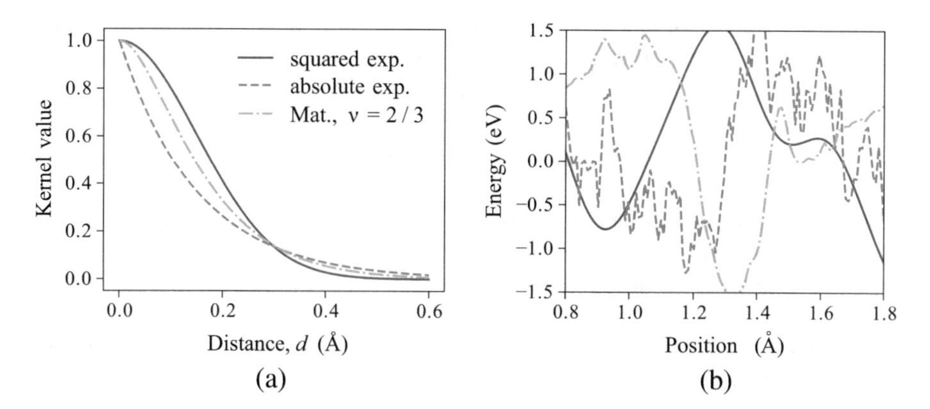

Fig. 5.3 Effect of three kernel functions on the smoothness of the corresponding stochastic processes

energy or force function is typically sought. For this reason, the absolute exponential is not appropriate and has never been used while the flexibility of the Matérn covariance has only found limited applicability [41]. In fact, the squared exponential has been almost always preferred, in conjunction with suitable representations ρ of the atomic environment, [14, 16, 31, 42], and will be used also in this work.

5.2.3.2 Physical Symmetries

Any energy or force function has to respect the symmetry properties listed below.

Translations Physical systems are invariant upon rigid translations of all their components. This basic property is relatively easy to enforce in any learning algorithm via a local representation of the atomic environments. In particular, it is customary to express a given local atomic environment as the unordered set of M vectors $\{\mathbf{r}_i\}_{i=1}^{M}$ going from the "central" atom to every neighbour lying within a given cutoff radius [14, 15, 21, 33]. It is clear that any representation ρ and any function learned within this space will be invariant upon translations.

Permutations Atoms of the same chemical species are indistinguishable, and any permutation $\mathcal{P}$ of identical atoms in a configuration necessarily leaves energy (as well as the force) invariant. Formally one can write $\varepsilon(\mathcal{P}\rho) = \varepsilon(\rho)\ \forall\mathcal{P}$. This property corresponds to the kernel invariance

$$k\left(\mathcal{P}\rho, \mathcal{P}'\rho'\right) = k\left(\rho, \rho'\right) \quad \forall\mathcal{P}, \mathcal{P}'. \tag{5.19}$$

Typically, the above equality has been enforced either by the use of invariant descriptors [13, 14, 42, 43] or via an explicit invariant summation of the kernel over the permutation group [15, 16, 44], with the latter choice being feasible only when the symmetrisation involves a small number of atoms.

Rotations The potential energy associated with a configuration should not change upon any rigid rotation $\mathcal{R}$ of the same (i.e., formally, $\varepsilon(\mathcal{R}\rho) = \varepsilon(\rho)\,\forall\mathcal{R}$). Similarly to permutation symmetry, this invariance is expressed via the kernel property

$$k\left(\mathcal{R}\rho, \mathcal{R}'\rho'\right) = k\left(\rho, \rho'\right) \quad \forall\mathcal{R}, \mathcal{R}'. \tag{5.20}$$

The use of rotation-invariant descriptors to construct the representation ρ immediately guarantees the above. Typical examples of such descriptors are the symmetry functions originally proposed in the context of neural networks [3,45], the internal vector matrix [13] or the set of distances between groups of atoms [15,42,43].

Alternatively, a "base" kernel k_b can be made invariant with respect to the rotation group via the following symmetrisation ("Haar integral" over the full 3D rotation group):

$$k\left(\rho, \rho'\right) = \int d\mathcal{R}\, k_b\left(\rho, \mathcal{R}\rho'\right). \tag{5.21}$$

Such a procedure (called "transformation integration" in the ML community [46]) was first used to build a potential energy kernel in Ref. [21].

When learning forces, as well as other tensorial physical quantities (e.g., a stress tensor, or the (hyper)polarisability of a molecule), the learnt function must be *covariant* under rotations. This property can be formally written as $\mathbf{f}(\mathcal{R}\rho) = \mathbf{R}\mathbf{f}(\rho)\,\forall\mathcal{R}$ and, as shown in [14], it translates at the kernel level to

$$\mathbf{K}\left(\mathcal{R}\rho, \mathcal{R}'\rho'\right) = \mathbf{R}\mathbf{K}(\rho, \rho')\mathbf{R}'^{T}. \tag{5.22}$$

Note that, since forces are three dimensional vectorial quantities, the corresponding kernels are 3×3 matrices [14,47,48], here denoted by $\mathbf{K}$.

Designing suitable covariant descriptors is arguably harder than finding invariant ones. For this reason, the automatic procedure proposed in Ref. [14] to build covariant descriptors can be particularly useful. Covariant matrix valued kernels are generated starting with an (easy to construct) scalar base kernel k_b through a "covariant integral"

$$\mathbf{K}\left(\rho, \rho'\right) = \int d\mathcal{R}\, \mathbf{R}k_b\left(\rho, \mathcal{R}\rho'\right). \tag{5.23}$$

This approach has been extended to learn higher order tensors in Refs. [49,50].

Using rotational symmetry crucially improves the efficiency of the learned model. A very simple illustrative example of the importance of rotational symmetry is shown in Fig. 5.4, addressing an atomic dimer in which force predictions coming from a non-covariant squared exponential kernel and its covariant counterpart (obtained using Eq. (5.23)) are compared. The figure reports the forces predicted to act on an atom, as a function of the position on the x-axis of the other atom, relative to the first. So that, for positive x values the figure reports the forces on the

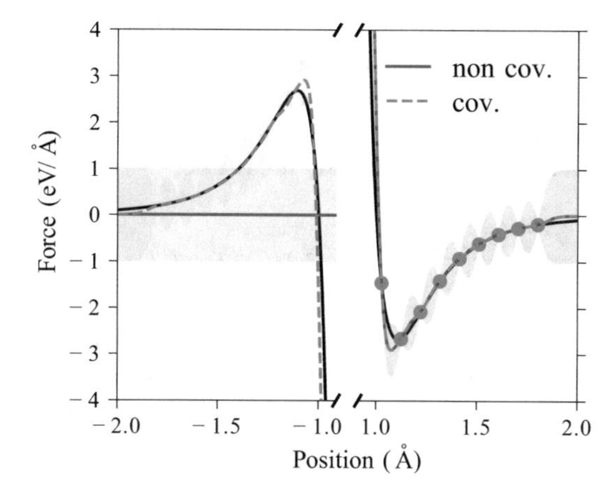

Fig. 5.4 Learning the force profile of a 1D LJ dimer using data (blue circle) coming from one atom only. It is seen that a non-covariant GP (solid red line) does not learn the symmetrically equivalent force acting on the other atom and it thus predicts a zero force and maximum error. If covariance is imposed to the kernel via Eq. (5.23) (dashed blue line), then the correct equivalent (inverted) profile is recovered. Shaded regions represent the predicted 1σ interval in the two cases

left atom as a function of the position of the right atom, while negative x values will be associated with forces acting on the right atom as a function of the position of the left atom. In the absence of the covariance force properties, training the model on a sample of nine forces acting on the left atom will populate correctly only the right side of the graph: a null force will be predicted to act on the right atom (solid red line on the left panel). However, the covariant transformation (in 1D, just a change of sign) will allow the transposition of the force field learned from one environment to the other, and thus the correct prediction of the (inverted) force profile in the left panel.

5.2.3.3 Interaction Order

Classical parametrised force fields are sometimes expressed as a truncated series of energy contributions of progressively higher n-body "interaction orders" [51–54]. The procedure is consistent with the intuition that, as long as the series converges rapidly, truncating the expansion reduces the amount of data necessary for the fitting, and enables a likely higher extrapolation power to unseen regions of configuration space. The lowest truncation order compatible with the target precision threshold is, in general, system dependent, as it will typically depend on the nature of the chemical interatomic bonds within the system. For instance, metallic bonding in a close-packed crystalline system might be described surprisingly well by a pairwise potential, while covalent bonding yielding a zincblende structure can never be, and it will always require three-body interactions terms to be present [14, 15]. Restricting the order of a machine learning force field has proven to be useful for both neural network [55] and Gaussian process regression [14,42]. In the particular context of GP-based ML-FFs, prior knowledge on the interaction order needs to be included in the form of an n-body kernel functions. A detailed and comprehensive exposition on how to do so was given in Ref. [15], and it will be summarised below and in the next subsection. The order of a kernel k_n can be defined as the smallest integer n for which the following property holds true

$$\frac{\partial^n k_n\left(\rho,\rho'\right)}{\partial \mathbf{r}_{i_1}\cdots\partial \mathbf{r}_{i_n}}=0 \qquad \forall\ \mathbf{r}_{i_1}\neq\mathbf{r}_{i_2}\neq\cdots\neq\mathbf{r}_{i_n}, \tag{5.24}$$

where $\mathbf{r}_{i_1},\ldots,\mathbf{r}_{i_n}$ are the positions of any choice of a set of n different surrounding atoms. By virtue of linearity, the predicted local energy in Eq. (5.6) will also satisfy the same property if k_n does. Thus, Eq. (5.24) implies that the central atom in a local configuration interacts with up to $n-1$ other atoms simultaneously, making the learned energy n-body.

5.2.4 Smooth, Symmetric Kernels of Finite Order n

In the previous subsection, we saw how the fundamental physical symmetries of energy and forces translate into the realm of kernels. Here, we show how to build n-body kernels that possess these properties.

We start by defining a smooth translation- and permutation-invariant 2-body kernel by summing all the squared exponential kernels calculated on the distances between the relative positions in ρ and those in ρ' [14–16]

$$k_2\left(\rho,\rho'\right)=\sum_{i\in\rho,j\in\rho'} e^{-\|\mathbf{r}_i-\mathbf{r}'_j\|^2/2\ell^2}. \tag{5.25}$$

As shown in [15], higher order kernels can be defined simply as integer powers of k_2

$$k_n\left(\rho,\rho'\right)=k_2\left(\rho,\rho'\right)^{n-1} \tag{5.26}$$

Note that, by building n-body kernels using Eq. (5.26), one can avoid the exponential cost of summing over all n-plets that a more naïve kernel implementation would involve. This makes it possible to model any interaction order paying only the quadratic computational cost of computing the 2-body kernel in Eq. (5.25).

Furthermore, one can at this point write the squared exponential kernel on the natural distance $d^2(\rho,\rho')=k_2(\rho,\rho)+k_2(\rho',\rho')-2k_2(\rho,\rho')$ induced by the ("scalar product") k_2 as a formal many-body expansion:

$$\begin{aligned} k_{MB}\left(\rho,\rho'\right) &= e^{-d^2(\rho,\rho')/2\ell^2} \\ &= e^{\frac{-k_2(\rho,\rho)-k_2\left(\rho',\rho'\right)}{2\ell^2}}\left[1+\frac{1}{\ell^2}k_2+\frac{1}{2!\ell^4}k_3+\frac{1}{3!\ell^6}k_4+\cdots\right]. \end{aligned} \tag{5.27}$$

So that, assuming a smooth underlying function, the completeness of the series and the "universal approximator" property of the squared exponential [35, 56] can be immediately seen to imply one another.

It is important to notice that the scalar kernels just defined are *not* rotation symmetric, i.e., they do not respect the invariance property of Eq. (5.20). This is due to the fact that the vectors $\mathbf{r}_i$ and $\mathbf{r}'_j$ featuring in Eq. (5.25) depend on the arbitrary reference frames with respect to which they are expressed. A possible solution would be given by carrying out the explicit symmetrisations provided by Eq. (5.21) (or Eq. (5.23) if the intent is to build a force kernel). The invariant integration Eq. (5.21) of k_3 is, for instance, a step in the construction of the (many-body) SOAP kernel [21], while an analytical formula for k_n (with arbitrary n) has been recently proposed [15]. The covariant integral (Eq. (5.23)) of finite-n kernels was also successfully carried out (see Ref. [14], which in particular contains a closed form expression for the $n = 2$ matrix valued two-body force kernel).

However, explicit symmetrisation via Haar integration invariably implies the evaluation of computationally expensive functions of the atomic positions. Motivated by this fact, one could take a different route and consider symmetric n-kernels defined, for any n, as functions of the effective rotation-invariant degrees of freedom of n-plets of atoms [15]. For $n = 2$ and $n = 3$, we can choose these degrees of freedom to be simply the interparticle distances occurring in atomic pairs and triplets (other equally simple choices are possible, and have been used before, see Ref. [42]). The resulting kernels read:

$$k_2^s(\rho, \rho') = \sum_{\substack{i \in \rho \\ j \in \rho'}} e^{-\left(r_i - r'_j\right)^2 / 2\ell^2}, \tag{5.28}$$

$$k_3^s(\rho, \rho') = \sum_{\substack{i_1 > i_2 \in \rho \\ j_1 > j_2 \in \rho'}} \sum_{\mathbf{P} \in \mathcal{P}} e^{-\|\left(r_{i_1}, r_{i_2}, r_{i_1 i_2}\right)^{\mathrm{T}} - \mathbf{P}\left(r'_{j_1}, r'_{j_2}, r'_{j_1 j_2}\right)^{\mathrm{T}}\|^2 / 2\ell^2}. \tag{5.29}$$

where r_i indicates the Euclidean norm of the relative position vector $\mathbf{r}_i$, and the sum over all permutations of three elements $\mathcal{P}$ ($|\ \mathcal{P}\ | = 6$) ensures the permutation invariance of the kernel (see Eq. (5.19)).

It was argued (and numerically tested) in [15] that these *direct* kernels are as accurate as the Haar-integrated ones, while their evaluation is very substantially faster. However, as is clear from Eqs. (5.28) and (5.29), even the construction of directly symmetric kernels becomes unfeasible for large values of n, since the number of terms in the sums grows exponentially. On the other hand, it is still possible to use Eq. (5.26) to increase the integer order of an already symmetric n'−body kernel by elevating it to an integer power. As detailed in [15], raising an already symmetric "input" kernel of order n' to a power ζ in general produces a symmetric "output" kernel

$$k_n^{\neg u}(\rho, \rho') = k_{n'}^s(\rho, \rho')^{\zeta} \tag{5.30}$$

of order $n = (n' - 1)\zeta + 1$. We can assume that the input kernel was built on the effective degrees of freedom of the n' particles in an atomic n'-plet (as is the case, e.g., the 2 and 3-kernels in Eqs. (5.28) and (5.29)). The number of these degrees of freedom is $(3n' - 6)$ for $n' > 2$ (or just 1 for $n' = 2$). Under this assumption, the output n-body kernel will depend on $\zeta(3n' - 6)$ variables (or just ζ variables for $n' = 2$). It is straightforward to check that this number is always smaller than the total number of degrees of freedom of n bodies (here, $3n - 6 = 3(n' - 1)\zeta - 3$). As a consequence, a rotation-symmetric kernel obtained as an integer power of an already rotation-symmetric kernel will *not* be able to learn an *arbitrary* n-body interaction even if fully trained: its convergence predictions upon training on a given n-body reference potential will not be in general exact, and the prediction errors incurred will be specific to the input kernel and ζ exponent used. For this reason, kernels obtained via Eq. (5.30) were defined *non-unique* in Ref. [15] (the superscript $\neg u$ in Eq. (5.30) stands for this).

In practice, the non-unicity issue appears to be a severe problem only when the input kernel is a two-body kernel, and as such it depends only on the radial distances from the central atoms occurring in the two atomic configurations (cf. Eq. (5.28)). In this case, the non-unique output n-body kernels will depend on ζ-plets of radial distances and will miss angular correlations encoded in the training data [15]. On the contrary, a symmetric 3-body kernel (Eq. (5.29)) contains angular information on all triplets in a configuration, and using this kernel as input will be able to capture higher interaction orders (as confirmed, e.g., by the numerical tests performed in Ref. [21]).

Following the above reasoning, one can define a many-body kernel invariant over rotations as a squared exponential on the 3-body invariant distance $d_s^2(\rho, \rho') = k_3^s(\rho, \rho) + k_3^s(\rho', \rho') - 2k_3^s(\rho, \rho')$, obtaining:

$$k_{MB}^{s}\left(\rho, \rho'\right) = e^{-\left(k_3^s(\rho,\rho)+k_3^s(\rho',\rho')-2k_3^s(\rho,\rho')\right)/2\ell^2}. \tag{5.31}$$

It is clear from the series expansion of the exponential function that this kernel is many-body in the sense of Eq. (5.24) and that the importance of high order contributions can be controlled by the hyperparameter ℓ. With $\ell \ll 1$ high order interactions become dominant, while for $\ell \gg 1$ the kernel falls back to a 3-body description.

For all values of ℓ, the above kernel will however always encompass an implicit sum over all contributions (no matter how suppressed), being hence incapable of pruning away irrelevant ones even when a single interaction order is clearly dominant. Real materials often possess dominant interaction orders, and the ionic or covalent nature of their chemical bonding makes the many-body expansion converge rapidly. In these cases, an algorithm which automatically selects the dominant contributions, truncating this way the many-body series in Eq. (5.27), would represent an attractive option. This is the subject of the following section.

5.2.5 Choosing the Optimal Kernel Order

In the previous sections, we analysed how prior information can be encoded in the kernel function. This brought us to designing kernels that implicitly define smooth potential energy surfaces and force fields with all the desired symmetries, corresponding to a given interaction order (Eqs. (5.29) and (5.30)). This naturally raises the problem of deciding the order n best suited to describe a given system. A good conceptual framework for a principled choice is that of Bayesian model selection, which we now briefly review.

We start by assuming we are given a set of models $\{\mathcal{M}_n^{\boldsymbol{\theta}}\}$ (each, e.g., defined by a kernel function of given order n). Each model will be equipped with a vector of *hyperparameters* $\boldsymbol{\theta}$ (typically associated with the covariance lengthscale ℓ, the data noise level σ_n and similar). A fully Bayesian treatment would involve calculating the posterior probability of each candidate model, formally expressed via Bayes' theorem as

$$p\left(\mathcal{M}_n^{\boldsymbol{\theta}} \mid \boldsymbol{\rho}, \boldsymbol{\varepsilon}^r\right) = \frac{p\left(\boldsymbol{\varepsilon}^r \mid \boldsymbol{\rho}, \mathcal{M}_n^{\boldsymbol{\theta}}\right) p\left(\mathcal{M}_n^{\boldsymbol{\theta}}\right)}{p\left(\boldsymbol{\varepsilon}^r \mid \boldsymbol{\rho}\right)}, \tag{5.32}$$

and selecting the model that maximises it. However, often little a priori information is available on the candidate models and their hyperparameters (or it is simply interesting to operate a selection unbiased by priors, and "let the data speak"). In such a case, the prior $p(\mathcal{M}_n^{\boldsymbol{\theta}})$ can be ignored as being flat and uninformative, and maximising the posterior becomes equivalent to maximising the *marginal likelihood* $p(\boldsymbol{\varepsilon}^r \mid \boldsymbol{\rho}, \mathcal{M}_n^{\boldsymbol{\theta}})$ (here equivalent to the *model evidence*.[2]), and the optimal selection tuple $(n, \boldsymbol{\theta})$ can be hence chosen as

$$\left(\hat{n}, \hat{\boldsymbol{\theta}}\right) = \underset{(n,\boldsymbol{\theta})}{\operatorname{argmax}}\; p\left(\boldsymbol{\varepsilon}^r \mid \boldsymbol{\rho}, \mathcal{M}_n^{\boldsymbol{\theta}}\right). \tag{5.33}$$

The marginal likelihood is an analytically computable normalised multivariate distribution, and it was given in Eq. (5.4).

The maximisation in Eq. (5.33) can be thought of as a formalisation of the Occam's razor principle in our particular context. This is illustrated in Fig. 5.5, which contains a cartoon of the marginal likelihood of three models of increasing complexity/flexibility (a useful analogy is to think of polynomials $P_n(x)$ of increasing order n, the likelihood representing how well these would fit a set of measurements ε^r of an unknown function $\varepsilon(x)$). By definition, the most complex model in the figure is the green one, as it assigns a non-zero probability to the

[2]The model evidence is conventionally defined as the integral over the hyperparameter space of the marginal likelihood times the hyperprior (cf. [35]). We here simplify the analysis by jointly considering the model and its hyperparameters.

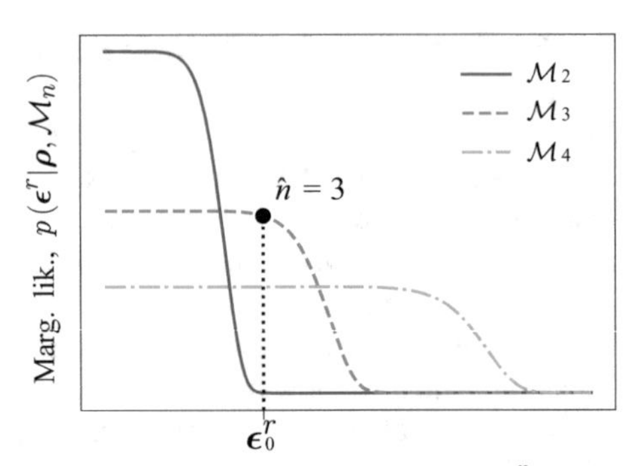

Fig. 5.5 Cartoon of the marginal likelihood profile of three models of increasing complexity. More complex models can fit very different datasets $\boldsymbol{\varepsilon}^r$, this is illustrated by the fact that their marginal likelihood is non-zero for a broader region of the dataset space (here pictorially one dimensional)

largest domain of possible outcomes, and would thus be able to explain the widest range of datasets. Consistently, the simplest model is the red one, which is instead restricted to the smallest dataset range (in our analogy, a straight line will be able to fit well fewer datasets than a fourth order polynomial). Once a reference database ε_0^r is collected, it is immediately clear that the $\mathcal{M}_3$ model with highest likelihood $p(\boldsymbol{\varepsilon} \mid \boldsymbol{\rho}, \mathcal{M}_n^{\boldsymbol{\theta}})$ at $\varepsilon^r = \varepsilon_0^r$ is the simplest that is still able to explain it (the blue one in Fig. 5.5). Indeed, the even simpler model $\mathcal{M}_2$ is not likely to explain the data, the more complex model $\mathcal{M}_4$ can explain more than is necessary for compatibility with the ε_0^r data at hand, and thus produces a lower likelihood value, due to normalisation.

To see how these ideas work in practice, we first test them on a simple system with controllable interaction order, while real materials are analysed in the next section. We here consider a one dimensional chain of atoms interacting via an *ad hoc* potential of order n^t (t standing for "true").[3]

For each value of n^t, we generate a database of N randomly sampled configurations and associated energies. To test Bayesian model selection, for different reference n^t and N values and for fixed $\sigma_n \approx 0$ (noiseless data), we selected the optimal lengthscale parameter ℓ and interaction order n of the n-kernel in Eq. (5.26) by solving the maximisation problem of Eq. (5.33). This procedure was repeated 10 times to obtain statistically significant conclusions; the results were however found to be very robust in the sense that they did not depend significantly on the specific realisation of the training dataset.

The results are reported in Fig. 5.6, where we graph the logarithm of the maximum marginal likelihood (MML), divided by the number of training points N, as a function of N for different combinations of true orders n^t and kernel order n. The model selected in each case is the one corresponding to the line achieving

[3]The n-body toy model used was set up as a hierarchy of two-body interactions defined via the negative Gaussian function $\varepsilon^g(d) = -e^{-\frac{(d-1)^2}{2}}$. This pairwise interaction, depending only on the distance d between two particles, was then used to generate n-body local energies as $\varepsilon_n(\rho) = \sum_{i_1 \neq \cdots \neq i_{n-1}} \varepsilon^g(x_{i_1})\varepsilon^g(x_{i_2} - x_{i_1}) \ldots \varepsilon^g(x_{i_{n-2}} - x_{i_{n-1}})$ where $x_{i_1}, \ldots, x_{i_n - 1}$ are the positions, relative to the central atom, of $n - 1$ surrounding neighbours.

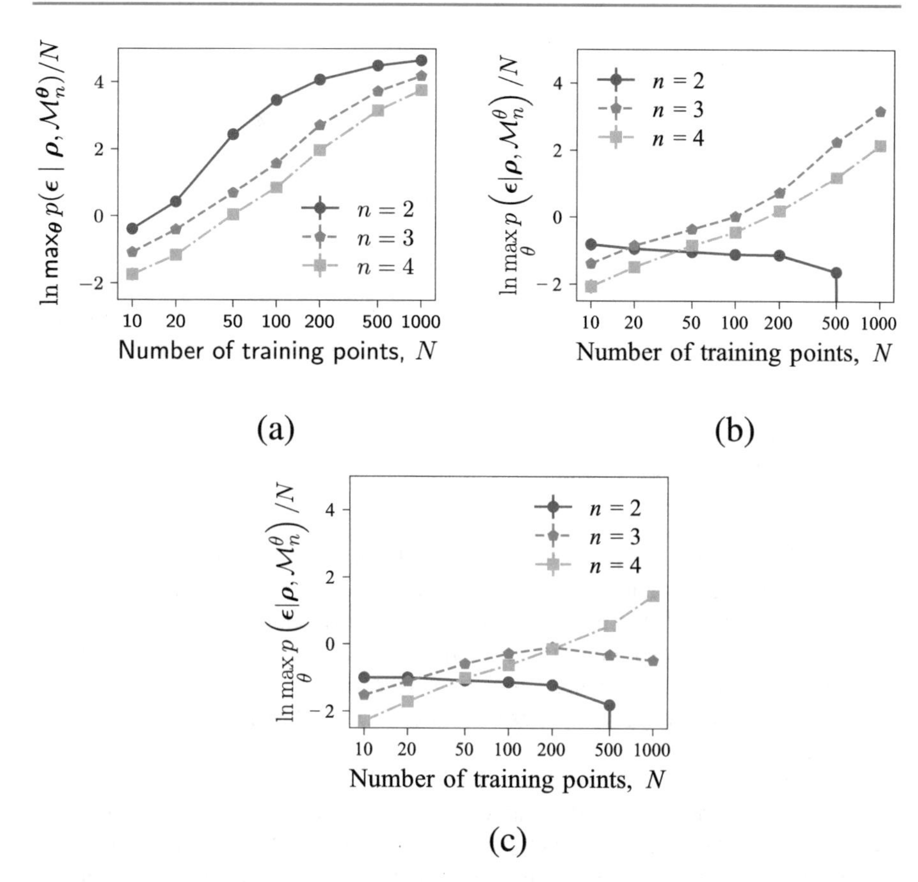

Fig. 5.6 Scaled log maximum marginal likelihood as a function of the number of training points for different kernel models n and true interaction orders n^t. **(a)** $n^t = 2$. **(b)** $n^t = 3$. **(c)** $n^t = 4$

the maximum value of this quantity. It is interesting to notice that, when the kernels order is lower than the true order (i.e., for $n < n^t$), the MML can be observed to *decreases* as a function of N (as, e.g., the red and blue lines in Fig. 5.6c). This makes the gap between the true model and the other models increase substantially as N becomes sufficiently large.

Figure 5.7 summarises the results of model selection. In particular, Fig. 5.7a illustrates the model-selected order $\hat{n}$ as a function of the true order n^t, for different training set sizes N. The graph reveals that, when the dataset is large enough ($N = 1000$ in this example) maximising the marginal likelihood always yields the true interaction order (green line). On the contrary, for smaller database sizes, a lower interaction order value n is selected (blue and red lines). This is consistent with the intuitive notion that smaller databases may simply not contain enough information to justify the selection of a complex model, so that a simpler one should be chosen. More insight can be obtained by observing Fig. 5.7b, reporting the model-selected order as a function of the training dataset size for different true interaction

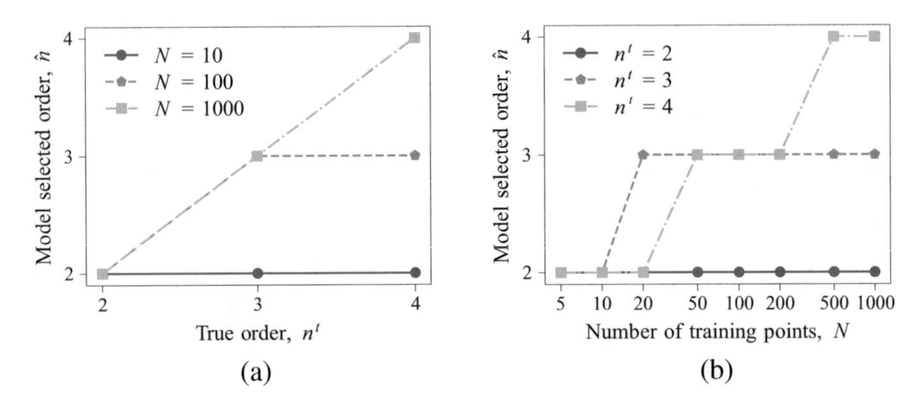

Fig. 5.7 Model-selected order $\hat{n}$ as a function of the true order n^t (left) and as a function of the number of training data points N (right)

orders. While the order of a simple 2-body model is always recovered (red line), to identify as optimal a higher order interaction model a minimum number of training points is needed, and this number grows with the system complexity. Although not immediately obvious, choosing a simpler model when only limited databases are available also leads to smaller prediction errors on unseen configurations, since overfitting is ultimately prevented, as illustrated in Refs. [14, 15] and further below in Sect. 5.3.1.

The picture emerging from these observations is one in which, although the quantum interactions occurring in atomistic systems will in principle involve all atoms in the system, there is never going to be sufficient data to select/justify the use of interaction models beyond the first few terms of the many-body expansion (or any similar expansion based on prior physical knowledge). At the same time, in many likely scenarios, a realistic target threshold for the average error on atomic forces (typically of the order of 0.1 eV/A) will be met by truncating the series at a complexity order that is still practically manageable. Hence, in practice *a small finite-order model will always be optimal.*

This is in stark contrast with the original hope of finding a single many-body "universal approximator" model to be used in every context, which has been driving a lot of interest in the early days of the ML-FF research field, producing, for instance, reference methods [3, 12]. Furthermore, the observation that it may be possible to use models of finite-order complexity without ever recurring to universal approximators suggests alternative routes for increasing the accuracy of GP models without increasing the kernels' complexity. These are worth a small digression.

Imagine a situation as the one depicted in Fig. 5.8, where we have an heterogeneous dataset composed of configurations that cluster into groups. This could be the case, for instance, if we imagine collecting a database which includes several relevant phases of a given material. Given the large amount of data and the complexity of the physical interactions within (and between) several phases, we can imagine the model selected when training on the full dataset to be a relatively

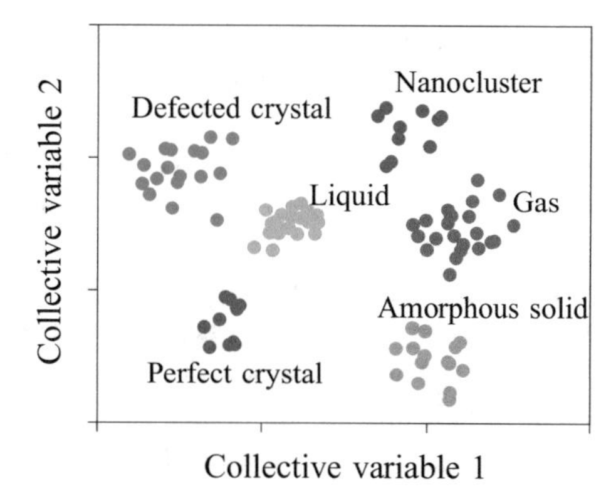

Fig. 5.8 An illustrative representation of a heterogeneous database composed of configurations which "cluster" around specific centroids in an arbitrary two dimensional space. The different clusters can be imagined to be different phases of the same material

complex one. On the other hand, each of the small datasets representative of a given phase may be well described by a model of much lower complexity. As a consequence, one could choose to train several GP, one for each of the phases, as well as a *gating function* $p(c|\rho)$ deciding, during an MD run, which of the clusters c to call at any given time. These GPs learners will effectively *specialise* on each particular phase of the material. This model can be considered a type of *mixture of experts* model [57, 58], and heavily relies on a viable partitioning of the configuration space into clusters that will comprise similar entries. This subdivision is far from trivially obtained in typical systems, and in fact obtaining "atlases" for real materials or molecules similar to the one in Fig. 5.8 is an active area of research [59–62]. However, another simpler technique to combine multiple learner is that of bootstrap aggregating ("Bagging") [63]. In our particular case, this could involve training multiple GPs on random subsections of the data and then averaging them to obtain a final prediction. While it should not be expected that the latter combination method will perform better than a GP trained on the full dataset, the approach can be very advantageous from a computational perspective since, similar to the mixture of experts model, it circumvents the $\mathcal{O}(N^3)$ computational bottleneck of inverting the kernel matrix in Eq. (5.5) by distributing the training data to multiple GP learners. ML algorithms based on the use of multiples learners belong to a broader class of *ensemble learning* algorithms [64, 65].

5.2.6 Kernels for Multiple Chemical Species

In this section, we briefly show how kernels for multispecies systems can be constructed, and provide specific expressions for the case of 2- and 3-body kernels.

It is convenient to show the reasoning behind multispecies kernel construction starting from a simple example. Defining by s_j the chemical species of atom j, a generic 2-body decomposition of the local energy of an atom i surrounded by the configuration ρ_i takes the form

$$\varepsilon(\rho_i) = \sum_{j\in\rho_i} \tilde{\varepsilon}_2^{s_i s_j}(r_{ij}), \tag{5.34}$$

where a pairwise function $\tilde{\varepsilon}_2^{s_i s_j}(r_{ij})$ is assumed to provide the energy associated with each couple of atoms i and j which depends on their distance r_{ij} and on their chemical species s_i and s_j. These pairwise energy functions should be invariant upon re-indexing of the atoms, i.e., $\tilde{\varepsilon}_2^{s_i s_j}(r_{ij}) = \tilde{\varepsilon}_2^{s_j s_i}(r_{ji})$. The kernel for the function $\varepsilon(\rho_i)$ then takes the form

$$\begin{aligned} k_2^s\left(\rho_i, \rho_l'\right) &= \left\langle \varepsilon(\rho_i)\varepsilon\left(\rho_l'\right)\right\rangle \\ &= \sum_{jm} \left\langle \tilde{\varepsilon}_2^{s_i s_j}(r_{ij})\tilde{\varepsilon}_2^{s_l' s_m'}\left(r_{lm}'\right)\right\rangle \\ &= \sum_{jm} \tilde{k}_2^{s_i s_j s_l' s_m'}\left(r_{ij}, r_{lm}'\right). \end{aligned} \tag{5.35}$$

The problem of designing the kernel k_2^s for two configurations in this way reduced to that of choosing a suitable kernel $\tilde{k}_2^{s_i s_j s_l' s_m'}$ comparing couples of atoms. An obvious choice for this would include a squared exponential for the radial dependence and a delta correlation for the dependence on the chemical species, giving rise to $\delta_{s_i s_l'}\delta_{s_j s_m'} k_{SE}(r_{ij}, r_{lm}')$. This kernel is however still not symmetric upon the exchange of two atoms and it would hence not impose the required property $\tilde{\varepsilon}_2^{s_i s_j}(r_{ij}) = \tilde{\varepsilon}_2^{s_j s_i}(r_{ji})$ on the learned pairwise potential. Permutation invariance can be enforced by a direct sum over the permutation group, in this case simply an exchange of the two atoms l and m in the second configuration. The resulting 2-body multispecies kernel reads

$$k_2^s\left(\rho_i, \rho_l'\right) = \sum_{\substack{j\in\rho_i \\ m\in\rho_l'}} \left(\delta_{s_i s_l'}\delta_{s_j s_m'} + \delta_{s_i s_m'}\delta_{s_j s_l'}\right) e^{-\left(r_{ij}-r_{lm}'\right)^2/2\ell^2}. \tag{5.36}$$

This can be considered the natural generalisation of the single species 2-body kernel in Eq. (5.28). A very similar sequence of steps can be followed for the 3-body kernel. By defining the vector containing the chemical species of an ordered triplet as $\mathbf{s}_{ijk} = (s_i s_j s_k)^{\mathrm{T}}$, as well as the vector containing the corresponding three distances $\mathbf{r}_{ijk} = (r_{ij} r_{jk} r_{ki})^{\mathrm{T}}$, a multispecies 3-body kernel can be compactly written down as

$$k_3^s\left(\rho_i, \rho_l'\right) = \sum_{\substack{j>k\in\rho_i \\ m>n\in\rho_l'}} \sum_{\mathbf{P}\in\mathcal{P}} \delta_{\mathbf{s}_{ijk}, \mathbf{P}\mathbf{s}_{lmn}'} e^{-\left\|\mathbf{r}_{ijk}^{\mathrm{T}} - \mathbf{P}\mathbf{r}_{lmn}'\right\|^2/2\ell^2}, \tag{5.37}$$

where the group $\mathcal{P}$ contains six permutations of three elements, represented by the matrices $\mathbf{P}$. The above can be considered the direct generalisation of the 3-body kernel in Eq. (5.29). It is simple to see how the reasoning can be extended to an arbitrary n-body kernel. Importantly, the computational cost of evaluating the multispecies kernels described above *does not* increase with the number of species present in a given environment, and the kernels' interaction order could be increased arbitrarily at no extra computational cost using Eqs. (5.30) and (5.31).

5.2.7 Summary

In this section, we first went through the basics of GP regression, and emphasised the importance of a careful design of the kernel function, which ideally should encode any available prior information on the (energy or force) function to be learned (Sect. 5.2.1). In Sect. 5.2.2, we detailed how a local energy function (which is not a quantum observable) can be learned in practice starting from a database containing solely total energies and atomic forces. We then discussed how fundamental properties of the target force field, such as the interaction order, smoothness, as well as its permutation, translation and rotation symmetries, can be included into the kernel function (Sect. 5.2.3). We next proceeded to the construction of a set of computationally affordable kernels that implicitly define smooth, fully symmetric potential energy functions with tunable "complexity" given a target interaction order n. In Sect. 5.2.5, we looked at the problem of choosing the order n best suited for predictions based on the information available in a given set of QM calculations. Bayesian theory for model selection prescribes in this case to choose the n-kernel yielding the largest marginal likelihood for the dataset, which is found to work very well in a 1D model system where the interaction order can be tuned and is correctly identified upon sufficient training. Finally, in Sect. 5.2.6 we showed how the ideas presented can be generalised to systems containing more than one chemical species.

5.3 Practical Considerations

We next focus on the application of the techniques described in the previous sections. In Sect. 5.3.1, we apply the model selection methodology described in Sect. 5.2.5 to two atomic systems described using density functional theory (DFT) calculations. Namely, we consider a small set of models with different interaction order n, and recast the optimal model selection problem into an optimal kernel order selection problem. This highlights the connections between the optimal kernel order n and the physical properties of the two systems, revealing how novel physical insight can be gained via model selection. We then present a more heuristic approach to kernel order selection and compare the results with the ones obtained from the MML procedure. The comparison reveals that typically the kernel selected via the Bayesian approach also incurs into lower average error for force prediction on a provided test set. In Sect. 5.3.2, we discuss computational efficiency of GPs. We argue that an important advantage of using GP kernels of known finite order is the possibility of “mapping” the kernel’s predictions onto the values of a compact approximator function of the same set of variables. This keeps all the advantages of the Bayesian framework, while removing the need of lengthy sums over the database and expensive kernel evaluations typical of GP predictions. For this we introduce a method that can be used to “map” the GP predictions for finite-body kernels and therefore increase the computational speed up to a factor of 10^4 when compared with the original 3-body kernel, while effectively producing identical interatomic forces.

5.3.1 Applying Model Selection to Nickel Systems

We consider two Nickel systems: a bulk face centred cubic (FCC) system described using periodic boundary conditions (PBC), and a defected double icosahedron nanocluster containing 19 atoms, both depicted in Fig. 5.9a. We note that all atoms in the bulk system experience a similar environment, their local coordination involving 12 nearest neighbours, as the system contains no surfaces, edges or vertexes. The atom-centred configurations $\boldsymbol{\rho}$ are therefore very similar in this system. The nanocluster system is instead exclusively composed by surface atoms, involving a different number of nearest neighbours for different atoms. The GP model is thus here required to learn the reference force field for a significantly more complex and more varied set of configurations. It is therefore expected that the GP model selected for the nanocluster systems will be more complex (have a higher kernel order n) than the one selected for the bulk system, even if the latter system is kept at an appreciably higher temperature.

The QM databases used here were extracted from first principles MD simulations carried out at 500 K in the case of bulk Ni, and at 300 K for the Ni nanocluster.

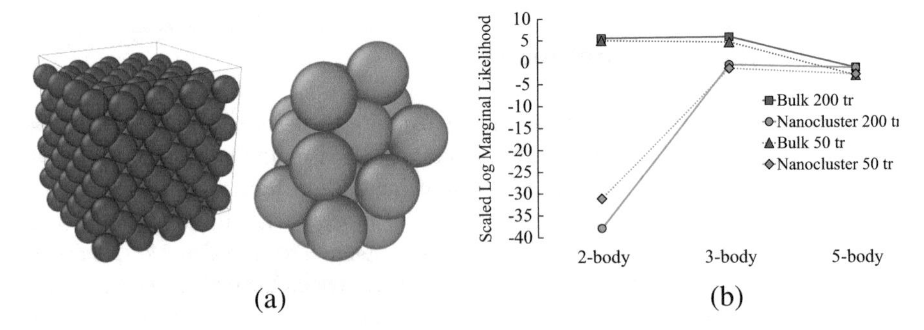

Fig. 5.9 Panel (**a**): the two Nickel systems used in this section as examples, with bulk FCC Nickel in periodic boundary conditions on the left (purple) and a Nickel nanocluster containing 19 atoms on the right (orange). Panel (**b**): maximum log marginal likelihood divided by the number of training points for the 2-, 3- and 5-body kernels in the bulk Ni (purple) and Ni nanocluster (orange) systems, using 50 (dotted lines) and 200 (full lines) training configurations

All atoms within a 4.45 Å cutoff from the central one were included in the atomic configurations $\boldsymbol{\rho}$ for the bulk Ni system, while no cutoff distance was set for the nanocluster configurations, which therefore all include 19 atoms. In this example, we perform model selection on a restricted, yet representative, model set $\{\mathcal{M}_2^{\boldsymbol{\theta}}, \mathcal{M}_3^{\boldsymbol{\theta}}, \mathcal{M}_5^{\boldsymbol{\theta}}\}$ containing, in increasing order of complexity, a 2-body kernel (see Eq. (5.28)), a 3-body kernel (see Eq. (5.29)) and a non-unique 5-body kernel obtained by squaring the 3-body kernel [14] (see Eq. (5.30)). Every kernel function depends on only two hyperparameters $\boldsymbol{\theta} = (\ell, \sigma_n)$, representing the characteristic lengthscale of the kernel ℓ and the modelled uncertainty of the reference data σ_n. While the value of σ_n is kept the same for all kernels, we optimise the lengthscale parameter ℓ for each kernel via marginal likelihood maximisation (Eq. (5.33)). We then select the optimal kernel order n as the one associated with the highest marginal likelihood.

Figure 5.9b reports the optimised marginal likelihood of the three models ($n = 2, 3, 5$) for the two systems while using 50 and 200 training configurations. The 2- and 3-body kernels reach comparable marginal likelihoods in the bulk Ni system, while a 3-body kernel is instead always optimal for the Ni nanocluster system. While intuitively correlated with the relative complexity of the two systems, these results yield further interesting insight. For instance, the occurrence of angular-dependent forces must have a primary role in small Ni clusters since a 3-body kernel is necessary and sufficient to accurately describe the atomic forces in the nanocluster. Meanwhile, the 5-body kernel does not yield a higher likelihood, suggesting that the extra correlation it encodes is not significant enough to be resolved at this level of training. On the other hand, the forces on atoms occurring in a bulk Ni environment at a temperature as high as 500 K are well described by a function of radial distance only, suggesting that angular terms play little to no role, as long as the bonding topology remains everywhere that of undefected FCC crystal.

The comparable maximum log marginal likelihoods the 2- and 3-body kernels produce on bulk environment suggest that the two kernels will achieve similar accuracies. In particular, the 2-body kernel produces the higher log marginal likelihood when the models are trained using $N = 50$ configurations, while the 3-body kernel has a better performance when N increases to 200. This result resonates with the results shown on the toy model in Fig. 5.7: the model selected following the MML principle is a function of the number of training points N used.

For this reason, when using a restricted dataset we should prefer the 2-body kernel to model bulk Ni and a 3-body kernel to model the Ni cluster, as these provide the simplest models that are able to capture sufficiently well the interactions of the two systems. Notice that the models selected in the two cases are different and this reflects the different nature of the chemical interactions involved. This is reassuring, as it shows that the MML principle is able to correctly identify the minimum interaction order needed for a fundamental characterisation of a material even with very moderate training set sizes. For most inorganic material, this minimum order can be expected to be low (typically either 2 or 3) as a consequence of the ionic or covalent nature of the chemical bonds involved, while for certain organic molecules,

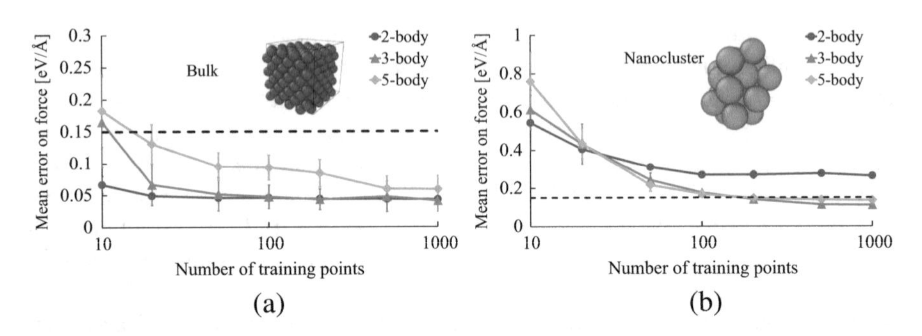

Fig. 5.10 Learning curves for bulk Ni (**a**) and Ni nanocluster (**b**) systems displaying the mean error incurred by the 2-body, 3-body and 5-body kernels as the number of training points used varies. The "error on force" reported here is defined as the norm of the difference vector between predicted and reference force. The error bars in the graphs show the standard deviation when five tests were repeated using different randomly chosen training and testing configuration sets. The black dashed line corresponds to the same target accuracy in the two cases (here 0.15 eV/Å), much more easily achieved in the bulk system

one can expect this to be higher (think, e.g., at the importance of 4-body dihedral terms).

Overall, this example showcases how the maximum marginal likelihood principle can be used to automatically select the simplest model which accurately describes the system, meanwhile providing some insight on the nature of the interactions occurring in the system. In the following, we will compare this procedure with a more heuristic approach based on comparing the kernels' generalisation error, which is commonly employed in the literature [14–16,43,66] for its ease of use.

Namely, let us assume that all of the hyperparameters $\boldsymbol{\theta}$ have been optimised for each kernel in our system of interest, either via maximum likelihood optimisation or via manual tuning. We then measure the error incurred by each kernel on a *test set*, i.e., a set of randomly chosen configurations and forces different from those used to train the GP. Tracing this error as the number of training points increases, we obtain a learning curve (Fig. 5.10). The selected model will be the lowest-complexity one that is capable of reaching a target accuracy (chosen by the user, here set to 0.15 eV/Å, cf. black dotted line in Fig. 5.10). Since lower-complexity kernels are invariably faster learners, if they can reach the target accuracy, they will do so using a smaller number of training points, consistent with all previous discussions and findings. More importantly, lower-complexity kernels are computationally faster and more robust extrapolators than higher-complexity ones—a property that derives from the low order interaction they encode. Furthermore, they can be straightforwardly mapped as described in the next section. For the bulk Ni system of the present example, all three kernels reach the target error threshold, so the 2-body kernel is the best choice for the bulk Ni system. In the Ni nanocluster case, the 2-body kernel is not able to capture the complexity of force field experienced by the atoms in the system, while both the 3- and 5-body kernels reach the threshold. Here the 3-body kernel is thus preferred.

In conclusion, marginal likelihood and generalisation error offer different approaches to the problem of optimal model selection. While their outcomes are generally consistent, these two methods differ in spirit, e.g., because the marginal likelihood distribution naturally incorporates information on the underlying model's variance when measured on the training target data and this will reflect into selecting the best model also on this basis (see Fig. 5.5, in which the target data ε_0^r select the model with $n = 3$). This is not true when using the generalisation error, where all that counts is the model's prediction, i.e., the predicted mean of the posterior GP. Moreover, while model selection according to the marginal likelihood is a function of the training set only, the generalisation error is also dependent on the choice of the test set, whose sampling uncertainty can be reduced through repeated tests, as reported in Fig. 5.10. Regardless of the model selection method, simpler models may perform better when the available data is limited, i.e., higher model complexity does not necessarily imply higher prediction accuracy: whether this is the case will each time depend on the target physical system, the desired accuracy threshold, and the amount data available for training. Due to the lower dimensionality of the feature spaces used to construct the kernels, the predictions of simpler models will also be easier to re-express into a more computationally efficient way than carrying out the summation in Eq. (5.6). For the examples described in this chapter, this means re-expressing the trained GPs based on n-body kernels as functions of $3n$-6 variables which can be evaluated directly, without using a database. These functions can be viewed as the nonparametric n-body classical force fields (here named "MFFs") that the n-body kernels' predictions *exactly* correspond to. Exploiting this correspondence allows us to achieve force fields as fast-executing as determined by the complexity of the physical problem at hand (which will determine the lowest n that can be used). Examples of MFF constructions and tests on their computational efficiency are provided in the next section.

5.3.2 Speeding Up Predictions by Building MFFs

In Sect. 5.2.4, we described how simple n-body kernels of any order n could be constructed. Force prediction based on these kernels effectively produces nonparametric classical n-body force fields: typically depending on distances (2-body) as well as on angles (3-body), dihedrals (4-body) and so on, but not bound by design to any particular functional form.

In this section, we describe a mapping technique (first presented in Ref. [15]) that faithfully encodes forces produced by n-body GP regression into classical tabulated force fields. This procedure can be carried out with arbitrarily low accuracy loss, and always yields a substantial computational speed gain.

We start from the expression of the GP energy prediction in Eq. (5.6), where we substitute k with a specific n-body kernel (in this example, the 2-body kernel of Eq. (5.25) for simplicity). Rearranging the sums, we obtain:

$$\hat{\varepsilon}(\rho) = \sum_{i \in \rho} \left(\sum_{d}^{N} \sum_{j \in \rho_d} e^{-(r_i - r_j)^2/2\ell^2} \alpha_d \right). \tag{5.38}$$

The expression within the parentheses in the above equation is a function of the single distance r_i in the target configuration ρ and the training dataset, and it will not change once the dataset is chosen and the model is trained (the covariance matrix is computed and inverted to give the coefficient α_d for each dataset entry). We can thus rewrite Eq. (5.38) as

$$\hat{\varepsilon}(\rho) = \sum_{i \in \rho} \tilde{\varepsilon}_2(r_i), \tag{5.39}$$

where the function $\tilde{\varepsilon}(r_i)$ can be now thought to be nonparametric 2-body potential expressing the energy associated with an atomic pair (a "bond") as a function of the interatomic distance, so that the energy associated with a local configuration ρ is simply the sum over all atoms surrounding the central one of this 2-body potential. It is now possible to compute the values of $\tilde{\varepsilon}_2(r_i)$ for a set of distances r_i, store them in an array, and from here on *interpolate* the value of the function for any other distances rather than using the GP to compute this function for every atomic configuration during an MD simulation. In practice, a spline interpolation of the so-tabulated potential can be very easily used to predict any $\hat{\varepsilon}(\rho)$ or its negative gradient $\hat{\mathbf{f}}(\rho)$ (analytically computed to allow for a constant of motion in MD runs). The interpolation approximates the GP predictions with arbitrary accuracy, which increases with the density of the grid of tabulated values, as illustrated in Fig. 5.11a.

The computational speed of the resulting "mapped force field" (MFF) is independent of the number of training points N and depends linearly, rather than quadratically, on the number of distinct atomic n-plets present in a typical atomic environment ρ including M atoms plus the central one (this is the number of combinations $\binom{M}{n-1} = M!/(n-1)!(M-n+1)!$, yielding, e.g., M pairs and $M(M-1)/2$ triplets). The resulting overall $N\binom{M}{n-1}$ speedup factor is typically several orders of magnitude over the original n-body GP, as shown in Fig. 5.11b.

The method just described can in principle be used to obtain n-body MFFs from any n-body GPs, for every finite n. In practice, however, while mapping 2-body or 3-body predictions on a 1D or 3D spline is straightforward, the number of values to store grows exponentially for n, consistent with the rapidly growing dimensionality associated with atomic n-plets. This makes the procedure quickly not viable for higher n values which would require ($3n$-6)-dimensional mapping grids and interpolation splines. On a brighter note, flexible 3-body force fields were shown to capture most of the features for a variety of inorganic materials [15, 16, 20, 42].

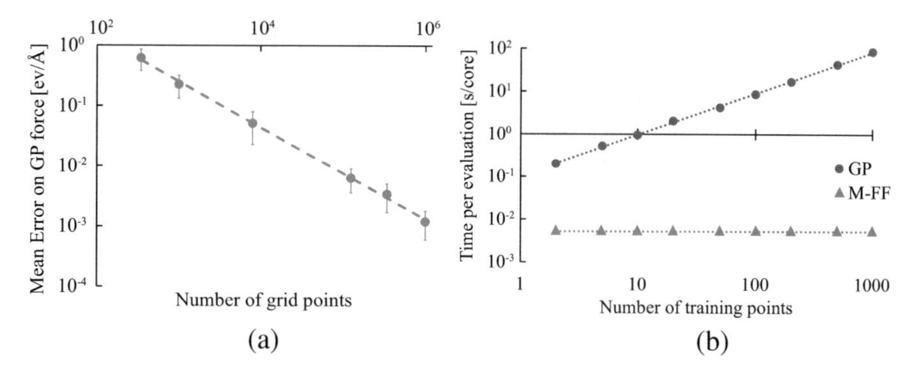

Fig. 5.11 Panel (**a**): error incurred by a 3-body MFF w.r.t. the predictions of the original GP used to build it as a function of the number of points in the MFF grid. Panel (**b**): computational time needed for the force prediction on an atom in a 19-atoms Ni nanocluster as a function of the number of training points for a 3-body GP (red dots) and for the MFF built from the same 3-body GP (blue dots)

Increasing the order of the kernel function beyond 3 might be unnecessary for many systems (and if only few training data are available, it could be still advantageous to use a low-n model to *improve* prediction accuracy, as discussed in Sect. 5.2.5).

MFFs can be built for systems containing any number of atomic species. As already described in Sect. 5.2.6, the cost of constructing a multispecies GP does not increase with the number of species modelled. On the other hand, the number of n-body MFFs that need to be constructed when k atomic species are present grows as the multinomial factor $\frac{(k+n-1)!}{n!(k-1)!}$ (just as any classical force field of the same order). Luckily, constructing multiple MFFs is an embarrassingly parallel problem as different MFFs can be assigned to different processors. This means that the MFF construction process can be considered affordable also for high values of k, especially when using a 3-body model (which can be expected to achieve sufficient accuracy for a large number of practical applications).

We finally note that the variance of a prediction $\hat{\sigma}^2(\rho)$ (third term in Eq. (5.5)) could also be mapped similarly to its mean. However, it is easy to check that the mapped variance will have twice as many arguments as the mapped mean, which again makes the procedure rather cumbersome for $n > 2$. For instance, for $n = 2$ one would have to store the function of two variables $\tilde{\sigma}^2(r_i, r_j)$ providing the variance contribution from any two distances within a configuration, and the final variance can be computed as a sum over all contributions. A more affordable estimate of the error could also be obtained by summing up only the contributions coming from single n-plets (i.e., $\tilde{\sigma}^2(r_i, r_i)$ in the $n = 2$ example). This alternative measure could again be mapped straightforwardly also for $n = 3$ and its accuracy in modelling the uncertainty in the real materials should be investigated.

MFFs obtained as described above have already been used to perform MD simulations on very long timescales while tracking with very good accuracy their reference ab initio DFT calculations for a set of Ni_{19} nanoclusters [16]. In this example application, a total of $1.2 \cdot 10^8$ MD time steps were performed, requiring the use of 24 CPUs for ~3.75 days. The same simulation would have taken ~80 years before mapping, and indicatively ~2000 years using the full DFT-PBE (Perdew–Burke–Ernzerhof) spin-orbit coupling method which was used to build the training database. A Python implementation for training and mapping two- and three-body nonparametric force fields for one or two chemical species is freely available within the MFF package [29].

5.4 Conclusions

In this chapter, we introduced the formalism of Gaussian process regression for the construction of force fields. We analysed a number of relevant properties of the kernel function, namely its smoothness and its invariance with respect to permutation of identical atoms, translation and rotation. The concept of interaction order, traditionally useful in constructing classical parametrised force fields and recently imported into the context of machine learning force fields, was also discussed. Examples on how to construct smooth and invariant n-body energy kernels have been given, with explicit formulas for the cases of $n = 2$ and $n = 3$. We then focused on the Bayesian model selection approach, which prescribes the maximisation of the marginal likelihood, and applied it to a set of standard kernels defined by an integer order n. In a 1D system where the target interaction order could be exactly set, explicit calculations exemplified how the optimal kernel order choice depends on the number of training points used, so that larger datasets are typically needed to resolve the appropriateness of more complex models to a target physical system. We next reported an example of application of the marginal likelihood maximisation approach to kernel order selection for two Nickel systems: face centred cubic crystal and a Ni_{19} nanocluster. In this example, prior knowledge about the system provides hints on the optimal kernel order choice which is *a posteriori* confirmed by the model selection algorithm based on the maximum marginal likelihood strategy. To complement the Bayesian approach to kernel order selection, we briefly discussed the use of learning curves based on the generalisation error to select the simplest model that reaches a target accuracy. We finally introduced the concept of "mapping" GPs onto classical MFFs, and exemplified how mapping of mean and variance of a GP energy prediction can be carried out, providing explicit expressions for the case of a 2-body kernel. The construction of MFFs allows for an accurate calculation of GP predictions while reducing the computational cost by a factor $\sim 10^4$ in most operational scenarios of interest in materials science applications, allowing for molecular dynamics simulations that are as fast as classical ones but with an accuracy that approaches ab initio calculations.

Acknowledgments The authors acknowledge funding by the Engineering and Physical Sciences Research Council (EPSRC) through the Centre for Doctoral Training "Cross Disciplinary Approaches to Non-Equilibrium Systems" (CANES, Grant No. EP/L015854/1) and by the Office of Naval Research Global (ONRG Award No. N62909-15-1-N079). The authors thank the UK

Materials and Molecular Modelling Hub for computational resources, which is partially funded by EPSRC (EP/P020194/1). ADV acknowledges further support by the EPSRC HEmS Grant No. EP/L014742/1 and by the European Union's Horizon 2020 research and innovation program (Grant No. 676580, The NOMAD Laboratory, a European Centre of Excellence). We, AG, CZ and AF, are immensely grateful to Alessandro De Vita for having devoted, with inexhaustible energy and passion, an extra-ordinary amount of his time and brilliance towards our personal and professional growth.

References

1. D.H. Wolpert, Neural Comput. **8**(7), 1341 (1996)
2. A.J. Skinner, J.Q. Broughton, Modell. Simul. Mater. Sci. Eng. **3**(3), 371 (1995)
3. J. Behler, M. Parrinello, Phys. Rev. Lett. **98**(14), 146401 (2007)
4. R. Kondor (2018). Preprint. arXiv:1803.01588
5. M. Gastegger, P. Marquetand, J. Chem. Theory Comput. **11**(5), 2187 (2015)
6. S. Manzhos, R. Dawes, T. Carrington, Int. J. Quantum Chem. **115**(16), 1012 (2014)
7. P. Geiger, C. Dellago, J. Chem. Phys. **139**(16), 164105 (2013)
8. N. Kuritz, G. Gordon, A. Natan, Phys. Rev. B **98**(9), 094109 (2018)
9. K.T. Schütt, F. Arbabzadah, S. Chmiela, K.R. Müller, A. Tkatchenko, Nat. Commun. **8**, 13890 (2017)
10. N. Lubbers, J.S. Smith, K. Barros, J. Chem. Phys. **148**(24), 241715 (2018)
11. K.T. Schütt, H.E. Sauceda, P.J. Kindermans, A. Tkatchenko, K.R. Müller, J. Chem. Phys. **148**(24), 241722 (2018)
12. A.P. Bartók, M.C. Payne, R. Kondor, G. Csányi, Phys. Rev. Lett. **104**(13), 136403 (2010)
13. Z. Li, J.R. Kermode, A. De Vita, Phys. Rev. Lett. **114**(9), 096405 (2015)
14. A. Glielmo, P. Sollich, A. De Vita, Phys. Rev. B **95**(21), 214302 (2017)
15. A. Glielmo, C. Zeni, A. De Vita, Phys. Rev. B **97**(18), 1 (2018)
16. C. Zeni, K. Rossi, A. Glielmo, Á. Fekete, N. Gaston, F. Baletto, A. De Vita, J. Chem. Phys. **148**(24), 241739 (2018)
17. W.J. Szlachta, A.P. Bartók, G. Csányi, Phys. Rev. B **90**(10), 104108 (2014)
18. A.P. Thompson, L.P. Swiler, C.R. Trott, S.M. Foiles, G.J. Tucker, J. Comput. Phys. **285**(C), 316 (2015)
19. A.V. Shapeev, Multiscale Model. Simul. **14**(3), 1153 (2016)
20. A. Takahashi, A. Seko, I. Tanaka, J. Chem. Phys. **148**(23), 234106 (2018)
21. A.P. Bartók, R. Kondor, G. Csányi, Phys. Rev. B **87**(18), 184115 (2013)
22. W.H. Jefferys, J.O. Berger, Am. Sci. **80**(1), 64 (1992)
23. C.E. Rasmussen, Z. Ghahramani, in *Proceedings of the 13th International Conference on Neural Information Processing Systems (NIPS'00)* (MIT Press, Cambridge, 2000), pp. 276–282
24. Z. Ghahramani, Nature **521**(7553), 452 (2015)
25. V.N. Vapnik, A.Y. Chervonenkis, in *Measures of Complexity* (Springer, Cham, 2015), pp. 11–30
26. V.N. Vapnik, *Statistical Learning Theory* (Wiley, Hoboken, 1998)
27. M.J. Kearns, U.V. Vazirani, *An Introduction to Computational Learning Theory* (MIT Press, Cambridge, 1994)
28. T. Suzuki, in *Proceedings of the 25th Annual Conference on Learning Theory*, ed. by S. Mannor, N. Srebro, R.C. Williamson. Proceedings of Machine Learning Research, vol. 23 (PMLR, Edinburgh, 2012), pp. 8.1–8.20
29. C. Zeni, F. Ádám, A. Glielmo, MFF: a Python package for building nonparametric force fields from machine learning (2018). https://doi.org/10.5281/zenodo.1475959
30. R.P. Feynman, Phys. Rev. **56**(4), 340 (1939)
31. V. Botu, R. Ramprasad, Phys. Rev. B **92**(9), 094306 (2015)

32. I. Kruglov, O. Sergeev, A. Yanilkin, A.R. Oganov, Sci. Rep. **7**(1), 1–7 (2017)
33. G. Ferré, J.B. Maillet, G. Stoltz, J. Chem. Phys. **143**(10), 104114 (2015)
34. A.P. Bartók, G. Csányi, Int. J. Quantum Chem. **115**(16), 1051 (2015)
35. C.K.I. Williams, C.E. Rasmussen, *Gaussian Processes for Machine Learning* (MIT Press, Cambridge, 2006)
36. I. Macêdo, R. Castro, *Learning Divergence-Free and Curl-Free Vector Fields with Matrix-Valued Kernels* (Instituto Nacional de Matematica Pura e Aplicada, Rio de Janeiro, 2008)
37. C.M. Bishop, in *Pattern Recognition and Machine Learning*. Information Science and Statistics (Springer, New York, 2006)
38. M. Rupp, A. Tkatchenko, K.R. Müller, O.A. von Lilienfeld, Phys. Rev. Lett. **108**(5), 058301 (2012)
39. M. Rupp, Int. J. Quantum Chem. **115**(16), 1058 (2015)
40. K. Hansen, G. Montavon, F. Biegler, S. Fazli, M. Rupp, M. Scheffler, O.A. von Lilienfeld, A. Tkatchenko, K.R. Müller, J. Chem. Theory Comput. **9**(8), 3404 (2013)
41. S. Chmiela, A. Tkatchenko, H.E. Sauceda, I. Poltavsky, K.T. Schütt, K.R. Müller, Sci. Adv. **3**(5), e1603015 (2017)
42. V.L. Deringer, G. Csányi, Phys. Rev. B **95**(9), 094203 (2017)
43. H. Huo, M. Rupp (2017). Preprint. arXiv:1704.06439
44. A.P. Bartók, M.J. Gillan, F.R. Manby, G. Csányi, Phys. Rev. B **88**(5), 054104 (2013)
45. J. Behler, J. Chem. Phys. **134**(7), 074106 (2011)
46. B. Haasdonk, H. Burkhardt, Mach. Learn. **68**(1), 35 (2007)
47. C.A. Micchelli, M. Pontil, in *Advances in Neural Information Processing Systems* (University at Albany State University of New York, Albany, 2005)
48. C.A. Micchelli, M. Pontil, Neural Comput. **17**(1), 177 (2005)
49. T. Bereau, R.A. DiStasio, A. Tkatchenko, O.A. von Lilienfeld, J. Chem. Phys. **148**(24), 241706 (2018)
50. A. Grisafi, D.M. Wilkins, G. Csányi, M. Ceriotti, Phys. Rev. Lett. **120**, 036002 (2018). https://doi.org/10.1103/PhysRevLett.120.036002
51. S.K. Reddy, S.C. Straight, P. Bajaj, C. Huy Pham, M. Riera, D.R. Moberg, M.A. Morales, C. Knight, A.W. Götz, F. Paesani, J. Chem. Phys. **145**(19), 194504 (2016)
52. G.A. Cisneros, K.T. Wikfeldt, L. Ojamäe, J. Lu, Y. Xu, H. Torabifard, A.P. Bartók, G. Csányi, V. Molinero, F. Paesani, Chem. Rev. **116**(13), 7501 (2016)
53. F.H. Stillinger, T.A. Weber, Phys. Rev. **B31**(8), 5262 (1985)
54. J. Tersoff, Phys. Rev. B **37**(12), 6991 (1988)
55. K. Yao, J.E. Herr, J. Parkhill, J. Chem. Phys. **146**(1), 014106 (2017)
56. K. Hornik, Neural Netw. **6**(8), 1069 (1993)
57. R.A. Jacobs, M.I. Jordan, S.J. Nowlan, G.E. Hinton, Neural Comput. **3**(1), 79 (1991)
58. C.E. Rasmussen, Z. Ghahramani, in *Advances in Neural Information Processing Systems* (UCL, London, 2002)
59. S. De, A.P. Bartók, G. Csányi, M. Ceriotti, Phys. Chem. Chem. Phys. **18**, 13754 (2016)
60. L.M. Ghiringhelli, J. Vybiral, S.V. Levchenko, C. Draxl, M. Scheffler, Phys. Rev. Lett. **114**(10), 105503 (2015)
61. J. Mavračić, F.C. Mocanu, V.L. Deringer, G. Csányi, S.R. Elliott, J. Phys. Chem. Lett. **9**(11), 2985 (2018)
62. S. De, F. Musil, T. Ingram, C. Baldauf, M. Ceriotti, J. Cheminf. **9**(1), 1–14 (2017)
63. L. Breiman, Mach. Learn. **24**(2), 123 (1996)
64. O. Sagi, L. Rokach, Wiley Interdiscip. Rev. Data Min. Knowl. Disc. **8**(4), e1249 (2018)
65. M. Sewell, Technical Report RN/11/02 (Department of Computer Science, UCL, London, 2008)
66. I. Kruglov, O. Sergeev, A. Yanilkin, A.R. Oganov, Sci. Rep. **7**(1), 8512 (2017)

Machine-Learning of Atomic-Scale Properties Based on Physical Principles

6

Gábor Csányi, Michael J. Willatt, and Michele Ceriotti

Abstract

We briefly summarize the kernel regression approach, as used recently in materials modelling, to fitting functions, particularly potential energy surfaces, and highlight how the linear algebra framework can be used to both predict and train from linear functionals of the potential energy, such as the total energy and atomic forces. We then give a detailed account of the smooth overlap of atomic positions (SOAP) representation and kernel, showing how it arises from an abstract representation of smooth atomic densities, and how it is related to several popular density-based representations of atomic structure. We also discuss recent generalizations that allow fine control of correlations between different atomic species, prediction and fitting of tensorial properties, and also how to construct structural kernels—applicable to comparing entire molecules or periodic systems—that go beyond an additive combination of local environments. (This chapter is adapted with permission from Ceriotti et al. (Handbook of materials modeling. Springer, Cham, 2019).)

G. Csányi
Engineering Laboratory, University of Cambridge, Cambridge, UK

M. J. Willatt · M. Ceriotti (✉)
Laboratory of Computational Science and Modelling, Institute of Materials, École Polytechnique Fédérale de Lausanne, Lausanne, Switzerland
e-mail: michele.ceriotti@epfl.ch

K. T. Schütt et al. (eds.), *Machine Learning Meets Quantum Physics*,
Lecture Notes in Physics 968, https://doi.org/10.1007/978-3-030-40245-7_6

6.1 Introduction[1]

There has been a surge of activity during the last couple of years in applying machine-learning methods to materials and molecular modelling problems, that was largely fuelled by the evident success of these techniques in what can loosely be called artificial intelligence. These successes have followed from the collective experience that the scientific community has gained in fitting high volumes of data with very complex functional forms that involve a large number of free parameters, while still keeping control of the regularity and thus avoiding catastrophic overfitting. In the context of molecular modelling, empirical fitting of potential energy surfaces has of course been used for many decades. Indeed it is generally held that this is the only practical way to simulate very large systems (many thousands of atoms) over long time scales (millions of time steps) [2].

Traditionally, when fitting empirical models of atomic interactions, regularity was ensured by writing functional forms that are expressed in terms of one-dimensional functions, e.g. pair potentials, spherically symmetric atomic electron densities, bond orders (as a function of number of neighbours), etc. Such functions are easy to inspect visually to ensure that they are physically and chemically meaningful, e.g. that pair potentials go to zero at large distances and are strongly repulsive at close approach, that atomic electron densities are decreasing with distance, that electron density embedding functions are convex, etc. Moreover, these natural properties are easy to build into the one-dimensional functional forms or enforced as constraints in the parameter optimization. It is widely held that employing such "physically meaningful" functional forms is key to achieving good *transferability* of the empirical models [3].

It is also recognized, however, that the limited functional forms that can be built from these one-dimensional functions ultimately limit the accuracy that these empirical models can achieve. In trying to replace them by high dimensional fits using much more flexible functional forms, two things immediately have to change. The first is the target data. When fitting only a few parameters, it is natural to demand that important observables that are deemed to be central to the scientific questions being addressed are reproduced correctly, and it is easiest to do this if they are part of the fit: e.g. melting points and other phase boundaries, radial distribution functions, etc. But in the case of very many parameters, their optimization also takes a significant number of evaluations, and it becomes impractical to use complex observables as targets. Moreover, there is a drive towards using a "first principles" approach, i.e. that the potentials should actually reproduce the real Born–Oppenheimer potential energy surface with sufficient accuracy *and therefore* the scientifically relevant observables also. The hope is that this will result in transferability in the sense that a wide array of macroscopic observables will be correctly predicted without any of them being part of the fit explicitly, *and therefore*, the corresponding microscopic mechanisms that are also dependent on the potential

[1]This chapter is adapted with permission from Ref. [1].

energy surface will also be correct. So it is natural to take values of the potential energy, computed by the electronic structure method of choice, as the target data. The large number of free parameters can then easily be counterbalanced by a large amount of calculated target data.

The second thing that has to change is how the smooth physically meaningful behaviour of the potential is controlled. It is not practical to inspect manually high dimensional functions to ensure that their predictions are physically and chemically meaningful for all possible configurations. Therefore it becomes even more important to build into the functional forms as much prior information as possible about limiting behaviour and *regularity* (the technical word for the kind of smoothness we are interested in). Reviewing recent work, this paper sets out an example framework for how to do this. The key goals are to create functional forms that preserve the (1) invariance of the properties over permutation of like atoms, (2) invariance of scalar and covariance of tensorial properties with three-dimensional rotations, (3) continuity and regularity with respect to changes in atomic coordinates, including compact support of atomic interactions by including finite cutoffs.

Evidence is accumulating that strictly enforcing these physically motivated properties is enormously beneficial, and many of the most successful machine-learning schemes for atomic-scale systems are built around symmetry arguments. One possible approach is to describe the system in terms of internal coordinates—that automatically satisfy rotational invariance—and then symmetrize explicitly the vector of representations or the functional relation between the representations and the properties. Permutationally invariant polynomials are an example that have been very effective to model the potential energy surfaces of small molecules (see, e.g. the work of Braams and Bowman [4]). Sorting the elements of the representation vector according to interatomic distances has also been used as a way of obtaining permutation invariance at the cost of introducing derivative discontinuities [5–7]. Another possibility, which we will focus on in this paper, starts from a representation of each structure in terms of atomic densities—that are naturally invariant to atom permutations—and then builds a representation that is further invariant to translations and rotations also.

Either way, once an appropriate description of each structure has been obtained, further regularization can be achieved at the level of the regression scheme. To this end, two prominent techniques are the use of artificial neural networks and kernel ridge regression [8]. We use the latter formalism here, and many further details about these techniques can be found in the rest of this volume. The kernel approach starts with the definition of a kernel function, which will be combined with a set of representative atomic configurations to construct the basis functions for the fit. It is a scalar function—at least when learning scalar quantities—with two input arguments, in the present case two atomic structures. Its value should quantify the similarity of the atomic configurations represented by its two arguments, and it can (but does not have to) be defined starting from their associated representations. The value should be largest when its two arguments are equal (or equivalent up to symmetry operations) and smallest for maximally different configurations. The degree to which the kernel is able to capture the variation of the function when varying the atomic configuration will determine how efficient the fit is. The better the correspondence, the fewer basis functions that are needed to achieve a given accuracy of fit.

6.2 Kernel Fitting

We start by giving a concise account of the kernel regression fitting approach, for more details see Refs. [8–10]. A function defined on an atomic structure is represented as a linear sum over kernel basis functions,

$$f(\mathcal{A}) = \sum_{\mathcal{B}\in M} x_{\mathcal{B}} K(\mathcal{A}, \mathcal{B}), \tag{6.1}$$

where the sum runs over a *representative set* of configurations M, selected from the total set N of input configurations. The set of coefficients, combined into a vector $\mathbf{x}$, are determined by solving the linear system that is obtained when the available data (e.g. values of the target function evaluated for a set of structures) are substituted into Eq. (6.1). In the simplest case, there is one input data value corresponding to each atomic configuration. Let $\mathbf{y}$ be the vector of all available input data, and $\mathbf{K}$ be the kernel matrix with rows and columns corresponding to atomic structures, so that the element of $\mathbf{K}$ with row and column corresponding to structures $\mathcal{A}$ and $\mathcal{B}$, respectively, is $K(\mathcal{A}, \mathcal{B})$. The fit is then obtained by solving a linear system in the least squares sense, i.e. minimizing the quadratic loss function,

$$\ell(\mathbf{x}) = \|\mathbf{K}\mathbf{x} - \mathbf{y}\|^2. \tag{6.2}$$

The text book case is when the set of all configurations for which we have target data available is used in its entirety as the representative set (i.e. $N = M$), $\mathbf{K}$ is square, and as long as it is invertible, the optimal solution is

$$\mathbf{x} = \mathbf{K}^{-1}\mathbf{y}. \tag{6.3}$$

In practice, for large data sets, using all the configurations in the data set as representatives is unnecessary. In this case, $M \subset N$, the solution is given by the pseudoinverse,

$$\mathbf{x}_M = (\mathbf{K}_{MN}\mathbf{K}_{NM})^{-1}\mathbf{K}_{MN}\mathbf{y}_N, \tag{6.4}$$

where we used subscripts to emphasize the set that the vector elements correspond to, e.g. $\mathbf{y} \equiv \mathbf{y}_N$ is the data vector with one element for each input data structure and $\mathbf{x} \equiv \mathbf{x}_M$ is the vector of coefficients, one for each representative configuration. The subscripts on the kernel matrix denote array slices, i.e. $\mathbf{K}_{MN} = \mathbf{K}_{NM}^\top$ is the rectangular matrix whose elements correspond to the kernel values between the representative configurations and the input configurations.

Using a representative set much smaller than the total number of structures has significant advantages in terms of computational cost, often with no reduction in fitting accuracy. The training cost is dominated by computing the pseudoinverse, which scales as $O(NM^2)$, which is linear in the size of the training data, N, and evaluating the model scales as $O(M)$, now independent of the size of the training data. These cost scalings are analogous to those of artificial neural networks with a fixed number of nodes.

While the above solutions are formally correct, it is widely recognized that they lead to numerical instability and *overfitting*, i.e. they are solutions that attempt to maximize the fit to the input data, even when this might not be desirable, which is almost always the case. At first sight, this might sound surprising, since electronic structure calculations can be made deterministic, with precise convergence behaviour in terms of its parameters, such as k-point sampling, SCF tolerance, etc. However, practical calculations are never converged to machine precision, and the resulting inconsistencies between the potential energy values for different configurations are not something that is desirable to propagate to a fitted potential energy surface. The magnitude of such inconsistencies can be easily assessed before the fit is made. Previous experience [11, 12] suggests that for large databases for materials applications using plane wave density functional theory, the error due to k-point sampling is dominant, and difficult to reduce below a few meV/atom due to the associated computational cost.

In case we are fitting a potential energy surface with a representation that does not characterize the atomic positions of the whole system completely due to, e.g., a finite cutoff, or some other choices made to gain computational efficiency, the fit is not expected to be exact, irrespective of the amount of input data. Sometimes, such *model error* can also be assessed a priori, e.g. in the case of a finite cutoff by measuring the contribution made to forces on an atom by other atoms beyond the cutoff [13–15].

These two considerations suggest that allowing some "looseness" in the linear system might be beneficial, because it can be exploited to allow smaller linear coefficients, making the fit more regular and thus better at extrapolation. We collect the errors we expect in the fit of each target data value on the diagonal of an $N \times N$ matrix, $\mathbf{\Lambda}$. The common procedure to regularizing the problem is due to Tikhonov [16]. Specifically, in "kernel ridge regression" (and the equivalent "Gaussian process regression", a Bayesian view of the same) the Tikhonov matrix is chosen to be the kernel matrix between the M representative points, $\mathbf{K}_{MM}$. With highly regular ("smooth") kernel functions, this regularization leads to smooth fits, and the sizes of the elements of $\mathbf{\Lambda}$ control the trade-off between the accuracy of the fit and smoothness. The corresponding solutions are

$$\mathbf{x} = (\mathbf{K} + \mathbf{\Lambda})^{-1}\mathbf{y}, \tag{6.5}$$

for the square problem, and

$$\mathbf{x}_M = \left(\mathbf{K}_{MM} + \mathbf{K}_{MN}\mathbf{\Lambda}^{-1}\mathbf{K}_{NM}\right)^{-1}\mathbf{K}_{MN}\mathbf{\Lambda}^{-1}\mathbf{y}_N, \tag{6.6}$$

for the rectangular problem, where we again emphasized the index sets. This solution is equivalent to minimizing

$$\|\mathbf{Kx} - \mathbf{y}\|^2_{\mathbf{\Lambda}^{-1}} + \|\mathbf{x}\|^2_{\mathbf{K}}, \tag{6.7}$$

which shows that the inverse of the tolerances in $\mathbf{\Lambda}$ is equivalent to regression weights on the different data points. With the solution of the linear system in hand, the value of the fitted function for a new structure C can be written as

$$f(C) = \mathbf{K}_{CM}\mathbf{x}_M. \tag{6.8}$$

Note that the $\mathbf{K}_{CM}$ slice is just a vector, with elements given by the kernel between the new structure C and the structures in the representative set M.

6.2.1 Selection of a Representative Set

Next we describe some ways to choose the set of representative environments over which the sum in Eq. (6.1) is taken. This can be done by simple random sampling, but we find it advantageous to use this freedom to optimize interpolation accuracy. Among the many strategies that have been proposed [17–20], we discuss two that have been used successfully in the context of potential energy fitting. One approach to this is to maximize the dissimilarity between the elements of the representative set. A greedy algorithm to select the configurations for the representative set is "farthest point sampling", in which we start with a randomly selected structure, and then iteratively pick as the next structure to include the one which is farthest away from any of the structures already in the set [21–23]. The distance between two structures is measured using the "kernel metric" [10], defined as

$$d^2(\mathcal{A}, \mathcal{B}) = K(\mathcal{A}, \mathcal{A}) + K(\mathcal{B}, \mathcal{B}) - 2K(\mathcal{A}, \mathcal{B}). \tag{6.9}$$

This algorithm performed well for selecting molecules in regression tasks, enabling the significant reduction of the data set sizes needed to achieve a given level of accuracy [24].

Another technique that has been successfully used is based on matrix factorization, which is particularly appealing when the kernel function is linear or a low order polynomial of the representation vector. Consider the matrix of feature vectors, $\mathbf{D}$, in which each row is the representation vector of an input atomic configuration, such that a linear kernel is $\mathbf{K} = \mathbf{DD}^\top$. We are looking to select rows, many fewer than the total number, which span as much of the space as all rows span. This is a problem of matrix representation, specifically the representative set should serve as a low-rank approximation of **K** and/or **D**. One solution to this is called CUR matrix decomposition [25], which can be applied to either **K** or **D**, the latter being much cheaper if the length of the representation vectors is less than the number of data points.

To determine the optimal set of representative configurations, we start with a singular value decomposition of **D**,

$$\mathbf{D} = USV^{\top}. \tag{6.10}$$

For each data point, a *leverage score* is calculated, essentially the weight that the top singular vectors have on that configuration,

$$\pi_{\mathcal{A}} = \frac{1}{k}\sum_{\xi=1}^{k}\left(u_{\mathcal{A}}^{\xi}\right)^{2}, \tag{6.11}$$

where $u_{\mathcal{A}}^{\xi}$ is the element of the ξ-th left singular vector that corresponds to structure $\mathcal{A}$. The sum runs over the first k singular vectors, e.g. $k = 20$ is typical. The configuration $\mathcal{A}$ is included in the representative set with a probability that is proportional to its leverage score, $\pi_{\mathcal{A}}$. A deterministic variant is to select one structure $\mathcal{A}$ at a time—the one with the highest leverage score—delete the associated row from the representation matrix and orthogonalize the remaining rows of **D** relative to it. The next data point can then be selected repeating the same procedure on the now smaller feature matrix [26].

Note that in the Gaussian process literature, using a subset of the data to construct the basis is called *sparsification* [27, 28], even though the approximation relies on a low-rank matrix reconstruction rather than the kernel matrix being sparse.

6.2.2 Linear Combination of Kernels

When fitting interatomic potentials for materials, a model is constructed for the *atomic energy*, sometimes called the “site energy”. This is both for computational efficiency and to reduce the complexity of the functional relation between structures and properties: each atomic energy is only a function of a limited number of degrees of freedom corresponding to the coordinates of the neighbouring atoms and can therefore be evaluated independently from any other atomic energy. In fact this is the defining characteristic of an interatomic potential, in contrast to a quantum mechanical model that explicitly includes delocalized electrons. Going from atomic energies to the total energy is trivial, the latter being the sum of the former. However, going in the other direction is not unambiguous. The total energy can be calculated from a quantum mechanical model, but the atomic energies are not defined uniquely, and it becomes part of the fitting task to find the best possible decomposition of the total energy into atomic energies. Treating these two transformations on the same footing helps. Suppose we want to predict the sum of function values for two (or more) configurations. For the simple case of the sum of two energies for structures $\mathcal{A}$ and $\mathcal{B}$, the prediction is, trivially, just the sum of the individual function value predictions, e.g.

$$E_{\text{tot}} = E(\mathcal{A}) + E(\mathcal{B}) = \mathbf{K}_{\mathcal{A}M}\mathbf{x}_M + \mathbf{K}_{\mathcal{B}M}\mathbf{x}_M. \tag{6.12}$$

If we define a new "sum-kernel" to be the sum of kernel values between a number of new configurations and the representative set, the expression for the above total energy prediction takes the same form as the prediction of the individual function values. For some set I of new configurations, let

$$^{\Sigma}\mathbf{K}_M = \sum_{\mathcal{A}\in I} \mathbf{K}_{\mathcal{A}M}, \tag{6.13}$$

where $^{\Sigma}\mathbf{K}_M$ is the *vector* of sum-kernel values, each element of which is the sum of the kernel between all the configurations in I and a given configuration in the representative set M. The predicted total energy of the configurations in I is then

$$E_{\text{tot}} = {}^{\Sigma}\mathbf{K}_M\mathbf{x}_M. \tag{6.14}$$

This same sum-kernel can be used to fit the model to sum data, rather than to individual function values. This is necessary in order to fit interatomic potentials for materials systems, since only total energies, and not the atomic energies themselves, are available from electronic structure calculations. At the same time, in order to enforce a finite short range in the interatomic potential, we must express the potential as an atomic energy. Using the sum-kernel, this is straightforward, the original functional form in Eq. (6.1) can be retained, and we now minimize (omitting the regularization term for clarity)

$$\left\| {}^{\Sigma}\mathbf{K}\mathbf{x} - \mathbf{E}_{\text{tot}} \right\|^2, \tag{6.15}$$

where $^{\Sigma}\mathbf{K}$ is a matrix containing the sum-kernel values for all configurations in the input database and the representative set, and the vector $\mathbf{E}_{\text{tot}}$ is the collection of corresponding total energy data.

6.2.3 Derivatives

The explicit analytic functional form of Eq. (6.1) leads to analytic derivates with respect to the atomic coordinates, e.g. forces in the case of fitting an energy. Considering for the moment the simpler case in which we are computing the derivatives of an atom-centred quantity $f(\mathcal{A})$, we define $\nabla_{\mathcal{A}}$ as the vector of derivatives with respect to all the atomic coordinates in structure $\mathcal{A}$. We use the notation $\overleftarrow{\nabla}$ to indicate a derivative operator that applies to the first argument of the kernel, and $\overrightarrow{\nabla}$ to indicate a derivative that applies to the second argument. The derivatives of $f(\mathcal{A})$ are non-zero only for atoms that belong to the structure $\mathcal{A}$ and are then given by differentiating Eq. (6.1)

$$\nabla_{\mathcal{A}} f(\mathcal{A}) = \sum_{\mathcal{B} \in M} x_{\mathcal{B}} \overleftarrow{\nabla}_{\mathcal{A}} K(\mathcal{A}, \mathcal{B}) = \mathbf{K}_{\nabla \mathcal{A} M} \mathbf{x}_M, \tag{6.16}$$

where we introduced the notation $\mathbf{K}_{\nabla \mathcal{A} M}$ to indicate the matrix that contains the derivatives of the kernels relative to all the relevant atomic coordinates. Similarly to the case of sums above, the gradient of the kernel function can also be used for fitting the model not to target values, but to *gradient data* [29]. This is especially useful when the target represents a potential energy surface. When using typical electronic structure methods, the cost of computing the gradient with respect to all atomic positions is only a little bit more than the cost of computing the energy, but yields much more information, $3n$ pieces of data for an n-atom structure. There are two approaches one can take to incorporate gradient information. In the first one, used in Ref. [30] and subsequent work of that group [11, 12, 14, 15, 31–37], the functional form for the energy is again retained to be the same as in Eq. (6.1). The corresponding loss function (again without regularization) is

$$\|\mathbf{K}_{\nabla N M} \mathbf{x}_M - \mathbf{y}_{\nabla N}\|^2, \tag{6.17}$$

where $\mathbf{y}_{\nabla N}$ refers to the concatenated vector of gradients on all atoms in the set of input structures and $\mathbf{K}_{\nabla N M}$ to the corresponding matrix of kernel derivatives. The form of the solution for the coefficients is unchanged from Eq. (6.5) or (6.6) with $\mathbf{K}_{\nabla N M}$ taking the role of $\mathbf{K}_{NM}$.

In the second approach, used recently in Ref. [38], derivatives of the kernel are the basis functions in the functional form of the fit,

$$f(\mathcal{A}) = \sum_{\mathcal{B} \in M} \mathbf{x}_{\nabla \mathcal{B}} \cdot \overrightarrow{\nabla}_{\mathcal{B}} K(\mathcal{A}, \mathcal{B}), \tag{6.18}$$

where $\mathbf{x}_{\nabla \mathcal{B}}$ contains one weight for each of the derivatives relative to the atoms in structure $\mathcal{B}$. The number of basis functions and corresponding coefficients is now much larger, $3nM$, for n-atom structures. Since the model is fitted to the derivatives, given by gradients of Eq. (6.18), the loss is

$$\|\mathbf{K}_{\nabla N \nabla M} \mathbf{x}_{\nabla M} - \mathbf{y}_{\nabla N}\|^2, \tag{6.19}$$

the target properties can be computed as

$$f(\mathcal{A}) = \mathbf{K}_{\mathcal{A} \nabla M} \mathbf{x}_{\nabla M}, \tag{6.20}$$

and their derivatives as

$$\nabla_{\mathcal{A}} f(\mathcal{A}) = \mathbf{K}_{\nabla\mathcal{A}\nabla M} \mathbf{x}_{\nabla M}. \tag{6.21}$$

The original motivation for this approach is apparent from Eq. (6.19) in which the matrix can be understood as a kernel directly between atomic forces (and in case of $M = N$, between the input data forces).

Both approaches constitute valid ways of learning a function from data representing its gradients, differing only in the choice of the kernel basis. The kernel-derivative basis functions could also be used in conjunction with a reduced representative set, and it is not yet clear which approach is better, or indeed a combination: one could choose different basis functions (kernels or their derivatives) depending on the amount and kind of data available and on the size and choice of the representative set.

6.2.4 Learning from Linear Functionals

We can combine the sum-kernel and the derivative kernel naturally, and write a single least squares problem for the coefficients in Eq. (6.1) that is solved to fit an interatomic potential to all available total energy, force, and virial stress data (the only condition being that the input data has to be expressible using a linear operator on function values). We define $\mathbf{y}$ as the vector with L components containing all the input data: all total energies, forces, and virial stress components in the training database, and $\mathbf{y}'$ as the vector with N components containing the *unknown* atomic energies of the N atomic environments in the database, and $\hat{\mathbf{L}}$ as the linear differential operator of size $L \times N$ which connects $\mathbf{y}$ with $\mathbf{y}'$ such that $\hat{\mathbf{L}}\mathbf{y}' = \mathbf{y}$ (note that the definition of $\hat{\mathbf{L}}$ we use here is the transpose of that in Ref. [32]). The regularized least squares problem is now to minimize

$$\left\| \hat{\mathbf{L}}\mathbf{K}\mathbf{x} - \mathbf{y} \right\|^2_{\mathbf{\Lambda}^{-1}} + \|\mathbf{x}\|^2_{\mathbf{K}}, \tag{6.22}$$

and the expression for the coefficients is given by

$$\mathbf{x} = \left[\mathbf{K}_{MM} + \left(\hat{\mathbf{L}}\mathbf{K}_{NM}\right)^{\top} \mathbf{\Lambda}^{-1} \hat{\mathbf{L}}\mathbf{K}_{NM} \right]^{-1} \left(\hat{\mathbf{L}}\mathbf{K}_{NM}\right)^{\top} \mathbf{\Lambda}^{-1} \mathbf{y}, \tag{6.23}$$

where the sizes of the elements of $\mathbf{\Lambda}$ control the trade-off between the accuracy of the fit and smoothness.

It is instructive to write down the above matrices for the simple case when the system consists of just two atoms, A and B, with position vectors $\mathbf{r}_A, \mathbf{r}_B$, target total energy E, and target forces $\mathbf{f}_A \equiv (f_{Ax}, f_{Ay}, f_{Az})$ and $\mathbf{f}_B \equiv (f_{Bx}, f_{By}, f_{Bz})$. The data vector is then given by

$$\mathbf{y} = \left[E \; f_{Ax} \; f_{Ay} \; f_{Az} \; f_{Bx} \; f_{By} \; f_{Bz} \right]^{\top}. \tag{6.24}$$

The aim of the fit is to determine two unknown atomic energy functions ε_A and ε_B as a function of the atomic environments centred around the two atoms, $\mathcal{A}$ and $\mathcal{B}$, respectively. The total energy is their sum, $E = \varepsilon_A + \varepsilon_B$, and the forces need to include the cross terms,

$$\begin{aligned} \mathbf{f}_A &= \frac{\partial \varepsilon_A}{\partial \mathbf{r}_A} + \frac{\partial \varepsilon_B}{\partial \mathbf{r}_A}, \\ \mathbf{f}_B &= \frac{\partial \varepsilon_A}{\partial \mathbf{r}_B} + \frac{\partial \varepsilon_B}{\partial \mathbf{r}_B}. \end{aligned} \tag{6.25}$$

The representative set in this case consists of the same two atoms, so $N = M$, and the kernel matrix is square,

$$\mathbf{K} = \begin{bmatrix} K(\mathcal{A}, \mathcal{A}) & K(\mathcal{A}, \mathcal{B}) \\ K(\mathcal{B}, \mathcal{A}) & K(\mathcal{B}, \mathcal{B}) \end{bmatrix}, \tag{6.26}$$

and the linear operator $\hat{\mathbf{L}}$ is a 7×2 matrix and is given by

$$\hat{\mathbf{L}} = \begin{bmatrix} 1 & 1 \\ \overleftarrow{\nabla}_{\mathbf{r}_A} & \overleftarrow{\nabla}_{\mathbf{r}_A} \\ \overleftarrow{\nabla}_{\mathbf{r}_B} & \overleftarrow{\nabla}_{\mathbf{r}_B} \end{bmatrix}, \tag{6.27}$$

so the $\hat{\mathbf{L}}\mathbf{K}$ matrix to be substituted into Eq. (6.23) is

$$\hat{\mathbf{L}}\mathbf{K} = \begin{bmatrix} K(\mathcal{A}, \mathcal{A}) + K(\mathcal{A}, \mathcal{B}) & K(\mathcal{B}, \mathcal{A}) + K(\mathcal{B}, \mathcal{B}) \\ \overleftarrow{\nabla}_{\mathbf{r}_A} K(\mathcal{A}, \mathcal{A}) + \overleftarrow{\nabla}_{\mathbf{r}_A} K(\mathcal{B}, \mathcal{A}) & \overleftarrow{\nabla}_{\mathbf{r}_A} K(\mathcal{A}, \mathcal{B}) + \overleftarrow{\nabla}_{\mathbf{r}_A} K(\mathcal{B}, \mathcal{B}) \\ \overleftarrow{\nabla}_{\mathbf{r}_B} K(\mathcal{A}, \mathcal{A}) + \overleftarrow{\nabla}_{\mathbf{r}_B} K(\mathcal{B}, \mathcal{A}) & \overleftarrow{\nabla}_{\mathbf{r}_B} K(\mathcal{A}, \mathcal{B}) + \overleftarrow{\nabla}_{\mathbf{r}_B} K(\mathcal{B}, \mathcal{B}) \end{bmatrix} \tag{6.28}$$

Note that terms such as $\overleftarrow{\nabla}_{\mathbf{r}_A} K(\mathcal{B}, \mathcal{B})$ or $\overleftarrow{\nabla}_{\mathbf{r}_A} K(\mathcal{B}, \mathcal{A})$ are not zero because atom A is present in the environment $\mathcal{B}$ of atom B, and so $K(\mathcal{B}, \mathcal{A})$, and also $K(\mathcal{B}, \mathcal{B})$, depend on $\mathbf{r}_A$ explicitly.

Using the approach of Ref. [38] for the dimer, the kernel matrix is 6×6 and is given by

$$\mathbf{K}_{\nabla\mathcal{A}\nabla\mathcal{B}} = \begin{bmatrix} \overleftarrow{\nabla}_{\mathbf{r}_A} \overrightarrow{\nabla}_{\mathbf{r}_A} K(\mathcal{A}, \mathcal{A}) & \overleftarrow{\nabla}_{\mathbf{r}_A} \overrightarrow{\nabla}_{\mathbf{r}_B} K(\mathcal{A}, \mathcal{B}) \\ \overleftarrow{\nabla}_{\mathbf{r}_B} \overrightarrow{\nabla}_{\mathbf{r}_A} K(\mathcal{B}, \mathcal{A}) & \overleftarrow{\nabla}_{\mathbf{r}_B} \overrightarrow{\nabla}_{\mathbf{r}_B} K(\mathcal{B}, \mathcal{B}) \end{bmatrix}. \tag{6.29}$$

In practice it is always worth using *all* available data, even though once the fit is converged in the limit of infinite amount of data, the information from derivatives (forces) is the same as from energies. With finite amount of data, however, choosing the weights corresponding to energies and forces via the diagonal regularizer allows control of the fit, in the sense of its relative accuracy in reproducing energies and forces.

6.2.5 Learning Multiple Models Simultaneously

Being able to fit to sums of function values has an interesting consequence. It enables in a very natural way the fitting of a model that is explicitly and a priori written as a sum of terms, each using a different kernel function, perhaps even using a different representation. That this is a good idea for potential energy functions is shown by the relative success of empirical force fields both for materials and molecules, in which the total energy is written as a sum of body-ordered terms, i.e. an atomic term, a pair potential, and a three-body (angle-dependent) term, etc.

$$E_{\mathrm{tot}} = \sum_{i} E^{(1)} + \sum_{i,j} E^{(2)}(r_{ij}) + \sum_{i,j,k} E^{(3)}(r_{ij}, r_{ik}, r_{jk}) + \cdots \tag{6.30}$$

It is notable that while pair potentials and three-body potentials using various simple parametrizations are widely used in the materials modelling literature, there are few models that take advantage of the full three-dimensional flexibility of the three-body term. Four-body terms are almost always restricted to one-dimensional parametrizations such as a dihedral angle. The reason for this is presumably because there is little intuition about what kinds of functional forms would be appropriate—kernel fitting avoids this problem. Such a framework was introduced [32] and is beginning to be used for low body order model fitting [14, 15, 39, 40]. Furthermore, by bringing everything together under the kernel formalism, the above expansion can also be augmented with a many-body term which enables the systematic convergence to the true Born–Oppenheimer potential energy surface, but with the many-body term having a relatively small magnitude (because the low body order terms account for most of the energy already), which helps transferability and stability.

The two-body term could be represented as a linear combination of kernels whose arguments are simply the interatomic distances, the three-body term is again a linear combination of kernels whose arguments are some representation of the geometry of three atoms, e.g. the one above using the three distances, but two distances and an angle are equally viable. The fit is then made to the target data of total energies and forces of atomic configurations, in complete analogy with Eq. (6.12), and now the value of the sum-kernel is the sum of pair- and triplet-kernel values between all pairs and triplets present in the two atomic configurations. A stringent test of this scheme is that in case of a target potential energy surface that is explicitly the sum of two- and three-body terms, the fit recovers these terms explicitly from just the total energies and forces [32].

6.3 Density-Based Representations and Kernels

Having summarized the algorithms that can be used to perform kernel ridge regression using atomic-scale properties and their derivatives as inputs, we now proceed to describe a framework for defining physics-based representations of local atomic environments and the kernels built from them. In kernel ridge regression, the representations do not necessarily need to be expressed explicitly, but can also be defined implicitly by means of the kernel function $K(\mathcal{A}, \mathcal{B})$ that corresponds to the scalar product of representation vectors that span a (possibly infinite-dimensional) Hilbert space [10]. Vectors $|\mathcal{A}\rangle$ in this "reproducing kernel Hilbert space" do correspond to atomic structures, and one can write formally $K(\mathcal{A}, \mathcal{B}) \equiv \langle\mathcal{A}|\mathcal{B}\rangle$ even if the kernel might be computed without ever determining the vectors explicitly.

The reader trained in quantum mechanics will recognize an isomorphism between representations and the state vectors on the one hand, and kernels and expectation values on the other. This analogy suggests that it may be beneficial to formulate atomic-scale representations using a formalism that mimics Dirac notation. Whereas in a quantum mechanical setting the physical symmetries of the problem are built into the Hamiltonian, in a machine-learning setting they are more conveniently included in the representation itself, that should be made invariant to basic symmetries such as atom labelling, rigid translations, and rotations. In this section we show how starting from these intuitions one can build a very abstract description of a molecular structure that is naturally invariant with respect to the physical symmetries, based on a representation of the atom density.

Translational and rotational symmetries can be included by decomposing the structure into a collection of local environments, and by explicit symmetrization over the $SO(3)$ group. In fact, it has been recently shown [41] how this construction leads naturally to the SOAP representation and kernel [42], and to several other popular choices of density-based representations—from Behler–Parrinello symmetry functions [43], to voxel density representations [44] to the binning of the pair correlation function [45]—that can be regarded as different projections of the same smooth atomic amplitude. A peculiarity of the SOAP framework is that it is formulated very naturally in terms of a kernel, that corresponds to the symmetrized overlap of atomic densities, and that it allows one to explicitly compute the representations whose scalar product constitutes the kernel function, which allows one to go back and forth between a kernel and a representation language. The atomic environmental representations can then be modified to generate non-linear kernels, as well as combined into global structural kernels. We will briefly discuss different possible approaches to the latter, either by simple linear combination of the local representations or by a more sophisticated procedure that takes into account the most effective matching between pairs of environments in the two structures that are being compared.

6.3.1 A Dirac Notation for Structural Representations

Let us introduce an abstract notation to describe atomistic structures in terms of the positions and chemical nature of the atoms that compose them [41]. Taking inspiration from Dirac notation for quantum mechanical states, we associate a ket $|\mathcal{A}\rangle$ with each configuration. Let us start with a simple example to see how such a formalism can be introduced and used. Much like in the case of quantum states, we can define a concrete representation of the ket associated with a structure in terms of positions and chemical species, e.g.

$$\left\langle \mathbf{r} \middle| \mathcal{A} \right\rangle = \sum_i g_i(\mathbf{r} - \mathbf{r}_i) \left| \alpha_i \right\rangle, \tag{6.31}$$

where the position of each atom is represented by a smooth density g_i (that in principle could depend on the nuclear charge and the position of atom i) and the kets $|\alpha_i\rangle$ contain the information on the nuclear charge of each atom. The Dirac notation lends itself naturally to the definition of overlap kernels between structures, $\langle \mathcal{A}|\mathcal{B}\rangle$. To compute such an integral, one can use the position representation and assume that the kets associated with different elements are orthonormal:

$$\begin{aligned} \langle \mathcal{A}|\mathcal{B}\rangle &= \int \mathrm{d}\mathbf{r}\, \langle \mathcal{A}|\mathbf{r}\rangle \langle \mathbf{r}|\mathcal{B}\rangle \\ &= \sum_{ij} \int \mathrm{d}\mathbf{r}\, g_i^A \left(\mathbf{r} - \mathbf{r}_i^A\right)^\star g_j^B \left(\mathbf{r} - \mathbf{r}_j^B\right) \left\langle \alpha_i^A \middle| \alpha_j^B \right\rangle \\ &= \sum_\alpha \sum_{i,j\in\{\alpha\}} \int \mathrm{d}\mathbf{r}\, g_i^A \left(\mathbf{r} - \mathbf{r}_i^A\right)^\star g_j^B \left(\mathbf{r} - \mathbf{r}_j^B\right). \end{aligned} \tag{6.32}$$

This density-based representation would not be in itself very useful, as the kernel is not invariant to relative rotations of the structures, and not even to the absolute position of the two structures in space, or their periodic representation. Nevertheless, it can be taken as the starting point to introduce many of the most successful feature representations that have been used in recent years for machine-learning of materials and molecules.

To see how, one can take inspiration from linear-scaling electronic structure methods, and the nearsightedness principle for electronic matter [46–49]. We then shift the attention from the description of complete structures to that of spherical atomic environments, that one can conveniently centre on top of each atom. An atom-centred representation arises naturally from the symmetrization over the translation group of tensor products of the representation Eq. (6.31) [41] and is also consistent with the atom-centred potentials that have been discussed in the previous section as an obvious application of this framework.

We will use the notation $|\mathcal{X}_j\rangle$ to indicate an environment centred around the j-th atom in a structure and express it in the position representation as

$$\langle \mathbf{r}|\mathcal{X}_j\rangle = \sum_i f_c(r_{ij}) g_{ij}(\mathbf{r} - \mathbf{r}_{ij})|\alpha_i\rangle, \tag{6.33}$$

where $f_c(r_{ij})$ is a cutoff function that restricts the environment to a spherical region centred on the atom, for the sake of computational efficiency and/or localization of the density information. The atom-centred smoothing functions are typically taken to be uniform-width Gaussians, but it would be easy to generalize the expression to include a dependency on the atomic species and/or the distance of an atom from the centre of the environment, which could be used to, e.g., reduce the resolution of the representation at the periphery of the environment, or adapt the smoothing length scale to each atomic species.

Note that one could also combine the density contributions from atoms of the same species into a species-dependent atomic amplitude,

$$\langle \alpha \mathbf{r} | \mathcal{X}_j \rangle = \psi^{\alpha}_{\mathcal{X}_j}(\mathbf{r}) = \sum_{i \in \alpha} f_c(r_{ij}) g_{ij}(\mathbf{r} - \mathbf{r}_{ij}), \tag{6.34}$$

and then write

$$\langle \mathbf{r} | \mathcal{X}_j \rangle = \sum_{\alpha} \psi^{\alpha}_{\mathcal{X}_j}(\mathbf{r}) | \alpha \rangle. \tag{6.35}$$

This notation is very useful to reveal how different representations can be seen as alternative representations of the same abstract ket. For instance, one can expand the atom density in orthogonal radial functions $R_n(r)$ and spherical harmonics. The coefficients in such an expansion can be written as

$$\begin{aligned} \langle \alpha n l m | \mathcal{X}_j \rangle &= \int \mathrm{d}\mathbf{r} \langle n l m | \mathbf{r} \rangle \langle \alpha \mathbf{r} | \mathcal{X}_j \rangle \\ &= \int \mathrm{d}r \mathrm{d}\hat{\mathbf{r}}\, r^2 R_n(r) Y^l_m(\hat{\mathbf{r}}) \psi^{\alpha}_{\mathcal{X}_j}(r\hat{\mathbf{r}}). \end{aligned} \tag{6.36}$$

As another example, Behler–Parrinello atom-centred symmetry functions that have been used in the construction of artificial neural network based interatomic potentials for materials [43,50–52] and molecules [53] can be written by setting the basis functions to be delta distributions $g_{ij}(\mathbf{r} - \mathbf{r}_{ij}) = \delta(\mathbf{r} - \mathbf{r}_{ij})$, and averaging the atom density with an appropriate pair weighting function G_2, e.g.

$$\begin{aligned} \langle \alpha \beta G_2 | \mathcal{X}_j \rangle &= \langle \alpha | \alpha_j \rangle \int \mathrm{d}\mathbf{r}\, G_2(r) \langle \beta \mathbf{r} | \mathcal{X}_j \rangle \\ &= \delta_{\alpha_j \alpha} \sum_{i \in \{\beta\}} f_c(r_{ij}) G_2(r_{ij}) \end{aligned} \tag{6.37}$$

The basis functions of the spectral neighbour analysis potential [54] also start with the same density and expands it in hyperspherical harmonics as introduced in Ref. [30].

6.3.2 Smooth Overlap of Atomic Positions

It is clear that a density-based representation such as Eq. (6.33) is invariant to translations of the entire structure, but not to rotations that would change the orientation of the atomic neighbour amplitude. This reflects the fact that scalar products of the form $\langle \mathcal{X}_j | \mathcal{X}_k \rangle$ depend on the relative orientation of the environments being compared. In the smooth overlap of atomic positions (SOAP) framework, we define a symmetrized version of the overlap kernel, using the Haar integral [55] of the rotation group,

$$K^{(\nu)}(\mathcal{X}_j, \mathcal{X}_k) = \int \mathrm{d}\hat{R} \left| \left\langle \mathcal{X}_j \middle| \hat{R} \middle| \mathcal{X}_k \right\rangle \right|^{\nu} = \left\langle \mathcal{X}_j{}^{(\nu)} \middle| \mathcal{X}_k{}^{(\nu)} \right\rangle \tag{6.38}$$

where the integral is performed over all possible rotation matrices. If the base kernel is raised to the ν-th power, the average preserves information on the correlations between atoms up to the $(\nu + 1)$-th order [39]. As we will show below, a crucial feature of the SOAP framework is that an explicit expression for the symmetrized representation vectors $|\mathcal{X}_j{}^{(\nu)}\rangle$ can be given. In fact, an alternative derivation of the SOAP framework can be achieved by symmetrizing directly tensor products of the translationally invariant ket Eq. (6.36) [41].

The complexity of the SOAP features is quite manageable for $\nu = 1, 2$, but becomes increasingly cumbersome for higher ν. An effective description of higher-order interactions, that does not increase too much the complexity of the analytical evaluation of Eq. (6.38), can be obtained by manipulating the $\nu = 2$ kernel, e.g. by taking a non-linear function of it. In practice it has been found that raising it to a power ζ and normalizing it to one

$$\left\langle \mathcal{X}_j{}^{(2)} \middle| \mathcal{X}_k{}^{(2)} \right\rangle_{\zeta} = \frac{\langle \mathcal{X}_j{}^{(2)} | \mathcal{X}_k{}^{(2)} \rangle^{\zeta}}{\sqrt{\langle \mathcal{X}_j{}^{(2)} | \mathcal{X}_j{}^{(2)} \rangle^{\zeta} \langle \mathcal{X}_k{}^{(2)} | \mathcal{X}_k{}^{(2)} \rangle^{\zeta}}} \tag{6.39}$$

is sufficient to include many-body contributions in the final kernel.

Using the Dirac notation, it is easy to see how one can give an explicit representation of the $SO(3)$ symmetrized ket for the case with $\nu = 1, 2$. Using a spherical harmonics expansion of $|\mathcal{X}_j\rangle$ it is very natural to perform the rotational average analytically by introducing the Wigner matrix associated with the rotation, $\langle lm | \hat{R} | l'm' \rangle = \delta_{ll'} D^l_{mm'}(\hat{R})$

$$\begin{aligned} &\int \mathrm{d}\hat{R} \sum_{\alpha nlm} \langle \mathcal{X}_j | \alpha nlm \rangle \langle \alpha nlm | \hat{R} | \mathcal{X}_k \rangle \\ &\quad = \sum_{\alpha nlmm'} \langle \mathcal{X}_j | \alpha nlm \rangle \langle \alpha nlm' | \mathcal{X}_k \rangle \int \mathrm{d}\hat{R}\, D^l_{mm'}(\hat{R}) \end{aligned} \tag{6.40}$$

which simplifies greatly due to the properties of the Wigner matrices. Only the term with $l = 0$ survives, which makes it possible to write explicitly the $\nu = 1$ symmetrized SOAP representations in terms of the spherical harmonics coefficients

$$\left\langle \alpha n \middle| \mathcal{X}_k{}^{(1)} \right\rangle = \sqrt{8\pi^2}\langle \alpha n00|\mathcal{X}_k\rangle, \tag{6.41}$$

which corresponds to the simple kernel

$$\left\langle \mathcal{X}_j{}^{(1)} \middle| \mathcal{X}_k{}^{(1)} \right\rangle = \sum_{\alpha n} \left\langle \mathcal{X}_j{}^{(1)} \middle| \alpha n \right\rangle \left\langle \alpha n \middle| \mathcal{X}_k{}^{(1)} \right\rangle. \tag{6.42}$$

A position representation of the $\nu = 1$ representation $\langle r|\mathcal{X}_k{}^{(1)}\rangle$ yields naturally the rotational average of $\langle \mathbf{r}|\mathcal{X}_k\rangle$. This can be seen by expressing $K^{(1)}(\mathcal{X}_j, \mathcal{X}_k)$ in a position basis

$$\begin{aligned} \left\langle \alpha \mathcal{X}_j{}^{(1)} \middle| \alpha \mathcal{X}_k{}^{(1)} \right\rangle &= \int \mathrm{d}\hat{R} \int \mathrm{d}\mathbf{r}\, \psi^{\alpha}_{\mathcal{X}_j}(\mathbf{r}) \psi^{\alpha}_{\mathcal{X}_k}(\hat{R}\mathbf{r}) \\ &= 32\pi^3 \int \mathrm{d}r\, r^2 \bar{\psi}^{\alpha}_{\mathcal{X}_j}(r) \bar{\psi}^{\alpha}_{\mathcal{X}_k}(r) \end{aligned} \tag{6.43}$$

where we have defined the rotationally averaged atom density

$$\bar{\psi}^{\alpha}_{\mathcal{X}_j}(r) = \frac{1}{4\pi} \int \mathrm{d}\hat{\mathbf{r}}\, \psi^{\alpha}_{\mathcal{X}_j}(r\hat{\mathbf{r}}) = \frac{1}{\sqrt{32\pi r^3}} \left\langle \alpha r \middle| \mathcal{X}_j{}^{(1)} \right\rangle, \tag{6.44}$$

which is thus closely related to the pair correlation function around the tagged atom. Similar representations have been used for machine-learning of molecules and materials [6, 45], revealing once more the intimate relationships between different atom-density based representations.

The $\nu = 1$ representation integrates away all angular correlations and therefore does not provide a unique representation of an environment. The representations with $\nu = 2$ provide information on 3-body correlations and can also be obtained relatively easily in closed form. The Haar integral now contains the product of two Wigner matrices. Exploiting their orthogonality relations, one obtains

$$\int \mathrm{d}\hat{R} \left| \sum_{\alpha n l m} \left\langle \mathcal{X}_j \middle| \alpha n l m \right\rangle \left\langle \alpha n l m \middle| \hat{R} \middle| \mathcal{X}_k \right\rangle \right|^2 = \sum_{\alpha n \alpha' n' l} \left\langle \mathcal{X}_j{}^{(2)} \middle| \alpha n \alpha' n' l \right\rangle \left\langle \alpha n \alpha' n' l \middle| \mathcal{X}_k{}^{(2)} \right\rangle \tag{6.45}$$

where the $\nu = 2$ symmetrized SOAP representations read

$$\left\langle \alpha n \alpha' n' l \middle| \mathcal{X}_j{}^{(2)} \right\rangle = \sqrt{\frac{8\pi^2}{2l+1}} \sum_m \left\langle \mathcal{X}_j \middle| \alpha n l m \right\rangle \left\langle \alpha' n' l m \middle| \mathcal{X}_j \right\rangle. \tag{6.46}$$

This notation corresponds to the power spectrum components introduced in Refs. [24, 42], $\langle\alpha n\alpha' n' l|\mathcal{X}_j{}^{(2)}\rangle \equiv p^{\alpha\alpha'}_{nn'l}(\mathcal{X}_j)$. Note also that, while the representation of the symmetrized kets in terms of the nlm expansion is very convenient, it is not the only possibility. Similar to Eq. (6.44), an explicit position representation can be obtained for $\langle\alpha\mathbf{r}_1\alpha'\mathbf{r}_2|\mathcal{X}_k{}^{(2)}\rangle$, that provides a complete representation of the 3-body rotationally invariant correlations. The 3-body symmetry functions of the Behler–Parrinello kind can be seen as projections of this representation, similarly to the case of 2-body functions in Eq. (6.37).

The case of $\nu = 3$ leads to an explicit representation of the ket that is proportional to the bispectrum of the environment [42]

$$\begin{aligned}\left\langle\alpha_1 n_1 l_1 \alpha_2 n_2 l_2 \alpha n l\middle|\mathcal{X}_j{}^{(3)}\right\rangle &\propto \sum_{m m_1 m_2}\left\langle\mathcal{X}_j\middle|\alpha n l m\right\rangle \\ &\times \langle\alpha_1 n_1 l_1 m_1|\mathcal{X}_j\rangle\langle\alpha_2 n_2 l_2 m_2|\mathcal{X}_j\rangle\langle l_1\, m_1\, l_2\, m_2|l\, m\rangle .\end{aligned} \tag{6.47}$$

While the dimensionality of this representation makes it impractical unless somehow sparsified, it does give direct access to higher-order correlations. An interesting detail is that $|\mathcal{X}_j{}^{(3)}\rangle$, contrary to the $\nu = 1, 2$ cases, is not invariant to mirror symmetry, which makes it capable of distinguishing enantiomers.

Finally, one should note that the normalization of the kernel Eq. (6.39) can be achieved by normalizing the SOAP vector, so that an explicit representation of the normalized feature vector is possible. While in principle one could write out an explicit representation that yields the kernel for $\zeta > 1$, it would contain an exponentially increasing number of terms. As in the case of $|\mathcal{X}_j{}^{(3)}\rangle$, this only makes sense if combined with a sparsification procedure.

6.3.3 Body-Order Potentials

Representations based on a symmetrized atom-density correspond to $(\nu + 1)$ body-order correlations. As a consequence, they can be used as a linear basis to expand $(\nu + 1)$-body potentials [39], drawing a connection with a long-standing, physically inspired tradition in the construction of empirical force fields as a sequence of terms of increasing complexity [56–58]. This connection has been recently discussed in great detail in Ref. [59], which reaches a conclusion similar to Ref. [41] on the existence of a deep connection among all of the most widespread feature sets for atomistic machine-learning. In Ref. [59], smoothness in the atomic cluster expansion is effectively imposed through the choice of a smooth set of functions that are evaluated at atomic positions. In contrast, the SOAP representation is smooth regardless of the choice of basis for the density expansion because the density itself is a superposition of smooth functions (e.g. Gaussians).

When discussing the connection between density-based representations, atomic cluster expansion, and traditional body-order potentials, it is important to consider the computational complexity associated with the evaluation of different body-order terms. An explicit $(\nu + 1)$-body potential requires summing over every group of ν of the N atoms within each neighbourhood, leading to an $N^{(\nu)}$ scaling for each atom-centred term. In a density-based representation, or in the atomic cluster expansion, this cost is mitigated by computing expansion coefficients of the density (with a linear scaling in N), and then computing $(\nu + 1)$-order invariant representations. This second step still entails an exponential scaling with body order, e.g. for $\nu = 2$ SOAP written in a radial basis with $n_{\max}$ functions and $l_{\max}$ angular channels, one has to perform $n_{\max}^2(l_{\max} + 1)^2$ multiplications to evaluate all invariant coefficients. In general, the computational complexity of the complete set of invariants scales with the power of ν. Computational savings come through the choice of an optimized basis set (keeping $n_{\max}$ and $l_{\max}$ low), the low prefactor associated with the evaluation of the invariant coefficients, and potentially through the use of data-driven compression techniques [26], that can beat the exponential scaling by only retaining a small number of $(\nu + 1)$-body order invariant features. Note that apart from computing high body order terms explicitly by increasing ν, non-linear transformations (e.g. Eq. (6.39)) also increase the effective body order of the models within the kernel fitting framework.

6.3.4 Kernel Operators and Feature Optimization

Provided one takes a long-range environmental cutoff, and chooses a kernel that can represent high orders of many-body interactions, a density-based representation of atomic structures should provide a complete description of any atomic structure and—given a sufficiently complete training set—predict any atomistic property with arbitrary accuracy. In practice, obviously, the accuracy of a model depends on the details of the representation, which is why different representations or kernels provide different levels of accuracy for the same training and test set [60]. The performance of a set of representations can be improved by modifying them so that they represent more efficiently the relations between structure and properties.

This kind of optimizations are best understood in terms of changes to the translationally invariant environmental ket Eq. (6.33) and can be described in an abstract and basis-set independent manner as a Hermitian operator acting on the ket,

$$|\mathcal{X}_j\rangle \to \hat{U}|\mathcal{X}_j\rangle. \tag{6.48}$$

The most general form of this operator that makes it rotationally invariant—so that it commutes with the rotation matrix in the definition of the SOAP kernel Eq. (6.38)—is readily expressed in the $\{|\alpha nlm\rangle\}$ basis [41]:

$$\langle \alpha nlm|\hat{U}|\alpha' n' l' m'\rangle = \delta_{ll'}\delta_{mm'}\langle \alpha nl|\hat{U}|\alpha' n' l'\rangle. \tag{6.49}$$

While this is the most general form of the operator that is consistent with $SO(3)$ symmetry, one can use simpler forms to represent feature space transformations that can be easily understood. For instance, taking

$$\langle \alpha n l | \hat{U} | \alpha' n' l' \rangle = u_n \delta_{\alpha\alpha'} \delta_{nn'} \delta_{ll'} \quad (6.50)$$

corresponds to a scaling of the smooth atom density according to the distance from the centre. This kind of scaling has been shown to improve significantly the performance of SOAP [61], as well as other density-based representations [62, 63].

Another simple form of the transformation matrix involves only the "chemical" channels

$$\langle \alpha n l | \hat{U} | \alpha' n' l' \rangle = u_{\alpha\alpha'} \delta_{nn'} \delta_{ll'}. \quad (6.51)$$

This operator amounts to a change of representation for the elemental space. It is easy to see that $\langle \alpha | \hat{U}^\dagger \hat{U} | \alpha' \rangle = \kappa_{\alpha\alpha'}$ is nothing but the "alchemical similarity matrix" between elements that has been shown to improve the accuracy of SOAP in the presence of multiple atomic species [23, 24]. What is more, by writing a low-rank approximation of $\hat{U} = \sum_{J\alpha} = u_{J\alpha} | J \rangle \langle \alpha$ one can express atomic density in terms of a small number of "chemical archetypes", improving dramatically both the accuracy and the computational cost of machine-learning models that involve more than a handful of elements [61]. Note that this transformation can be applied at the level of the translationally invariant representation, where one can write

$$\psi^J_{\mathcal{X}_j}(\mathbf{r}) = \langle J\mathbf{r} | \mathcal{X}_j \rangle = \sum_\alpha u_{J\alpha} \langle \alpha \mathbf{r} | \mathcal{X}_j \rangle \quad (6.52)$$

that makes it evident how the action of this particular $\hat{U}$ operator amounts to using linear combination of atomic densities in which each species is given weights that can be optimized by cross-validation.

6.3.5 λ-SOAP: Symmetry-Adapted Gaussian Process Regression

When building a machine-learning model for a tensorial property **T**, one should consider that the target is not invariant under the action of a symmetry operation (e.g. a rotation) but transforms covariantly. The most effective strategy to encode the appropriate covariance properties in the model involves the decomposition of the tensor into its irreducible spherical components, i.e. combinations of the elements of the tensor that transform as the spherical harmonics of order λ [64]. For these irreducible components,

$$T_{\lambda\mu}(\hat{R}\mathcal{X}_j) = \sum_{\mu'} D^\lambda_{\mu\mu'}(\hat{R}) T_{\lambda\mu'}(\mathcal{X}_j) \quad (6.53)$$

As shown in Ref. [65] for the case of vectors and in Ref. [66] for tensors of arbitrary order, one has to consider a matrix-valued kernel that describes the geometric relationship between the different components of $\mathbf{T}_\lambda$, which can be obtained by including an additional Wigner matrix $D^\lambda_{\mu\mu'}(\hat{R})$ in the Haar integral

$$\left\langle \mathcal{X}^{(\nu)}_{j,\lambda\mu} \middle| \mathcal{X}^{(\nu)}_{k,\lambda\mu'} \right\rangle = \int \mathrm{d}\hat{R}\, D^\lambda_{\mu\mu'}(\hat{R}) \left| \langle \mathcal{X}_j | \hat{R} | \mathcal{X}_k \rangle \right|^\nu . \tag{6.54}$$

For the case with $\nu = 2$ the symmetrized kets can be written explicitly based on a $\alpha n l m$ expansion of the atom density

$$\begin{aligned} \left\langle \alpha n l \alpha' n' l' \middle| \mathcal{X}^{(2)}_{j,\lambda\mu} \right\rangle = \sqrt{\frac{8\pi^2}{2l+1}} \sum_{mm'} \langle \mathcal{X}_j | \alpha n l m \rangle \\ \times \langle \alpha' n' l' m' | \mathcal{X}_j \rangle \langle l\, m\, l'\, -m' | \lambda\, -\mu \rangle \end{aligned} \tag{6.55}$$

We write Eq. (6.55) in this form because it is somewhat symmetric, but the properties of the Clebsch-Gordan coefficients require that $m' = m + \mu$ so the expression can be evaluated with a single sum. Furthermore, the expression evaluates to zero whenever $|l - l'| < \lambda$, which reduces the number of elements that must be evaluated and stored, and makes it clear that Eq. (6.55) reduces to the scalar SOAP Eq. (6.46) when $\lambda = 0$.

When using a linear model, each of the symmetry-adapted representations Eq. (6.55) can be used to represent tensorial components that transform as Y^λ_μ. Linearity, in this case, is necessary for preserving the symmetry properties of the λ-SOAP. A non-linear model, however, can be obtained by scaling each $\langle \alpha n l \alpha' n' l' | \mathcal{X}^{(2)}_{j,\lambda\mu} \rangle$ by a (in principle different) non-linear function of some $\lambda = 0$ representations [67]. In the kernel language, a high-order version of the λ-SOAP kernel can be introduced with an expression analogous to Eq. (6.39):

$$\left\langle \mathcal{X}^{(2)}_{j,\lambda\mu} \middle| \mathcal{X}^{(2)}_{k,\lambda\mu'} \right\rangle_\zeta = \frac{\left\langle \mathcal{X}^{(2)}_{j,\lambda\mu} \middle| \mathcal{X}^{(2)}_{k,\lambda\mu'} \right\rangle \left\langle \mathcal{X}_j{}^{(2)} \middle| \mathcal{X}_k{}^{(2)} \right\rangle_{\zeta-1}}{\left\| \left\langle \mathcal{X}^{(2)}_{j,\lambda\boldsymbol{\mu}} \middle| \mathcal{X}^{(2)}_{j,\lambda\boldsymbol{\mu}^\top} \right\rangle \right\|_F \left\| \left\langle \mathcal{X}^{(2)}_{k,\lambda\boldsymbol{\mu}} \middle| \mathcal{X}^{(2)}_{k,\lambda\boldsymbol{\mu}^\top} \right\rangle \right\|_F}, \tag{6.56}$$

where $\|\cdot\|_F$ indicates the Frobenius norm and $\langle \mathcal{X}_j{}^{(2)} | \mathcal{X}_k{}^{(2)} \rangle_{\zeta-1}$ is a (scalar) SOAP kernel. This second term makes the overall kernel non-linear, without affecting the symmetry properties of the overall tensorial kernel.

6.3.6 Computing SOAP Representations Efficiently

A practical calculation of both scalar and tensorial $\nu = 2$ SOAP representations $\langle \alpha n l \alpha' n' l' | \mathcal{X}^{(2)}_{j,\lambda\mu} \rangle$ requires the evaluation of the expansion coefficients $\langle \alpha n l m | \mathcal{X}_j \rangle$. Let us start with the atom density written in the position representation, according to Eq. (6.36), and consider the case in which $\psi_{\mathcal{X}}^{\alpha}(\mathbf{r})$ is written as a superposition of spherical Gaussian functions of width σ placed at the positions of the atoms of type α. Then, the spherical harmonics projection in Eq. (6.36) can be carried out analytically, leading to:

$$\begin{aligned}\langle \alpha nlm|\mathcal{X}_j\rangle = &\sum_{i\in\alpha} Y_{lm}(\hat{\mathbf{r}}_{ij})\, e^{-\frac{r_{ij}^2}{2\sigma^2}} \\ &\times \int_0^\infty \mathrm{d}r\, r^2\, R_n(r) e^{-\frac{r^2}{2\sigma^2}} \iota_l\left(\frac{r r_{ij}}{\sigma^2}\right),\end{aligned} \tag{6.57}$$

where the sum runs over all neighbouring atoms of type α and ι_l indicates a modified spherical Bessel function of the first kind. It is convenient to choose a form for the orthogonal radial basis functions $R_n(r)$ that makes it possible to perform the radial integration analytically.

One possible choice starts by using Gaussian type orbitals as non-orthogonal primitive functions $\tilde{R}_k(r)$

$$\tilde{R}_k(r) = \mathcal{N}_k\, r^k \exp\left\{-\frac{1}{2}\left(\frac{r}{\sigma_k}\right)^2\right\}, \tag{6.58}$$

where $\mathcal{N}_k$ is a normalization factor, such that $\int_0^\infty dr r^2 \tilde{R}_k^2(r) = 1$. The set of Gaussian widths $\{\sigma_k\}$ can be chosen to span effectively the radial interval involved in the environment definition. Assuming that the smooth cutoff function approaches one at a distance $r_{\mathrm{cut}} - \delta r_{\mathrm{cut}}$, one could take $\sigma_k = (r_{\mathrm{cut}} - \delta r_{\mathrm{cut}}) \max(\sqrt{k}, 1)/n_{\max}$, that gives functions that are peaked at equally spaced positions in the range between 0 and $r_{\mathrm{cut}} - \delta r_{\mathrm{cut}}$.

While the $\tilde{R}_k(r)$ are not themselves orthogonal, they can be used to write orthogonal basis functions $R_n(r) = \sum_k S_{nk}^{-1/2} \tilde{R}_k(r)$, where the overlap matrix $S_{kk'} = \int \mathrm{d}r r^2 \tilde{R}_k(r) \tilde{R}_{k'}(r)$ can be computed analytically. The full decomposition of the translationally invariant environmental ket can then be obtained without recourse to numerical integration.

Once the spherical decomposition of the atomic density has been obtained, the coefficients can be combined to give the SOAP representations of order 1 and 2. Particularly in the presence of many different chemical species, the number of components can become enormous. Ignoring for simplicity a few symmetries, and the fact that if all species do not appear in every environment it is possible to store a sparse representation of the representation, the power spectrum contains a number of components of the order of $n_{\mathrm{species}}^2 n_{\max}^2 l_{\max}$, which can easily reach into the tens of thousands. In the case of the tensorial λ-SOAP the number increases further to $\lambda^2 n_{\mathrm{species}}^2 n_{\max}^2 l_{\max}$. It is however not necessary to compute and store all of these representations: each of them, or any linear combination, is a spherical invariant (covariant) description of the environment and can be used separately as a representation. This can be exploited to reduce dramatically the computational cost and the memory footprint of a SOAP calculation, determining a low-rank approximation of the representation. One can use dimensionality reduction techniques similar to those discussed in Sect. 6.2.1 to identify the most suitable reference structures. As shown in Ref. [26], both CUR decomposition and a greedy selection strategy based on farthest point sampling make it possible to reduce by more than 95% the number of SOAP representations that are needed to predict the energy of small organic molecules with chemical accuracy.

6.3.7 Back to the Structures

Whenever one is interested in computing properties that are associated to individual atoms (for instance, their NMR chemical shieldings [68], or the forces) one can use directly the representations corresponding to each environment, or the kernel between two environments, as the basis for a linear or non-linear regression model. As discussed in Sect. 6.2, it is often the case that one is interested in using as structure labels some properties that are instead associated with the entirety of a structure—e.g. its cohesive energy, its dielectric constant, etc. In these cases a ridge regression model should be used that is based on "global" kernels between the structures, $K(\mathcal{A}, \mathcal{B})$, rather than those between individual atom-centred environments. This is reflected in how the kernels between environments should be combined to give a kernel that is suitable to represent the relation between local environments and the overall property of a structure. When the target property can be seen as an additive combination of local, atom-centred contributions, the most natural (and straightforward) choice, that is consistent with Eq. (6.13), is

$$K(\mathcal{A}, \mathcal{B}) = \sum_{j\in\mathcal{A},k\in\mathcal{B}} K(\mathcal{X}_j, \mathcal{X}_k). \tag{6.59}$$

It is worth stressing that in the case where the environment kernel is a linear kernel based on SOAP representations, this sum-kernel can be written in terms of a global representation associated with the entire structure,

$$K(\mathcal{A}, \mathcal{B}) = \left\langle \mathcal{A}^{(\nu)} \middle| \mathcal{B}^{(\nu)} \right\rangle, \tag{6.60}$$

where we introduced

$$|\mathcal{A}^{(\nu)}\rangle = \sum_{j\in\mathcal{A}} |\mathcal{X}_j{}^{(\nu)}\rangle. \tag{6.61}$$

An alternative way to combine the information from individual environments in a symmetrized global kernel corresponds to averaging the Fourier coefficients of each environment,

$$\langle \alpha nlm|\mathcal{A}\rangle = \sum_{j\in\mathcal{A}} \langle \alpha nlm|\mathcal{X}_j\rangle \tag{6.62}$$

and then taking the Haar integral of the resulting sum. For instance, for $\nu = 2$,

$$\left\langle \alpha n \alpha' n' l \middle| \bar{\mathcal{A}}^{(2)} \right\rangle = \sum_m \langle \alpha n l m | \mathcal{A} \rangle \langle \mathcal{A} | \alpha' n' l m \rangle. \tag{6.63}$$

The form Eq. (6.59) is more general, and one can readily introduce non-linear kernels such as $\langle \mathcal{X}_j{}^{(\nu)} | \mathcal{X}_k{}^{(\nu)} \rangle_\zeta$ for which an explicit expression for the representations would be too cumbersome. Equation (6.59) also suggests that the combination of environment kernels could be generalized by introducing a weighting matrix

$$K_W(\mathcal{A}, \mathcal{B}) = \sum_{j \in \mathcal{A}, k \in \mathcal{B}} W_{jk}(\mathcal{A}, \mathcal{B}) K(\mathcal{X}_j, \mathcal{X}_k). \tag{6.64}$$

One could, for instance, determine the importance of each environment within a structure, and set $W_{jk}(\mathcal{A}, \mathcal{B}) = w_j(\mathcal{A}) w_k(\mathcal{B})$. Alternatively, one can use techniques from optimal transport theory [69] to define an entropy-regularized matching (REMatch) procedure [24], in which W_{jk} is a doubly stochastic matrix that matches the most similar environments in the two structures, disregarding the environmental kernels between very dissimilar environments

$$\mathbf{W}(\mathcal{A}, \mathcal{B}) = \underset{\mathbf{W} \in \mathcal{U}(N_\mathcal{A}, N_\mathcal{B})}{\operatorname{argmin}} \sum_{jk} W_{jk} \left[d^2(\mathcal{X}_j, \mathcal{X}_k) + \gamma \ln W_{jk} \right], \tag{6.65}$$

where d^2 indicates the kernel-induced squared distance Eq. (6.9). The parameter γ weights the entropy regularization and makes it possible to interpolate between strict matching of the most similar pairs of environments ($\gamma \to 0$) to an average kernel that weights all pairs equally ($\gamma \to \infty$). Although this construction complicates considerably the combination of local kernels, it provides a strategy to introduce an element of non-locality in the comparison between structures. Given the cost of computing the REMatch kernel, and the fact that it prevents using some sparsification strategies that act at the level of individual environments, this method should be used when the target property is expected to exhibit very strong non-additive behaviour, e.g. when just one portion of the system is involved—for instance, when determining the activity of a drug molecule, a problem for which REMatch has been shown to improve dramatically the accuracy of the ML model [23].

6.3.8 Multi-Kernel Learning

We have shown that SOAP representations can be seen as just one possible embodiment of a general class of rotationally symmetrized density-based representations, that also encompasses other popular representations for atomic-scale machine learning, and that can be tuned to a great extent, e.g. by changing the way different components are weighted. The fact that different representations can be computed within the same formalism does not imply they are fully equivalent: each expression or kernel emphasizes different components of the structure/property relations. For instance, kernels with varying radial scaling or cutoff distance focus the machine-learning model on short-, mid-, or long-range interactions. It is then natural to consider whether a better overall model can be constructed by combining representations that are associated with different cutoff distances, or different levels of body-order expansions. This can be achieved by a weighted combination of kernels of the form

$$K_{\text{tot}}(\mathcal{A}, \mathcal{B}) = \sum_{\aleph} w_{\aleph} K_{\aleph}(\mathcal{A}, \mathcal{B}), \tag{6.66}$$

where each $K_{\aleph}$ corresponds to a distinct model.

This is equivalent to an additive model for a property, similar to the construction of an atom-centred decomposition of the total energy in Eq. (6.12). In this case, instead, the property y associated with each structure is written as the sum of contributions $y_{\aleph}(\mathcal{A})$ that are associated with the various kernels $K_{\aleph}$

$$y(\mathcal{A}) = \sum_{\aleph} y_{\aleph}(\mathcal{A}) = \sum_{\aleph, \mathcal{B}} x_{\mathcal{B}} w_{\aleph} K_{\aleph}(\mathcal{A}, \mathcal{B}), \tag{6.67}$$

where $x_{\mathcal{B}}$ are the kernel regression weights for each of the representative structures $\mathcal{B}$. The weights $w_{\aleph}$ correspond to the estimated contribution that each model will give to the final property and can be obtained by cross-validation, or by physical intuition. For instance, in the case of multiple radial cutoffs, it is found that much smaller weights should be associated with long-range kernels, consistent with the fact that distant interactions contribute a small (although often physically relevant) contribution to the total energy [23]. It should also be noted that, provided that the representations corresponding to the kernels are linearly independent, Eq. (6.66) effectively corresponds to a feature space of increased dimensionality, obtained by concatenating the representations that are—implicitly or explicitly—associated with each kernel.

6.4 Conclusions

We have laid out a mathematical framework, based on the concept of the atomic density, for building representations of atomic environments that preserve the geometric symmetries, and chemically sensible limits. Coupled with kernel regression, this allows the fitting of complex models of physical properties on the atomic scale, both scalars like interatomic potentials (force fields), and tensors such as multipole moments and quantum mechanical operators. We discuss in general terms how kernel regression can be extended to include a sparse selection of reference structures and to predict and learn from linear functionals of the target property. To leverage the many formal similarities between kernel regression and quantum mechanics, we use a Dirac bra-ket notation to formulate the main results concerning the SOAP representations. This notation also helps making apparent the relationship between SOAP representations and other popular density-based approaches to represent atomic structures. The framework can be extended and tuned in many different ways to incorporate insight about the relations between properties, structures, and representations. With physical principles such as symmetry and nearsightedness of interactions at its core, we believe this formulation is ideally suited to provide a unified framework to machine learn atomic-scale properties.

Nomenclature

$\mathcal{A}$	An item—structure or atomic environment for which one wants to predict a property
$K(\mathcal{A}, \mathcal{B})$	The kernel function computed between items $\mathcal{A}$ and $\mathcal{B}$
N	Number of input structures in the training set
M	Number of structures in the representative set
$\mathbf{x}$	The vector of KRR weights, also written as $\mathbf{x}_M$; the weight associated with a structure $\mathcal{B}$ is indicated as $x_{\mathcal{B}}$
$\mathbf{y}$	The vector containing the values of the target property, also written as $\mathbf{y}_N$. $y_{\mathcal{B}}$ indicates the value for the item $\mathcal{B}$
$\Vert \cdot \Vert$	The 2-norm of the quantity $\cdot$
$\Vert \cdot \Vert_F$	The Frobenius norm of the quantity $\cdot$
$\mathbf{K}$	Kernel matrix
$\mathbf{K}_{MN}$	Slice of the kernel matrix $\mathbf{K}$, corresponding to rows in set M and columns in set N
$\Vert \cdot \Vert_{\mathbf{A}}$	The 2-norm of the quantity $\cdot$, in the metric given by $\mathbf{A}$
${}^{\Sigma}\mathbf{K}$	Sum-kernel, defined as the sum of the regular kernel over a set of configurations
$\overleftarrow{\nabla}$	Derivative operator applying to the first argument of the kernel matrix
$\overrightarrow{\nabla}$	Derivative operator applying to the second argument of the kernel matrix
$\mathbf{K}_{\nabla\mathcal{A}\mathcal{B}}$	Derivative of the kernel matrix, applying to its first argument, with respect to the coordinates of atoms in structure $\mathcal{A}$, with structure $\mathcal{B}$ as its second argument
$\hat{\mathbf{L}}$	Linear operator connecting the observed values $\mathbf{y}$ with the unobserved atomic energies $\mathbf{y}'$
$\lvert\mathcal{A}\rangle$	An abstract vector that describes the input $\mathcal{A}$

(continued)

$\langle \mathcal{A} \vert \mathcal{B} \rangle$	The scalar product between the features associated with $\mathcal{A}$ and $\mathcal{B}$. Could be either an explicit scalar product, or an abstract notation equivalent to $K(\mathcal{A}, \mathcal{B})$
$g_i(\mathbf{r})$	A smooth function—typically a Gaussian that is used to represent the density associated with atom i
$\vert\alpha\rangle$	An abstract vector that represents the chemical species α
$\mathbf{r}$	Position in 3D Cartesian coordinates
r	The modulus of the vector $\mathbf{r}$
$\hat{\mathbf{r}}$	The unit vector $\mathbf{r}/r$
$\mathbf{r}_i$	Position of the i-th atom
$\mathbf{r}_{ij}$	Displacement vector $\mathbf{r}_i - \mathbf{r}_j$ between the i-th and j-th atoms
$\psi^{\alpha}_{\mathcal{X}_j}(\mathbf{r})$	The atom density of species α centred around the j-th atom
$Y^l_m(\hat{\mathbf{r}})$	The l, m-th spherical harmonic
$R_n(r)$	The n-th orthogonal radial basis function
$\vert\mathcal{X}_j{}^{(\nu)}\rangle$	The spherically averaged SOAP representation of order ν
$\langle \mathcal{X}_j{}^{(\nu)} \vert \mathcal{X}_k{}^{(\nu)} \rangle_\zeta$	The normalized SOAP kernel of order ν and non-linear exponent ζ
$D^l_{mm'}(\hat{R})$	The Wigner rotation matrix associated with the rotation $\hat{R}$
$\langle \alpha n \alpha' n' l \vert \mathcal{X}_j{}^{(2)} \rangle$	The radial/spherical representation of the SOAP $\nu = 2$ vector, corresponding to the power spectrum between species α and α'
$\langle l_1\, m_1\, l_2\, m_2 \vert l\, m \rangle$	A Clebsch–Gordan coefficient
$T_{\lambda\mu}$	The μ-th component of the irreducible spherical component of order λ for the tensorial quantity $\mathbf{T}$
$\vert\mathcal{X}^{(\nu)}_{j,\lambda\mu}\rangle$	The λ-SOAP representation of order ν, corresponding to the irreducible spherical component $\lambda\mu$ centred on atom j

References

1. M. Ceriotti, M.J. Willatt, G. Csányi, Machine-learning of atomic-scale properties based on physical principles, in *Handbook of Materials Modeling*, ed. by W. Andreoni, S. Yip (Springer, Cham, 2019)
2. M.W. Finnis, *Interatomic Forces in Condensed Matter* (Oxford University Press, Oxford, 2004)
3. D.W. Brenner, Phys. Status Solidi B **217**, 23 (2000)
4. B.J. Braams, J.M. Bowman, Int. Rev. Phys. Chem. **28**(4), 577–606 (2009)
5. M. Rupp, A. Tkatchenko, K.R. Müller, O.A. von Lilienfeld, Phys. Rev. Lett. **108**, 058301 (2012)
6. F. Faber, A. Lindmaa, O.A. von Lilienfeld, R. Armiento, Int. J. Quantum Chem. **115**(16), 1094–1101 (2015)

7. L. Zhang, J. Han, H. Wang, R. Car, E. Weinan, Phys. Rev. Lett. **120**(14), 143001 (2018)
8. C.M. Bishop, *Pattern Recognition and Machine Learning* (Springer, Berlin, 2016)
9. C.E. Rasmussen, C.K.I. Williams, *Gaussian Processes for Machine Learning* (MIT Press, Cambridge, 2006)
10. B. Schölkopf, A.J. Smola, *Learning with Kernels: Support Vector Machines, Regularization, Optimization, and Beyond* (MIT Press, Cambridge, 2002)
11. W.J. Szlachta, A.P. Bartók, G. Csányi, Phys. Rev. B Condens. Matter **90**(10), 104108 (2014). https://doi.org/10.1103/PhysRevB.90.104108
12. D. Dragoni, T.D. Daff, G. Csányi, N. Marzari, Phys. Rev. Mater. **2**(1), 013808 (2018)
13. N. Bernstein, J.R. Kermode, G. Csányi, Rep. Prog. Phys. **72**(2), 026501 (2009). https://doi.org/10.1088/0034-4885/72/2/026501
14. V.L. Deringer, G. Csányi, Phys. Rev. B **95**(9), 094203 (2017). https://doi.org/10.1103/physrevb.95.094203
15. S. Fujikake, V.L. Deringer, T.H. Lee, M. Krynski, S.R. Elliott, G. Csányi, J. Chem. Phys. **148**, 241714 (2018)
16. A.N. Tikhonov, A. Goncharsky, V.V. Stepanov, A.G. Yagola, *Numerical Methods for the Solution of Ill-Posed Problems* (Kluwer Academic, Dordrecht, 1995)
17. J.A. Hartigan, M.A. Wong, J. R. Stat. Soc. Ser. C Appl. Stat. **28**(1), 100 (1979)
18. S. Prabhakaran, S. Raman, J.E. Vogt, V. Roth, *Joint DAGM (German Association for Pattern Recognition) and OAGM Symposium* (Springer, Berlin, 2012), pp. 458–467
19. E.V. Podryabinkin, A.V. Shapeev, Comput. Mater. Sci. **140**, 171 (2017). https://doi.org/10.1016/j.commatsci.2017.08.031.
20. B. Huang, O.A. von Lilienfeld (2017). arxiv:1707.04146 . http://arxiv.org/abs/1707.04146v3
21. T.F. Gonzalez, Theor. Comput. Sci. **38**, 293 (1985)
22. M. Ceriotti, G.A. Tribello, M. Parrinello, J. Chem. Theory Comput. **9**, 1521 (2013)
23. A.A.P. Bartók, S. De, C. Poelking, N. Bernstein, J.R.J. Kermode, G. Csányi, M. Ceriotti, Sci. Adv. **3**, e1701816 (2017)
24. S. De, A.A.P. Bartók, G. Csányi, M. Ceriotti, Phys. Chem. Chem. Phys. **18**, 13754 (2016)
25. M.W. Mahoney, P. Drineas, Proc. Natl. Acad. Sci. USA **106**, 697 (2009)
26. G. Imbalzano, A. Anelli, D. Giofré, S. Klees, J. Behler, M. Ceriotti, J. Chem. Phys. **148**, 241730 (2018)
27. J.Q. Quinonero-Candela, C.E. Rasmussen, J. Mach. Learn. Res. **6**, 1939–1959 (2005)
28. E. Snelson, Z. Ghahramani, *Advances in Neural Information Processing Systems* (2005)
29. E. Solak, C.E. Rasmussen, D.J. Leith, R. Murray-Smith, W.E. Leithead, *Advances in Neural Information Processing Systems* (2003)
30. A.P. Bartók, M.C. Payne, R. Kondor, G. Csányi, Phys. Rev. Lett. **104**(13), 136403 (2010)
31. A.P. Bartók, M.J. Gillan, F.R. Manby, G. Csányi, Phys. Rev. B **88**(5), 054104 (2013). https://doi.org/10.1103/PhysRevB.88.054104
32. A.P. Bartók, G. Csányi, Int. J. Quant. Chem. **116**(13), 1051 (2015). https://doi.org/10.1002/qua.24927
33. S.T. John, G. Csányi, J. Phys. Chem. B **121**(48), 10934 (2017). https://doi.org/10.1021/acs.jpcb.7b09636
34. V.L. Deringer, C.J. Pickard, G. Csányi, Phys. Rev. Lett. **120**(15), 156001 (2018). https://doi.org/10.1103/PhysRevLett.120.156001
35. M.A. Caro, V.L. Deringer, J. Koskinen, T. Laurila, G. Csányi, Phys. Rev. Lett. **120**(16), 166101 (2018). https://doi.org/10.1103/PhysRevLett.120.166101
36. P. Rowe, G. Csányi, D. Alfè, A. Michaelides, Phys. Rev. B **97**(5), 054303 (2018). https://doi.org/10.1103/PhysRevB.97.054303
37. T.T. Nguyen, E. Szekely, G. Imbalzano, J. Behler, G. Csányi, M. Ceriotti, A.W. Götz, F. Paesani, J. Chem. Phys. **148**, 241725 (2018)
38. S. Chmiela, A. Tkatchenko, H.E. Sauceda, I. Poltavsky, K.T. Schütt, K.R. Müller, Sci. Adv. **3**(5), e1603015 (2017). https://doi.org/10.1126/sciadv.1603015
39. A. Glielmo, C. Zeni, A.D. Vita, Phys. Rev. B **97**(18) (2018). https://doi.org/10.1103/physrevb.97.184307

40. C. Zeni, K. Rossi, A. Glielmo, A. Fekete, N. Gaston, F. Baletto, A. Dr Vita, J. Chem. Phys. **148**(23), 234106 (2018)
41. M.J. Willatt, F. Musil, M. Ceriotti, J. Chem. Phys. **150**, 154110 (2019)
42. A.P. Bartók, R. Kondor, G. Csányi, Phys. Rev. B **87**, 184115 (2013)
43. J. Behler, M. Parrinello, Phys. Rev. Lett. **98**, 146401 (2007)
44. S. Kajita, N. Ohba, R. Jinnouchi, R. Asahi, Sci. Rep. **7**, 1 (2017)
45. K.T. Schütt, H. Glawe, F. Brockherde, A. Sanna, K.R. Müller, E.K.U. Gross, Phys. Rev. B **89**, 205118 (2014)
46. W. Yang, Phys. Rev. Lett. **66**, 1438 (1991)
47. G. Galli, M. Parrinello, Phys. Rev. Lett. **69**, 3547 (1992)
48. S. Goedecker, Rev. Mod. Phys. **71**, 1085 (1999)
49. E. Prodan, W. Kohn, Proc. Natl. Acad. Sci. USA **102**, 11635 (2005)
50. H. Eshet, R.Z. Khaliullin, T.D. Kühne, J. Behler, M. Parrinello, Phys. Rev. Lett. **108**, 115701 (2012)
51. T. Morawietz, A. Singraber, C. Dellago, J. Behler, Proc. Natl. Acad. Sci. USA **113**, 8368 (2016)
52. B. Cheng, J. Behler, M. Ceriotti, J. Phys. Chem. Lett. **7**, 2210 (2016)
53. J.S. Smith, O. Isayev, A.E. Roitberg, Chem. Sci. **8**, 3192 (2017)
54. A.P. Thompson, L.P. Swiler, C.R. Trott, S.M. Foiles, G.J. Tucker, J. Comput. Phys. **285**, 316 (2015)
55. A. Haar, Ann. Math. **34**, 147 (1933)
56. J. Tersoff, Phys. Rev. B **39**, 5566 (1989)
57. G.R. Medders, V. Babin, F. Paesani, J. Chem. Theory Comput. **10**, 2906 (2014)
58. J.A. Moriarty, Phys. Rev. B **42**, 1609 (1990)
59. R. Drautz, Phys. Rev. B **99**, 014104 (2019)
60. F.A. Faber, L. Hutchison, B. Huang, J. Gilmer, S.S. Schoenholz, G.E. Dahl, O. Vinyals, S. Kearnes, P.F. Riley, O.A. von Lilienfeld, J. Chem. Theory Comput. **13**, 5255 (2017)
61. M.J. Willatt, F. Musil, M. Ceriotti, Phys. Chem. Chem. Phys. **20**, 29661 (2018)
62. B. Huang, O.A. von Lilienfeld, J. Chem. Phys. **145**(16), 161102 (2016). https://doi.org/10.1063/1.4964627
63. F.A. Faber, A.S. Christensen, B. Huang, O.A. von Lilienfeld, J. Chem. Phys. **148**, 241717 (2018)
64. D.A. Varshalovich, A.N. Moskalev, V.K. Khersonskii, *Quantum Theory of Angular Momentum* (World Scientific, Singapore, 1988)
65. A. Glielmo, P. Sollich, A. De Vita, Phys. Rev. B **95**, 214302 (2017)
66. A. Grisafi, D.D.M. Wilkins, G. Csányi, M. Ceriotti, Phys. Rev. Lett. **120**, 036002 (2018)
67. D.M. Wilkins, A. Grisafi, Y. Yang, K.U. Lao, R.A. DiStasio, M. Ceriotti, Proc. Natl. Acad. Sci. USA **116**, 3401 (2019)
68. F.M. Paruzzo, A. Hofstetter, F. Musil, S. De, M. Ceriotti, L. Emsley, Nat. Commun. **9**, 4501 (2018)
69. M. Cuturi, in *Advances in Neural Information Processing Systems*, vol. 26, ed. by C.J.C. Burges, L. Bottou, M. Welling, Z. Ghahramani, K.Q. Weinberger (Curran Associates, Inc., Red Hook, 2013), pp. 2292–2300

Accurate Molecular Dynamics Enabled by Efficient Physically Constrained Machine Learning Approaches

7

Stefan Chmiela, Huziel E. Sauceda, Alexandre Tkatchenko, and Klaus-Robert Müller

Abstract

We develop a combined machine learning (ML) and quantum mechanics approach that enables data-efficient reconstruction of flexible molecular force fields from high-level ab initio calculations, through the consideration of fundamental physical constraints. We discuss how such constraints are recovered and incorporated into ML models. Specifically, we use conservation of energy—a fundamental property of closed classical and quantum mechanical systems—to derive an efficient gradient-domain machine learning (GDML) model. The challenge of constructing conservative force fields is accomplished by learning in a Hilbert space of vector-valued functions that obey the law of energy conservation. We proceed with the development of a multi-partite matching algorithm that enables a fully automated recovery of physically relevant point group and fluxional symmetries from the training dataset into a symmetric variant of our model. The symmetric GDML (sGDML) approach is able to faithfully reproduce global force fields at the accuracy high-level ab initio methods, thus enabling sample intensive tasks like molecular dynamics simulations at that level of accuracy. (This chapter is adapted with permission from Chmiela (Towards exact molecular dynamics simulations with invariant machine-learned models, PhD thesis. Technische Universität, Berlin, 2019).)

S. Chmiela (✉) · H. E. Sauceda · K.-R. Müller
Machine Learning Group, Technische Universität Berlin, Berlin, Germany
e-mail: stefan@chmiela.com

A. Tkatchenko
Physics and Materials Science Research Unit, University of Luxembourg, Luxembourg, Luxembourg

K. T. Schütt et al. (eds.), *Machine Learning Meets Quantum Physics*, Lecture Notes in Physics 968, https://doi.org/10.1007/978-3-030-40245-7_7

7.1 Introduction

Molecular dynamics (MD) simulations have become an indispensable atomistic modeling tool, revealing the equilibrium thermodynamic and dynamical properties of a system, while simultaneously providing atomic-scale insight. The predictive power of such simulations is crucially determined by the accuracy of the underlying description of inter-atomic forces, which is typically the limiting factor. Most commonly, the inter-atomic forces are obtained from classical potentials, which provide a mechanistic description in terms of fixed interaction patterns between bonds and bond angles within a molecule. In contrast to exact ab initio methods, classical force fields are computationally affordable enough to allow for long simulation time scales, but their simplistic characterization of atomic interactions prevents them from capturing wide range of important effects, such as the anharmonic nature of atomic bonds, charge transfer, and many-body effects [1].

As an alternative, a series of methodological advances in the field of machine learning (ML) [2–55] have yielded universal approximators with virtually no inherent flexibility restrictions. Parametrized from high-level ab initio reference calculations, they are theoretically able to represent arbitrarily accurate force fields that faithfully capture key quantum effects, as represented in the training data. However, the major handicap of such off-the-shelf solutions is that they require extraordinary amounts of computationally expensive reference calculations until convergence to a useful level of predictive robustness. This limits their practicality, as the full procedure of data generation, training, and inference has to outpace the reference method that the ML model is based on. Only then do ML models represent a meaningful addition to existing approaches.

This predicament suggests that a tight integration between ML and fundamental concepts from physics is necessary to close the gap between efficient FFs and accurate high-level ab initio methods. To this end, we review ideas of how to take advantage of conserved quantities in dynamical processes in addition to other physical laws in order to inform universal approximators without compromising their generality. In doing so, we focus statistical inference on the challenging aspects of the problem, while readily available a priori knowledge about the atomic interactions is represented exactly and artifact-free through explicit constraints. A key philosophical difference distinguishes this approach from classical FFs: we aim to exclude physically impossible interactions, as opposed to (approximately) parametrizing known behavior. Inherent modeling biases can thus be avoided. It is a fascinating prospect, that the necessary constraints can often be expressed in simple terms, although they originate from complex interactions, as many regularities can be exploited without an explicit concept of the underlying principles that cause them. For example, Noether's theorem [56] allows us to derive symmetry constraints as conservation laws from the Lagrangian.

We approach this challenge using principles of probabilistic inference, which define a set of hypotheses and conditions them on the made observations. The resulting predictions are particularly robust to overfitting, because all viable hypotheses that agree with the data are taken into account. The incorporation of strong priors restricts this set of hypotheses, which carries data-efficiency advantages over approaches that start from a more general set of assumptions. The so-called Hilbert space learning algorithms [57] will enable us to rigorously incorporate fundamental temporal and spatial symmetries of atomic systems to create models for which parametrization from highly accurate, but costly coupled-cluster reference data becomes viable. In the end, we will be able to perform sampling intensive, long-time scale path-integral MD at that level of accuracy.

7.2 Hilbert Space Learning

Supervised ML infers a relationship between pairs of inputs $\mathbf{x} \in \mathcal{X}$ and associated outputs $y \in \mathcal{Y}$ from a finite *training set* of M examples. The objective is to formulate a hypothesis that generalizes beyond these known data points, which is estimated by measuring the prediction error of the model on an independent *test set*. While data efficiency, in the sense of a quickly converging generalization error with growing training set size, is certainly a consideration, it has rarely been the primary focus in typical ML application domains. These fields are typically fortuned with an abundance of data and the need for ML purely arises from the lack of theory to describe the sought-after mapping. For example, there is no rigorous way of deriving the contents of an image from its pixel values and the best results are currently achieved by deep neural networks that match each input against hundreds of thousands of characteristic patterns for the respective image class. Here, it does not matter that several millions of example images were needed until the model was able to give useful predictions, because it is the best tool available.

However, the data efficiency demands are much more stringent in the sciences where the baseline is an essentially exact theory and the role of ML algorithms is to forego some of the computational complexity by means of empirical inference. Naturally, this also means that a considerable computational cost is associated with the generation of high-level ab initio training data in that case, making it rare. Practically obtainable datasets of that quality are thus often too small to enable deep learning architectures to play up their strengths.

A more efficient alternative is provided by Hilbert space learning algorithms, as they operate in spaces of functions that match prior beliefs about the observed process. While a small number of training points is not enough to condition general estimators adequately, it might be sufficient to constrain a well-behaved, physically meaningful function space. This is alluring, because even complex physical processes involve quantities with well understood properties that can be exploited to define the structure of those Hilbert spaces. In the following, we briefly review the fundamentals of Hilbert space learning and highlight different ways of incorporating prior knowledge.

7.2.1 Hilbert Spaces

A *Hilbert space* $\mathcal{H}$ is a vector space over $\mathbb{R}$ with an inner product that yields a complete metric space. The inner product gives rise to a norm $\|\mathbf{x}\| = \sqrt{\langle \mathbf{x}, \mathbf{x} \rangle}$, which induces a distance metric $d(\mathbf{x}, \mathbf{x}') = \|\mathbf{x} - \mathbf{x}'\|$, for $\mathbf{x}, \mathbf{x}' \in \mathcal{H}$. Although any N-dimensional Euclidean space $\mathbb{R}^N$ is technically a Hilbert space, this formalism becomes particularly interesting in infinite dimension, where $\mathcal{H}$ is a space of functions, while retaining almost all of linear algebra from vector spaces [57,58].

7.2.1.1 Reproducing Kernels

Many ML algorithms make use of infinite-dimensional Hilbert spaces indirectly via the so-called *kernel-trick*, which allows to express inner products of mappings $\Phi : \mathcal{X} \to \mathcal{H}$ in terms of inputs $\mathbf{x} \in \mathcal{X}$ via a *kernel function* $k : \mathcal{X} \times \mathcal{X} \to \mathbb{R}$:

$$k(\mathbf{x}, \mathbf{x}') = \langle \Phi(\mathbf{x}), \Phi(\mathbf{x}') \rangle_{\mathcal{H}}. \tag{7.1}$$

Equation (7.1) holds true for any symmetric and positive semidefinite kernel, i.e. it is required that $k(\mathbf{x}, \mathbf{x}') = k(\mathbf{x}', \mathbf{x})$ and any linear combination $f = \sum_i \alpha_i \Phi(\mathbf{x}_i)$ with $\alpha_i \in \mathbb{R}$ must satisfy

$$\langle f, f \rangle_{\mathcal{H}} = \sum_{ij} \alpha_i \alpha_j k(\mathbf{x}_i, \mathbf{x}_j) \geq 0. \tag{7.2}$$

These two properties guarantee the *reproducing property* of $\mathcal{H}$

$$f(\mathbf{x}) = \langle k(\cdot, \mathbf{x}), f \rangle_{\mathcal{H}}, \tag{7.3}$$

due to which any evaluation of f corresponds to an inner product evaluation in $\mathcal{H}$ between the representer $k(\cdot, \mathbf{x}) = \Phi(\mathbf{x})$ of $\mathbf{x}$ and the function itself. We say that k is reproducing for a subset of $\mathcal{H}$, the *reproducing kernel Hilbert space* (RKHS). Intuitively, this means that the feature maps $\Phi(\mathbf{x}_i)$ for all training points $i \in [1, \ldots, M]$ provide an over-complete basis for the RKHS [58].

7.2.1.2 Representer Theorem

The computational tractability of Hilbert space learning algorithms is afforded by the *representer theorem* which states that in an RKHS $\mathcal{H}$, the minimizer $\hat{f} \in \mathcal{H}$ of a loss function $\mathcal{L} : \mathcal{Y} \times \mathcal{Y} \to \mathbb{R}$ in a regularized risk functional with $\lambda > 0$,

$$\hat{f} = \underset{f \in \mathcal{F}}{\arg\min} \left[\frac{1}{M} \sum_i^M \mathcal{L}(f(\mathbf{x}_i), y_i) + \lambda \|f\|^2 \right], \tag{7.4}$$

admits a representation of the form

$$f(\cdot) = \sum_{i}^{M} \alpha_i k(\cdot, \mathbf{x}_i) \tag{7.5}$$

for any α_i. It therefore reduces the infinite-dimensional minimization problem in a function space to finding the optimal values for an M-dimensional vector of coefficients $\boldsymbol{\alpha}$ [57, 59, 60]. Because we are not fitting a model with a fixed number of predetermined parameters, Hilbert space algorithms are generally regarded as non-parametric methods, i.e. the complexity of the model is able to grow with the amount of available data.

7.2.2 Gaussian Process Models

When formulated in terms of the squared loss $\mathcal{L}(\hat{f}(\mathbf{x}), y) = (\hat{f}(\mathbf{x}) - y)^2$, the regularized risk functional in Eq. (7.4) can be interpreted as the maximum a posteriori estimate of a Gaussian process (GP) [57]. One common perspective on GPs is that they specify a prior distribution over a function space. GPs are defined as a collection of random variables that jointly represent the distribution of the function $f(\mathbf{x})$ at each location $\mathbf{x}$ and thus as a generalization of the Gaussian probability distribution from vectors to functions. This conceptual extension makes it possible to model complex beliefs.

At least in part, the success of GPs—in contrast to other stochastic processes—can be attributed to the fact that they are completely defined by only the first- and second-order moments, the mean $\mu(\mathbf{x})$, and covariance $k(\mathbf{x}, \mathbf{x}')$ for all pairs of random variables [61]:

$$f(\mathbf{x}) \sim \mathcal{GP}\left[\mu(\mathbf{x}), k(\mathbf{x}, \mathbf{x}')\right]. \tag{7.6}$$

Any symmetric and positive definite function is a valid covariance that specifies the prior distribution over functions we expect to observe and want to capture by a GP. Altering this function can change the realizations of the GP drastically: e.g. the squared exponential kernel $k(\mathbf{x}, \mathbf{x}') = \exp(-\|\mathbf{x} - \mathbf{x}'\|^2\sigma^{-1})$ (with a freely selectable length-scale parameters σ) defines a smooth, infinitely differentiable function space, whereas the exponential kernel $k(\mathbf{x}, \mathbf{x}') = \exp(-\|\mathbf{x} - \mathbf{x}'\|\sigma^{-1})$ produces a non-differentiable realizations. The ability to define a prior explicitly gives us the opportunity to express a wide range of hypotheses like boundary conditions, coupling between variables or different symmetries like periodicity or group invariants. Most critically, the prior characterizes the generalization behavior of the GP, defining how it extrapolates to previously unseen data. Furthermore, the closure properties of covariance functions allow many compositions, providing additional flexibility to encode complex domain knowledge from existing simple priors [62].

The challenge in applying GP models lies in finding a kernel that represents the structure in the data that is being modeled. Many kernels are able to approximate universal continuous functions on a compact subset arbitrarily well, but a strong prior restricts the hypothesis space and drastically improves the convergence of a GP while preventing overfitting [63]. Each training point conditions the GP, which allows increasingly accurate predictions from the posterior distribution over functions with growing training set size.

A number of attractive properties beyond their expressivity make GPs particularly useful in the physical sciences:

- There is a unique and exact closed form solution for the predictive posterior, which allows GPs to be trained analytically. Not only is this faster and more accurate than numerical solvers, but also more robust. For example, choosing the hyper-parameters of the numerical solver for NNs often involves intuition and time-consuming trial and error.
- Because a trained model is the average of *all* hypotheses that agree with the data, GPs are less prone to overfit, which minimizes the chance of artifacts in the reconstruction [64]. Other types of methods that start from a more general hypothesis space require more complex regularization schemes.
- Lastly, their simple linear form makes GPs easier to interpret, which simplifies analysis of the modeled phenomena and supports theory building.

7.2.2.1 Gaussian Process Regression

It is straightforward to use GPs for regression: Given a sample $(\mathbf{X}, \mathbf{y}) = \{(\mathbf{x}_i, y_i)\}_i^M$, we compute the sample covariance matrix $(\mathbf{K})_{ij} = k(\mathbf{x}_i, \mathbf{x}_j)$ and use the posterior mean

$$\mu(\mathbf{x}) = \mathbb{E}[f(\mathbf{x})] = k_{\mathbf{X}}(\mathbf{x})^\top (\mathbf{K} + \lambda \mathbb{I})^{-1} \mathbf{y} \tag{7.7}$$

to make predictions about new points $\mathbf{x}$. Here, $k_{\mathbf{X}}(\mathbf{x}) = [k(\mathbf{x}, \mathbf{x}_1), \ldots, k(\mathbf{x}, \mathbf{x}_M)]^\top$ is the vector of covariances between the new point $\mathbf{x}$ and all training points. In the frequentist interpretation, this algorithm is also referred to as *kernel ridge regression*.

We can also calculate the variability of the hypotheses at every point via the posterior variance

$$\sigma^2(\mathbf{x}) = \mathbb{E}\left[(f(\mathbf{x}) - \mu(\mathbf{x}))^2\right] = k(\mathbf{x}, \mathbf{x}) - k_{\mathbf{X}}(\mathbf{x})^\top (\mathbf{K} + \lambda \mathbb{I})^{-1} k_{\mathbf{X}}(\mathbf{x}), \tag{7.8}$$

which gives us an idea about the uncertainty of the prediction. We remark here, that the posterior variance is generally not a measure for the accuracy of the prediction. It rather describes how well the hypothesized space of solutions is conditioned by the observations and whether the made assumptions are correct.

7.3 Encoding Prior Knowledge

Prior knowledge about the problem at hand is an essential ingredient to the learning task, as it can drastically increase the efficiency of the training process and robustness of the reconstruction. An ML model that starts from weak assumptions will require more training data to achieve the same performance, compared to one that is restricted to solutions with certain known properties. A unique feature of GPs is that they provide a direct way to incorporate such constraints on the hypothesis space [58].

In the context of this chapter, we are particularly interested in regularities that arise from invariances and symmetries of physical systems. Sure enough, the idea to reduce equations in a way that leaves them invariant is not new in physics. In fact, Jacobi already developed a procedure to simplify Hamilton's dynamical equations of mechanics based on the conserved quantities of dynamical systems [65] in the middle of the eighteenth century. Heisenberg was the first to apply group theory to quantum mechanics, where he exploited the permutational symmetry of indistinguishable quantum particles in 1926. Even in modern physics, new symmetries are still routinely discovered [66].

Here, we will review the three most important ways to include prior knowledge into GPs: via the representation of the input, the construction of suitable mean, and the covariance functions. As the choice of covariance function is especially important, which is why we will describe several distinct ways to construct them.

7.3.1 Representation

Once the data is captured, it needs to be represented in terms of features that are considered to be particularly informative, i.e. well-correlated with the predicted quantity. For example, parametrizing a molecular graph in terms of dihedral angles instead of pairwise distances might be advantageous when modeling complex transition paths.

The representation of the data also provides the first opportunity to incorporate known invariances of the task at hand. Especially in physical systems, there are transformations that leave its properties invariant, which introduces redundancies that can be exploited with a representation that shares those symmetries.

For example, physical systems can generally be translated and rotated in space without affecting their attributes. Often, the invariances extend to more interesting group of transformations like rotations, reflections, or permutations, providing further opportunities to reformulate the learning problem into a simpler, but equivalent one. Conveniently, any non-linear map $\mathbf{D} : \mathcal{X} \to \mathcal{D}$ of the input to a covariance function yields another valid covariance function, providing a direct way to incorporate desired invariances into existing kernels [67].

7.3.2 Covariance Function

Symmetries in the input data naturally translate to symmetries in the output. If a molecular graph is mirror symmetric, so will be its potential energy surface. However, sometimes there is structure in the output that is not tied to the input at all. This is the case when the predicted property is subject to a conservation law, e.g. the energy of a system is conserved as its geometry propagates through time. There is no representation of individual data points that would be able to capture this kind of symmetry.

Instead, conservation laws have to be incorporated as constraints into the predictor, to restrict the space of eligible solutions. This is achieved elegantly in GPs, via modification of the covariance function in a way that gives rise to a prior that obeys the desired symmetry. Any function drawn from that prior will then inherit the same invariances [58, 61]. Before developing a covariance function that fits our problem, we will briefly highlight different ways to construct them. After all, arbitrary functions of two inputs $\mathbf{x}, \mathbf{x}' \in \mathcal{X}$ are not necessarily valid covariance functions. For that purpose we will switch away from the probabilistic view that we held so far and provide a perspective that is more intuitive in the physics context.

7.3.2.1 Integral Transforms

We can think of the covariance function as a kernel of a linear integral transform that defines an operator

$$\hat{T}_k f(\mathbf{x}) = \int_{\mathcal{X}} k(\mathbf{x}, \mathbf{x}') f(\mathbf{x}')\, d\mathbf{x}', \tag{7.9}$$

which maps a function $f(\mathbf{x})$ from one domain to another [61, 68]. In this view, $\hat{T}_k f(\mathbf{x}) = \hat{f}(\mathbf{x})$ corresponds to the posterior mean of our GP. Note that $\hat{T}_k f(\mathbf{x})$ remains a continuous function even if we discretize the integration domain. This is the case in the regression setting, when we are only able to observe our target function partially, i.e. when $f(\mathbf{x}') = \sum_{i=1}^{M} f(\mathbf{x}_i)\delta(\mathbf{x}' - \mathbf{x}_i)$. With that in mind, an integral operator can be regarded as a continuous generalization of the matrix-vector product using a square matrix with entries $(\mathbf{K})_{ij} = k(\mathbf{x}_i, \mathbf{x}_j)$ and a vector $\boldsymbol{\alpha}$. Then,

$$(\mathbf{K}\boldsymbol{\alpha})_i = \sum_{j}^{M} k(\mathbf{x}_i, \mathbf{x}_j)\alpha_j \tag{7.10}$$

is the discrete analogon to $\hat{T}_k f(\mathbf{x})$ [69]. Note that this expression is closed under linear transformation: any linear constraint $\hat{G}[\hat{T}_k]$ propagates into the integral and gives rise to a new covariance function.

However, there are several alternative construction options, one of them through explicit definition of the frequency spectrum of $\hat{T}_k$. Due to the translational symmetry of physical systems, we are particularly interested in stationary covariance functions that only depend on pairwise distances $\boldsymbol{\delta} = \mathbf{x} - \mathbf{x}'$ between points. In that setting, *Bochner's theorem* says that symmetric, positive definite kernels can be constructed via the inverse Fourier transform of a probability density function $p(\boldsymbol{\delta})$ in frequency space [61, 67, 68, 70]:

$$k(\boldsymbol{\delta}) = \mathcal{F}(p(\boldsymbol{\delta})) = \int p(\omega)e^{-i\omega^{\top}\boldsymbol{\delta}}\mathrm{d}\omega. \tag{7.11}$$

The following perspective might however be more intuitive when approaching this problem form a physics background: Since $\hat{T}_k f(\mathbf{x})$ is the reconstruction from pointwise observations $y_i = f(\mathbf{x}_i)$, we are ideally looking for an operator that leaves our unknown target function invariant, such that $\hat{T}_k f(\mathbf{x}) = f(\mathbf{x})$. This is another way of saying that our estimate $\hat{f}(\mathbf{x})$ lives in the space spanned by the eigenfunctions $\varphi_i \in \mathcal{F}$ of the operator defined by the kernel function (with coefficients $c_i \in \mathbb{R}$), giving

$$\hat{f}(\mathbf{x}) = \sum_i c_i \varphi_i(\mathbf{x}) \quad \text{with} \quad \hat{T}_k \varphi_i = \lambda_i \varphi_i. \tag{7.12}$$

It is impossible to overlook that there is a strong analogy between the covariance function in a GP process and the Hamiltonian in the Schrödinger equation (SE). Both operators formulate constraints that give rise to Hilbert space of possible states of the modeled object, whether it is the wave-function or the hypothesis space of the GP. Although this is where the similarities end, this connection certainly illustrates that GPs are particularly suitable to reconstruct physical processes in a principled way.

7.3.3 Mean Function

In most applications, the GP prior mean function $\mu(\mathbf{x}) = 0$ is set to zero, which leads to predictions $\hat{f}(\mathbf{x}) \approx 0$ as $\|\mathbf{x} - \mathbf{x}'\| \to 0$ for stationary kernels. Convergence to a constant outside of the training regime is desirable for data-driven models, because it means that the prediction degrades gracefully in the limit, instead of producing unforeseeable results. However, if a certain *asymptotic* behavior of the modeled function is known, the prior mean function offers the possibility to prescribe it. For example, we could introduce a log barrier function

$$\mu(\mathbf{x}) = -\log(\mathbf{b} - \mathbf{x}) \tag{7.13}$$

that ramps up the predicted quantity towards infinity for $\mathbf{x} \geq 0$. In a molecular PES model, such a barrier would represent an atom dissociation limit, which could be useful to ensure that a dynamical process stays confined to the data regime as it moves around the PES.

In the spirit of how the Slater determinant in quantum mechanics accounts for the average affect of electron repulsion without explicit correlation, the mean of a GP can be used to prescribe a sensible predictor response outside of the data regime.

7.4 Energy-Conserving Force Field Reconstructions

A fundamental property that any molecular force field $\mathbf{F}(\mathbf{r}_1, \mathbf{r}_2, \ldots, \mathbf{r}_N)$ must satisfy is the conservation of total energy, which implies that

$$\mathbf{F}(\mathbf{r}_1, \mathbf{r}_2, \ldots, \mathbf{r}_N) = -\nabla E(\mathbf{r}_1, \mathbf{r}_2, \ldots, \mathbf{r}_N). \tag{7.14}$$

While any analytically derivable expression for the potential energy satisfies energy conservation by definition, a direct reconstruction of a force field is more involved. It requires mapping to an explicitly conservative vector field and thus special constraints on the hypothesis space that the ML model navigates in. Before we discuss how this can be implemented in practice, we will briefly review a number of advantages that make this direct approach worthwhile.

7.4.1 Forces Are Quantum-Mechanical Observables

A major reason is the fact that atomic forces are true quantum-mechanical observables within the BO approximation by virtue of the *Hellmann–Feynman theorem*. It provides a way to obtain analytical derivatives by relating changes in the total energy δE with respect to any variation $\delta\lambda$ of the Hamiltonian H through the expectation value

$$\frac{\partial E}{\partial \lambda} = \left\langle \Psi_\lambda \left| \frac{\partial \hat{H}_\lambda}{\partial \lambda} \right| \Psi_\lambda \right\rangle. \tag{7.15}$$

It thus allows the direct computation of forces $\mathbf{F} = -\partial E/\partial \mathbf{R}$ as derivatives with respect to nuclei positions $\mathbf{R}$. Similar analogous expressions for density-based approaches exist as well [71, 72]. Once the SE is solved for a particular atomic configuration to compute the energy, this theorem makes the additional computation of forces relatively cheap, by reusing some of the results (most importantly, the parametrization of the wave-function Ψ).

The appealing aspect about (analytic) force observations is that they are considerably more informative, as they represent a linearization of the PES in all directions of the $3N$-dimensional configuration space, instead of a single point evaluation. Gathering a similar amount of insight about the PES numerically via energy examples would require solutions of the SE for at least $3N + 1$ perturbations $E(r_1, \ldots, r_i + \epsilon, \ldots, r_{3N})$ of the original geometry at each point. Even then, the obtained force would be subject to approximation error. In contrast, computing analytical forces using the Hellman–Feynman theorem only requires about one to seven times the computational effort of a single energy calculation. Effectively, this theorem thus offers a more efficient way to sample PESs.

7.4.2 Differentiation Amplifies Noise

When force estimates are obtained via differentiation of an *approximate* PES reconstruction, there are no guarantees regarding their quality, since forces are neither constrained nor directly regularized within the loss function of an energy-based model. Inevitably, this can lead to artifacts.

Reconstructions of functions based on a limited number of observations will almost always not be error-free, either due to aliasing effects, non-ideal choice of hypothesis space, or noisy training data [60]. Furthermore, the regularization term in the loss function will reduce the variance of the model and thus promote any approximation errors into the high-frequency band of the residual $f - \hat{f}$. Unfortunately, the application of the derivative operator amplifies high frequencies ω with increasing gain [73], drastically magnifying these errors. This phenomenon can be easily understood when looking at the derivative of a model $\hat{f}'$ in the frequency domain

$$\mathcal{F}\left[\hat{f}'\right] = \mathrm{i}\omega\mathcal{F}\left[\hat{f}\right], \tag{7.16}$$

where ω is a factor on the Fourier transform $\mathcal{F}[\hat{f}]$ of the original model (see Fig. 7.1). A low test error in the energy prediction task does therefore not necessarily imply that the model also reconstructs the forces of the target function reliably.

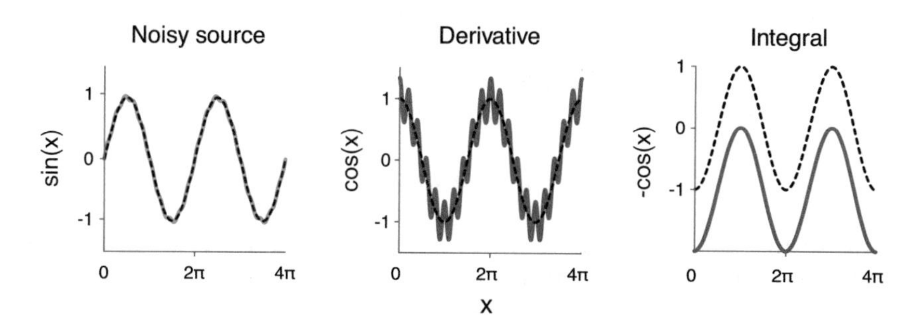

Fig. 7.1 A noisy approximation of a sine wave (blue). Although all instantaneous values are represented well, the derivative of the approximation is a poor estimator for the true derivative. This is because differentiation amplifies the high-frequency noise component within the approximation (middle). Integration on the other hand acts as a low-pass filter (right) that attenuates noise. It is therefore easier to approximate a function with accurate first derivatives from derivative examples instead of function values. Note that integrals are only defined up to an additive constant, which needs to be recovered separately. Figure taken from Chmiela [74]

Conversely, however, high-frequency noise is attenuated upon integration of a force field estimate, which gives corresponding potentials with controlled variance.

7.4.3 Constructing Conservative Vector-Valued GPs

In the simplest, and by far most prevalent regression setting, a *single* output variable $y \in \mathbb{R}$ is predicted from an input vector $\mathbf{x} \in \mathbb{R}^N$. Being a scalar field, the PES reconstruction problem fits this template; however, we intend to pursue the reconstruction of the associated force field, i.e. the negative gradient of the PES, instead. In that reformulation of the problem, the output $\mathbf{y} \in \mathbb{R}^N$ becomes vector-valued as well, thus requiring a mapping $\hat{\mathbf{f}} : \mathbb{R}^N \rightarrow \mathbb{R}^N$.

Naively, we could model each partial force separately and treat them as independent, implicitly assuming that the individual components of the force vector do not affect each other. Then, a straightforward formulation of a vector-valued estimator takes the form

$$\hat{\mathbf{f}}(\mathbf{x}) = \left[\hat{f}_1(\mathbf{x}), \ldots, \hat{f}_N(\mathbf{x})\right]^\top, \tag{7.17}$$

where each component $\hat{f}_i : \mathbb{R}^N \rightarrow \mathbb{R}$ is a separate scalar-valued GP [75]. However, this assumed independence of the individual outputs is hard to justify in many practical scenarios. Especially, since correlations between the individual noise processes associated with each output channel could introduce dependencies in the posterior process, even if they were independent a priori [61]. Bypassing this dependence would therefore ignore valuable information and yield sub-optimal estimates.

Instead of mapping to scalar outputs, we can alternatively model the covariance function as a matrix $\mathbf{k} : \mathcal{X} \times \mathcal{X} \rightarrow \mathbb{R}^{N \times N}$ that expresses the interaction among multiple output components. Together with a vector-valued mean function $\boldsymbol{\mu} : \mathcal{X} \rightarrow \mathbb{R}^N$, we can then sample realizations of vector-valued functions from the GP

$$\mathbf{f}(\mathbf{x}) \sim \mathcal{GP}\left[\boldsymbol{\mu}(\mathbf{x}), \mathbf{k}(\mathbf{x}, \mathbf{x}')\right]. \tag{7.18}$$

In this setting, the corresponding RKHS is vector-valued and it has been shown that the representer theorem continues to hold [76]. Each component of the kernel function k_{ij} specifies a covariance between a pair of outputs $f_i(\mathbf{x})$ and $f_j(\mathbf{x})$, which makes it straightforward to impose linear constraints $\mathbf{g}(x) = \hat{G}\left[\mathbf{f}(\mathbf{x})\right]$ on the GP prior

$$\mathbf{g}(\mathbf{x}) \sim \mathcal{GP}\left[\hat{G}\boldsymbol{\mu}(\mathbf{x}), \hat{G}\,\mathbf{k}(\mathbf{x}, \mathbf{x}')\,\hat{G}'^{\top}\right], \tag{7.19}$$

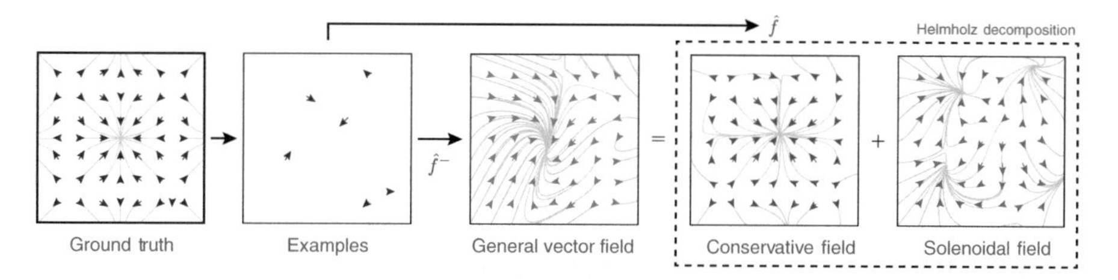

Fig. 7.2 Modeling gradient fields (leftmost subfigure) based on a small number of examples. With GDML, a conservative vector field estimate $\hat{\mathbf{f}}$ is obtained directly (purple). In contrast, a naïve estimator $\hat{\mathbf{f}}^-$ with no information about the correlation structure of its outputs is not capable to uphold the energy conservation constraint (blue). We perform a Helmholtz decomposition of the naïve non-conservative vector field estimate to show the error component due to violation of the law of energy conservation (red). This significant contribution to the overall prediction error is completely avoided with the GDML approach. Figure taken from Chmiela et al. [88]

and hence also the posterior [77–80]. Here, $\hat{G}$ and $\hat{G}'$ act on the first and second argument of the kernel function, respectively. Linear constraints are wide-spread in physics. They include simple conservation laws, but also operations like differential equations, allowing the construction of models that are consistent with the laws that underpin many physical processes [81–85].

Here, we aim to construct a GP that inherits the correct structure of a *conservative* force field to ensure integrability, so that the corresponding energy potential can be recovered from the same model. We start by considering, that the force field estimator $\hat{\mathbf{f}}_{\mathbf{F}}(\mathbf{x})$ and the PES estimator $\hat{f}_E(\mathbf{x})$ are related via some operator $\hat{G}$. To impose energy conservation, we require that the curl vanishes (see Fig. 7.2) for every input to the transformed energy model[1]:

$$\nabla \times \hat{G}\left[\hat{f}_E\right] = \mathbf{0}. \tag{7.20}$$

As expected, this is satisfied by the derivative operator $\hat{G} = \nabla$ or, in the case of energies and forces, the negative gradient operator

$$\hat{\mathbf{f}}_{\mathbf{F}}(\mathbf{x}) = \hat{G}\left[\hat{f}_E\right](\mathbf{x}) = -\nabla \hat{f}_E(\mathbf{x}). \tag{7.21}$$

[1]For illustrative purposes, we use the definition of curl in three dimensions here, but the theory directly generalizes to arbitrary dimension. One way to prove this is via path-independence of conservative vector fields: the circulation of a gradient along any closed curve is zero and the curl is the limit of such circulations.

As outlined previously, we can directly apply this transformation to a standard scalar-valued "energy" GP with realizations $f_E : \mathcal{X}^{3N} \rightarrow \mathbb{R}$. Since differentiation is a linear operator, the result is another GP with realizations $\mathbf{f_F} : \mathcal{X}^{3N} \rightarrow \mathbb{R}^{3N}$:

$$\hat{\mathbf{f}}_{\mathbf{F}} \sim \mathcal{GP}\left[-\nabla\mu(\mathbf{x}), \nabla_{\mathbf{x}} k(\mathbf{x}, \mathbf{x}')\nabla_{\mathbf{x}'}^{\top}\right]. \tag{7.22}$$

Note, that this gives the second derivative of the original kernel (with respect to each of the two inputs) as the (co-)variance structure, with entries

$$k_{ij} = \frac{\partial^2 k}{\partial \mathbf{x}_i \partial \mathbf{x}'_j}. \tag{7.23}$$

It is equivalent (up to sign) to the Hessian $\nabla k \nabla^{\top} = \text{Hess}_{\mathbf{x}}(k)$ (i.e., second derivative with respect to one of the inputs), provided that the original covariance function k is stationary. A GP using this covariance enables inference based on the distribution of partial derivative observations, instead of function values [86, 87]. Effectively, this allows us to train GP models in the gradient domain.

This Hessian kernel gives rise to the following *gradient domain machine learning* [88, 89] force model as the posterior mean of the corresponding GP:

$$\hat{\mathbf{f}}_{\mathbf{F}}(\mathbf{x}) = \sum_{i}^{M} \sum_{j}^{3N} (\boldsymbol{\alpha}_{\mathbf{i}})_j \frac{\partial}{\partial x_j} \nabla k(\mathbf{x}, \mathbf{x}_i) \tag{7.24}$$

Because the trained model is a (fixed) linear combination of kernel functions, integration only affects the kernel function itself. The corresponding expression for the energy predictor

$$\hat{f}_E(\mathbf{x}) = \sum_{i}^{M} \sum_{j}^{3N} (\boldsymbol{\alpha}_{\mathbf{i}})_j \frac{\partial}{\partial x_j} k(\mathbf{x}, \mathbf{x}_i) + c \tag{7.25}$$

is therefore neither problem-specific nor does it require retraining. It is however only defined up to an integration constant

$$c = \frac{1}{M} \sum_{i}^{M} E_i + \hat{f}_E(\mathbf{x}_i), \tag{7.26}$$

that we recover separately in the least-squares sense. Here, E_i are the energy labels for each training example. We remark that this reconstruction approach yields two models at the same time, by correctly implementing the fundamental physical connection between $\mathbf{k_F}$ and k_E (Fig. 7.3).

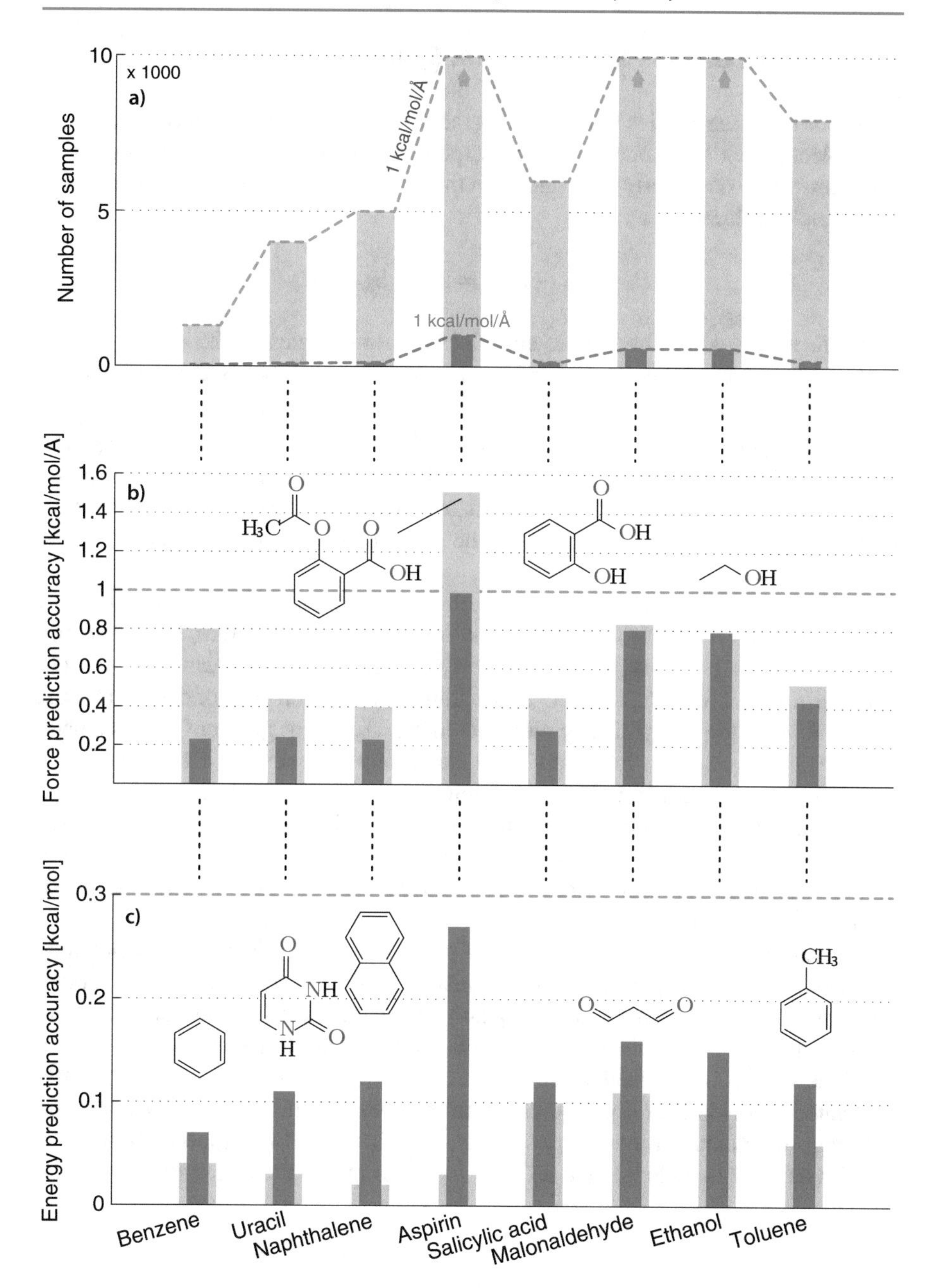

Fig. 7.3 Efficiency of GDML predictor versus a model that has been trained on energies. (**a**) Required number of samples for a force prediction performance of MAE ($1\,\mathrm{kcal\,mol^{-1}\,\AA^{-1}}$) with the energy-based model (gray) and GDML (blue). The energy-based model was not able to achieve the targeted performance with the maximum number of 63,000 samples for aspirin. (**b**) Force prediction errors for the converged models (same number of partial derivative samples and energy samples). (**c**) Energy prediction errors for the converged models. All reported prediction errors have been estimated via cross-validation. Figure taken from Chmiela et al. [88]

7.5 Point Groups and Fluxional Symmetries

Atoms of the same species are interchangeable within a numerical representation of the molecule, without affecting the corresponding energy or forces. This permutational invariance is directly inherited from the indistinguishability of identical atoms in the nuclear Hamiltonian

$$H_n = T_n + E_{tot} = T_n + E_{nn} + E_e. \tag{7.27}$$

Here, T_n and E_{nn} are the kinetic and electrostatic energies of the nuclei and E_e is the eigenvalue of the electronic Hamiltonian H_e. E_e inherits the parametrical dependence on the nuclear coordinates of H_e, which is invariant with respect to atomic permutations given its analytical form: $H_e = T_e + E_{ee} + \sum_{i=1}^{N_n} \sum_{j=1}^{N_e} \frac{1}{|\mathbf{R}_i - \mathbf{r}_j|}$, where the T_e and E_{ee} are the electronic kinetic and electrostatic energies. Since the dependence of H_e on the nuclear coordinates appears only as a sum, we can freely permute the nuclear positions, defining the permutations symmetry over the whole symmetric group.

For example, the benzene molecule with six carbon and six hydrogen atoms can be indexed (and therefore represented) in $6!6! = 518{,}400$ different, but physically equivalent ways. However, not all of these symmetric variants are actually important to define a kernel function that gauges the similarity between pairs of configurations. To begin with, the 24 symmetry elements in the $\mathbf{D}_{6h}$ point group of this molecule are relevant. In addition to these rigid space group symmetries (e.g., reflections), there are additional dynamic non-rigid symmetries [90] (e.g., rotations of functional groups or torsional displacements) that appear as the structure transforms over time. Fortunately, we can safely ignore the massive amount of the remaining configurations in the full symmetric group of factorial size if we manage to identify this relevant subset. While methods for identifying molecular point groups for polyatomic rigid molecules are readily available [91], dynamical symmetries are usually not incorporated in traditional force fields and electronic structure calculations. This is because extracting non-rigid symmetries requires chemical and physical intuition about the system at hand, which is hard to automate.

Since this is impractical in an ML setting, we will now review a physically motivated algorithm for purely data-driven recovery of this particularly important subgroup of molecular symmetries. This will allow us to incorporate these symmetries as prior knowledge into a GP to further improve the data-efficiency of the model.

7.5.1 Positive-Semidefinite Assignment

We begin with the basic insight that MD trajectories consist of smooth consecutive changes in nearly isomorphic molecular graphs. When sampling from these trajectories the combinatorial challenge is to correctly identify the same atoms across the examples such that the learning method can use consistent information for comparing two molecular conformations in its kernel function. While the so-called bi-partite matching allows to locally assign atoms $\mathbf{R} = \{\mathbf{r}_1, \ldots, \mathbf{r}_N\}$ for each pair of molecules in the training set, this strategy alone is not sufficient as the assignment needs to be made globally consistent by multi-partite matching in a second step [92–94]. The reason is that optimal bi-partite assignment yields indefinite functions in general, which are problematic in combination with kernel methods. They give rise to indefinite kernel functions, which do not define a Hilbert space [95]. Practically, there will not exist a metric space embedding of the complete set of approximate pairwise similarities defined in the kernel matrix and the learning problem becomes ill-posed. A multi-partite correction is therefore necessary to recover a non-contradictory notion of similarity across the whole training set. A side benefit of such a global matching approach is that it can robustly establish correspondence between distant transformations of a geometry using intermediate pairwise matchings, even if the direct bi-partite assignment is not unambiguously possible.

7.5.1.1 Solving the Multi-Way Matching Problem

We start by defining the bi-partite matching problem in terms of adjacency matrices as representation for the molecular graph. To solve the pairwise matching problem we therefore seek to find the assignment τ which minimizes the squared Euclidean distance between the adjacency matrices $\mathbf{A}$ of two isomorphic graphs G and H with entries $(\mathbf{A})_{ij} = \|\mathbf{r}_i - \mathbf{r}_j\|$, where $\mathbf{P}(\tau)$ is the permutation matrix that realizes the assignment:

$$\arg\min_{\tau} \mathcal{L}(\tau) = \left\| \mathbf{P}(\tau)\mathbf{A}_G\mathbf{P}(\tau)^\top - \mathbf{A}_H \right\|^2 . \tag{7.28}$$

Notably, most existing ML potentials use representations based on adjacency matrices as input [2–5, 7–49, 52–54]. An optimal assignment in terms of Eq. (7.28) therefore transfers to almost any other model and the GDML model in particular.

Adjacency matrices of isomorphic graphs have identical eigenvalues and eigenvectors, only their assignment differs. Following the approach of Umeyama [96], we identify the correspondence of eigenvectors $\mathbf{U}$ by projecting both sets $\mathbf{U}_G$ and $\mathbf{U}_H$ onto each other to find the best overlap. We use the overlap matrix

$$\mathbf{M} = \text{abs}(\mathbf{U}_G)\text{abs}(\mathbf{U}_H)^\top \tag{7.29}$$

after sorting eigenvalues and overcoming sign ambiguity. Then $-\mathbf{M}$ is provided as the cost matrix for the Hungarian algorithm [97], maximizing the overall overlap which finally returns the approximate assignment $\tilde{\tau}$ that minimizes Eq. (7.28) and thus provides the results of step one of the procedure. As indicated, global inconsistencies may arise, observable as violations of the transitivity property $\tau_{jk} \circ \tau_{ij} = \tau_{ik}$ of the assignments [92]. Therefore a second step is necessary which is based on the composite matrix $\tilde{\mathcal{P}}$ of all pairwise assignment matrices $\tilde{\mathbf{P}}_{ij} \equiv \mathbf{P}(\tilde{\tau}_{ij})$

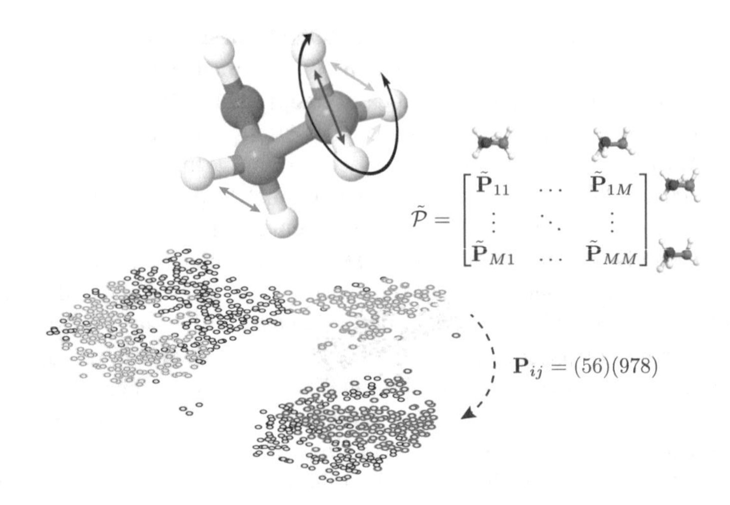

Fig. 7.4 T-SNE [98] embedding of all molecular geometries in an ethanol training set. Each data point is color coded to show the permutation transformations that align it with the arbitrarily chosen canonical reference state (gray points). These permutations are recovered by restricting the rank of the pairwise assignment matrix $\tilde{\mathcal{P}}$ to obtain a consistent multi-partite matching $\mathcal{P}$

within the training set. We propose to reconstruct a rank-limited $\mathcal{P}$ via the transitive closure of the minimum spanning tree (MST) that minimizes the bi-partite matching cost (see Eq. (7.28)) over the training set. The MST is constructed from the most confident bi-partite assignments and represents the rank N skeleton of $\tilde{\mathcal{P}}$, defining also $\mathcal{P}$ (see Fig. 7.4). Finally, the resulting *multi-partite matching* $\mathcal{P}$ is a consistent set of atom assignments across the whole training set.

As a first test, we apply our algorithm to a diverse set of non-rigid molecules that have been selected by Longuet-Higgins [90] to illustrate the concept of dynamic symmetries. Each of the chosen examples changes easily from one conformation to another due to internal rotations that cannot be described by point groups. Those molecules require the more complete *permutation-inversion group* of symmetry operations that include energetically feasible permutations of identical nuclei. Our multi-partite matching algorithm successfully recovers those symmetries from short MD trajectories (see Table 7.1), giving us the confidence to proceed.

7.5.1.2 Symmetric Kernels

The resulting consistent multi-partite matching $\mathcal{P}$ enables us to construct symmetric kernel-based ML models of the form

$$\hat{f}(\mathbf{x}) = \sum_{ij}^{M} \alpha_{ij} k(\mathbf{x}, \mathbf{P}_{ij}\mathbf{x}_i), \tag{7.30}$$

by augmenting the training set with the symmetric variations of each molecule [99]. A particular advantage of our solution is that it can fully populate all recovered

Table 7.1 Recovering the permutation-inversion (PI) group of symmetry operations of fluxional molecules from short MD trajectories

	Molecule	PG order	PI group order	Recovered
	Hydrazine	2	8	8
	Ammonia	6	6	6
	(Difluoromethyl)borane	2	12	12
	Cyclohexane	6	12	12
	Trimethylborane	2	324	339
	Dimethylacetylene	6	36	39
	Ethane	6	16	36

We used our multi-partite matching algorithm to recover the symmetries of the molecules used in Longuet-Higgins [90]. Our algorithm identifies PI group symmetries (a superset that also includes the PG), as well as additional symmetries that are an artifact of the metric used to compare molecular graphs in our matching algorithm. Each dataset consists of a MD trajectory of 5000 time steps. Figure taken from Chmiela [74]

permutational configurations even if they do not form a symmetric group, severely reducing the computational effort in evaluating the model. Even if we limit the range of j to include all S unique assignments only, the major downside of this approach is that a multiplication of the training set size leads to a drastic increase in the complexity of the cubically scaling GP regression algorithm. We overcome this drawback by exploiting the fact that the set of coefficients $\boldsymbol{\alpha}$ for the symmetrized training set exhibits the same symmetries as the data, hence the linear system can be contracted to its original size, while still defining the full set of coefficients exactly.

Without affecting the pairwise similarities expressed by the kernel, we transform all training geometries into a canonical permutation $\mathbf{x}_i \equiv \mathbf{P}_{i1}\mathbf{x}_i$, enabling the use of uniform symmetry transformations $\mathbf{P}_j \equiv \mathbf{P}_{1j}$. Simplifying Eq. (7.30) accordingly, gives rise to the symmetric kernel that we originally set off to construct

$$\begin{aligned}\hat{f}(\mathbf{x}) &= \sum_i^M \alpha_i \sum_q^S k(\mathbf{x}, \mathbf{P}_q \mathbf{x}_i) \\ &= \sum_i^M \alpha_i k_{\text{sym}}(\mathbf{x}, \mathbf{x}_i),\end{aligned} \tag{7.31}$$

Table 7.2 Relative increase in accuracy of the sGDML@DFT vs. the non-symmetric GDML model: the benefit of a symmetric model is directly linked to the number of permutational symmetries in the system

Molecule	# Sym. in k_{sym}	Δ MAE (%)	
		Energy	Forces
Benzene	12	−1.6	−62.3
Uracil	1	0.0	0.0
Naphthalene	4	0.0	−52.2
Aspirin	6	−29.6	−31.3
Salicylic acid	1	0.0	0.0
Malonaldehyde	4	−37.5	−48.8
Ethanol	6	−53.4	−58.2
Toluene	12	−16.7	−67.4
Paracetamol	12	−40.7	−52.9
Azobenzene	8	−74.3	−47.4

All symmetry counts in this table include the identity transformation

and yields a model with the exact same number of parameters as the original, non-symmetric one. This ansatz is known as *invariant integration* and frequently applied to symmetrize ML potentials [17, 24, 100]. However, our solution, motivated by the concept of permutation-inversion groups [90], is able to truncate the sum over potentially hundreds of thousands permutations in the full symmetric group of the molecule to a few physically reasonable ones. We remark that this step is essential in making invariant integration practical beyond systems with five or six identical atoms (with $5! = 120$ and $6! = 720$ permutations, respectively). For example, the molecules benzene, toluene, and azobenzene each only have 12 physically relevant symmetries, whereas the associated symmetric groups have orders $6!6!$, $7!8!$, and $12!10!2!$, respectively. Our multi-partite matching algorithm is therefore able to shorten the sum over S in Eq. (7.31) by up to 15 orders of magnitude, without significant loss of accuracy.

The data-driven symmetry adaptation approach outlined above can be applied universally, but in particular to the energy-conserving force field kernel that we have derived in the previous section. It improves the data-efficiency of the model in proportion to the number of symmetries that are present in molecules (see Table 7.2) to the point that costly high-level coupled-cluster CCSD(T) calculations become viable as reference data. We have shown that such a model effectively allows converged MD simulations with fully quantized electrons and nuclei for molecules with up to a few dozen atoms [101].

7.6 Conclusion

Typically, the parametrization of ML potentials relies on the availability of large reference datasets to obtain accurate results, which prevents the construction of ML models using costly high-level ab initio methods due to the exploding computational cost. In this chapter, we have shown how to overcome this restrictive requirement by informing the model with fundamental physical invariances and conservation laws. Not only does this make the models more data-efficient, it also guarantees that the incorporated physics are represented without artifacts. In particular, we have successively developed a theoretical framework for construction of ML potentials that include the full set of temporal and spatial symmetries of molecules. Homogeneity of time implies energy conservation and global spatial symmetries include rotational and translational invariance of the energy.

Using a generalization of GPs to vector-valued Hilbert spaces, we have defined a predictor that explicitly maps to energy-conserving solutions and thus allows the simultaneous prediction of accurate forces and energies at the same time. We have then extended this model to additionally incorporate all relevant rigid space group symmetries as well as dynamic non-rigid symmetries. Typically, the identification of symmetries requires chemical and physical intuition about the system at hand, which is impractical in an ML setting. Through a data-driven multi-partite matching approach, we have automated the discovery of permutation matrices of molecular graph pairs in different permutational configurations and thus between symmetric transformations undergone within the scope of a dataset. This allowed us to define a compact symmetric model that can be parametrized from very small training datasets of just a few hundreds of examples, enabling the direct construction of flexible molecular force fields from expensive high-level ab initio calculations.

While the resulting sGDML model constitutes a substantial step towards enabling highly accurate and thus truly predictive MD simulations, there is a number of challenges that remain to be solved in terms of its applicability and scaling to larger molecular systems. For example, a fragmentation of large atomistic systems would allow scaling up and transferable predictions across different molecules with similar atom types. The well-separated inter- and intramolecular correlation scales within molecular solids suggest that a hierarchical decomposition is possible with limited degradation of prediction accuracy. Furthermore, advanced sampling strategies could be employed to combine forces from different levels of theory to minimize the need for computationally intensive ab initio calculations even further.

7.7 Data and Software

The ML potentials described in this chapter are implemented in the sGDML software package—see www.sgdml.org for details. Reconstruct FFs from your own datasets today!

References

1. M.E. Tuckerman, Ab initio molecular dynamics: basic concepts, current trends and novel applications. J. Phys. Condens. Matter **14**(50), R1297 (2002)
2. M. Rupp, A. Tkatchenko, K.-R. Müller, O.A. Von Lilienfeld. Fast and accurate modeling of molecular atomization energies with machine learning. Phys. Rev. Lett. **108**(5), 58301 (2012)

3. K. Hansen, G. Montavon, F. Biegler, S. Fazli, M. Rupp, M. Scheffler, O.A. von Lilienfeld, A. Tkatchenko, K.-R. Müller, Assessment and validation of machine learning methods for predicting molecular atomization energies. J. Chem. Theory Comput. **9**(8), 3404–3419 (2013)
4. K. Hansen, F. Biegler, R. Ramakrishnan, W. Pronobis, O.A. von Lilienfeld, K.-R. Müller, A. Tkatchenko, Machine learning predictions of molecular properties: accurate many-body potentials and nonlocality in chemical space. J. Phys. Chem. Lett. **6**(12), 2326–2331 (2015)
5. M. Rupp, R. Ramakrishnan, O.A. von Lilienfeld, Machine learning for quantum mechanical properties of atoms in molecules. J. Phys. Chem. Lett. **6**(16), 3309–3313 (2015)
6. V. Botu, R. Ramprasad, Adaptive machine learning framework to accelerate ab initio molecular dynamics. Int. J. Quantum Chem. **115**(16), 1074–1083 (2015)
7. M. Hirn, N. Poilvert, S. Mallat, Quantum energy regression using scattering transforms. CoRR, abs/1502.02077 (2015)
8. R. Ramakrishnan, P.O. Dral, M. Rupp, O.A. von Lilienfeld, Big data meets quantum chemistry approximations: the δ-machine learning approach. J. Chem. Theory Comput. **11**(5), 2087–2096 (2015)
9. S. De, A.P. Bartók, G. Csányi, M. Ceriotti, Comparing molecules and solids across structural and alchemical space. Phys. Chem. Chem. Phys. **18**(20), 13754–13769 (2016)
10. N. Artrith, A. Urban, G. Ceder, Efficient and accurate machine-learning interpolation of atomic energies in compositions with many species. Phys. Rev. B **96**(1), 14112 (2017)
11. A.P. Bartók, S. De, C. Poelking, N. Bernstein, J.R. Kermode, G. Csányi, M. Ceriotti, Machine learning unifies the modeling of materials and molecules. Sci. Adv. **3**(12), e1701816 (2017)
12. A. Glielmo, P. Sollich, A. De Vita, Accurate interatomic force fields via machine learning with covariant kernels. Phys. Rev. B **95**, 214302 (2017)
13. K. Yao, J.E. Herr, J. Parkhill, The many-body expansion combined with neural networks. J. Chem. Phys. **146**(1), 14106 (2017)
14. S.T. John, G. Csányi, Many-body coarse-grained interactions using Gaussian approximation potentials. J. Phys. Chem. B **121**(48), 10934–10949 (2017)
15. F.A. Faber, L. Hutchison, B. Huang, J. Gilmer, S.S. Schoenholz, G.E. Dahl, O. Vinyals, S. Kearnes, P.F. Riley, O.A. von Lilienfeld, Prediction errors of molecular machine learning models lower than hybrid DFT error. J. Chem. Theory Comput. **13**(11), 5255–5264 (2017)
16. M. Eickenberg, G. Exarchakis, M. Hirn, S. Mallat, L. Thiry, Solid harmonic wavelet scattering for predictions of molecule properties. J. Chem. Phys. **148**(24), 241732 (2018)
17. A. Glielmo, C. Zeni, A. De Vita, Efficient nonparametric n-body force fields from machine learning. Phys. Rev. B **97**(18), 184307 (2018)
18. Y.-H. Tang, D. Zhang, G. Em Karniadakis, An atomistic fingerprint algorithm for learning ab initio molecular force fields. J. Chem. Phys. **148**(3), 34101 (2018)
19. A. Grisafi, D.M. Wilkins, G. Csányi, M. Ceriotti, Symmetry-adapted machine learning for tensorial properties of atomistic systems. Phys. Rev. Lett. **120**, 36002 (2018)
20. W. Pronobis, A. Tkatchenko, K.-R. Müller, Many-body descriptors for predicting molecular properties with machine learning: analysis of pairwise and three-body interactions in molecules. J. Chem. Theory Comput. **14**(6), 2991–3003 (2018)
21. J. Behler, M. Parrinello, Generalized neural-network representation of high-dimensional potential-energy surfaces. Phys. Rev. Lett. **98**(14), 146401 (2007)
22. A.P. Bartók, M.C. Payne, R. Kondor, G. Csányi, Gaussian approximation potentials: the accuracy of quantum mechanics, without the electrons. Phys. Rev. Lett. **104**(13), 136403 (2010)
23. K.V. Jovan Jose, N. Artrith, J. Behler, Construction of high-dimensional neural network potentials using environment-dependent atom pairs. J. Chem. Phys. **136**(19), 194111 (2012)
24. A.P. Bartók, R. Kondor, G. Csányi, On representing chemical environments. Phys. Rev. B **87**(18), 184115 (2013)
25. G. Montavon, M. Rupp, V. Gobre, A. Vazquez-Mayagoitia, K. Hansen, A. Tkatchenko, K.-R. Müller, O.A. von Lilienfeld, Machine learning of molecular electronic properties in chemical compound space. New J. Phys. **15**(9), 95003 (2013)

26. A.P. Bartók, G. Csányi, Gaussian approximation potentials: a brief tutorial introduction. Int. J. Quantum Chem. **115**(16), 1051–1057 (2015)
27. V. Botu, R. Ramprasad, Learning scheme to predict atomic forces and accelerate materials simulations. Phys. Rev. B **92**, 94306 (2015)
28. T. Bereau, D. Andrienko, O.A. von Lilienfeld, Transferable atomic multipole machine learning models for small organic molecules. J. Chem. Theory Comput. **11**(7), 3225–3233 (2015)
29. Z. Li, J.R. Kermode, A. De Vita, Molecular dynamics with on-the-fly machine learning of quantum-mechanical forces. Phys. Rev. Lett. **114**, 96405 (2015)
30. J. Behler, Perspective: machine learning potentials for atomistic simulations. J. Chem. Phys. **145**(17), 170901 (2016)
31. F. Brockherde, L. Vogt, L. Li, M.E. Tuckerman, K. Burke, K.-R. Müller, Bypassing the Kohn-Sham equations with machine learning. Nat. Commun. **8**, 872 (2017)
32. M. Gastegger, J. Behler, P. Marquetand, Machine learning molecular dynamics for the simulation of infrared spectra. Chem. Sci. **8**, 6924–6935 (2017)
33. K.T. Schütt, F. Arbabzadah, S. Chmiela, K.-R. Müller, A. Tkatchenko, Quantum-chemical insights from deep tensor neural networks. Nat. Commun. **8**, 13890 (2017)
34. K. Schütt, P.-J. Kindermans, H.E. Sauceda, S. Chmiela, A. Tkatchenko, K.-R. Müller, SchNet: a continuous-filter convolutional neural network for modeling quantum interactions, in *Advances in Neural Information Processing Systems*, vol. 31, pp. 991–1001 (2017)
35. K.T. Schütt, H.E. Sauceda, P.-J. Kindermans, A. Tkatchenko, K.-R. Müller, SchNet—A deep learning architecture for molecules and materials. J. Chem. Phys. **148**(24), 241722 (2018)
36. B. Huang, O.A. von Lilienfeld, The "DNA" of chemistry: scalable quantum machine learning with "amons". arXiv preprint:1707.04146 (2017)
37. T.D. Huan, R. Batra, J. Chapman, S. Krishnan, L. Chen, R. Ramprasad, A universal strategy for the creation of machine learning-based atomistic force fields. NPJ Comput. Mater. **3**(1), 37 (2017)
38. E.V. Podryabinkin, A.V. Shapeev, Active learning of linearly parametrized interatomic potentials. Comput. Mater. Sci. **140**, 171–180 (2017)
39. P.O. Dral, A. Owens, S.N. Yurchenko, W. Thiel, Structure-based sampling and self-correcting machine learning for accurate calculations of potential energy surfaces and vibrational levels. J. Chem. Phys. **146**(24), 244108 (2017)
40. L. Zhang, J. Han, H. Wang, R. Car, E. Weinan, Deep potential molecular dynamics: a scalable model with the accuracy of quantum mechanics. Phys. Rev. Lett. **120**(14), 143001 (2018)
41. N. Lubbers, J.S. Smith, K. Barros, Hierarchical modeling of molecular energies using a deep neural network. J. Chem. Phys. **148**(24), 241715 (2018)
42. K. Ryczko, K. Mills, I. Luchak, C. Homenick, I. Tamblyn, Convolutional neural networks for atomistic systems. Comput. Mater. Sci. **149**, 134–142 (2018)
43. K. Kanamori, K. Toyoura, J. Honda, K. Hattori, A. Seko, M. Karasuyama, K. Shitara, M. Shiga, A. Kuwabara, I. Takeuchi, Exploring a potential energy surface by machine learning for characterizing atomic transport. Phys. Rev. B **97**(12), 125124 (2018)
44. T.S. Hy, S. Trivedi, H. Pan, B.M. Anderson, R. Kondor, Predicting molecular properties with covariant compositional networks. J. Chem. Phys. **148**(24), 241745 (2018)
45. J. Wang, S. Olsson, C. Wehmeyer, A. Pérez, N.E. Charron, G. De Fabritiis, F. Noé, C. Clementi, Machine learning of coarse-grained molecular dynamics force fields. ACS Cent. Sci. **5**(5), 755–767 (2019)
46. T. Bereau, R.A. DiStasio Jr., A. Tkatchenko, O.A. Von Lilienfeld, Non-covalent interactions across organic and biological subsets of chemical space: physics-based potentials parametrized from machine learning. J. Chem. Phys. **148**(24), 241706 (2018)
47. A. Mardt, L. Pasquali, H. Wu, F. Noé, VAMPnets for deep learning of molecular kinetics. Nat. Commun. **9**(1), 5 (2018)
48. F. Noé, S. Olsson, J. Köhler, H. Wu, Boltzmann generators: Sampling equilibrium states of many-body systems with deep learning. Science **365**(6457), eaaw1147 (2019)

49. N. Thomas, T. Smidt, S. Kearnes, L. Yang, L. Li, K. Kohlhoff, P. Riley, Tensor field networks: rotation-and translation-equivariant neural networks for 3D point clouds. arXiv preprint:1802.08219 (2018)
50. J.S. Smith, B. Nebgen, N. Lubbers, O. Isayev, A. Roitberg, Less is more: sampling chemical space with active learning. J. Chem. Phys. **148**(24), 241733 (2018)
51. K. Gubaev, E.V. Podryabinkin, A.V. Shapeev, Machine learning of molecular properties: locality and active learning. J. Chem. Phys. **148**(24), 241727 (2018)
52. F.A. Faber, A.S. Christensen, B. Huang, O.A. von Lilienfeld, Alchemical and structural distribution based representation for universal quantum machine learning. J. Chem. Phys. **148**(24), 241717 (2018)
53. A.S. Christensen, F.A. Faber, O.A. von Lilienfeld, Operators in quantum machine learning: response properties in chemical space. J. Phys. Chem. **150**(6), 64105 (2019)
54. R. Winter, F. Montanari, F. Noé, D.-A. Clevert, Learning continuous and data-driven molecular descriptors by translating equivalent chemical representations. Chem. Sci. **10**(6), 1692–1701 (2019)
55. K. Gubaev, E.V. Podryabinkin, G.L.W. Hart, A.V. Shapeev, Accelerating high-throughput searches for new alloys with active learning of interatomic potentials. Comput. Mater. Sci. **156**, 148–156 (2019)
56. E. Noether, Invarianten beliebiger Differentialausdrücke. Gött. Nachr. Mathematisch-Physikalische Klasse **1918**, 37–44 (1918)
57. K.-R. Müller, S. Mika, G. Rätsch, K. Tsuda, B. Schölkopf, An introduction to kernel-based learning algorithms. IEEE Trans. Neural Netw. Learn. Syst. **12**(2), 181–201 (2001)
58. B. Schölkopf, A.J. Smola, *Learning with Kernels: Support Vector Machines, Regularization, Optimization, and Beyond* (MIT Press, Cambridge, 2002)
59. G. Wahba, *Spline Models for Observational Data*, vol. 59 (SIAM, Philadelphia, 1990)
60. B. Schölkopf, R. Herbrich, A.J. Smola, A generalized representer theorem, in *International Conference on Computational Learning Theory* (Springer, Berlin, 2001), pp. 416–426
61. C.E. Rasmussen, Gaussian processes in machine learning, in *Advanced Lectures on Machine Learning* (Springer, Berlin, 2004), pp. 63–71
62. D. Duvenaud, *Automatic Model Construction with Gaussian Processes*, PhD thesis, University of Cambridge, Cambridge, 2014
63. C.A. Micchelli, Y. Xu, H. Zhang, Universal kernels. J. Mach. Learn. Res. **7**(Dec), 2651–2667 (2006)
64. A. Damianou, N. Lawrence, Deep Gaussian processes, in *Artificial Intelligence and Statistics* (2013), pp. 207–215
65. C. Lanczos, *The Variational Principles of Mechanics* (University of Toronto Press, Toronto, 1949)
66. K. Brading, E. Castellani, *Symmetries in Physics: Philosophical Reflections* (Cambridge University Press, Cambridge, 2003)
67. D.J.C. MacKay, Introduction to Gaussian processes, in *NATO ASI Series F: Computer and Systems Sciences*, vol. 168 (Springer, Berlin, 1998)
68. A.J. Smola, B. Schölkopf, K.-R. Müller, The connection between regularization operators and support vector kernels. Neural Netw. **11**(4), 637–649 (1998)
69. C. Heil, *Metrics, Norms, Inner Products, and Operator Theory* (Birkhäuser, Basel, 2018)
70. A. Rahimi, B. Recht, Random features for large-scale kernel machines, in *Advances in Neural Information Processing Systems* (2008), pp. 1177–1184
71. P. Politzer, J.S. Murray, The Hellmann-Feynman theorem: a perspective. J. Mol. Model. **24**(9), 266 (2018)
72. R.P. Feynman, Forces in molecules. Phys. Rev. **56**(4), 340 (1939)
73. C.E. Shannon, Communication in the presence of noise. Proc. IEEE **86**(2), 447–457 (1998)
74. S. Chmiela, *Towards Exact Molecular Dynamics Simulations with Invariant Machine-Learned Models*, PhD thesis. Technische Universität, Berlin, 2019
75. T. Hastie, R. Tibshirani, J.H. Friedman, *The Elements of Statistical Learning: Data Mining, Inference, and Prediction*. Springer Series in Statistics (Springer, Berlin, 2009)

76. M.A. Alvarez, L. Rosasco, N.D. Lawrence, et al., Kernels for vector-valued functions: a review. Found. Trends Mach. Learn. **4**(3), 195–266 (2012)
77. P. Boyle, M. Frean, Dependent Gaussian processes, in *Advances in Neural Information Processing Systems* (2005), pp. 217–224
78. C.A. Micchelli, M. Pontil, On learning vector-valued functions. Neural Comput. **17**(1), 177–204 (2005)
79. C.A. Micchelli, M. Pontil, Kernels for multi-task learning, in *Advances in Neural Information Processing Systems* (2005), pp. 921–928
80. L. Baldassarre, L. Rosasco, A. Barla, A. Verri, Multi-output learning via spectral filtering. Mach. Learn. **87**(3), 259–301 (2012)
81. T. Graepel, Solving noisy linear operator equations by Gaussian processes: application to ordinary and partial differential equations, in *International Conference on Machine Learning* (2003), pp. 234–241
82. S. Särkkä, Linear operators and stochastic partial differential equations in Gaussian process regression, in *International Conference on Artificial Neural Networks* (Springer, Berlin, 2011), pp. 151–158
83. E.M. Constantinescu, M. Anitescu, Physics-based covariance models for Gaussian processes with multiple outputs. Int. J. Uncertain. Quantif. **3**(1) (2013)
84. N.C. Nguyen, J. Peraire, Gaussian functional regression for linear partial differential equations. Comput. Methods Appl. Mech. Eng. **287**, 69–89 (2015)
85. C. Jidling, N. Wahlström, A. Wills, T.B. Schön, Linearly constrained Gaussian processes, in *Advances in Neural Information Processing Systems* (2017), pp. 1215–1224
86. F.J. Narcowich, J.D. Ward, Generalized Hermite interpolation via matrix-valued conditionally positive definite functions. Math. Comput. **63**(208), 661–687 (1994)
87. E. Solak, R. Murray-Smith, W.E. Leithead, D.J. Leith, C.E. Rasmussen, Derivative observations in Gaussian process models of dynamic systems, in *Advances in Neural Information Processing Systems* (2003), pp. 1057–1064
88. S. Chmiela, A. Tkatchenko, H.E. Sauceda, I. Poltavsky, K.T. Schütt, K.-R. Müller, Machine learning of accurate energy-conserving molecular force fields. Sci. Adv. **3**(5), e1603015 (2017)
89. S. Chmiela, H.E. Sauceda, I. Poltavsky, K.-R. Müller, A. Tkatchenko, sGDML: constructing accurate and data efficient molecular force fields using machine learning. Comput. Phys. Commun. **240**, 38–45 (2019)
90. H.C. Longuet-Higgins, The symmetry groups of non-rigid molecules. Mol. Phys. **6**(5), 445–460 (1963)
91. E.B. Wilson, *Molecular Vibrations: The Theory of Infrared and Raman Vibrational Spectra* (McGraw-Hill Interamericana, New York, 1955)
92. D. Pachauri, R. Kondor, V. Singh, Solving the multi-way matching problem by permutation synchronization, in *Advances in Neural Information Processing Systems* (2013), pp. 1860–1868
93. M. Schiavinato, A. Gasparetto, A. Torsello, *Transitive Assignment Kernels for Structural Classification* (Springer, Cham, 2015), pp. 146–159
94. N.M. Kriege, P.-L. Giscard, R.C. Wilson, On valid optimal assignment kernels and applications to graph classification, in *Advances in Neural Information Processing Systems*, vol. 30 (2016), pp. 1623–1631
95. J.-P. Vert, The optimal assignment kernel is not positive definite. CoRR, abs/0801.4061 (2008)
96. S. Umeyama, An eigendecomposition approach to weighted graph matching problems. IEEE Trans. Pattern Anal. Mach. Intell. **10**(5), 695–703 (1988)
97. H.W. Kuhn, The Hungarian method for the assignment problem. Nav. Res. Logist. **2**(1–2), 83–97 (1955)
98. L. van der Maaten, G. Hinton, Visualizing data using t-SNE. J. Mach. Learn. Res. **9**(2579–2605), 85 (2008)

99. T. Karvonen, S. Särkkä, Fully symmetric kernel quadrature. SIAM J. Sci. Comput. **40**(2), A697–A720 (2018)
100. B. Haasdonk, H. Burkhardt, Invariant kernel functions for pattern analysis and machine learning. Mach. Learn. **68**(1), 35–61 (2007)
101. S. Chmiela, H.E. Sauceda, K.-R. Müller, A. Tkatchenko, Towards exact molecular dynamics simulations with machine-learned force fields. Nat. Commun. **9**(1), 3887 (2018)

Quantum Machine Learning with Response Operators in Chemical Compound Space

8

Felix Andreas Faber, Anders S. Christensen, and O. Anatole von Lilienfeld

Abstract

The choice of how to represent a chemical compound has a considerable effect on the performance of quantum machine learning (QML) models based on kernel ridge regression (KRR). A carefully constructed representation can lower the prediction error for out-of-sample data by several orders of magnitude with the same training data. This is a particularly desirable effect in data scarce scenarios, such as they are common in first principles based chemical compound space explorations. Unfortunately, representations which result in KRR models with low and steep learning curves for extensive properties, for example, energies, do not necessarily lead to well performing models for response properties. In this chapter we review the recently introduced FCHL18 representation (Faber et al., J Chem Phys 148(24):241717, 2018), in combination with kernel-based QML models to account for response properties by including the corresponding operators in the regression (Christensen et al., J Chem Phys 150(6):064105, 2019). FCHL18 was designed to describe an atom in its chemical environment, allowing to measure distances between elements in the periodic table, and consequently providing a metric for both structural and chemical similarities between compounds. The representation does not decouple the radial and angular degrees of freedom, which makes it well-suited for comparing atomic environments. QML models using FCHL18 display low and steep learning curves for energies of molecules, clusters, and crystals. By contrast, the same QML models exhibit less favorable learning for other properties, such as forces, electronic eigenvalues, or dipole moments. We discuss the use of the electric field differential operator within a kernel-based operator QML (OQML) approach. Using OQML results in

F. A. Faber · A. S. Christensen · O. A. von Lilienfeld (✉)
Institute of Physical Chemistry, Department of Chemistry, University of Basel, Basel, Switzerland
e-mail: anatole.vonlilienfeld@uibas.ch

K. T. Schütt et al. (eds.), *Machine Learning Meets Quantum Physics*,
Lecture Notes in Physics 968, https://doi.org/10.1007/978-3-030-40245-7_8

the same predictive accuracy for molecular dipole norm, but with approximately 20× less training data, as directly learning the dipole norm with a KRR model.

8.1 Introduction

The specific architecture of quantum machine learning models (QML) plays a large role in the predictive accuracy (or learning efficiency) of a model. For example, it has now become common practice to decompose a model into a sum of atomic contributions when learning and predicting atomization energies, as it has been shown to give better and more scalable performance. This is because the atomization energy is an extensive property, meaning that it tends to grow with system size. It is therefore natural to formulate an additive model of atomic contributions. These additive machine learning models have been shown to display remarkable accuracies for learning extensive properties such as atomization energies [1–5].

A recent benchmark study [6] demonstrates that an efficient model for one property is not necessarily as efficient for others. For example, the HDAD based KRR model in Ref. [6] produces less favorable learning curves for intensive response properties, such as dipole moments, electronic eigenvalues, or forces, than for atomization energies, due to the fact that the extensive nature of the kernel does not necessarily reflect the behavior of response properties correctly.

This chapter is divided into two parts: First, we will discuss a way to represent the surrounding chemistry and structure of an atomic environment, using a set of many-body distributions [1]. We then proceed with the discussion on how to learn response properties, and provide examples on how to do this for energy derivatives, as exemplified for dipole moments [7].

8.2 Representing an Atomic Environment

There are many ways of encoding the composition and structure of a chemical compound into a representation suitable for machine learning. It has become common practice to decompose the kernel into contributions from the environment of each atom in the compound. Representations based on atomic densities [8] have yielded machine learning models with remarkable predictive power. They are also easy to understand and modify to the learning problem. For deriving the FCHL18 representation (Ref. [1]), we have therefore chosen to use atomic density based representations. In essence, the FCHL2018 representation encodes an atom I in its environment by a set of distributions $\mathcal{A}_M(I) = \{A_1(I), A_2(I), \ldots, A_M(I)\}$, based on many body expansions with up to M atoms.

Each distribution $A_m(I)$ with m interatomic contributions encodes both the structural and chemical environment of atom I. This is done by placing scaled Gaussian functions, centered on structural and chemical degrees of freedom. The structural degrees of freedom are among other quantities, distances d, angles θ, or dihedrals between atoms, and the extension to contributions going beyond four bodies would also be straightforward. The elemental identity is encoded by the atom's group G and period P in Mendeleev's table. This information was already demonstrated to be helpful in QML models of crystalline properties [9], and was subsequently also used by others [10]. However, any feature which uniquely differentiates between the elements in the periodic table would suffice to ensure uniqueness. Uniqueness of the representation is a necessary condition for representations in order to avoid absurd results [11, 12]. Note that the FCHL representation is constructed such that the mth distribution contains m-body information. For example, $A_1(I)$ only encodes G and P of atom I, while $A_2(I)$ encodes the chemical identities of atom I and neighboring atoms, as well as the distances to neighboring atoms.

A given N-dimensional Gaussian distribution with means μ and standard deviations σ along all dimensions is denoted by $\mathcal{N}(\mathbf{x})$ where $\mathbf{x} = \{\mu_1, \sigma_1; \mu_2, \sigma_2; \ldots; \mu_N, \sigma_N\}$. Within FCHL we multiply the interatomic 2 and 3-body Gaussians with a scaling function ξ_m, which is a function of internal degrees of freedom, such as distances and/or angles between atom I and its neighboring atoms. The scaling functions both serve as a way to modulate the relative importance of the contribution from each atom, depending on vicinity, and to help with the introduction of a smooth damping function if need be. We will discuss more details regarding the choice of possible cutoff functions later in this chapter. Before we do this, however, we will go through and explicitly define each of the distributions $A(I)_m$, starting from the first-order expansion, and going up to the third order expansion. Discussion of higher-order terms have been omitted for FCHL18 since preliminary testing indicated that learning curves do not improve significantly beyond $\mathcal{A}_3(I)$.

8.2.1 First-Order Term A_1

The first-order expansion $A_1(I)$ encodes the elemental identity of atom I. This is done by placing a Gaussian on the period P_I and group G_I in the periodic table of the element I. This is seen in Eq. (8.1), with $\mathbf{x}_I^{(1)} = \{P_I, \sigma_P; G_I, \sigma_G\}$ where σ_P and σ_G are hyperparameters, corresponding to the width of the Gaussians along the respective dimensions. χ_i are the dummy variables in the dimensions in which the Gaussians are placed, and will be integrated out when calculating distances between any two atoms (also discussed in the next section).

$$A_1(I) = \mathcal{N}\left(\mathbf{x}_I^{(1)}\right) = e^{-\frac{(P_I-\chi_1)^2}{2\sigma_P^2} - \frac{(G_I-\chi_2)^2}{2\sigma_G^2}} \tag{8.1}$$

The drawback of using Gaussians to measure elemental distances consists of its locality. It can only be used efficiently as a metric for elements that overlap due to elemental smearing, i.e., that they are sufficiently close-by in the periodic table and, for example, will not be able to distinguish between two elements at opposite sides of the periodic table. Na and Ne, for example, are close to each other in the sense that they differ in nuclear charge by only one proton, yet within our usual parameterization of FCHL they will overlap negligibly. Of course, this could be rectified by selecting more appropriate widths in Eq. (8.1) but we have not yet studied this effect in detail. Alternatively, other dimensions or other basis functions than Gaussians could be studied just as well.

8.2.2 Second-Order Term A_2

The second-order term $A_2(I)$, in addition to encoding the chemical element of I, also includes the distance information with respect to all neighbors of atom I. $A_2(I) = \mathcal{N}(\mathbf{x}_I^{(1)}) \sum_{i \neq I} \mathcal{N}(\mathbf{x}_{iI}^{(2)}) \xi_2(d_{iI})$ consists of a product between $\mathcal{N}(\mathbf{x}_I^{(1)})$ and a sum that runs over all neighboring atoms $\{i\}$. The entries of the sum are Gaussians $\mathbf{x}_{iI}^{(2)} = \{d_{iI}, \sigma_d; P_i, \sigma_P; G_i, \sigma_G\}$, placed at the distance d_{iI} between atom i and I, and σ_d is the corresponding Gaussian width. Here, $\xi_2(d_{iI})$ is the aforementioned scaling function. Figure 8.1 depicts $A_2(I)$ for a carbon with sp^1, sp^2, and sp^3 electron configuration as encountered in ethane, ethylene, and ethyne, respectively. Note that neither elemental smearing is used in this case ($\sigma_P \to 0$ and $\sigma_G \to 0$),

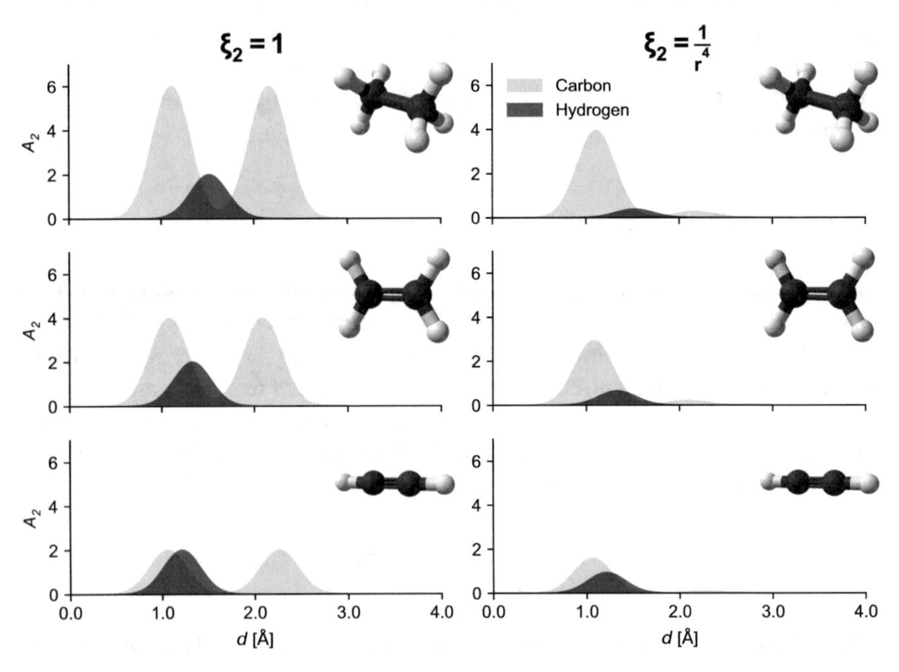

Fig. 8.1 The two-body term (A_2) for one of the carbon atoms in ethane, ethylene, and ethyne, as a function of radial (d) degrees of freedom. The scaling function ξ_2 is set to unity (left) and $\frac{1}{r^4}$ (right). The elemental smearing is set to zero ($\sigma_P, \sigma_G \to 0$) so that the five-dimensional distribution can be reduced to two sets two-dimensional distributions, one for each element triplet

which reduces $A_2(I)$ to a one-dimensional function for each element pair, nor have we multiplied in the scaling function yet.

8.2.3 Third Order Term A_3

$A_3(I) = \mathcal{N}(\mathbf{x}^{(1)}) \sum_{i \neq I} \mathcal{N}(\mathbf{x}_{iI}^{(2)}) \sum_{j \neq i,I} \mathcal{N}(\mathbf{x}_{ijI}^{(3)}) \xi_3(d_{iI}, d_{jI}, \theta_{ij}^I)$, extends upon $A_2(I)$ by containing an additional sum, running over all the other remaining neighboring atoms $\{j \neq i\}$. Again, the elements in the sum consist of Gaussians, $\mathbf{x}_{ijI}^{(3)} = \{\theta_{ij}^I, \sigma_\theta; P_j, \sigma_P; G_j, \sigma_G\}$, placed at angles θ_{ij}^I, with independent width σ_θ. θ is the principal angle between the two distance vectors spanned from atoms I to i and I to j. P_j and G_j, with accompanying widths σ_P and σ_G, are the group and period of the element of atom j, respectively. $\xi_3(d_{iI}, d_{jI}, \theta_{ij}^I)$ is the scaling function for $A_3(I)$ and will be discussed in the next paragraph. Figure 8.2 depicts how $A_3(I)$ looks like for the oxygen, carbon, and hydrogen atoms in ethanol. Again, neither elemental smearing is used in the figure ($\sigma_P \to 0$ and $\sigma_G \to 0$), so $A_3(I)$ reduces to a two-dimensional function for each element triplet, nor has the scaling function been multiplied in.

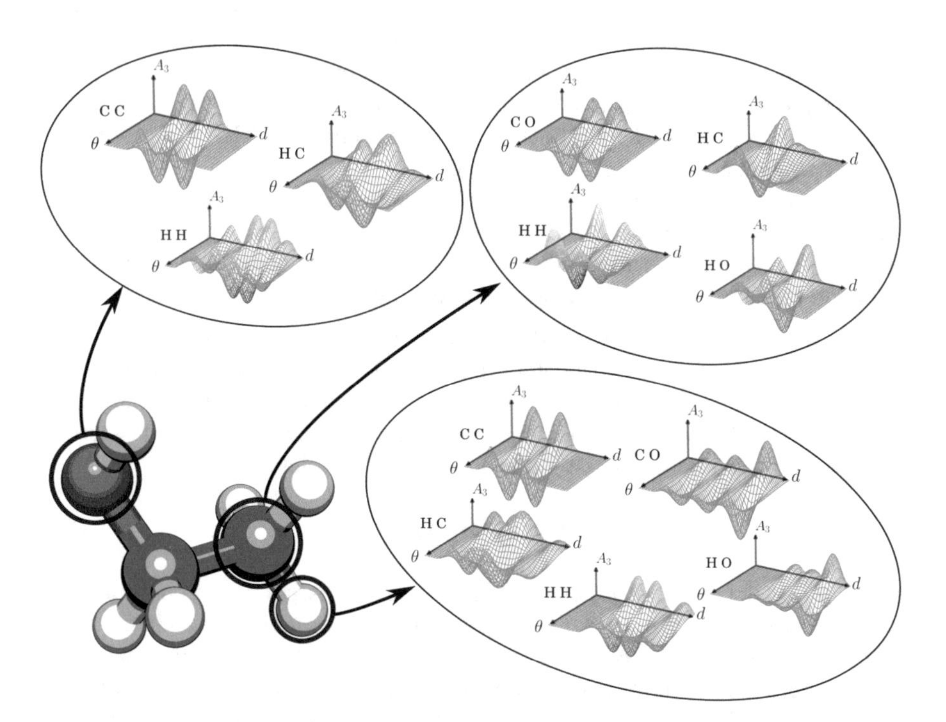

Fig. 8.2 Figure taken from Ref. [1]. The three-body term (A_3) for the atomic environments of C, H, and O (circled) in ethanol, as a function of radial (d) and angular (θ) degrees of freedom. For simplicity, the scaling function is set to unity and elemental smearing is set to zero. It then reduces to a set of two-dimensional distributions, one for each element triplet

8.2.4 Scaling Function

As mentioned earlier, the scaling function ξ_m is used to weight the importance of the Gaussians based on internal distances and angles. We use power-law scaling throughout this chapter since resulting ML models have been shown to yield numerically accurate results. Inspired by BAML [11] and SLATM [5], we have modified the London dispersion interaction $\xi_2(d_{iI}) = \frac{1}{d_{iI}^{n_2}}$ by optimizing the exponent on various data-sets: $n_2 = 4$. For the angular scaling function we rely on the Axilrod–Teller–Muto [13, 14] formula $\xi_3(d_{iI}, d_{jI}, \theta_{ij}^{I}) = \frac{1+3\cos(\theta_{ij}^{I})\cos(\theta_{Ij}^{i})\cos(\theta_{iI}^{j})}{(d_{iI}d_{jI}d_{ij})^{n_3}}$ with $n_3 = 2$.

8.2.5 Electric Field-Dependent Representation

While any response property, such as atomic forces, chemical potentials, or alchemical derivatives, could be considered, in this chapter we focus on the energy response resulting from an externally applied electric field. This approach can in principle be used as long as the derivative of the QML model predicting the energy with respect to the response operator is non-zero. This is in most cases trivial for forces, since representations that can recover potential energy surfaces already include an explicit dependence on the atomic coordinates. For other properties, this dependency can, for example, be obtained by making the representation dependent on the perturbing field. In this section, we demonstrate this by modifying the FCHL18 representation to be dependent on an external electric field.

While an electric field can be included in the representation in multiple ways, we have chosen to include it via the aforementioned two- and three-body scaling functions. The importance of the features in the representations is weighted by applying appropriate scaling factors to those features. For energies a factor of $\frac{1}{d_{iI}^{n_2}}$ is well-suited, since it is expected that atoms that are closer contribute more to the energy. Similarly, we expect atom pairs with large, opposite partial charges to contribute more to the dipole moment compared to pairs of atoms that are close to neutral. Building on this principle, the scaling function for the two-body term is modified (denoted by an asterisk) to include the electric field response:

$$\xi_2^{*IJ} = \xi_2^{IJ} - \epsilon(\boldsymbol{\mu}_{IJ} \cdot \mathbf{E}) \tag{8.2}$$

where $\boldsymbol{\mu}_{IJ}$ is the dipole vector due to fictitious partial charges placed on the atoms I and J, that is $\boldsymbol{\mu}_{IJ} = \mathbf{r}_I q_I + \mathbf{r}_J q_J$ with the coordinate system centered in the center of nuclear charge, and $\mathbf{E}$ is an external electric field, and ϵ is a scaling parameter that balances the two terms in the scaling function. Similarly, the three-body term is modified by including the dipole due to partial charges placed on three atoms I,J, and K.

$$\xi_3^{*IJK} = \xi_3^{IJK} - \epsilon\left(\boldsymbol{\mu}_{IJK} \cdot \mathbf{E}\right) \tag{8.3}$$

where $\boldsymbol{\mu}_{IJK}$ is the dipole due to fictitious partial charges placed on the three atoms, calculated similarly to the two-body dipole, that is $\boldsymbol{\mu}_{IJK} = \mathbf{r}_I q_I + \mathbf{r}_J q_J + \mathbf{r}_K q_K$. The model seems to be insensitive to the exact value of the fictitious partial charges, so long as they qualitatively describe a reasonable interaction with the external field. We have relied on partial charges taken from the Gasteiger charge model [15], but we found that learning curves are quite insensitive with respect to the specific charges used, and therefore we think that any reasonable charge model could be used just as well.

In the absence of an electric field, the kernel elements resulting from the representation with the modified scaling functions are unchanged, but the derivative with respect to the electric field is now non-zero. Furthermore, the kernel elements now *can* change whenever an electric field is applied. Rotational invariance is preserved since the model outputs a scalar field, any vectorial properties are obtained from the gradient of a scalar quantity. The reader should note that the model does not learn the partial charges of the model nor does it use them as a proxy to learn the dipole moment. Rather, they serve as dummy variables that enforce a physically motivated weighting of the molecular features, allowing the energy (and its responses) to be regressed with improved transferability.

8.3 Kernel-Based Regression Model

Many important observables can be formulated as a derivative of the energy. For example, Table 8.1 contains 11 observables that are partial derivatives of the energy with respect to external electric field, external magnetic field, internal magnetic moments, and the nuclear coordinates. These measurable quantities are important in many domains of chemistry, e.g., when used in various types of spectroscopy. The OQML formalism is capable of treating these properties simultaneously. As noted in Ref. [6], many machine learning models struggle to simultaneously describe multiple of these properties accurately. The numerical evidence presented in Sect. 8.4 suggests that it is advantageous to explicitly exploit these linear dependencies within the OQML formalism.

This section presents the general frame work of OQML, and the derivative of the machine learning model of the energy with respect to nuclear coordinates and external electric field. Some modifications to the representation are still necessary in order to be able to treat properties resulting from other vector potentials, such as external magnetic fields and internal magnetic moments.

8.3.1 General Response Formalism

In kernel ridge regression (KRR) the energy of a set of molecules, $\mathbf{U}$, is calculated as

$$\mathbf{U} = \mathbf{K}\boldsymbol{\alpha} \tag{8.4}$$

Table 8.1 The table shows order of derivatives of the energy with respect to the external electric field (**E**), external magnetic field (**B**), internal magnetic moments (**I**), and the nuclear coordinates (**R**), and their corresponding response properties [16]

E	**B**	**I**	**R**	Property
0	0	0	0	Energy
1	0	0	0	Electric dipole moment
0	1	0	0	Magnetic dipole moment
0	0	1	0	Hyperfine coupling constant
0	0	0	1	Molecular (nuclear) gradient
2	0	0	0	Electric polarizability
0	2	0	0	Magnetizability
0	0	2	0	Nuclear spin–spin coupling
0	0	0	2	Harmonic vibrational frequencies
1	0	0	1	Infrared absorption intensities
1	1	0	0	Optical rotation, circular dichroism
0	1	1	0	Nuclear magnetic shielding

where $\mathbf{K}$ is the kernel matrix and $\boldsymbol{\alpha}$ is the set of regression coefficients. Most commonly, basis functions are placed on each of the molecules in the training set, such that the kernel matrix $\mathbf{K}$ is a square symmetric matrix.

By extension, a response property ω with the corresponding response operator O can be calculated by applying the response operator to the energy calculated in Eq. (8.4):

$$\omega = O[\mathbf{U}] \approx O[\mathbf{K}]\boldsymbol{\alpha} \tag{8.5}$$

The implication of this relation is that a single set of regression coefficients can describe both the energy and the energy response simultaneously. This is similar to the GDML [17] model and Gaussian process regression with covariant kernels [18], with the main difference being the choice of basis functions in the kernel.

This also exploits the fact that the kernel is well-suited for energy learning, which then extends the kernels capability to predict energy responses with higher accuracy, as will be shown later in this chapter.

By training on a set of reference values of $O[\mathbf{U}^{\text{ref}}]$, the set of regression coefficients can be obtained by minimizing the following Lagrangian:

$$J(\boldsymbol{\alpha}) = \sum_{\gamma} \beta_{\gamma} \left\| O_{\gamma}\left[\mathbf{U}^{\text{ref}}\right] - O_{\gamma}[\mathbf{K}\boldsymbol{\alpha}] \right\|^{2}_{L_2(\Omega_{\gamma})} \tag{8.6}$$

$$\equiv \sum_{\gamma} \beta_{\gamma} \int_{\Omega_{\gamma}} \left[O_{\gamma}\left[\mathbf{U}^{\text{ref}}\right] - O_{\gamma}[\mathbf{K}\boldsymbol{\alpha}]\right]^{T} \left[O_{\gamma}\left[\mathbf{U}^{\text{ref}}\right] - O_{\gamma}[\mathbf{K}\boldsymbol{\alpha}]\right] \tag{8.7}$$

where γ denotes the perturbation, so that it is possible to train on multiple response properties simultaneously, and β_{γ} is a hyperparameter that can balance the weight of the terms. To avoid a very overdetermined regression problem, the set of basis functions can be increased by placing the kernel function on each of the atoms in the training set. One consequence of this is that the kernel matrix is no longer a square matrix, and Eq. (8.6) cannot be solved by simply inverting the kernel, as is commonly done in KRR and Gaussian Process Regression [19]. The analytical solution to Eq. (8.6) is given by

$$\boldsymbol{\alpha} = \Big[\sum_{\gamma} \beta_{\gamma} \int_{\Omega_{\gamma}} O_{\gamma}[\mathbf{K}]^{T} O_{\gamma}[\mathbf{K}]\Big]^{-1} \Big[\sum_{\gamma} \beta_{\gamma} \int_{\Omega_{\gamma}} O_{\gamma} \left[\mathbf{U}^{\text{ref}}\right]^{T} O_{\gamma}[\mathbf{K}]\Big] \tag{8.8}$$

As an alternative to this normal-equation solution, the regression coefficients can be obtained by using an orthogonal decomposition such as a singular-value (SVD) or a QR decomposition.

Note that the number of training labels in the regression (i.e., the length of $O_{\gamma}[\mathbf{U}^{\text{ref}}]$) may exceed the number of molecules in the training set, for example, when training a model for force components. Consequently, the training error is non-zero, which is in contrast to conventional KRR where the training error is very small for noise-less data and unique representations.

8.3.2 Kernel Derivatives in the Basis of Atomic Environments

Here we introduce the kernel derivatives that correspond to force and dipole moment operators in the basis of kernel functions placed on atomic environment. Firstly, the kernel that correspond to the unperturbed kernel (corresponding to energy learning) is given as

$$(\mathbf{K})_{iJ} = \sum_{I \in i} k(q_J, q_I^*) \tag{8.9}$$

where I runs over the atoms in the i'th molecule and J is the index of an atomic environment in the basis.

From this it follows—by taking the derivative—that the kernel elements that correspond to the force, i.e., minus the nuclear gradient operator acting on the kernel, are given by

$$-\frac{\partial}{\partial x_I^*}(\mathbf{K})_{IJ} = -\sum_{K \in i} \frac{\partial k(q_J, q_K^*)}{\partial x_I^*} \qquad \text{where} \qquad I \in i \tag{8.10}$$

The kernel elements used to learn dipole moments, corresponding to the kernel's response to the external electric field $\mathbf{E}$, and are given by

$$-\frac{\partial}{\partial E_{\nu}^*}(\mathbf{K})_{i_{\nu}J} = -\sum_{K \in i} \frac{\partial k(q_J, q_K^*)}{\partial E_{\nu}^*} \qquad \text{where} \qquad \nu \in \{x, y, z\} \tag{8.11}$$

Similarly, higher-order kernel derivatives are also possible. For example, the nuclear Hessian kernel is given by

$$\frac{\partial^2}{\partial x^*_{I'}\partial x^*_I}(\mathbf{K})_{I'IJ} = \sum_{K\in i} \frac{\partial k(q_J, q^*_K)}{\partial x^*_{I'}\partial x^*_I} \quad \text{where} \quad I', I \in i \tag{8.12}$$

Lastly, it is also possible to define higher-order, mixed derivatives of the kernel as long as the representation has a response to both perturbations. The kernel that yields the dipole derivatives necessary for the infrared intensities is written as the following mixed second-order derivative:

$$\frac{\partial^2}{\partial E^*_\nu \partial x^*_I}(\mathbf{K})_{i_\nu IJ} = \sum_{K\in i} \frac{\partial k(q_J, q^*_K)}{\partial E^*_\nu \partial x^*_I}$$
$$\text{where } I \in i \text{ and } \nu \in \{x, y, z\} \tag{8.13}$$

8.4 Numerical Results

To measure the predictive power of the representation we now consider learning curves, resulting from using our model on the QM9 data set [20] which was generated from the SMILES strings stored in GDB-17 [21]. QM9 is built up from ~134k drug-like molecules with up to nine heavy atoms (C, N, O or F) not counting hydrogen, and consists of relaxed structures as well as 13 electronic ground-state properties, obtained using the B3LYP functional in DFT. We begin by examining how well the model can predict atomization energies. The learning curves of FCHL18 based KRR models are shown in Fig. 8.3, and show plausible trends and

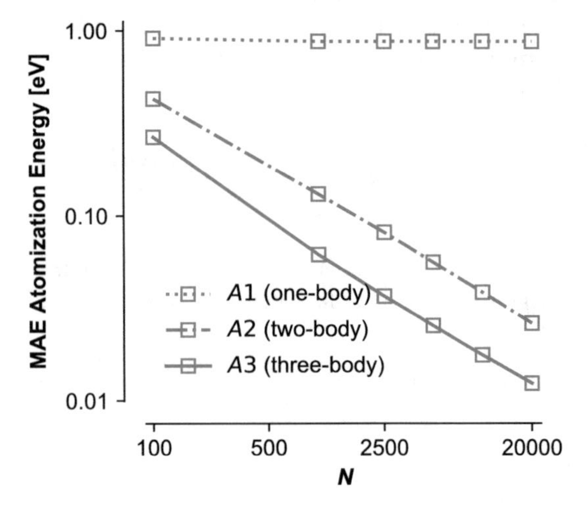

Fig. 8.3 Learning curves of atomization energies form the QM9 dataset, resulting from including the first, second, and third order terms in the representation ($\mathcal{A}_m$, $m = 1, 2, 3$)

behavior: Including only the first-order term (amounting to stoichiometry) yields hardly any improvement beyond 100 training instances. This is not surprising, since the dataset contains conformational as well as compositional degrees of freedom, making this model incapable of distinguishing structures beyond their elemental composition. Including second-order terms drastically improves the off-set and the learning rate. Inclusion of three-body effects results in further lowering of the learning curve, reaching chemical accuracy (∼0.05 eV) already for training set sizes with ∼1000 molecules. Such physics based understanding and control of learning efficiency is expected to be important when dealing with data scarce problems where the quantum reference calculations are very costly.

Figure 8.4 provides a comparison of FCHL based QML models to various alternative representations and regressors (including neural networks). Overall, KRR based models appear to outperform neural networks. And structural distribution based representations, such as SOAP, (a)SLATM, and FCHL fare most favorably.

As demonstrated above, QML models with the FCHL representation can easily reach chemical accuracy for energetic properties with small training set sizes. However, as pointed out in Ref. [6] it can be very difficult for QML models to reach chemical accuracy for certain properties, even when the QML model shows very promising learning for extensive properties such as the energy. The reason for the difference in learning rates can be visualized by employing kernel principal

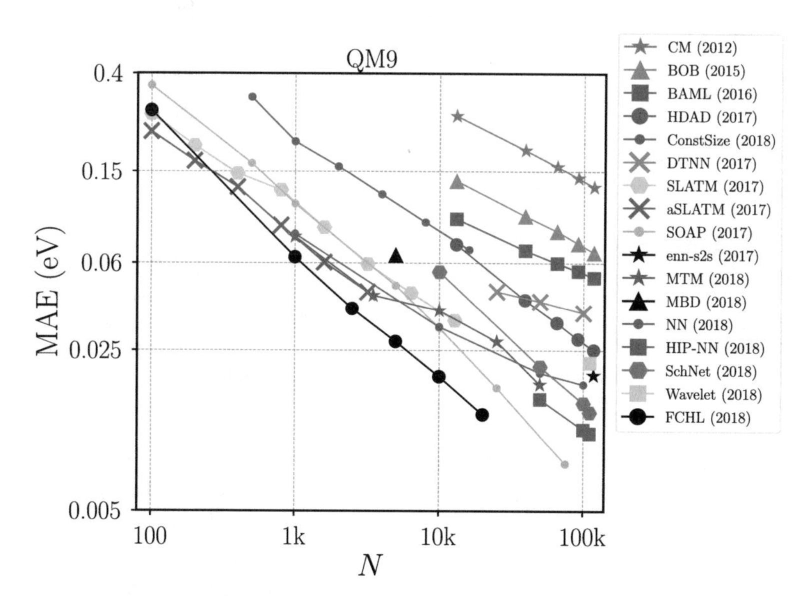

Fig. 8.4 Performance overview of various QML models published ever since Ref. [22]. Prediction errors of atomization energies in the QM9 [20] are shown as a function of training set size. The QML models included differ solely by representation, model architecture, and training/test set and cross-validation details. They correspond to CM [22], BOB [23], BAML [11], HDAD [6], constant size [24], DTNN [25], (a)SLATM [26], SOAP [2], enn [27], MTM [28], MBD [29], NN [30], HIP-NN [31], SchNet [32], Wavelet [33], and FCHL [1]

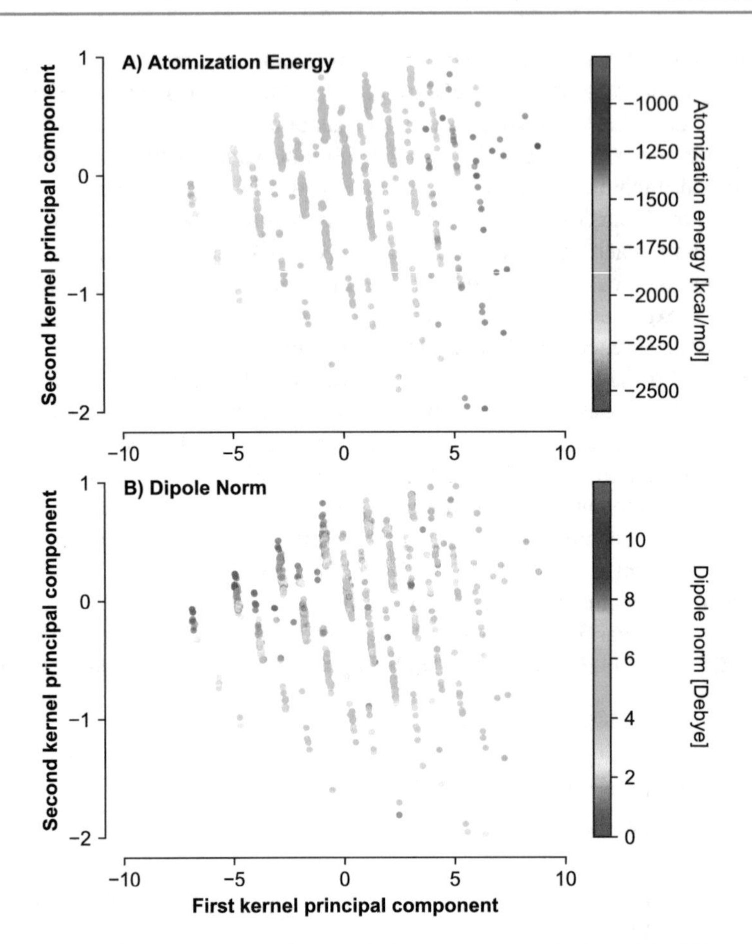

Fig. 8.5 Visualization of the two first kernel principal components of a kernel for 1000 randomly selected QM9 molecules [20], calculated with the FCHL representation [1]. Each dot corresponds to one molecule. In (**a**) the molecules are colored by their DFT atomization energy, showing how the kernel correlates well with the energy of the molecules. In (**b**) the same molecules are colored by their dipole norm which correlates worse with the kernel principal components, compared to the energy i (**a**), which causes a slower learning rate for the dipole norm

component analysis (PCA). An example is illustrated in Fig. 8.5, where a kernel PCA is performed for the FCHL kernel for 1000 randomly selected QM9 molecules with the elements HCNO. The first principal component for this kernel roughly correspond to the number of atoms in a molecule, while the second principal component further separates the molecules by chemical composition.

The smooth and monotonic changes in color suggest that this kernel allows the regressor to easily interpolate energies between the training points, as seen in Fig. 8.5a. By contrast, when it comes to dipole moments, the coloring of the same components is much less monotonic, as seen in Fig. 8.5b. The result is that the kernel is well-suited for learning the energy directly, as demonstrated by steep learning curves, while a comparatively slow learning rate is obtained for intensive response properties, such as the dipole moment [6].

8.4.1 Dipole Learning for QM9 Molecules

As alluded to before, the OQML formalism can now be used to greatly improve learning rates of molecular dipole norms. Here we compare a standard KRR approach to learning the dipole norms of the molecules in the QM9 dataset [20], obtained using DFT for the GDB-9 subset in GDB-17 [21], with the OQML approach introduced in Ref. [7], and outlined above. In order to quantify the improvement, two machines have been trained, one using conventional KRR trained directly on the QM9 dipole norms, the other using our response formalism trained on the dipole moment components of the same molecules.

The resulting learning curves for the two models are displayed in Fig. 8.6. As previously observed [6], the learning rate for most KRR models is rather poor, even for models that work well for energy learning. The same is true for a KRR model with the FCHL representation [1]. Even at 10,000 training molecules, the MAE on out-of-sample dipole norms is far from chemical accuracy and at almost 0.6 Debye. However, the machine trained on the energy response using the OQML formalism shows a substantial increase in learning rate. Around 20× less data is required to reach the same accuracy [7].

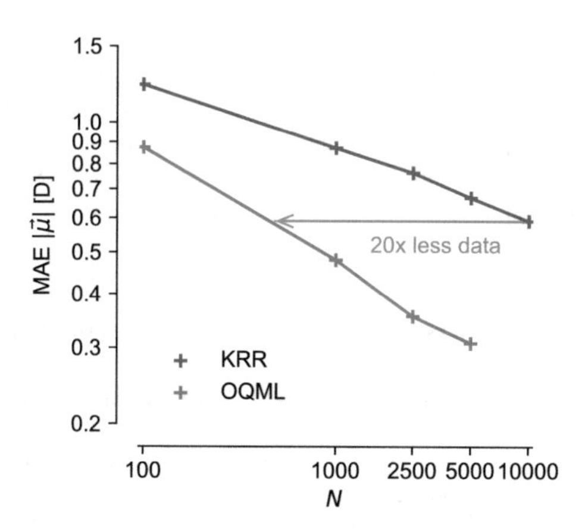

Fig. 8.6 The figure displays learning curves of the dipole norm of molecules in the QM9 dataset. The mean absolute error for the predicted dipole norms is plotted against the size of the training set. The two models are trained on either the dipole norms directly using conventional kernel ridge regression (KRR), or the dipole vectors using the operator quantum machine learning (OQML) approach described in this chapter. Despite using the same representation for both models, the OQML approach requires about 20× less data to reach the same accuracy compared to KRR

8.5 Outlook

We have discussed the FCHL representation and the generalization of KRR based QML models to account for response properties through Operator based QML. While there are still many improvements to be made, we consider these two developments to represent is a significant steps forward towards the general QML philosophy, as also recently discussed in Ref. [34]. For example, we only investigated the use of response operator on kernel-based models. However, OQML can in principle be applied to any differentiable regression model, including neural networks, which already have been proven to be well-suited for multitask learning [35, 36]. Furthermore, we only tested the formalism on dipole moment and forces, and only trained the model on one of them at a time. Training on multiple response properties simultaneously, as well as studying the formalism on other properties will be part of future work.

References

1. F.A. Faber, A.S. Christensen, B. Huang, O.A. von Lilienfeld, J. Chem. Phys. **148**(24), 241717 (2018)
2. A.P. Bartók, S. De, C. Poelking, N. Bernstein, J.R. Kermode, G. Csányi, M. Ceriotti, Sci. Adv. **3**(12) (2017). https://doi.org/10.1126/sciadv.1701816
3. J. Behler, J. Chem. Phys. **134**, 074106 (2011)
4. K.T. Schütt, H.E. Sauceda, P.J. Kindermans, A. Tkatchenko, K.R. Müller, J. Chem. Phys. **148**(24), 241722 (2018)
5. B. Huang, O.A. von Lilienfeld (2017). Preprint. arXiv:1707.04146
6. F.A. Faber, L. Hutchison, B. Huang, J. Gilmer, S.S. Schoenholz, G.E. Dahl, O. Vinyals, S. Kearnes, P.F. Riley, O.A. von Lilienfeld, J. Chem. Theory Comput. **13**, 5255 (2017)
7. A.S. Christensen, F.A. Faber, O.A. von Lilienfeld, J. Chem. Phys. **150**(6), 064105 (2019)
8. M.J. Willatt, F. Musil, M. Ceriotti (2018). Preprint. arXiv:1807.00408
9. F.A. Faber, A. Lindmaa, O.A. von Lilienfeld, R. Armiento, Phys. Rev. Lett. **117**, 135502 (2016). https://doi.org/10.1103/PhysRevLett.117.135502
10. J. Schmidt, J. Shi, P. Borlido, L. Chen, S. Botti, M.A. Marques, Chem. Mater. **29**(12), 5090 (2017)
11. B. Huang, O.A. von Lilienfeld, J. Chem. Phys. **145**(16) (2016). https://doi.org/10.1063/1.4964627
12. O.A. von Lilienfeld, R. Ramakrishnan, M. Rupp, A. Knoll, Int. J. Quantum Chem. **115**, 1084 (2015). https://arxiv.org/abs/1307.2918
13. B.M. Axilrod, E. Teller, J. Chem. Phys **11**(6), 299 (1943). https://doi.org/10.1063/1.1723844
14. Y. Muto, J. Phys. Math. Soc. Jpn. **17**, 629 (1943)
15. J. Gasteiger, M. Marsili, Tetrahedron **36**(22), 3219 (1980). https://doi.org/10.1016/0040-4020(80)80168-2
16. F. Jensen, *Introduction to Computational Chemistry* (Wiley, Chichester, 2007)
17. S. Chmiela, A. Tkatchenko, H.E. Sauceda, I. Poltavsky, K.T. Schütt, K.R. Müller, Sci. Adv. **3**(5), e1603015 (2017)
18. A. Glielmo, P. Sollich, A. De Vita, Phys. Rev. B **95**(21), 214302 (2017)
19. C.K. Williams, C.E. Rasmussen, *Gaussian Processes for Machine Learning*, vol. 2 (MIT Press, Cambridge, 2006)
20. R. Ramakrishnan, P. Dral, M. Rupp, O.A. von Lilienfeld, Sci. Data **1**, 140022 (2014)
21. L. Ruddigkeit, R. van Deursen, L. Blum, J.L. Reymond, J. Chem. Inf. Model. **52**, 2684 (2012)

22. M. Rupp, A. Tkatchenko, K.R. Müller, O.A. von Lilienfeld, Phys. Rev. Lett. **108**, 058301 (2012)
23. K. Hansen, F. Biegler, O.A. von Lilienfeld, K.R. Müller, A. Tkatchenko, J. Phys. Chem. Lett. **6**, 2326 (2015)
24. C.R. Collins, G.J. Gordon, O.A. von Lilienfeld, D.J. Yaron, J. Chem. Phys. **148**(24), 241718 (2018)
25. K.T. Schütt, F. Arbabzadah, S. Chmiela, K.R. Müller, A. Tkatchenko, Nat. Comm. **8**, 13890 (2017). https://doi.org/10.1038/ncomms13890
26. B. Huang, O.A. von Lilienfeld, Nature (2017). arXiv:1707.04146
27. J. Gilmer, S.S. Schoenholz, P.F. Riley, O. Vinyals, G.E. Dahl, in *Proceedings of the 34th International Conference on Machine Learning, ICML 2017* (2017)
28. K. Gubaev, E.V. Podryabinkin, A.V. Shapeev, J. Chem. Phys. **148**(24), 241727 (2018)
29. W. Pronobis, A. Tkatchenko, K.R. Müller, J. Chem. Theory Comput. **14**(6), 2991–3003 (2018)
30. O.T. Unke, M. Meuwly, J. Chem. Phys. **148**(24), 241708 (2018)
31. B. Nebgen, N. Lubbers, J.S. Smith, A.E. Sifain, A. Lokhov, O. Isayev, A.E. Roitberg, K. Barros, S. Tretiak, J. Chem. Theory Comput. **9**(16), 4495–4501 (2018)
32. H.E. Sauceda, S. Chmiela, I. Poltavsky, K.R. Müller, A. Tkatchenko, Molecular force fields with gradient-domain machine learning: Construction and application to dynamics of small molecules with coupled cluster forces. J. Chem. Phys. **150**(11), 114102 (2019)
33. M. Eickenberg, G. Exarchakis, M. Hirn, S. Mallat, L. Thiry, J. Chem. Phys. **148**(24), 241732 (2018)
34. O.A. von Lilienfeld, Angew. Chem. Int. Ed. **57**, 4164 (2018). https://doi.org/10.1002/anie.201709686
35. G. Montavon, M. Rupp, V. Gobre, A. Vazquez-Mayagoitia, K. Hansen, A. Tkatchenko, K.R. Müller, O.A. von Lilienfeld, New J. Phys. **15**(9), 095003 (2013)
36. M. Tsubaki, T. Mizoguchi, J. Phys. Chem. Lett. **9**(19), 5733 (2018)

Physical Extrapolation of Quantum Observables by Generalization with Gaussian Processes

9

R. A. Vargas-Hernández and R. V. Krems

Abstract

For applications in chemistry and physics, machine learning is generally used to solve one of three problems: interpolation, classification or clustering. These problems use information about physical systems in a certain range of parameters or variables in order to make predictions at unknown values of these variables within the same range. The present work illustrates the application of machine learning to prediction of physical properties outside the range of the training parameters. We define 'physical extrapolation' to refer to accurate predictions $y(\boldsymbol{x}^*)$ of a given physical property at a point $\boldsymbol{x}^* = \left[x_1^*, \ldots, x_{\mathcal{D}}^*\right]$ in the $\mathcal{D}$-dimensional space, if, at least, one of the variables $x_i^* \in \left[x_1^*, \ldots, x_{\mathcal{D}}^*\right]$ is *outside* of the range covering the training data. We show that Gaussian processes can be used to build machine learning models capable of physical extrapolation of quantum properties of complex systems across quantum phase transitions. The approach is based on training Gaussian process models of variable complexity by the evolution of the physical functions. We show that, as the complexity of the models increases, they become capable of predicting new transitions. We also show that, where the evolution of the physical functions is analytic and relatively simple (one example considered here is $a+b/x+c/x^3$), Gaussian process models with simple kernels already yield accurate generalization results, allowing for accurate predictions of quantum properties in a different quantum phase. For more complex problems, it is necessary to build models with complex kernels. The complexity of the kernels is increased using the Bayesian Information Criterion (BIC). We illustrate the importance of the BIC by comparing the results with random kernels of various complexity. We discuss strategies to minimize

R. A. Vargas-Hernández (✉) · R. V. Krems
Department of Chemistry, University of British Columbia, Vancouver, BC, Canada
e-mail: ravh011@chem.ubc.ca; rkrems@chem.ubc.ca

K. T. Schütt et al. (eds.), *Machine Learning Meets Quantum Physics*, Lecture Notes in Physics 968, https://doi.org/10.1007/978-3-030-40245-7_9

overfitting and illustrate a method to obtain meaningful extrapolation results without direct validation in the extrapolated region.

9.1 Introduction

As described throughout this book, machine learning has in recent years become a powerful tool for physics research. A large number of machine learning applications in physics can be classified as supervised learning, which aims to build a model $\mathcal{F}(\cdot)$ of the $\boldsymbol{x} \mapsto y$ relation, given a finite number of $\boldsymbol{x}_i \mapsto y_i$ pairs. Here, $\boldsymbol{x}$ is a vector of (generally multiple) parameters determining the physical problem of interest and y is a physics result of relevance. For example, $\boldsymbol{x}$ could be a vector of coordinates specifying the positions of atoms in a polyatomic molecule and y the potential energy of the molecule calculated by means of a quantum chemistry method [1–11]. In this case, $\mathcal{F}(\boldsymbol{x})$ is a model of the potential energy surface constructed based on n energy calculations $\boldsymbol{y} = (y_1, \ldots, y_n)^\top$ at n points $\boldsymbol{x}_i$ in the configuration space of the molecule. To give another example, $\boldsymbol{x}$ could represent the parameters entering the Hamiltonian of a complex quantum system (e.g., the tunnelling amplitude, the on-site interaction strength and/or the inter-site interaction strength of an extended Hubbard model) and y some observable such as the energy of the system. Trained by a series of calculations of the observable at different values of the Hamiltonian parameters, $\mathcal{F}(\boldsymbol{x})$ models the dependence of the observable on the Hamiltonian parameters [12–44], which could be used to map out the phase diagram of the corresponding system.

The ability of a machine learning model to predict previously unseen data is referred to as 'generalization'. These previously unseen data must usually come from the same distribution as the training data, but may also come from a different distribution. Most of the applications of machine learning in physics aim to make predictions within the range of training data. In the present work, we discuss a method for building machine learning models suitable for physical extrapolation. We define 'physical extrapolation' to refer to accurate predictions $y(\boldsymbol{x}^*)$ of a given physical property at a point $\boldsymbol{x}^* = \left[x_1^*, \ldots, x_{\mathcal{D}}^*\right]$ in the $\mathcal{D}$-dimensional input space, if, at least, one of the variables $x_i^* \in \left[x_1^*, \ldots, x_{\mathcal{D}}^*\right]$ is *outside* of the range covering the training data. Thus, in the present work, the training data and test data distributions are necessarily separated in input space. We will refer to the predictions of machine learning models as generalization and the physical problems considered here as extrapolation.

Our particular goal is to extrapolate complex physical behaviour without *a priori* knowledge of the physical laws governing the evolution of the system. For this purpose, we consider a rather challenging problem: prediction of physical properties of complex quantum systems with multiple phases based on training data entirely in one phase. The main goal of this work is schematically illustrated in Fig. 9.1. We aim to construct the machine learning models that, when trained by the calculations or experimental measurements within one of the Hamiltonian phases (encircled region in Fig. 9.1), are capable of predicting the physical properties of the system in the

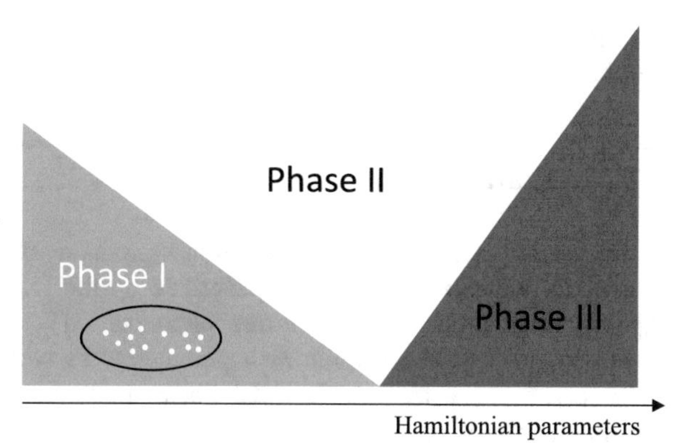

Fig. 9.1 Schematic diagram of a quantum system with three phases. The goal of the present work is to predict both of the phase transitions based on information about the properties of the system in the encircled region of phase I

other phases. Of particular interest is the prediction of the phase transitions, which are often challenging to find with rigorous quantum calculations.

This problem is challenging because the wave functions of the quantum systems—as well as the physical observables characterizing the phases—undergo sharp changes at the phase transitions. Most of the machine learning models used for interpolation/fitting are, however, smooth functions of $\boldsymbol{x}$. So, how can one construct a machine learning model that would capture the sharp and/or discontinuous variation of the physical properties? The method discussed here is based on the idea put forward in our earlier work [45].

We assume that the properties of a physical system within a given phase contain information about multiple transitions and that, when a system approaches a phase transition, the properties must change in a way that is affected by the presence of the transition as well as the properties in the other phase(s). In addition, a physical system is often characterized by some properties that vary smoothly through the transition. The goal is then to build a machine learning model that could be trained by such properties within a given phase, make a prediction of these properties in a different phase and predict the properties that change abruptly at the transition from the extrapolated properties. We will use Gaussian processes to build such models.

9.1.1 Organization of This Chapter

The remainder of this chapter is organized as follows. The next section describes the quantum problems considered here. Understanding the physics of these problems is not essential for understanding the contents of this chapter. The main purpose of Sect. 9.2 is to introduce the notation for the physical problems discussed here. These problems are used merely as examples. Section 9.3 briefly discusses the application of Gaussian process regression for interpolation in multi-dimensional spaces, mainly to set the stage and define the notation for the subsequent discussion. Section 9.4 describes the extension of Gaussian process models to the extrapolation problem. Section 9.5 presents the results and Sect. 9.6 concludes the present chapter.

We will abbreviate 'Gaussian process' as GP, 'Artificial Neural Network' as NN and 'machine learning' as ML throughout this chapter.

9.2 Quantum Systems

In this section, we describe the quantum systems considered in the present work. In general, we consider a system described by the Hamiltonian $\hat{H} = \hat{H}(\Gamma)$ that depends on a finite number of free parameters $\Gamma = \{\alpha, \beta, \dots\}$. The observables depend on these Hamiltonian parameters as well as the intrinsic variables $V = \{v_1, v_2, \dots\}$ such as the total linear momentum for few-body systems or thermodynamic variables for systems with a large number of particles. The set $\Gamma + V$ comprises the independent variables of the problems considered here. The ML models $\mathcal{F}$ will be functions of $\Gamma + V$.

More specifically, we will illustrate the extrapolation method using two completely different quantum models: the lattice polaron model and the mean-field Heisenberg model.

9.2.1 Lattice Polarons

The lattice polaron model describes low-energy excitations of a quantum particle hopping on a lattice coupled to the bosonic field provided by lattice phonons. We consider a quantum particle (often referred to as the 'bare' particle) in a one-dimensional lattice with $N \to \infty$ sites coupled to a phonon field:

$$\mathcal{H} = \sum_k \epsilon_k c_k^\dagger c_k + \sum_q \omega_q b_q^\dagger b_q + V_{\text{e-ph}}, \tag{9.1}$$

where c_k and b_q are the annihilation operators for the bare particle with momentum k and phonons with momentum q, $\epsilon_k = 2t\cos(k)$ is the energy of the bare particle and $\omega_q = \omega =$ const is the phonon frequency. The particle–phonon coupling is chosen to represent a combination of two qualitatively different polaron models:

$$V_{\text{e-ph}} = \alpha H_1 + \beta H_2, \tag{9.2}$$

where

$$H_1 = \sum_{k,q} \frac{2i}{\sqrt{N}} \left[\sin(k+q) - \sin(k)\right] c_{k+q}^\dagger c_k \left(b_{-q}^\dagger + b_q\right) \tag{9.3}$$

describes the Su–Schrieffer–Heeger (SSH) [46] particle–phonon coupling, and

$$H_2 = \sum_{k,q} \frac{2i}{\sqrt{N}} \sin(q) c^{\dagger}_{k+q} c_k \left(b^{\dagger}_{-q} + b_q \right) \tag{9.4}$$

is the breathing-mode model [47]. We will focus on two specific properties of the polaron in the ground state: the polaron momentum and the polaron effective mass. The ground state band of the model (9.1) represents polarons whose effective mass and ground-state momentum are known to exhibit two sharp transitions as the ratio α/β increases from zero to large values [48]. At $\alpha = 0$, the model (9.1) describes breathing-mode polarons, which have no sharp transitions [49]. At $\beta = 0$, the model (9.1) describes SSH polarons, whose effective mass and ground-state momentum exhibit one sharp transition in the polaron phase diagram [46]. At these transitions, the ground-state momentum and the effective mass of the polaron change abruptly.

9.2.2 The Heisenberg Model

The second model we consider here is the Heisenberg model

$$H = -\frac{J}{2} \sum_{\langle i,j \rangle} \mathbf{S}_i \cdot \mathbf{S}_j . \tag{9.5}$$

This model describes a lattice of interacting quantum spins S_i, which—depending on the strength of the interaction J—can be either aligned in the same direction (ferromagnetic phase) or oriented randomly leading to zero net magnetization (paramagnetic phase). The parameter J is the amplitude of the interaction and the $\langle .. \rangle$ brackets indicate that the interaction is non-zero only between nearest-neighbour spins.

Within a mean-field description, this many-body quantum system has free-energy density [50, 51]

$$f(T, m) \approx \frac{1}{2} \left(1 - \frac{T_c}{T} \right) m^2 + \frac{1}{12} \left(\frac{T_c}{T} \right)^3 m^4, \tag{9.6}$$

where m is the magnetization, T is the temperature and T_c is the critical temperature of the phase transition. At temperatures $T > T_c$, the model yields the paramagnetic phase, while $T < T_c$ corresponds to the ferromagnetic phase. The main property of interest here will be the order parameter. This property undergoes a sharp change at the critical temperature of the paramagnetic—ferromagnetic phase transition.

9.3 Gaussian Process Regression for Interpolation

The purpose of GP regression is to make a prediction of some quantity y at an arbitrary point $\boldsymbol{x} \in [\boldsymbol{x}_{\min}, \boldsymbol{x}_{\max}]$ of a $\mathcal{D}$-dimensional space, given a finite number of values $\boldsymbol{y} = (y_1, \ldots, y_n)^\top$, where y_i is the value of y at $\boldsymbol{x}_i$. Here, $\boldsymbol{x}_i$ is a $\mathcal{D}$-dimensional vector specifying a particular position in the input space and it is assumed that the values $\boldsymbol{x}_i$ sample the entire range $[\boldsymbol{x}_{\min}, \boldsymbol{x}_{\max}]$. If the training data are noiseless (as often will be the case for data coming from the solutions of physical equations), it is assumed that y is represented by a continuous function f that passes through the points y_i, so the vector of given results is $\boldsymbol{y} = (f(\boldsymbol{x}_1), \ldots, f(\boldsymbol{x}_n))^\top$. The goal is thus to infer the function $f(\boldsymbol{x})$ that interpolates the points $y_i \equiv f(x_i)$. The values y_i in the vector $\boldsymbol{y}$ represent the 'training data'.

GPs infer a distribution over functions $p(f|\boldsymbol{y})$ given the training data, as illustrated in Fig. 9.2. The left panel of Fig. 9.2 shows an example of the GP prior, i.e. the GP before the training. The right panel shows the GP conditioned by the training data (red dots). The GP is characterized by a mean function $\boldsymbol{\mu}(\boldsymbol{x})$ and covariance $\Sigma(\boldsymbol{x})$. The matrix elements of the covariance are defined as $\Sigma_{ij} = k(\boldsymbol{x}_i, \boldsymbol{x}_j)$, where $k(\cdot, \cdot)$ is a positively defined kernel function.

It is possible to derive the closed-form expressions for the conditional mean and variance of a GP [52], yielding

$$\mu(\boldsymbol{x}_*) = K(\boldsymbol{x}_*, \boldsymbol{x})^\top \left[K(\boldsymbol{x}, \boldsymbol{x}) + \sigma_n^2 I\right]^{-1} \boldsymbol{y} \tag{9.7}$$

$$\sigma(\boldsymbol{x}_*) = K(\boldsymbol{x}_*, \boldsymbol{x}_*) - K(\boldsymbol{x}_*, \boldsymbol{x})^\top \left[K(\boldsymbol{x}, \boldsymbol{x}) + \sigma_n^2 I\right]^{-1} K(\boldsymbol{x}_*, \boldsymbol{x}), \tag{9.8}$$

where $\boldsymbol{x}_*$ is a point in the input space where the prediction $\boldsymbol{y}_*$ is to be made; $K(\boldsymbol{x}, \boldsymbol{x})$ is the $n \times n$ square matrix with the elements $K_{i,j} = k(\boldsymbol{x}_i, \boldsymbol{x}_j)$ representing

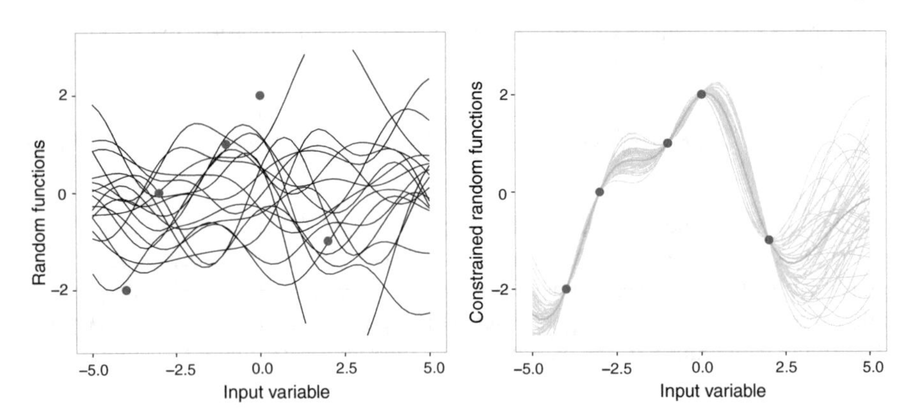

Fig. 9.2 Left: Gaussian process prior (grey curves). Right: Gaussian process (grey curves) conditioned by the training data (red dots). The green curve represents the mean of the GP posterior

the covariance between $y(\boldsymbol{x}_i)$ and $y(\boldsymbol{x}_j)$. The elements $k(\boldsymbol{x}_i, \boldsymbol{x}_j)$ are represented by the kernel function. Equation (9.7) can then be used to make the prediction of the quantity y at point $\boldsymbol{x}^*$, while Eq. (9.8) can be used to define the error of the prediction.

In this work, the GP models are trained by the results of quantum mechanical calculations. For the case of the polaron models considered here,

$$\boldsymbol{x}_i \Rightarrow \{\text{polaron momentum } K, \text{ Hamiltonian parameter } \alpha, \\ \text{Hamiltonian parameter } \beta, \text{phonon frequency } \omega\}.$$

For the case of the Heisenberg model considered here,

$$\boldsymbol{x}_i \Rightarrow \{\text{Temperature } T, \text{ magnetization } m\}$$

As already mentioned, $\boldsymbol{y} \Rightarrow f(\boldsymbol{x})$ is a vector of quantum mechanics results at the values of the parameters specified by $\boldsymbol{x}_i$. For the case of the polaron models considered here, $\boldsymbol{y} \Rightarrow$ polaron energy E. For the case of the Heisenberg model considered here, $\boldsymbol{y} \Rightarrow$ free energy density.

To train a GP model, it is necessary to assume some analytic form for the kernel function $k(\cdot, \cdot)$. In the present work, we will use the following analytic forms for the kernel functions:

$$k_{\text{LIN}}(\boldsymbol{x}_i, \boldsymbol{x}_j) = \boldsymbol{x}_i^\top \boldsymbol{x}_j + \ell \tag{9.9}$$

$$k_{\text{RBF}}(\boldsymbol{x}_i, \boldsymbol{x}_j) = \exp\left(-\frac{1}{2} r^2(\boldsymbol{x}_i, \boldsymbol{x}_j)\right) \tag{9.10}$$

$$k_{\text{MAT}}(\boldsymbol{x}_i, \boldsymbol{x}_j) = \left(1 + \sqrt{5}\, r(\boldsymbol{x}_i, \boldsymbol{x}_j) + \frac{5}{3}\, r^2(\boldsymbol{x}_i, \boldsymbol{x}_j)\right) \\ \times \exp\left(-\sqrt{5}\, r(\boldsymbol{x}_i, \boldsymbol{x}_j)\right) \tag{9.11}$$

$$k_{\text{RQ}}(\boldsymbol{x}_i, \boldsymbol{x}_j) = \left(1 + \frac{|\boldsymbol{x}_i - \boldsymbol{x}_j|^2}{2\alpha\ell^2}\right)^{-\alpha}, \tag{9.12}$$

where $r^2(\boldsymbol{x}_i, \boldsymbol{x}_j) = (\boldsymbol{x}_i - \boldsymbol{x}_j)^\top \times M \times (\boldsymbol{x}_i - \boldsymbol{x}_j)$ and M is a diagonal matrix with different length-scales ℓ_d for each dimension of $\boldsymbol{x}_i$. The unknown parameters of these functions are found by maximizing the log *marginal likelihood* function,

$$\log p(\boldsymbol{y}|X, \boldsymbol{\theta}) = -\frac{1}{2} \boldsymbol{y}^\top K^{-1} \boldsymbol{y} - \frac{1}{2} \log|K| - \frac{n}{2} \log(2\pi), \tag{9.13}$$

where $\boldsymbol{\theta}$ denotes collectively the parameters of the analytical function for $k(\cdot, \cdot)$ and $|K|$ is the determinant of the matrix K. $\boldsymbol{X}$ is known as the design matrix and contains the training points, $\{\boldsymbol{x}_i\}_{i=1}^N$. Given the kernel functions thus found, Eq. (9.7) is a GP model, which can be used to make a prediction by interpolation.

9.3.1 Model Selection Criteria

As Eq. (9.7) clearly shows, the GP models with different kernel functions will generally have a different predictive power. In principle, one could use the marginal likelihood as a metric to compare models with different kernels. However, different kernels have different numbers of free parameters and the second term of Eq. (9.13) directly depends on the number of parameters in the kernel. This makes the log marginal likelihood undesirable to compare kernels of different complexity.

As shown in Ref. [53], a better metric could be the Bayesian information criterion (BIC) defined as

$$\text{BIC}(\mathcal{M}_i) = \log p(\boldsymbol{y}|\boldsymbol{x}, \hat{\boldsymbol{\theta}}, \mathcal{M}_i) - \frac{1}{2}|\mathcal{M}_i| \log n, \tag{9.14}$$

where $|\mathcal{M}_i|$ is the number of kernel parameters of the kernel $\mathcal{M}_i$. In this equation, $p(\boldsymbol{y}|\boldsymbol{x}, \hat{\boldsymbol{\theta}}, \mathcal{M}_i)$ is the marginal likelihood for the optimized kernel $\hat{\boldsymbol{\theta}}$ which maximizes the logarithmic part. The assumption—one that will be tested in the present work for physics applications—is that more physical models have a larger BIC. The last term in Eq. (9.14) penalizes kernels with a larger number of parameters. The optimal BIC will thus correspond to the kernel yielding the largest value of the log marginal likelihood function with the fewest number of free parameters.

9.4 Physical Extrapolation by Generalization with Gaussian Processes

As shown Refs. [54, 55], one can use the BIC to increase the generalization power of GP models. The approach proposed in Refs. [54, 55] aims to build up the complexity of kernels, starting from the simple kernels (9.9)–(9.13), using a greedy search algorithm guided by the values of the BIC. Here, we employ this algorithm to extrapolate the quantum properties embodied in lattice models across phase transitions.

9.4.1 Learning with Kernel Combinations

The approach adopted here starts with the simple kernels (9.9)–(9.13). For each of the kernels, a GP model is constructed and the BIC is calculated. The kernel corresponding to the highest BIC is then selected as the best kernel. We will refer to such kernel as the 'base' kernel and denote it by k_0. The base kernel is then combined with each of the kernels (9.9)–(9.13). The new 'combined' kernels are chosen to be either of the sum form

$$c_0 k_0 + c_i k_i \tag{9.15}$$

or of the product form

$$c_i \times k_0 \times k_i, \tag{9.16}$$

where c_0 and c_i are treated as independent constants to be found by the maximization of the log marginal likelihood. The GP models with each of the new kernels are constructed and the BIC values are calculated. The kernel of the model with the largest BIC is then chosen as k_0 and the process is iterated. We thus have an 'optimal policy' algorithm [56] that selects the kernel assumed optimal based on the BIC at every step in the search.

We note that a similar procedure could be used to improve the accuracy of GP models for the interpolation problems. We have done this in one of our recent articles [57], where GP models were used to construct a six-dimensional potential energy surface for a chemically reactive complex with a very small number of training points. In the case of interpolation problems, it may also be possible to use cross-validation for kernel selection [58]. Cross-validation could also be applied to kernel selection for the extrapolation problems. We have not attempted to do this in the present work. We will compare the relative performance of the validation error and the BIC as the kernel selection metric in a future work.

9.5 Extrapolation of Quantum Properties

In this section, we present the results illustrating the performance of the algorithm described above for the prediction of the quantum properties of complex systems outside the range of the training data. Our particular focus is on predicting properties that undergo a sharp variation or discontinuity at certain values of the Hamiltonian parameters. Such properties cannot be directly modelled by GPs because the mean of a GP is a smooth, differentiable function.

The main idea [45] is to train a GP model with functions (obtained from the solutions of the Schrödinger equation) that vary smoothly across phase transitions and derive the properties undergoing sharp changes from such smoothly varying function. We thus expect this procedure to be generally applicable to extrapolation across second-order phase transitions. Here, we present two examples to illustrate this. The particular focus of the discussion presented below is on how the method converges to the accurate predictions as the complexity of the kernels increases.

9.5.1 Extrapolation Across Sharp Polaron Transitions

As discussed in Sect. 9.2, the Hamiltonian describing a quantum particle coupled to optical phonons through a combination of two couplings defined by Eq. (9.2) yields polarons with unusual properties. In particular, it was previously shown [48] that the ground-state momentum of such polarons undergoes two sharp transitions as the ratio α/β in Eq. (9.2) as well as the parameter $\lambda = 2\alpha^2/t\hbar\omega$ are varied. The dimensionless parameter λ is defined in terms of the bare particle hopping amplitude t and the phonon frequency ω. It quantifies the strength of coupling between the bare particle and the phonons. One can thus calculate the ground-state momentum or the effective mass of the polaron as a function of λ and α/β. The values of λ and α/β, where the polaron momentum and effective mass undergo sharp changes, separate the 'phases' of the Hamiltonian (9.1).

The GP models are trained by the *polaron energy dispersions* (i.e. the full curves of the dependence of the polaron energy on the polaron momentum) at different values of λ, α and β. These models are then used to generalize the full energy dispersions to values of λ, α and β outside the range of the training data and the momentum of the polaron with the lowest energy is calculated from these dispersion curves. The results are shown in Fig. 9.3. Each of the white dots in the phase diagrams depicted specifies the values of α, β and λ, for which the polaron dispersions were calculated and used as the training data. One can thus view the resulting GP models as four-dimensional, i.e. depending on α, β, λ and the polaron momentum.

Figure 9.3 illustrates two remarkable results:

- The left panel illustrates that the GP models are capable of predicting *multiple* new phase transitions by using the training data *entirely* in one single phase.

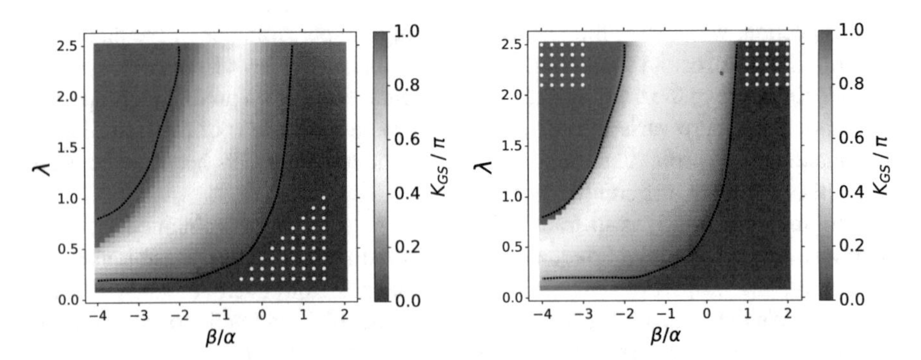

Fig. 9.3 Adapted with permission from Ref. [45], Copyright © APS, 2018. The polaron ground-state momentum K_{GS} for the mixed model (9.1) as a function of β/α for $\lambda = 2\alpha^2/t\hbar\omega$. The colour map is the prediction of the GP models. The curves are the quantum calculations from Ref. [48]. The models are trained by the polaron dispersions at the parameter values indicated by the white dots. The optimized kernel combination is $(k_{\text{MAT}} + k_{\text{RBF}}) \times k_{\text{LIN}}$ (left panel) and $(k_{\text{MAT}} \times k_{\text{LIN}} + k_{\text{RBF}}) \times k_{\text{LIN}}$ (right panel)

This proves our conjecture [45] that the evolution of physical properties with the Hamiltonian parameters in a single phase contains information about multiple phases and multiple phase transitions.

- While perhaps less surprising, the right panel illustrates that the accuracy of the predictions increases significantly and the predictions of the phase transitions become quantitative if the models are trained by data in two phases. The model illustrated in this panel extrapolates the polaron properties from high values of λ to low values of λ. Thus, the extrapolation becomes much more accurate if the models are trained by data in multiple phases.

In the following section we analyse how the kernel selection algorithm described in Ref. [45] and briefly above arrives at the models used for the predictions in Fig. 9.3.

9.5.2 Effect of Kernel Complexity

Figure 9.4 illustrates the performance of the models with kernels represented by a simple addition of two simple kernels, when trained by the data in two phases, as in the right panel of Fig. 9.3. The examination of this figure shows that the generalization accuracy, including the prediction of the number of the transitions, is sensitive to the kernel combination. For example, the models with the combination of the RBF and LIN kernels do not predict any transitions. Most of the other kernel combinations predict only one of the two transitions. Remarkably, the combination of two RBF kernels already leads to the appearance of the second transition, and allows the model to predict the location of the first transition quite accurately. The combination of Figs. 9.3 and 9.4 thus illustrates that the BIC is a meaningful metric to guide the kernel selection algorithm, as it rules out many of the kernels leading to incorrect phase diagrams shown in Fig. 9.4. The results in Fig. 9.4 also raise the question, how many combinations are required for kernels to allow quantitative predictions?

To answer this question, we show in Figs. 9.5 and 9.6 the convergence of the phase diagrams to the results in Fig. 9.3 with the number of iterations in the kernel selection algorithm. We use the following notation to label the figure panels: GPL-X, where X is the number of iteration. Thus, $X = 0$ corresponds to level zero of the kernel selection algorithm, i.e. GPL-0 is the phase diagram predicted by the model with the single simple kernel leading to the highest BIC. Level $X = 1$ corresponds to kernels constructed as the simple combinations (9.15) or (9.16). Level $X = 2$ corresponds to kernels of the form (9.15) or (9.16), where k_i is a combination of two kernels. As can be seen from Figs. 9.5 and 9.6, level $X = 2$ and $X = 3$ produce kernels with sufficient complexity for accurate predictions.

It must be noted that increasing the complexity of the kernels further (by increasing X) often decreases the accuracy of the predictions. This is illustrated in Fig. 9.7. We assume that this happens either due to overfitting or because the kernels become so complex that it is difficult to optimize them and the maximization of the log marginal likelihood gets stuck in a local maximum. To overcome this problem,

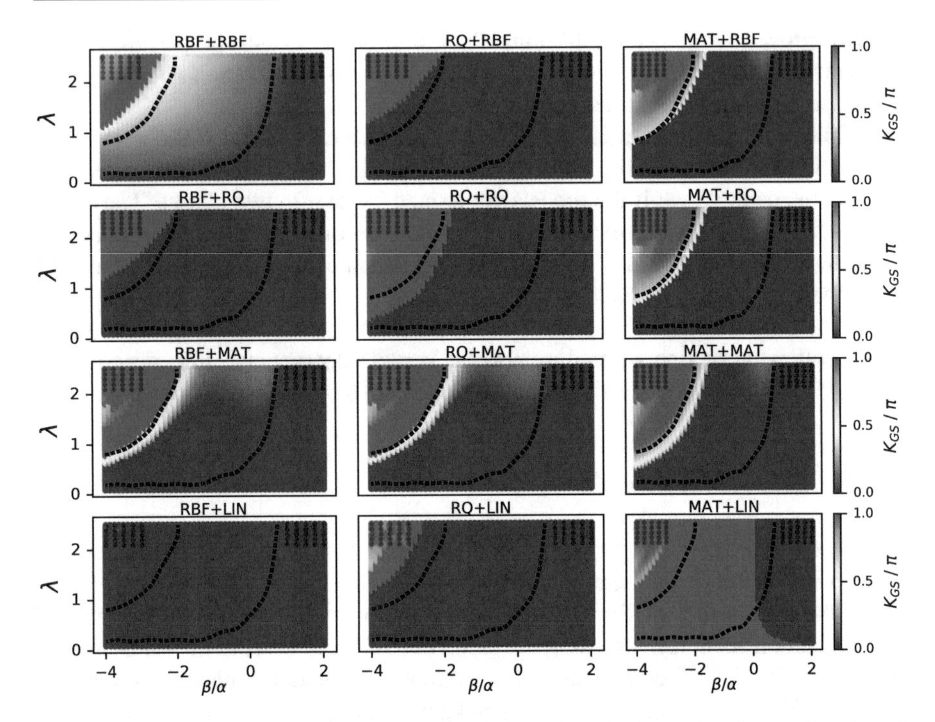

Fig. 9.4 The polaron ground-state momentum K_{GS} for the mixed model (9.1) as a function of β/α for $\lambda = 2\alpha^2/t\hbar\omega$. The black dashed curves are the calculations from Ref. [48]. The colour map is the prediction of the GP models with the fully optimized kernels. The models are trained by the polaron dispersions at the parameter values indicated by the black dots. The different kernels considered here are all possible pairwise additions (9.15) of two simple kernels from the family of kernels (k_{MAT}, k_{RQ} and k_{RBF})

one needs to optimize kernels multiple times starting from different conditions (either different sets of training data or different initial kernel parameters) and stop increasing the complexity of kernels when the optimization produces widely different results. Alternatively, the models could be validated by a part of the training data and the complexity of the kernels must be stopped at level X that corresponds to the minimal validation error, as often done to prevent overfitting with NNs [59].

9.5.3 Extrapolation Across Paramagnetic–Ferromagnetic Transition

In this section, we discuss the Heisenberg spin model described by the lattice Hamiltonian

$$\mathcal{H} = -\frac{J}{2}\sum_{\langle i,j\rangle} \mathbf{S}_i \cdot \mathbf{S}_j, \tag{9.17}$$

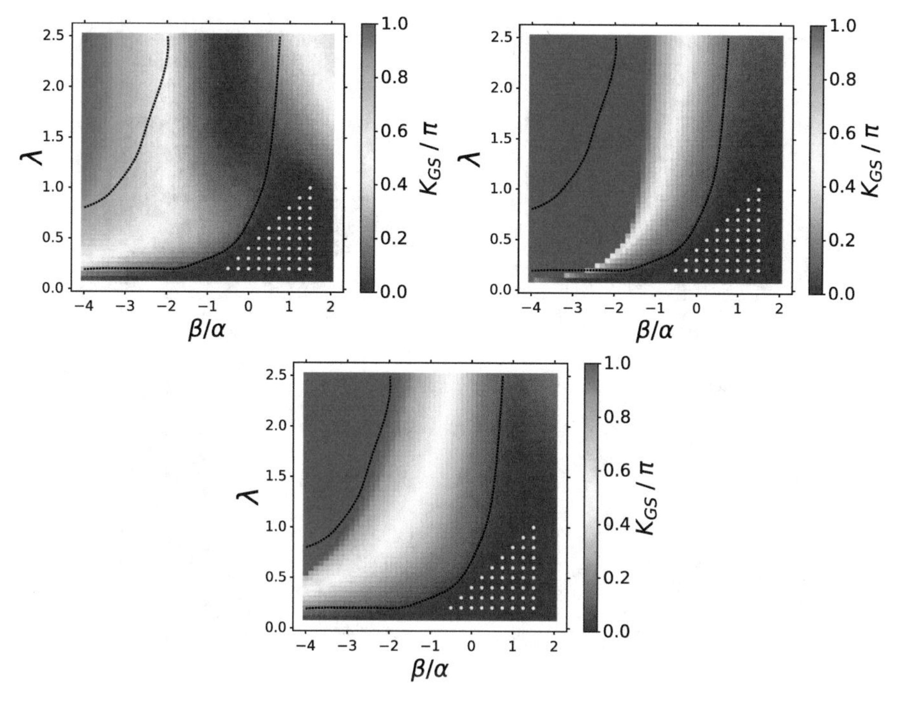

Fig. 9.5 Adapted from the supplementary material of Ref. [45]. Improvement of the phase diagram shown in Fig. 9.3 (upper panel) with the kernel complexity increasing as determined by the algorithm described in Sect. 9.4.1. The panels correspond to the optimized kernels GPL-0 (upper left), GPL-1 (upper right), GPL-2 (lowest panel), where "GPL-X" denotes the optimal kernel obtained after X depth levels

where $\langle i, j \rangle$ only account for nearest-neighbour interactions between different spins $\mathbf{S}_i$. The free energy of the system can be calculated within the mean-field approximation to yield

$$f(T, m) \approx \frac{1}{2}\left(1 - \frac{T_c}{T}\right) m^2 + \frac{1}{12}\left(\frac{T_c}{T}\right)^3 m^4, \tag{9.18}$$

where m is the magnetization and $T_c = 1.25$ is the critical temperature of the phase transition between the paramagnetic ($T > T_c$) and ferromagnetic ($T < T_c$) phase.

We train GP models by the entire free-energy curves at temperatures far above T_c. The free-energy curves are then predicted by the extrapolation models at temperatures decreasing to the other side of the transition. The order parameter m_0—defined as the value of magnetization that minimizes free energy—is then computed from the extrapolated predictions. The results are shown in Fig. 9.8.

As evident from Eq. (9.18), the free-energy curves have an analytic dependence on temperature T so this is a particularly interesting case for testing the generalization models. Can the kernel selection algorithm adopted here converge to a

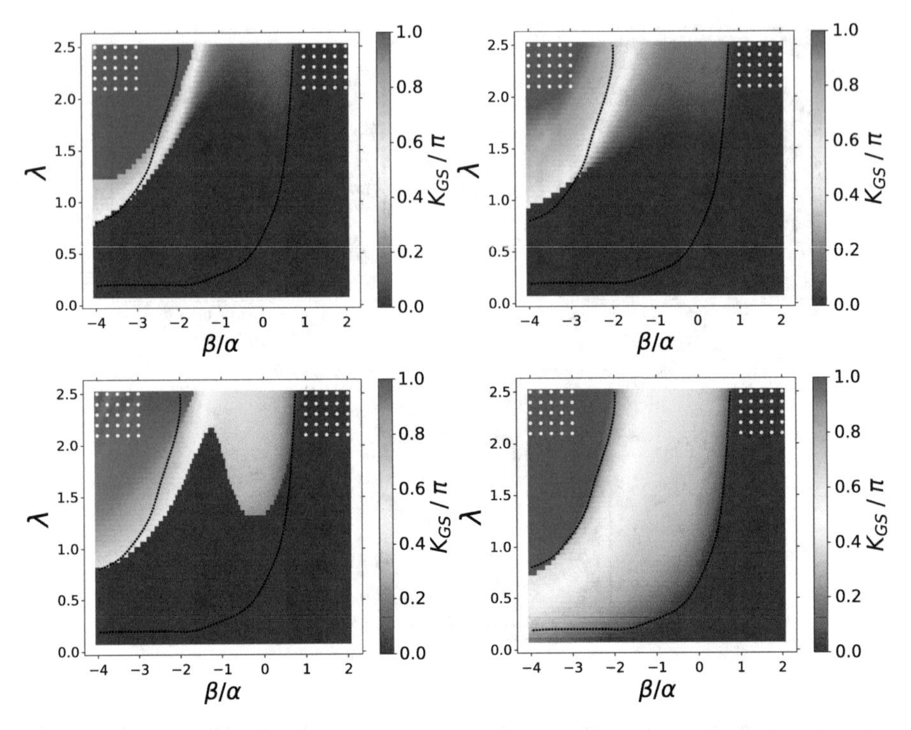

Fig. 9.6 Adapted from the supplementary material of Ref. [45]. Improvement of the phase diagram shown in Fig. 9.3 (lower panel) with the kernel complexity increasing as determined by the algorithm depicted in Sect. 9.4.1. The panels correspond to the optimized kernels GPL-0 (upper left), GPL-1 (upper right), GPL-2 (lower left), GPL-3 (lower right), where "GPL-X" denotes the optimal kernel obtained after X depth levels

model that will describe accurately the analytic dependence of the free energy (9.18) as well as the order parameter derived from it? We find that the temperature dependence of Eq. (9.18) can be rather well captured and accurately generalized by a model already with one simple kernel! However, this kernel must be carefully selected. As Fig. 9.9 illustrates, the accuracy of the free-energy prediction varies widely with the kernel. This translates directly into the accuracy of the order-parameter prediction illustrated by Fig. 9.10. Figure 9.10 illustrates that the RBF and RQ kernels capture the evolution of the order parameter quantitatively, while the LIN, MAT and quadratic kernels produce incorrect results.

Table 9.1 lists the BIC values for the models used to obtained the results depicted in Fig. 9.10, clearly demonstrating that the higher value of the BIC corresponds to the model with the better prediction power.

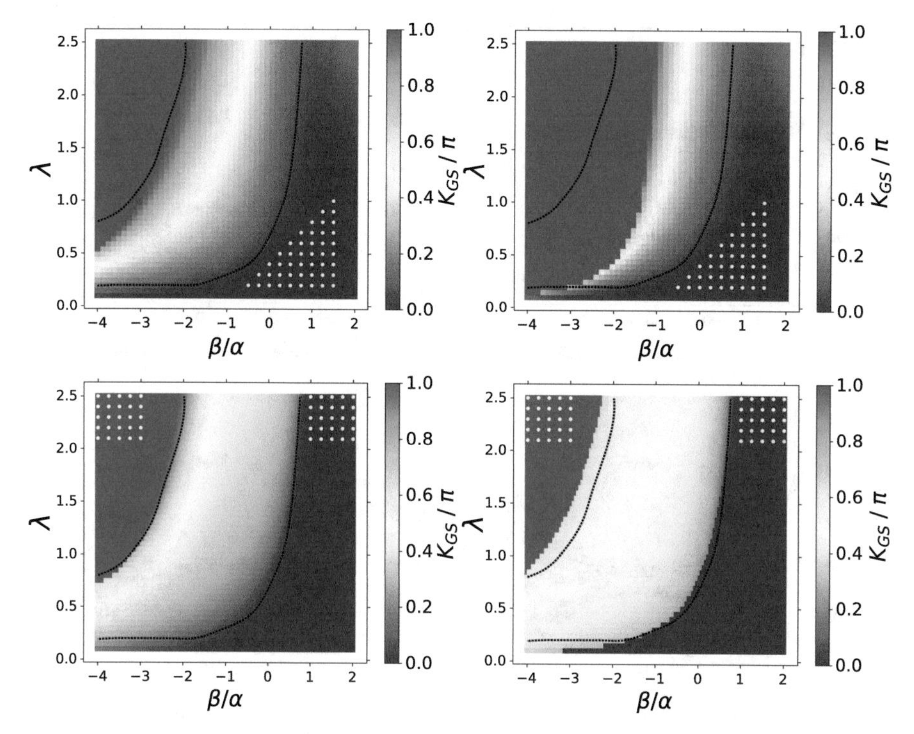

Fig. 9.7 Decrease of the prediction accuracy with increasing kernel complexity. Upper panels: left—GPL-2 (same as the right panel of Fig. 9.5), right—GPL-3. Lower panels: left—GPL-3 (same as the lower right panel of Fig. 9.6), right—GPL-4

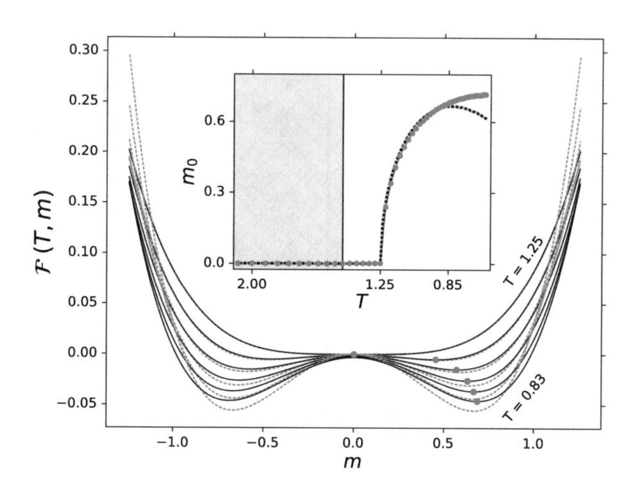

Fig. 9.8 Adapted with permission from Ref. [45], Copyright © APS, 2018. GP prediction (solid curves) of the free-energy density $f(T, m)$ of the mean-field Heisenberg model produced by Eq. (9.18) (dashed curves). Inset: the order parameter m_0 that minimizes $f(T, m)$: symbols—GP predictions, dashed curve—from Eq. (9.18). The GP models are trained with 330 points at $1.47 < T < 2.08$ (shaded area) and $-1.25 < m < 1.25$

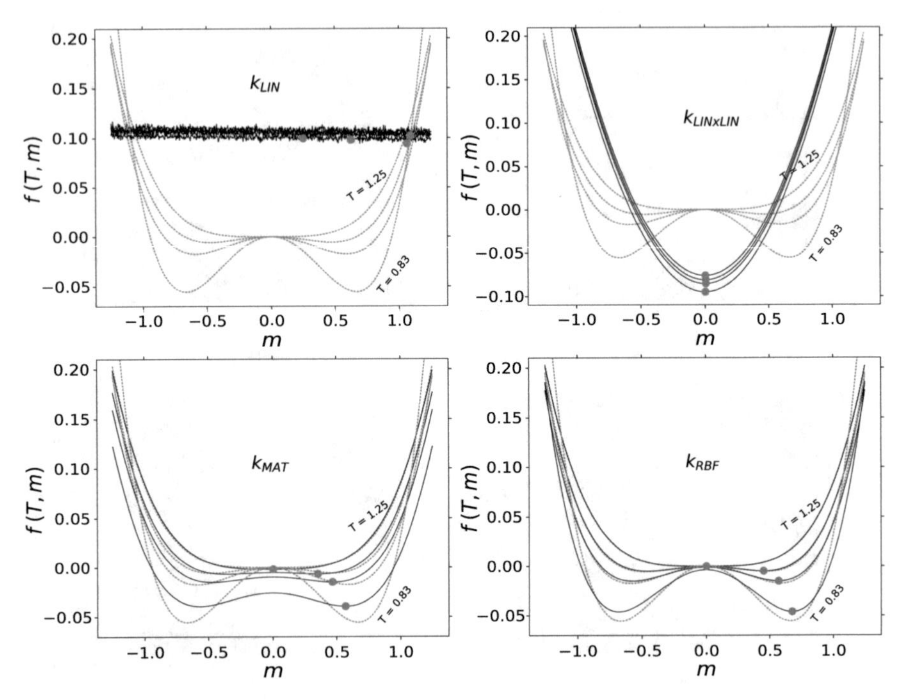

Fig. 9.9 GP prediction (solid curves) of the free-energy density $f(T, m)$ of the mean-field Heisenberg model produced by Eq. (9.18) (dashed curves). All GP models are trained with 330 points at $1.47 < T < 2.08$ (shaded area) and $-1.25 < m < 1.25$. The kernel function used in the GP models is indicated in each panel

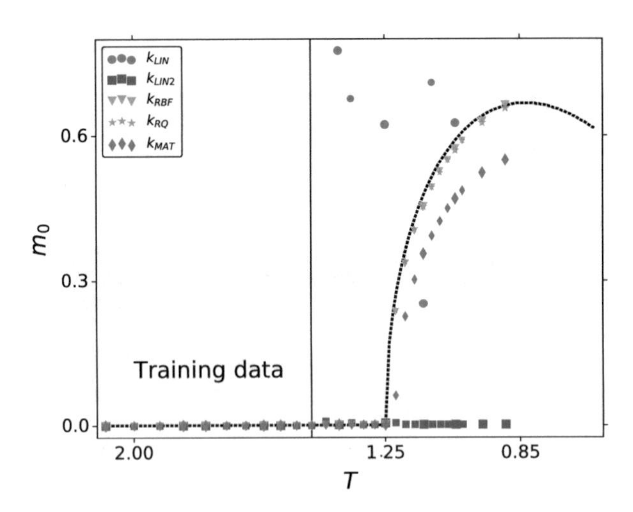

Fig. 9.10 The order parameter m_0 that minimizes $f(T, m)$: symbols—GP predictions, dashed curve—from Eq. (9.18). The order parameter m_0 is computed with the GP predictions using different kernels, illustrated in Fig. 9.9

Table 9.1 The numerical values of the BIC (9.14) for the models with different simple kernels (9.9)–(9.13) used for the predictions of the order parameter depicted in Fig. 9.10

Kernel type	BIC
RQ	8667.10
RBF	8657.13
MAT	7635.20
LIN	−104437128213.0
LIN × LIN	−10397873744.9

9.5.4 Validation of Extrapolation

Validation of the generalization predictions in the extrapolated region presents a major problem. By definition, there are no data in the extrapolated region. One can, of course, divide the given data into a training set and a validation set outside of the training data range. The validation set can then be used to verify the accuracy of the extrapolation. This is what is done throughout this work. However, this does not guarantee the accuracy of the predictions beyond the range of the validation data. Finding a proper method to validate the extrapolation predictions is particularly important for applications of the present approach to making predictions of observables at physical parameters, where no theoretical or experimental results are available.

A possible way to verify the accuracy of the extrapolation predictions without using data in the extrapolated region is to examine the sensitivity of the predictions to the *positions* and *number* of training points. If the predictions are stable to variations of the training data, one might argue that the predictions are valid. To illustrate this, we rebuild the models of the phase diagram depicted in Fig. 9.1 with a variable number of training points. Figure 9.11 shows the results obtained with models trained by the quantum calculations at different values of λ and α/β. The figure illustrates the following:

- The generalization models capture both transitions even when trained by the quantum calculations far removed from the transition line and with a random distribution of training points.
- The predictions of the transitions become more accurate as the distribution of the training points approaches the first transition line.

One may thus conclude that the predictions of the sharp transitions are physical. If possible, this can be further validated by training generalization models with data in a completely different part of the phase diagram. This is done in Fig. 9.12 that shows the same phase diagram obtained by the generalization models trained with quantum results entirely in the middle phase. Given the remarkable agreement of the results in the last panel of Fig. 9.11 and in the last panel of Fig. 9.12, one can argue that the predicted phase diagram is accurate.

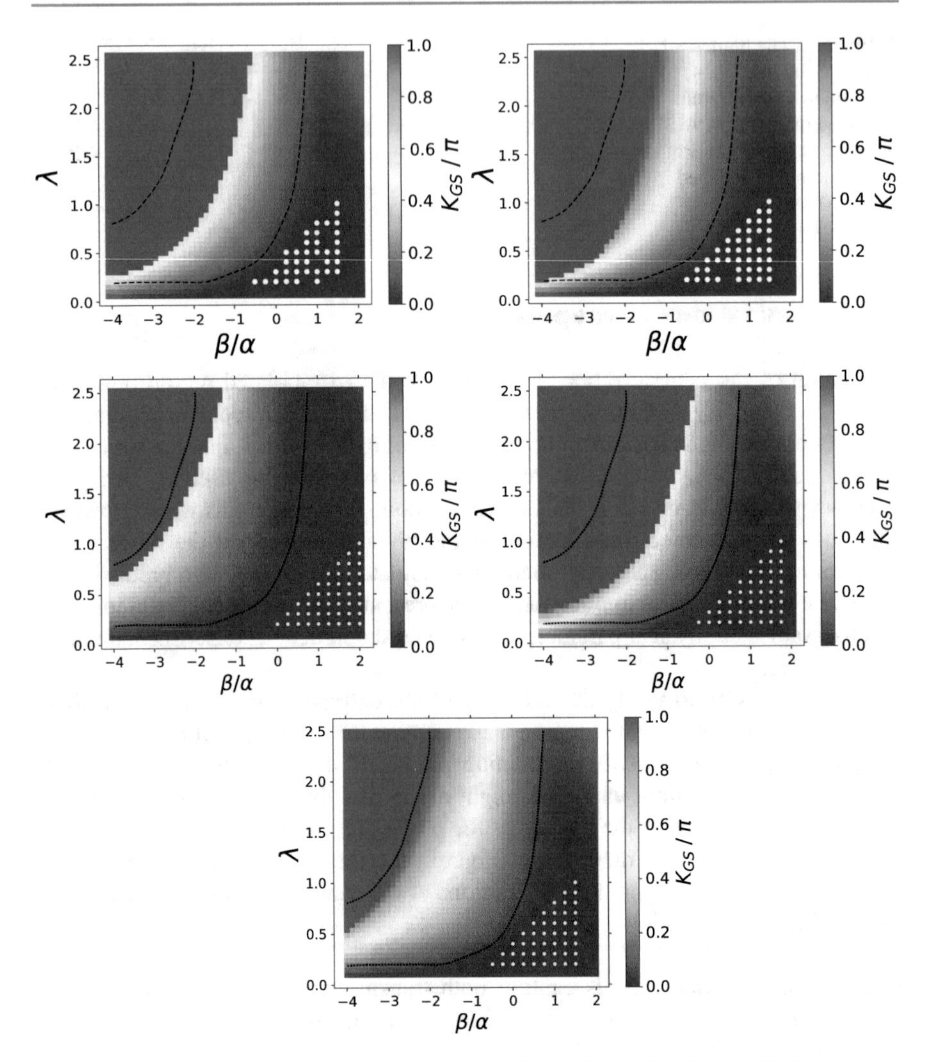

Fig. 9.11 Dependence of the prediction accuracy on the number and positions of training points. The white dots indicate the values of the parameters λ and α/β, at which the quantum properties were calculated for training the GP models. All results are computed with optimal kernels with the same complexity level GPL-2

Based on these results, we suggest the following algorithm to make stable predictions of unknown phase transitions by extrapolation:

1. Sample the phase diagram with a cluster of training points at random.
2. Identify the phase transitions by extrapolation in all directions.
3. Move the cluster of the training points towards any predicted transition.

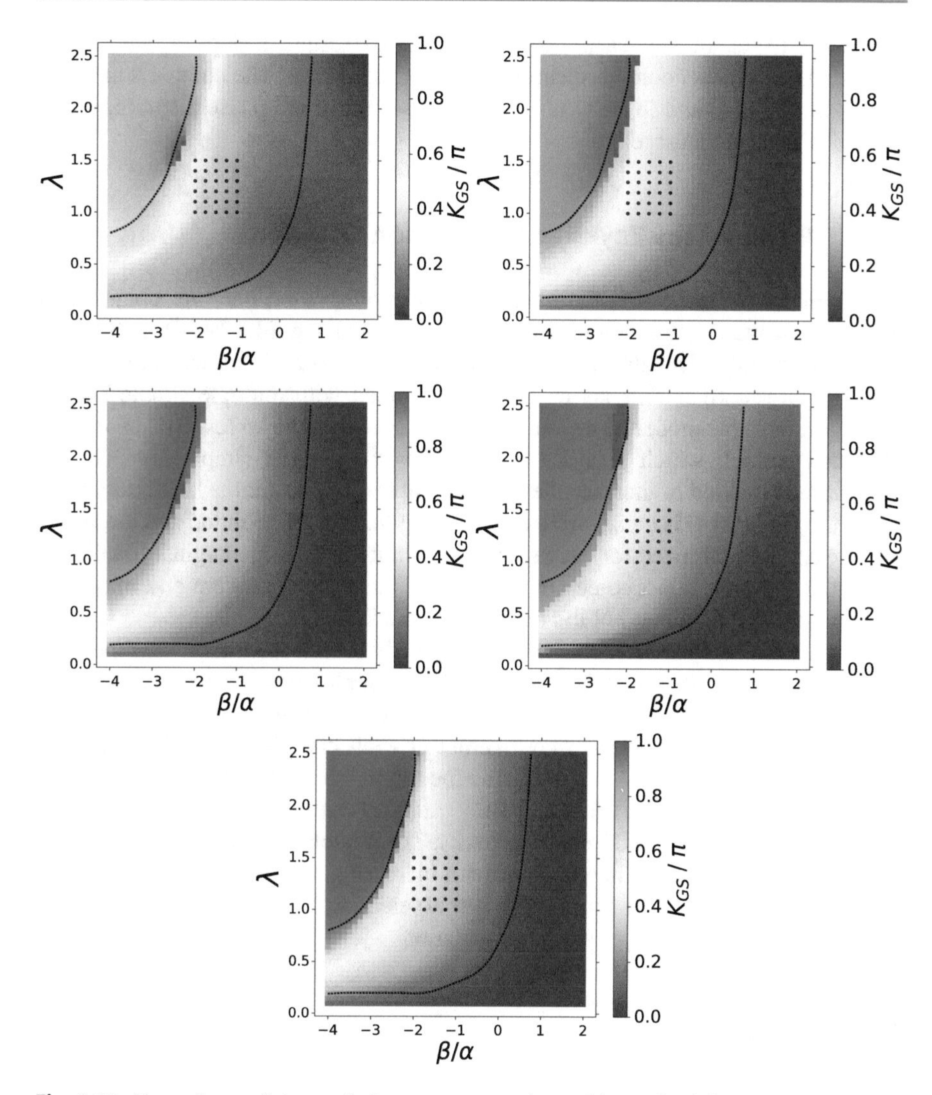

Fig. 9.12 Dependence of the prediction accuracy on the positions of training points. The black dots indicate the values of the parameters λ and α/β, at which the quantum properties were calculated for training the GP models. The different panels correspond to the optimal kernels with the complexity level ranging from GPL-0 (upper left) to GPL-4 (lowest panel)

4. Repeat the calculations until the predictions do not change with the change of the training point distributions.
5. If possible, rebuild the models with training points in a completely different part of the phase diagram.

While step (5) is not necessary, the agreement of the results in steps (4) and (5) can be used as an independent verification of the extrapolation. The comparison of the results in steps (4) and (5) may also point to the part of the phase diagram, where the predictions are least reliable.

9.5.5 Power of the Bayesian Information Criterion

As explained in Sect. 9.4.1, the generalization models used here are obtained by gradually increasing the complexity of kernels using the BIC (9.14) as a kernel selection criterion. The algorithm starts with a simple kernel that leads to a model with the largest BIC. This kernel is then combined with multiple simple kernels leading to multiple models. The kernel of the model with the largest BIC is selected as a new kernel, which is again combined with multiple simple kernels. The procedure is iterated to increase the kernel complexity, one simple kernel at a time. Since the BIC (9.14) is closely related to the log marginal likelihood and the kernel parameters are optimized for each step by maximizing the log marginal likelihood, why not to simply select some complex kernel function at random and maximize the log marginal likelihood of the model with this kernel?

To illustrate the power of the BIC in the greedy search algorithm, we repeat the calculations of Fig. 9.12 with kernels of various complexity selected at random. We mimic the iterative process used above, but instead of using the BIC as a kernel selection criterion, we select a new kernel at each complexity level at random. Every model is optimized by maximizing the log marginal likelihood as usual. The results are depicted in Fig. 9.13. The different panels of Fig. 9.13 are obtained with models using kernels of different complexity. The models are not physical and there is no evidence of model improvement with increasing kernel complexity. We thus conclude that the BIC is essential as the kernel selection criterion to build GP models for applications targeting physical extrapolation.

9.6 Conclusion

The present article presents clear evidence that Gaussian process models can be designed to predict the physical properties of complex quantum systems outside the range of the training data. As argued in Ref. [60], the generalization power of GP models in the extrapolated region is likely a consequence of the Bayesian approach to machine learning that underlies GP regression. For this reason, the authors believe that Bayesian machine learning has much potential for applications in physics and chemistry [60]. As illustrated here, it can be used as a new discovery tool of physical properties, potentially under conditions, where neither theory nor experiment are feasible.

The generalization models discussed here can also be used to guide rigorous theory in search of specific phenomena (such as phase transitions) and/or particular properties of complex systems. Generating the phase diagram, such as the one

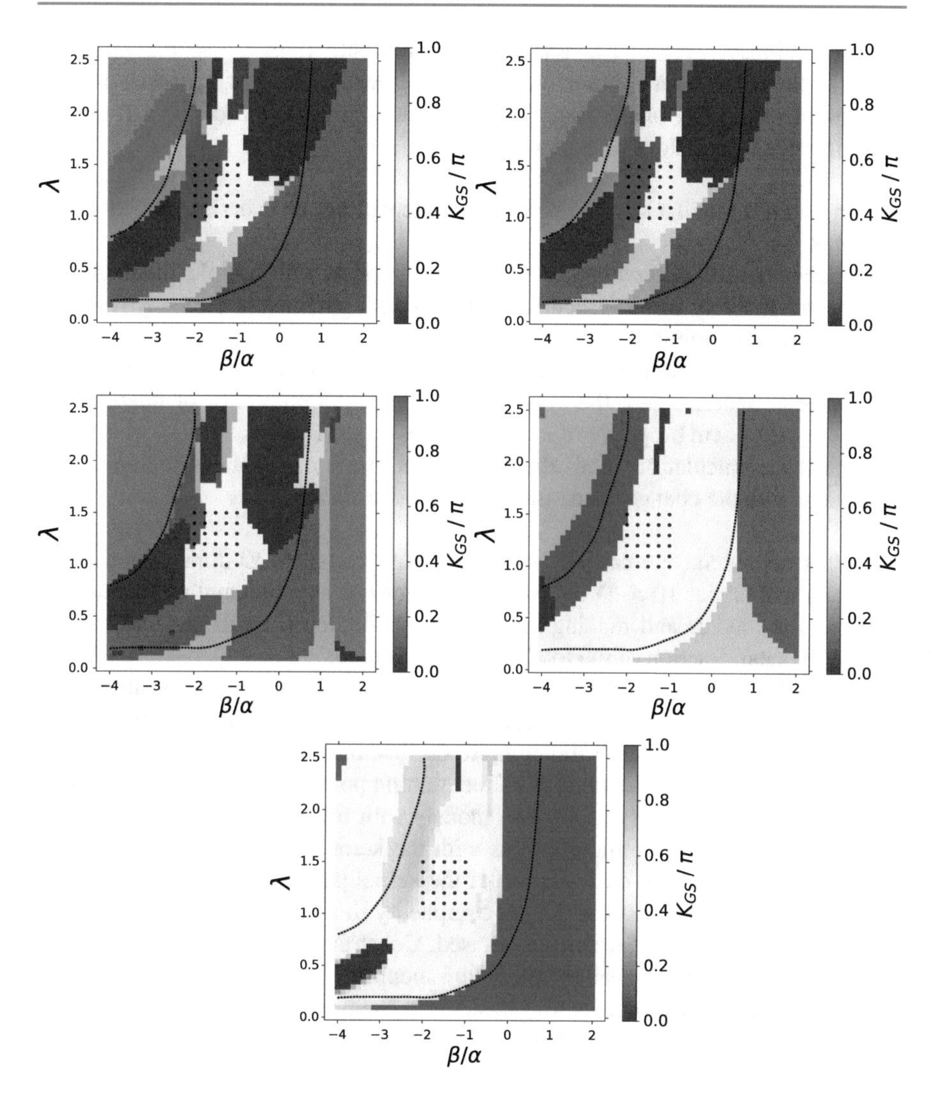

Fig. 9.13 Predictions obtained with randomly selected kernels. The black dots indicate the values of the parameters λ and α/β, at which the quantum properties were calculated for training the GP models. The different panels correspond to the optimal kernels with the complexity level ranging from GPL-0 (upper left) to GPL-4 (lowest panel). The initial kernel is selected at random. The kernel at the next complexity level GPL-X is obtained by combining the kernel from the previous complexity level with another randomly selected kernel. The parameters of the kernels thus obtained are optimized using the maximization of the log marginal likelihood. This procedure illustrates the importance of the BIC for the selection of the type of the kernel function

depicted in Fig. 9.1, presents no computational difficulty (taking essentially minutes of CPU time). One can thus envision the following efficient approach for the generation of the full phase diagrams based on a combination of the GP models with rigorous calculations or experiments:

1. Start with a small number of rigorous calculations or experimental measurements.
2. Generate the full phase diagram with the GP models with complex kernels. This diagram is likely to be inaccurate at the system parameters far away from the initial training points.
3. Use rigorous calculations or experiments to add training points in the parts of the parameter space, where the system exhibits desired properties of interest; and where the system properties undergo the most rapid change.
4. Repeat the calculations until the predictions in the extrapolated region do not change with the change of the training point distributions.

With this approach, one can envision generating complete $\mathcal{D}$-dimensional phase diagrams with about $10 \times \mathcal{D}$ rigorous calculations or experimental measurements. Training the models and making the predictions in step (2) will generally take a negligibly small fraction of the total computation time.

It should be pointed out that the results presented in this work suggest algorithms to construct complex GP models capable of meaningful predictions in the extrapolated region without direct validation. To do this, one can examine the sensitivity of the predictions to the distribution of the training points for models with the same level of kernel complexity as well as models with different complexity. Increase of the sensitivity to the training points with the kernel complexity would suggest overfitting or insufficient optimization of the kernel parameters. In such cases, the iterative process building up the kernel complexity should be stopped or the process of optimizing the kernel parameters revised. Constructing algorithms for physical extrapolation without the need for validation should be the ultimate goal of the effort aimed at designing ML models for physics and chemistry. Such models could then use all available chemistry and physics information to make meaningful discoveries.

Acknowledgments We thank Mona Berciu for the quantum results used for training and verifying the ML models for the polaron problem. We thank John Sous and Mona Berciu for the ideas that had led to work published in Ref. [45] and for enlightening discussions.

References

1. J.N. Murrell, S. Carter, S.C. Farantos, P. Huxley, A.J.C. Varandas, *Molecular Potential Energy Functions* (Wiley, Chichester, 1984)
2. T. Hollebeek, T.-S. Ho, H. Rabitz, Annu. Rev. Phys. Chem. **50**, 537 (1999)
3. B.J. Braams, J.M. Bowman, Int. Rev. Phys. Chem. **28**, 577 (2009)
4. M.A. Collins, Theor. Chem. Acc. **108**, 313 (2002)
5. C.M. Handley, P.L.A. Popelier, J. Phys. Chem. A **114**, 3371 (2010)

6. S. Manzhos, T. Carrington Jr., J. Chem. Phys. **125**, 194105 (2006)
7. J. Cui, R.V. Krems, Phys. Rev. Lett. **115**, 073202 (2015)
8. J. Cui, R.V. Krems, J. Phys. B **49**, 224001 (2016)
9. R.A. Vargas-Hernández, Y. Guan, D.H. Zhang, R.V. Krems, New J. Phys. **21**, 022001 (2019)
10. A. Kamath, R.A. Vargas-Hernández, R.V. Krems, T. Carrington Jr., S. Manzhos, J. Chem. Phys. **148**, 241702 (2018)
11. C. Qu, Q. Yu, B.L. Van Hoozen Jr., J.M. Bowman, R.A. Vargas-Hernández, J. Chem. Theory Comp. **14**, 3381 (2018)
12. L. Wang, Phys. Rev. B **94**, 195105 (2016)
13. J. Carrasquilla, R.G. Melko, Nat. Phys. **13**, 431 (2017)
14. E.P.L. van Nieuwenburg, Y.-H. Liu, S.D. Huber, Nat. Phys. **13**, 435 (2017)
15. P. Broecker, F. Assaad, S. Trebst, (2017). arXiv:1707.00663
16. S.J. Wetzel, M. Scherzer, Phys. Rev. B **96**, 184410 (2017)
17. S.J. Wetzel, Phys. Rev. E **96**, 022140 (2017)
18. Y.-H. Liu, E.P.L. van Nieuwenburg, Phys. Rev. Lett. **120**, 176401 (2018)
19. K. Chang, J. Carrasquilla, R.G. Melko, E. Khatami, Phys. Rev. X **7**, 031038 (2017)
20. P. Broecker, J. Carrasquilla, R.G. Melko, S. Trebst, Sci. Rep. **7**, 8823 (2017)
21. F. Schindler, N. Regnault, T. Neupert, Phys. Rev. B **95**, 245134 (2017)
22. T. Ohtsuki, T. Ohtsuki, J. Phys. Soc. Jpn **85**, 123706 (2016)
23. L.-F. Arsenault, A. Lopez-Bezanilla, O.A. von Lilienfeld, A.J. Millis, Phys. Rev. B **90**, 155136 (2014)
24. L.-F. Arsenault, O.A. von Lilienfeld, A.J. Millis, (2015). arXiv:1506.08858
25. M.J. Beach, A. Golubeva, R.G. Melko, Phys. Rev. B **97**, 045207 (2018)
26. E. van Nieuwenburg, E. Bairey, G. Refael, Phys. Rev. B **98**, 060301(R) (2018)
27. N. Yoshioka, Y. Akagi, H. Katsura, Phys. Rev. B **97**, 205110 (2018)
28. J. Venderley, V. Khemani, E.-A. Kim, Phys. Rev. Lett. **120**, 257204 (2018)
29. G. Carleo, M. Troyer, Science **355**, 602 (2017)
30. M. Schmitt, M. Heyl, SciPost Phys. **4**, 013 (2018)
31. Z. Cai, J. Liu, Phys. Rev. B **97**, 035116 (2017)
32. Y. Huang, J.E. Moore, (2017). arXiv:1701.06246
33. D.-L. Deng, X. Li, S.D. Sarma, Phys. Rev. B **96**, 195145 (2017)
34. Y. Nomura, A. Darmawan, Y. Yamaji, M. Imada, Phys. Rev. B **96**, 205152 (2017)
35. D.-L. Deng, X. Li, S.D. Sarma, Phys. Rev. X **7**, 021021 (2017)
36. X. Gao, L.-M. Duan, Nat. Commun. **8**, 662 (2017)
37. G. Torlai, G. Mazzola, J. Carrasquilla, M. Troyer, R. Melko, G. Carleo, Nat. Phys. **14**, 447 (2018)
38. T. Hazan, T. Jaakkola, (2015). arXiv:1508.05133
39. A. Daniely, R. Frostig, Y. Singer, NIPS **29**, 2253 (2016)
40. J. Lee, Y. Bahri, R. Novak, S.S. Schoenholz, J. Pennington, J. Sohl-Dickstein, Deep neural networks as Gaussian processes, in *ICLR* (2018)
41. K.T. Schütt, H. Glawe, F. Brockherde, A. Sanna, K.R. Müller, E.K.U. Gross, Phys. Rev. B **89**, 205118 (2014)
42. L.M. Ghiringhelli, J. Vybiral, S.V. Levchenko, C. Draxl, M. Scheffler, Phys. Rev. Lett. **114**, 105503 (2015)
43. F.A. Faber, A. Lindmaa, O.A, von Lilienfeld, R. Armient, Int. J. Quantum Chem. **115**, 1094 (2015)
44. F.A. Faber, A. Lindmaa, O.A, von Lilienfeld, R. Armient, Phys. Rev. Lett. **117**, 135502 (2016)
45. R.A. Vargas-Hernández, J. Sous, M. Berciu, R.V. Krems, Phys. Rev. Lett. **121**, 255702 (2018)
46. D.J.J. Marchand, G. De Filippis, V. Cataudella, M. Berciu, N. Nagaosa, N.V. Prokof'ev, A.S. Mishchenko, P.C.E. Stamp, Phys. Rev. Lett. **105**, 266605 (2010)
47. B. Lau, M. Berciu, G.A. Sawatzky, Phys. Rev. B **76**, 174305 (2007)
48. F. Herrera, K.W. Madison, R.V. Krems, M. Berciu, Phys. Rev. Lett. **110**, 223002 (2013)
49. B. Gerlach, H. Löwen, Rev. Mod. Phys. **63**, 63 (1991)

50. P.M. Chaikin, T.C. Lubensky, *Principles of Condensed Matter Physics* (Cambridge University Press, Cambridge, 1998)
51. S. Sachdev, *Quantum Phase Transitions* (Cambridge University Press, Cambridge, 1999)
52. C.E. Rasmussen, C.K.I. Williams, *Gaussian Process for Machine Learning* (MIT Press, Cambridge, 2006)
53. G. Schwarz, Ann. Stat. **6**(2), 461 (1978)
54. D.K. Duvenaud, H. Nickisch, C.E. Rasmussen, Adv. Neural Inf. Proces. Syst. **24**, 226 (2011)
55. D.K. Duvenaud, J. Lloyd, R. Grosse, J.B. Tenenbaum, Z. Ghahramani, *Proceedings of the 30th International Conference on Machine Learning Research*, vol. 28 (2013), p. 1166
56. R.S. Sutton, A.G. Barto, *Reinforcement Learning: An Introduction* (MIT Press, Cambridge, 2016)
57. J. Dai, R.V. Krems, J. Chem. Theory Comput. **16**(3), 1386–1395 (2020). arXiv:1907.08717
58. A. Christianen, T. Karman, R.A. Vargas-Hernández, G.C. Groenenboom, R.V. Krems, J. Chem. Phys. **150**, 064106 (2019)
59. N. Srivastava, G. Hinton, A. Krizhevsky, I. Sutskever, R. Salakhutdinov, J. Mach. Learn. Res. **15**, 1929 (2014)
60. R.V. Krems, Bayesian machine learning for quantum molecular dynamics. Phys. Chem. Chem. Phys. **21**, 13392 (2019)

Part III

Deep Learning of Atomistic Representations

Atomistic Representations: Preface

Deep learning has led to several breakthroughs in applications such as computer vision or natural language processing in recent years [1]. Neural networks can handle millions of training examples and can easily be parallelized on modern *graphics processing units* (GPUs) and distributed over large computer clusters. While the machine learning methods introduced in the last part of this book rely on the hand-crafted descriptors or kernels, neural networks owe a large part of their success to the ability to learn powerful multi-scale representations directly from structured data—a paradigm called *end-to-end learning*. Just like images or text, atomistic systems are structured data: their atoms correspond to pixels or words, and neighboring atoms generally tend to interact stronger than those at larger distances. Even though neural networks have been used to parametrize potential energy surfaces before [2–6], these methods still rely on manually crafted features. Only recently have end-to-end neural networks been applied to learn representations of molecules and materials.

This part reviews two popular frameworks—deep tensor neural networks (DTNN) [7] and message-passing neural networks (MPNN) [8]—which were both extended and generalized by later architectures [9–16]. While mathematically similar, both frameworks arrived from different underlying concepts: an encoding of the many-body expansion in a neural network [5, 17] for DTNNs versus the learning of graph representations [18, 19] in the case of MPNNs. Chapter 10 [20] reviews the MPNN framework including its numerous variants in the context of predicting molecular properties. It describes the building blocks of MPNNs in detail and covers both learning on atomic position and using only molecular graph connectivity. The deep tensor neural network SchNet [21], as presented in Chap. 11 [22], models atomic interactions using continuous, spatial convolution filters while encoding prior information, as described in Part II of this book, into an end-to-end neural network architecture. A focus of the chapter is the analysis of the learned representation to gain insights regarding what the model has learned about molecules and materials in the spirit of explainable AI [23]. Combined with the techniques to include prior knowledge presented in the previous part, these chapters provide the knowledge to design interpretable neural network architectures yielding accurate predictions for quantum simulations.

Berlin, Germany — Kristof T. Schütt
Berlin, Germany — Stefan Chmiela
Basel, Switzerland — O. Anatole von Lilienfeld
Luxembourg, Luxembourg — Alexandre Tkatchenko
Kashiwa, Japan — Koji Tsuda
Berlin, Germany — Klaus-Robert Müller
September 2019

References

1. Y. LeCun, Y. Bengio, G. Hinton, Nature **521**(7553), 436 (2015)
2. T.B. Blank, S.D. Brown, A.W. Calhoun, D.J. Doren, J. Chem. Phys. **103**(10), 4129 (1995)
3. S. Manzhos, T. Carrington Jr, J. Chem. Phys. **125**(8), 084109 (2006)
4. J. Behler, M. Parrinello, Phys. Rev. Lett. **98**(14), 146401 (2007)
5. M. Malshe, R. Narulkar, L.M. Raff, M. Hagan, S. Bukkapatnam, P.M. Agrawal, R. Komanduri, J. Chem. Phys. **130**(18), 184102 (2009)
6. A. Pukrittayakamee, M. Malshe, M. Hagan, L. Raff, R. Narulkar, S. Bukkapatnum, R. Komanduri, J. Chem. Phys. **130**(13), 134101 (2009)
7. K.T. Schütt, F. Arbabzadah, S. Chmiela, K.-R. Müller, A. Tkatchenko, Nat. Commun. **8**, 13890 (2017)
8. J. Gilmer, S.S. Schoenholz, P.F. Riley, O. Vinyals, G.E. Dahl, *Proceedings of the 34th International Conference on Machine Learning* (2017), pp. 1263–1272
9. N. Lubbers, J.S. Smith, K. Barros, J. Chem. Phys. **148**(24), 241715 (2018)
10. P.B. Jørgensen, K.W. Jacobsen, M.N. Schmidt (2018). Preprint. arXiv:1806.03146
11. T.S. Hy, S. Trivedi, H. Pan, B.M. Anderson, R. Kondor, J. Chem. Phys. **148**(24), 241745 (2018)
12. N. Thomas, T. Smidt, S. Kearnes, L. Yang, L. Li, K. Kohlhoff, P. Riley (2018). Preprint. arXiv:1802.08219
13. C. Chen, W. Ye, Y. Zuo, C. Zheng, S.P. Ong, Chem. Mater. **31**(9), 3564 (2019)
14. O.T. Unke, M. Meuwly, PhysNet: a neural network for predicting energies, forces, dipole moments, and partial charges. J. Chem. Theory Comput. **15**(6), 3678–3693 (2019)
15. N.W. Gebauer, M. Gastegger, K.T. Schütt (2019). Preprint. arXiv:1906.00957
16. K.T. Schütt, M. Gastegger, A. Tkatchenko et al., Unifying machine learning and quantum chemistry with a deep neural network for molecular wavefunctions. Nat. Commun. **10**, 5024 (2019). https://doi.org/10.1038/s41467-019-12875-2
17. B.J. Braams, J.M. Bowman, Int. Rev. Phys. Chem. **28**(4), 577 (2009)
18. F. Scarselli, M. Gori, A.C. Tsoi, M. Hagenbuchner, G. Monfardini, IEEE Trans. Neural Netw. **20**(1), 61 (2009)
19. D.K. Duvenaud, D. Maclaurin, J. Iparraguirre, R. Bombarell, T. Hirzel, A. Aspuru-Guzik, R.P. Adams, in *NIPS*, ed. by C. Cortes, N.D. Lawrence, D.D. Lee, M. Sugiyama, R. Garnett (Curran Associates, Red Hook, 2015), pp. 2224–2232
20. J. Gilmer, S.S. Schoenholz, P.F. Riley, O. Vinyals, G.E. Dahl, in *Machine Learning for Quantum Simulations of Molecules and Materials*, ed. by K.T. Schütt, S. Chmiela, A. von Lilienfeld, A. Tkatchenko, K. Tsuda, K.-R. Müller. Lecture Notes Physics (Springer, Berlin, 2019)
21. K.T. Schütt, H.E. Sauceda, P.-J. Kindermans, A. Tkatchenko, K.-R. Müller, J. Chem. Phys. **148**(24), 241722 (2018)

22. K.T. Schütt, A. Tkatchenko, K.-R. Müller, in *Machine Learning for Quantum Simulations of Molecules and Materials*, ed. by K.T. Schütt, S. Chmiela, A. von Lilienfeld, A. Tkatchenko, K. Tsuda, K.-R. Müller. Lecture Notes Physics (Springer, Berlin, 2019)
23. W. Samek, G. Montavon, A. Vedaldi, L.K. Hansen, K.-R. Müller, *Explainable AI: Interpreting, Explaining and Visualizing Deep Learning*, vol. 11700 (Springer, Cham, 2019)

Message Passing Neural Networks 10

Justin Gilmer, Samuel S. Schoenholz, Patrick F. Riley, Oriol Vinyals, and George E. Dahl

Abstract

Supervised learning on molecules has incredible potential to be useful in chemistry, drug discovery, and materials science. Luckily, several promising and closely related neural network models invariant to molecular symmetries have already been described in the literature. These models learn a message passing algorithm and aggregation procedure to compute a function of their entire input graph. In this chapter, we describe a general common framework for learning representations on graph data called message passing neural networks (MPNNs) and show how several prior neural network models for graph data fit into this framework. This chapter contains large overlap with Gilmer et al. (International Conference on Machine Learning, pp. 1263–1272, 2017), and has been modified to highlight more recent extensions to the MPNN framework.

10.1 Introduction

The past decade has seen remarkable success in the use of deep neural networks to understand and translate natural language [1], generate and decode complex audio

J. Gilmer (✉) · S. S. Schoenholz · G. E. Dahl
Google Brain, Mountain View, CA, USA
e-mail: gilmer@google.com; schsam@google.com; gdahl@google.com

P. F. Riley
Google, Mountain View, CA, USA
e-mail: pfr@google.com

O. Vinyals
DeepMind, London, UK
e-mail: vinyals@google.com

K. T. Schütt et al. (eds.), *Machine Learning Meets Quantum Physics*, Lecture Notes in Physics 968, https://doi.org/10.1007/978-3-030-40245-7_10

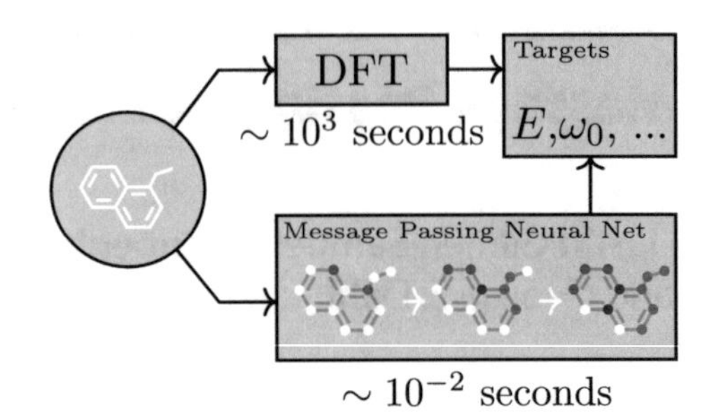

Fig. 10.1 A message passing neural network predicts quantum properties of an organic molecule by modeling a computationally expensive DFT calculation

signals [2], and infer features from real-world images and videos [3]. Although chemists have applied machine learning to many problems over the years, predicting the properties of molecules and materials using machine learning (and especially deep learning) is still in its infancy. A classic approach to applying machine learning to chemistry tasks [4–10] revolves around feature engineering. Recently, large scale quantum chemistry calculation and molecular dynamics simulations coupled with advances in high throughput experiments have begun to generate data at an unprecedented rate. Most classical techniques do not make effective use of the larger amounts of data that are now available. The time is ripe to apply more powerful and flexible machine learning methods to these problems, assuming we can find models with suitable inductive biases. The symmetries of atomic systems suggest neural networks that are invariant to the symmetries of graph structured data might also be appropriate for molecules. Sufficiently successful models could someday help automate challenging chemical search problems in drug discovery or materials science.

In this chapter, we describe a general framework for supervised learning on graphs called message passing neural networks (MPNNs) that simply abstracts the commonalities between several of the most promising prior neural models for graph structured data, in order to make it easier to understand the relationships between them and come up with novel variations. MPNNs have proven to have a strong inductive bias for graph data, with applications ranging from program synthesis [11], modeling citation networks [12], reinforcement learning [13], modeling physical systems, and predicting properties of molecules [14–17]. In this chapter, we describe the general MPNN framework before discussing specific applications of this framework in predicting the quantum mechanical properties of small organic molecules (see task schematic in Fig. 10.1).

In general, the search for practically effective machine learning (ML) models in a given domain proceeds through a sequence of increasingly realistic and interesting benchmarks. Here, we focus on the QM9 dataset as such a benchmark [18]. QM9 consists of 130k molecules with 13 properties for each molecule which are approximated by an expensive[1] quantum mechanical simulation method (DFT), to

[1]By comparison, the inference time of the neural networks discussed in this work is 300k times faster.

yield 13 corresponding regression tasks. These tasks serve as a useful benchmark for developing architectures with a strong inductive bias in the chemical domain. Additionally, QM9 also includes complete spatial information for the single low energy conformation of the atoms in the molecule that was used in calculating the chemical properties. QM9 therefore lets us consider both the setting where the complete molecular geometry is known (atomic distances, bond angles, etc.) and the setting where we need to compute properties that might still be *defined* in terms of the spatial positions of atoms, but where only the atom and bond information (i.e., graph) is available as input. In the latter case, the model must implicitly fit something about the computation used to determine a low energy 3D conformation and hopefully would still work on problems where it is not clear how to compute a reasonable 3D conformation.

When measuring the performance of our models on QM9, there are two important benchmark error levels. The first is the estimated average error of the DFT approximation to nature, which we refer to as "DFT error." The second, known as "chemical accuracy," is a target error that has been established by the chemistry community. Estimates of DFT error and chemical accuracy are provided for each of the 13 targets in Faber et al. [16]. One important goal of this line of research is to produce a model which can achieve chemical accuracy with respect to the *true* targets as measured by an extremely precise experiment. The ability to fit the DFT approximation to within chemical accuracy would be an encouraging step in this direction. In the rest of this paper, when we talk about chemical accuracy we generally mean with respect to our available ground truth labels.

In this chapter, we discuss our original application of the MPNN framework [17] to the QM9 dataset as well as related work [15, 19, 20] which have improved upon our initial results. To date MPNNs have been shown to predict DFT to within chemical accuracy on 11 out of 13 targets in the QM9 benchmark, and are less than DFT error on all 13 targets. We also show that MPNNs can predict DFT to within chemical accuracy on 5 out of 13 targets while operating on the topology of the molecule alone (with no spatial information as input). In this sparse setting, we explore some simple graph preprocessing techniques which can improve performance. Finally, we discuss a general method to train MPNNs with larger node representations without a corresponding increase in computation time or memory, yielding a substantial savings for high dimensional node representations.

10.2 Message Passing Neural Networks

There are many notable examples of models from the literature that we can describe using the message passing neural networks (MPNN) framework; in this section, we discuss a few specific instances. For simplicity, we describe MPNNs which operate on undirected graphs G with node features x_v and edge features e_{vw}. It is trivial to extend the formalism to directed multigraphs. The forward pass has two phases, a message passing phase and a readout phase. The message passing phase runs for T time steps and is defined in terms of message functions M_t and vertex update functions U_t. During the message passing phase, hidden states h_v^t at each node in the graph are updated based on aggregated messages m_v^{t+1} according to

$$m_v^{t+1} = \sum_{w \in N(v)} M_t(h_v^t, h_w^t, e_{vw}) \tag{10.1}$$

$$h_v^{t+1} = U_t(h_v^t, m_v^{t+1}) \tag{10.2}$$

where in the sum, $N(v)$ denotes the neighbors of v in graph G. Note the sum in Eq. 10.1 can be replaced with any permutation invariant function α operating on the set of neighbors $N(v)$. Some instantiations of MPNNs use self-attention in this aggregation step [21]. It has also been shown [22] that for more general aggregation functions α, other model families such as *non local neural networks* [23] can be described in this framework. The readout phase computes a feature vector for the whole graph using some readout function R according to

$$\hat{y} = R(\{h_v^T \mid v \in G\}). \tag{10.3}$$

The message functions M_t, vertex update functions U_t, and readout function R are all learned differentiable functions. R operates on the set of node states and must be invariant to permutations of the node states in order for the MPNN to be invariant to graph isomorphism. Some MPNNs [19, 24] learn edge features by introducing hidden states for all edges in the graph $h_{e_{vw}}^t$ and updating them using a learned differentiable edge function E according to Eq. 10.4.

$$e_{vw}^{t+1} = E(h_v^t, h_w^t, e_{vw}^t) \tag{10.4}$$

In what follows, we describe several models in the literature as they fit into the MPNN framework by specifying the specific message, update, and readout functions used.

10.2.1 Convolutional Networks for Learning Molecular Fingerprints [14]

The message function used is

$$M(h_v, h_w, e_{vw}) = (h_w, e_{vw}), \tag{10.5}$$

where $(\cdot\,,\ \cdot)$ denotes concatenation. The vertex update function used is

$$U_t(h_v^t, m_v^{t+1}) = \sigma(H_t^{\deg(v)} m_v^{t+1}), \tag{10.6}$$

where σ is the sigmoid function, $\deg(v)$ is the degree of vertex v, and H_t^N is a learned matrix for each time step t and vertex degree N. The readout function has skip connections to all previous hidden states h_v^t and is

$$R = f\left(\sum_{v,t} \text{softmax}(W_t h_v^t)\right), \tag{10.7}$$

where f is a neural network and W_t are learned readout matrices, one for each time step t. This message passing scheme may be problematic since the resulting message vector is

$$m_v^{t+1} = \left(\sum h_w^t, \sum e_{vw}\right), \tag{10.8}$$

which separately sums over connected nodes and connected edges. It follows that the message passing implemented in Duvenaud et al. [14] is unable to identify correlations between edge states and node states.

10.2.2 Gated Graph Neural Networks (GG-NN) [25]

The message function used is

$$M_t(h_v^t, h_w^t, e_{vw}) = A_{e_{vw}} h_w^t, \tag{10.9}$$

where $A_{e_{vw}}$ is a learned matrix, one for each edge label e (the model assumes discrete edge types). The update function is

$$U_t = \text{GRU}(h_v^t, m_v^{t+1}), \tag{10.10}$$

where GRU is the gated recurrent unit introduced by Cho et al. [26]. This work used weight tying, so the same update function is used at each time step t. Finally, the readout function is

$$R = \sum_{v \in V} \sigma\left(i(h_v^T, h_v^0)\right) \odot \left(j(h_v^T)\right), \tag{10.11}$$

where i and j are neural networks, and $\odot$ denotes element-wise multiplication.

10.2.3 Interaction Networks [27]

This work considered both the case where there is a target at each node in the graph and where there is a graph level target. It also considered the case where there are node level effects applied at each time step, in such a case the update function takes as input the concatenation (h_v, x_v, m_v) where x_v is an external vector representing some outside influence on the vertex v. The message function

$$M(h_v, h_w, e_{vw}) \tag{10.12}$$

is a neural network which takes the concatenation (h_v, h_w, e_{vw}). The vertex update function

$$U(h_v, x_v, m_v) \tag{10.13}$$

is a neural network which takes as input the concatenation (h_v, x_v, m_v). Finally, in the case where there is a graph level output, the readout function is

$$R = f\left(\sum_{v\in G} h_v^T\right), \tag{10.14}$$

where f is a neural network which takes the sum of the final hidden states h_v^T. Note the original work only defined the model for $T = 1$.

10.2.4 Molecular Graph Convolutions [24]

To the best of our knowledge, this work was the first to consider learned edge representations e_{vw}^t which are updated during the message passing phase. The message function used for node messages is

$$M(h_v^t, h_w^t, e_{vw}^t) = e_{vw}^t. \tag{10.15}$$

The vertex update function is

$$U_t(h_v^t, m_v^{t+1}) = \alpha\left(W_1\left(\alpha\left(W_0 h_v^t\right), m_v^{t+1}\right)\right), \tag{10.16}$$

where $(\cdot, \cdot)$ denotes concatenation, α is the ReLU activation, and W_1, W_0 are learned weight matrices. The edge state update is defined by

$$e_{vw}^{t+1} = E(e_{vw}^t, h_v^t, h_w^t) = \alpha\left(W_4\left(\alpha\left(W_2, e_{vw}^t\right), \alpha\left(W_3\left(h_v^t, h_w^t\right)\right)\right)\right), \tag{10.17}$$

where the W_i are also learned weight matrices.

10.2.5 Deep Tensor Neural Networks [15]

The message from w to v is computed by

$$M_t(h_v^t, h_w^t, e_{vw}) = \tanh\left(W^{fc}((W^{cf}h_w^t + b_1) \odot (W^{df}e_{vw} + b_2))\right), \tag{10.18}$$

where W^{fc}, W^{cf}, W^{df} are matrices and b_1, b_2 are bias vectors. The update function used is

$$U_t(h_v^t, m_v^{t+1}) = h_v^t + m_v^{t+1}. \tag{10.19}$$

The readout function passes each node independently through a single hidden layer neural network and sums the outputs, in particular

$$R = \sum_v \mathrm{NN}(h_v^T). \tag{10.20}$$

10.2.6 SchNet with Edge Updates [19]

This work extends SchNet [20] to include an edge update function. The message function used is

$$M_t(h_v^t, h_w^t, e_{vw}^t) = (W_1^t h_w^t) \odot g(W_3^t g(W_2^t e_{vw}^t)) \tag{10.21}$$

where g is the softplus function, and $\odot$ denotes the element-wise multiplication operation. The update function used is

$$U_t(h_v^t, m_v^{t+1}) = h_v^t + W_5^t g(W_4^t m_v^{t+1}) \tag{10.22}$$

while the edge update is

$$E(h_v^t, h_w^t, e_{vw}^t) = g(W_{E2}^{t+1} g(W_{E1}^{t+1}(h_v^t; h_w^t; e_{vw}^t))) \tag{10.23}$$

and the readout is

$$R = \sum_{v \in G} W_7 g(W_6 h_v^T). \tag{10.24}$$

10.2.7 Laplacian-Based Methods [12, 28, 29]

These methods generalize the notion of the convolution operation typically applied to image datasets to an operation that operates on an arbitrary graph G with a real valued adjacency matrix A. The operations defined in Bruna et al. [28], Defferrard et al. [29] result in message functions of the form

$$M_t(h_v^t, h_w^t) = C_{vw}^t h_w^t, \tag{10.25}$$

where the matrices C_{vw}^t are parameterized by the eigenvectors of the graph Laplacian L, and the learned parameters of the model. The vertex update function used is

$$U_t(h_v^t, m_v^{t+1}) = \sigma(m_v^{t+1}), \tag{10.26}$$

where σ is some pointwise non-linearity (such as ReLU). One model [12] results in a message function

$$M_t(h_v^t, h_w^t) = c_{vw} h_w^t \tag{10.27}$$

where $c_{vw} = (\deg(v)\deg(w))^{-1/2} A_{vw}$. The vertex update function is

$$U_v^t(h_v^t, m_v^{t+1}) = \text{ReLU}(W^t m_v^{t+1}). \tag{10.28}$$

The exact expressions for the C_{vw}^t and the derivation of the reformulation of these models as MPNNs can be found in [17].

10.3 MPNNs for Modeling Molecules

We began our exploration of MPNNs around the GG-NN model which we believe to be a strong baseline. We focused on trying different message functions, output functions, finding the appropriate input representation, and properly tuning hyper-parameters.

For the rest of the chapter, we use d to denote the dimension of the internal hidden representation of each node in the graph, and n to denote the number of nodes in the graph. Our implementation of MPNNs in general operates on directed graphs with a separate message channel for incoming and outgoing edges, in which case the incoming message m_v is the concatenation of m_v^{in} and m_v^{out}, this was also used in Li et al. [25]. When we apply this to undirected chemical graphs, we treat the graph as directed, where each original edge becomes both an incoming and outgoing edge with the same label. Note there is nothing special about the direction of the edge, it is only relevant for parameter tying. Treating undirected graphs as directed means that the size of the message channel is $2d$ instead of d.

The input to our MPNN model is a set of feature vectors for the nodes of the graph, x_v, and an adjacency matrix A with vector valued entries to indicate different bonds in the molecule as well as pairwise spatial distance between two atoms. We experimented as well with the message function used in the GG-NN family, which assumes discrete edge labels, in which case the matrix A has entries in a discrete alphabet of size k. The initial hidden states h_v^0 are set to be the atom input feature vectors x_v and are padded up to some larger dimension d. All of our experiments used weight tying at each time step t, and a GRU [26] for the update function as in the GG-NN family.

10.3.1 Message Functions

Matrix Multiplication We started with the message function used in GG-NN which is defined by the equation

$$M(h_v, h_w, e_{vw}) = A_{e_{vw}} h_w.$$

Edge Network To allow vector valued edge features, we propose the message function

$$M(h_v, h_w, e_{vw}) = A(e_{vw}) h_w,$$

where $A(e_{vw})$ is a neural network which maps the edge vector e_{vw} to a $d \times d$ matrix.

Pair Message One property that the matrix multiplication rule has is that the message from node w to node v is a function only of the hidden state h_w and the edge e_{vw}. In particular, it does not depend on the hidden state h_v^t. In theory, a network may be able to use the message channel more efficiently if the node messages are allowed to depend on both the source and destination node. Thus, we also tried using a variant on the message function as described in [27]. Here, the message from w to v along edge e is

$$m_{wv} = f\left(h_w^t, h_v^t, e_{vw}\right),$$

where f is a neural network.

When we apply the above message functions to directed graphs, there are two separate functions used, M^{in} and an M^{out}. Which function is applied to a particular edge e_{vw} depends on the direction of that edge.

10.3.2 Virtual Graph Elements

We explored two different ways to change how the messages are passed throughout the model. The simplest modification involves adding a separate "virtual" edge type for pairs of nodes that are not connected. This can be implemented as a data preprocessing step and allows information to travel long distances during the propagation phase.

We also experimented with using a latent "master" node, which is connected to every input node in the graph with a special edge type. The master node serves as a global scratch space that each node both reads from and writes to in every step of message passing. We allow the master node to have a separate node dimension d_{master}, as well as separate weights for the internal update function (in our case a GRU). This allows information to travel long distances during the propagation phase. It also, in theory, allows additional model capacity (e.g., large values of d_{master}) without a substantial hit in performance, as the complexity of the master node model is $O(|E|d^2 + nd^2_{master})$.

10.3.3 Readout Functions

We experimented with two readout functions. First is the readout function used in GG-NN, which is defined by Eq. 10.11. Second is a set2set model from Vinyals et al. [30]. The set2set model is specifically designed to operate on sets and should have more expressive power than simply summing the final node states. This model first applies a linear projection to each tuple (h_v^T, x_v) and then takes as input the set of projected tuples $T = \{(h_v^T, x_v)\}$. Then, after M steps of computation, the set2set model produces a graph level embedding q_t^* which is invariant to the order of the tuples T. We feed this embedding q_t^* through a neural network to produce the output.

10.3.4 Multiple Towers

One issue with MPNNs is scalability. In particular, a single step of the message passing phase for a dense graph requires $O(n^2d^2)$ floating point multiplications. As n or d get large this can be computationally expensive. To address this issue, we break the d dimensional node embeddings h_v^t into k different d/k dimensional embeddings $h_v^{t,k}$ and run a propagation step on each of the k copies separately to get temporary embeddings $\{\tilde{h}_v^{t+1,k}, v \in G\}$, using separate message and update functions for each copy. The k temporary embeddings of each node are then mixed together according to the equation

$$\left(h_v^{t,1}, h_v^{t,2}, \ldots, h_v^{t,k}\right) = g\left(\tilde{h}_v^{t,1}, \tilde{h}_v^{t,2}, \ldots, \tilde{h}_v^{t,k}\right) \tag{10.29}$$

where g denotes a neural network and $(x, y, \ldots)$ denotes concatenation, with g shared across all nodes in the graph. This mixing preserves the invariance to permutations of the nodes, while allowing the different copies of the graph to communicate with each other during the propagation phase. This can be advantageous in that it allows larger hidden states for the same number of parameters, which yields a computational speedup in practice. On dense graphs, when the message function is matrix multiplication (as in GG-NN) a propagation step of a single copy takes

$O\left(n^2(d/k)^2\right)$ time, and there are k copies, therefore the overall time complexity is $O\left(n^2d^2/k\right)$, with some additional overhead due to the mixing network. For $k = 8$, $n = 9$, and $d = 200$, we see a factor of 2 speedup in inference time over a $k = 1$, $n = 9$, and $d = 200$ architecture. This variation would be most useful for larger molecules, for instance, molecules from GDB-17 [31].

10.4 Input Representation

There are a number of features available for each atom in a molecule which capture both properties of the electrons in the atom and the bonds that the atom participates in. For a list of all of the features, see Table 10.1. We experimented with making the hydrogen atoms explicit nodes in the graph (as opposed to simply including the count as a node feature), in which case graphs have up to 29 nodes. Note that having larger graphs significantly slows training time, in this case by a factor of roughly 10. For the adjacency matrix, there are three edge representations used depending on the model.

Chemical Graph In the absence of distance information, adjacency matrix entries are discrete bond types: single, double, triple, or aromatic.

Distance Bins The matrix multiply message function assumes discrete edge types, so to include distance information we bin bond distances into 10 bins, the bins are obtained by uniformly partitioning the interval [2, 6] into 8 bins, followed by adding a bin [0, 2] and [6, ∞]. These bins were hand chosen by looking at a histogram of all distances. The adjacency matrix then has entries in an alphabet of size 14, indicating bond type for bonded atoms and distance bin for atoms that are not bonded. We found the distance for bonded atoms to be almost completely determined by bond type.

Raw Distance Feature When using a message function which operates on vector valued edges, the entries of the adjacency matrix are then 5 dimensional, where the first dimension indicates the Euclidean distance between the pair of atoms, and the remaining four are a one-hot encoding of the bond type.

Table 10.1 Atom features

Feature	Description
Atom type	H, C, N, O, F (one-hot)
Atomic number	Number of protons (integer)
Acceptor	Accepts electrons (binary)
Donor	Donates electrons (binary)
Aromatic	In an aromatic system (binary)
Hybridization	sp, sp2, sp3 (one-hot or null)
Number of hydrogens	(integer)

10.5 Training

Each model and target combination was trained using a uniform random hyperparameter search with 50 trials. T was constrained to be in the range $3 \leq T \leq 8$ (in practice, any $T \geq 3$ works). The number of set2set computations M was chosen from the range $1 \leq M \leq 12$. All models were trained using SGD with the ADAM optimizer (Kingma and Ba [32]), with batch size 20 for 3 million steps (540 epochs). The initial learning rate was chosen uniformly between $1e^{-5}$ and $5e^{-4}$. We used a linear learning rate decay that began between 10 and 90% of the way through training and the initial learning rate l decayed to a final learning rate $l \cdot F$, using a decay factor F in the range $[0.01, 1]$.

The QM9 dataset has 130,462 molecules in it. We randomly chose 10,000 samples for validation, 10,000 samples for testing, and used the rest for training. We use the validation set to do early stopping and model selection and we report scores on the test set. All targets were normalized to have mean 0 and variance 1. We minimize the mean squared error between the model output and the target, although we evaluate mean absolute error.

10.6 Results

In all of our tables, we report the ratio of the mean absolute error (MAE) of our models with the provided estimate of chemical accuracy for that target. Thus, any model with error ratio less than 1 has achieved chemical accuracy for that target. A list of chemical accuracy estimates for each target can be found in Faber et al. [16]. In this way, the MAE of our models can be calculated as (Error Ratio) × (Chemical Accuracy). Note, unless otherwise indicated, all tables display result of models trained individually on each target (as opposed to training one model to predict all 13).

We performed numerous experiments in order to find the best possible MPNN on this dataset as well as the proper input representation. In our experiments, we found that including the complete edge feature vector (bond type, spatial distance) and treating hydrogen atoms as explicit nodes in the graph to be very important for a number of targets. We also found that training one model per target consistently outperformed jointly training on all 13 targets. In some cases, the improvement was up to 40%. Our best MPNN variant used the edge network message function, set2set output, and operated on graphs with explicit hydrogens.

In Table 10.2, we compare the performance of our best MPNN variant (denoted with **enn-s2s**) with several baselines which use feature engineering as reported in Faber et al. [16]. We also show the performance of several other MPNNs, the GG-NN architecture [25], graph convolutions [24], and two more recent MPNNs which have improved upon our initial results reported in [17]. These include the SchNet MPNN [20], which developed message functions better suited for chemical systems, as well as [19] which added edge updates to the SchNet architectures. As of writing, the current state of the art on this benchmark is the MPNN described in [19]. For the exact functions used in these two models, see Sect. 10.2. For clarity, the error

Table 10.2 Comparison of hand engineered approaches (left) with different MPNNs (right)

Target	BAML	BOB	CM	ECFP4	HDAD	GC	GG-NN	enn-s2s	SchNet	SchNetE
mu	4.34	4.23	4.49	4.82	3.34	0.70	1.22	**0.30**	0.33	**0.29**
alpha	3.01	2.98	4.33	34.54	1.75	2.27	1.55	0.92	2.35	**0.77**
HOMO	2.20	2.20	3.09	2.89	1.54	1.18	1.17	0.99	0.95	**0.85**
LUMO	2.76	2.74	4.26	3.10	1.96	1.10	1.08	0.87	0.79	**0.72**
gap	3.28	3.41	5.32	3.86	2.49	1.78	1.70	1.60	1.47	**1.35**
R2	3.25	0.80	2.83	90.68	1.35	4.73	3.99	0.15	**0.06**	**0.06**
ZPVE	3.31	3.40	4.80	241.58	1.91	9.75	2.52	1.44	**1.37**	**1.35**
U0	1.21	1.43	2.98	85.01	0.58	3.02	0.83	0.45	0.33	**0.24**
U	1.22	1.44	2.99	85.59	0.59	3.16	0.86	0.45	0.44	**0.25**
H	1.22	1.44	2.99	86.21	0.59	3.19	0.81	0.39	0.33	**0.26**
G	1.20	1.42	2.97	78.36	0.59	2.95	0.78	0.44	0.33	**0.28**
Cv	1.64	1.83	2.36	30.29	0.88	1.45	1.19	0.80	0.66	**0.64**
Omega	0.27	0.35	1.32	1.47	0.34	0.32	0.53	0.19	N/A	N/A

Table 10.3 Models trained without spatial information

Model	Average error ratio
GG-NN	3.47
GG-NN + virtual edge	2.90
GG-NN + master node	2.62
GG-NN + set2set	**2.57**

Table 10.4 Towers vs vanilla GG-NN (no explicit hydrogen)

Model	Average error ratio
GG-NN + joint training	1.92
Towers8 + joint training	**1.75**
GG-NN + individual training	1.53
Towers8 + individual training	**1.37**

ratios of the best non-ensemble models are shown in bold. Overall, the best MPNNs achieve chemical accuracy on 11 out of 13 targets.

Training Without Spatial Information We also experimented in the setting where spatial information is not included in the input. In general, we find that augmenting the MPNN with some means of capturing long range interactions between nodes in the graph greatly improves performance in this setting. To demonstrate this, we performed 4 experiments, one where we train the GG-NN model on the sparse graph, one where we add virtual edges, one where we add a master node, and one where we change the graph level output to a set2set output. The error ratios averaged across the 13 targets are shown in Table 10.3. Overall, these three modifications help on all 13 targets, and the Set2Set output achieves chemical accuracy on 5 out of 13 targets. The experiments shown in Tables 10.3 and 10.4 were run with a partial charge feature as a node input. This feature is an output of the DFT calculation and thus could not be used in an applied setting. The numbers we report in Table 10.2 do not use this feature.

Table 10.5 Results from training the edge network + set2set model on different sized training sets (N denotes the number of training samples)

Target	$N = 11$k	$N = 35$k	$N = 58$k	$N = 82$k	$N = 110$k
mu	1.28	0.55	0.44	0.32	0.30
alpha	2.76	1.59	1.26	1.09	0.92
HOMO	2.33	1.50	1.34	1.19	0.99
LUMO	2.18	1.47	1.19	1.10	0.87
gap	3.53	2.34	2.07	1.84	1.60
R2	0.28	0.22	0.21	0.21	0.15
ZPVE	2.52	1.78	1.69	1.68	1.27
U0	1.24	0.69	0.58	0.62	0.45
U	1.05	0.69	0.60	0.52	0.45
H	1.14	0.64	0.65	0.53	0.39
G	1.23	0.62	0.64	0.49	0.44
Cv	1.99	1.24	0.93	0.87	0.80
Omega	0.28	0.25	0.24	0.15	0.19

Towers Our original intent in developing the towers variant was to improve training time, as well as to allow the model to be trained on larger graphs. However, we also found some evidence that the multi-tower structure improves generalization performance. In Table 10.4, we compare GG-NN + towers + set2set output vs a baseline GG-NN + set2set output when distance bins are used. We do this comparison in both the joint training regime and when training one model per target. The towers model outperforms the baseline model on 12 out of 13 targets in both individual and joint target training. We believe the benefit of towers is that it resembles training an ensemble of models. Unfortunately, our attempts so far at combining the towers and edge network message function have failed to further improve performance, possibly because the combination makes training more difficult.[2]

Additional Experiments In preliminary experiments, we tried disabling weight tying across different time steps. However, we found that the most effective way to increase performance was to tie the weights and use a larger hidden dimension d. We also early on found the pair message function to perform worse than the edge network function. This included a toy pathfinding problem which was originally designed to benefit from using pair messages. Also, when trained jointly on the 13 targets the edge network function outperforms pair message on 11 out of 13 targets, and has an average error ratio of 1.53 compared to 3.98 for pair message. Given the difficulties with training this function we did not pursue it further. For performance on smaller sized training sets, see Table 10.5.

[2]As reported in Schütt et al. [15]. The model was trained on a different train/test split with 100k training samples vs 110k used in our experiments.

10.7 Conclusions and Future Work

Our results show that MPNNs with the appropriate message, update, and output functions have a useful inductive bias for predicting molecular properties, outperforming several strong baselines, and eliminating the need for complicated feature engineering. Moreover, our results also reveal the importance of allowing long range interactions between nodes in the graph with either the master node or the set2set output. The towers variation makes these models more scalable, but additional improvements will be needed to scale to much larger graphs.

An important future direction is to design MPNNs that can generalize effectively to larger graphs than those appearing in the training set or at least work with benchmarks designed to expose issues with generalization across graph sizes. Generalizing to larger molecule sizes seems particularly challenging when using spatial information. First of all, the pairwise distance distribution depends heavily on the number of atoms. Second, our most successful ways of using spatial information create a fully connected graph where the number of incoming messages also depends on the number of nodes. To address the second issue, we believe that adding an attention mechanism over the incoming message vectors could be an interesting direction to explore.

Acknowledgments We would like to thank Lukasz Kaiser, Geoffrey Irving, Alex Graves, and Yujia Li for helpful discussions. Thank you to Adrian Roitberg for pointing out an issue with the use of partial charges in an earlier version of this work.

References

1. Y. Wu, M. Schuster, Z. Chen, Q.V. Le, M. Norouzi, W. Macherey, M. Krikun, Y. Cao, Q. Gao, K. Macherey, et al., (2016, preprint). arXiv:1609.08144
2. G. Hinton, L. Deng, D. Yu, G.E. Dahl, A.-R. Mohamed, N. Jaitly, A. Senior, V. Vanhoucke, P. Nguyen, T.N. Sainath, et al., IEEE Signal. Proc. Mag. **29**(6), 82 (2012)
3. A. Krizhevsky, I. Sutskever, G.E. Hinton, *Advances in Neural Information Processing Systems* (The MIT Press, Cambridge, 2012), pp. 1097–1105
4. K. Hansen, F. Biegler, R. Ramakrishnan, W. Pronobis, O.A. von Lilienfeld, K.-R. Müller, A. Tkatchenko, J. Phys. Chem. Lett. **6**(12), 2326 (2015). https://doi.org/10.1021/acs.jpclett.5b00831
5. B. Huang, O.A. von Lilienfeld, J. Chem. Phys. **145**(16), 161102 (2016). https://doi.org/10.1063/1.4964627
6. M. Rupp, A. Tkatchenko, K.-R. Müller, O.A. von Lilienfeld, Phys. Rev. Lett. **108**(5), 058301 (2012)
7. D. Rogers, M. Hahn, J. Chem. Inf. Model. **50**(5), 742 (2010)
8. G. Montavon, K. Hansen, S. Fazli, M. Rupp, F. Biegler, A. Ziehe, A. Tkatchenko, O.A. von Lilienfeld, K.-R. Müller, *Advances in Neural Information Processing Systems* (Curran Associates, Red Hook, 2012), pp. 440–448
9. J. Behler, M. Parrinello, Phys. Rev. Lett. **98**, 146401 (2007). https://doi.org/10.1103/PhysRevLett.98.146401
10. S.S. Schoenholz, E.D. Cubuk, D.M. Sussman, E. Kaxiras, A.J. Liu, A structural approach to relaxation in glassy liquids. Nat. Phys. **12**(5), 469–471 (2016)
11. M. Allamanis, M. Brockschmidt, M. Khademi, (2017, preprint). arXiv:1711.00740

12. T.N. Kipf, M. Welling, ArXiv e-prints (2016)
13. V. Zambaldi, D. Raposo, A. Santoro, V. Bapst, Y. Li, I. Babuschkin, K. Tuyls, D. Reichert, T. Lillicrap, E. Lockhart, et al. (2018, preprint). arXiv:1806.01830
14. D.K. Duvenaud, D. Maclaurin, J. Iparraguirre, R. Bombarell, T. Hirzel, A. Aspuru-Guzik, R.P. Adams, *Advances in Neural Information Processing Systems* (2015), pp. 2224–2232
15. K.T. Schütt, F. Arbabzadah, S. Chmiela, K.R. Müller, A. Tkatchenko, Quantum-chemical insights from deep tensor neural networks. Nat. Commun. **8**(1), 1–8 (2017)
16. F. Faber, L. Hutchison, B. Huang, J. Gilmer, S.S. Schoenholz, G.E. Dahl, O. Vinyals, S. Kearnes, P.F. Riley, O.A. von Lilienfeld, (2017). https://arxiv.org/abs/1702.05532
17. J. Gilmer, S.S. Schoenholz, P.F. Riley, O. Vinyals, G.E. Dahl, *International Conference on Machine Learning* (2017), pp. 1263–1272
18. R. Ramakrishnan, P.O. Dral, M. Rupp, O.A. Von Lilienfeld, Quantum chemistry structures and properties of 134 kilo molecules. Sci. Data **1**, 140022 (2014)
19. P.B. Jørgensen, K.W. Jacobsen, M.N. Schmidt (2018, preprint). arXiv:1806.03146
20. K.T. Schütt, P.-J. Kindermans, H.E.S. Felix, S. Chmiela, A. Tkatchenko, K.-R. Müller, *Advances in Neural Information Processing Systems* (Curran Associates, Red Hook, 2017), pp. 991–1001
21. P. Veličković, G. Cucurull, A. Casanova, A. Romero, P. Lio, Y. Bengio, (2017, preprint). arXiv:1710.10903
22. P.W. Battaglia, J.B. Hamrick, V. Bapst, A. Sanchez-Gonzalez, V. Zambaldi, M. Malinowski, A. Tacchetti, D. Raposo, A. Santoro, R. Faulkner, et al., (2018, preprint). arXiv:1806.01261
23. X. Wang, R. Girshick, A. Gupta, K. He, (2017, preprint). arXiv:1711.07971
24. S. Kearnes, K. McCloskey, M. Berndl, V. Pande, P. Riley, J. Comput.-Aided Mol. Des. **30**(8), 595 (2016)
25. Y. Li, D. Tarlow, M. Brockschmidt, R. Zemel, *International Conference on Learning Representations, ICLR* (2016)
26. K. Cho, B. Van Merriënboer, D. Bahdanau, Y. Bengio, (2014, preprint). arXiv:1409.1259
27. P. Battaglia, R. Pascanu, M. Lai, D.J. Rezende, K. Kavukcuoglu, *Advances in Neural Information Processing Systems* (Curran Associates, Red Hook, 2016), pp. 4502–4510
28. J. Bruna, W. Zaremba, A. Szlam, Y. LeCun, (2013, preprint). arXiv:1312.6203
29. M. Defferrard, X. Bresson, P. Vandergheynst, *Advances in Neural Information Processing Systems* (Curran Associates, Red Hook, 2016), pp. 3837–3845
30. O. Vinyals, S. Bengio, M. Kudlur, (2015, preprint). arXiv:1511.06391
31. L. Ruddigkeit, R. Van Deursen, L.C. Blum, J.-L. Reymond, J. Chem. Inf. Model. **52**(11), 2864 (2012)
32. D. Kingma, J. Ba, (2014, preprint). arXiv:1412.6980

Learning Representations of Molecules and Materials with Atomistic Neural Networks 11

Kristof T. Schütt, Alexandre Tkatchenko, and Klaus-Robert Müller

Abstract

Deep learning has been shown to learn efficient representations for structured data such as images, text, or audio. In this chapter, we present neural network architectures that are able to learn efficient representations of molecules and materials. In particular, the continuous-filter convolutional network SchNet accurately predicts chemical properties across compositional and configurational space on a variety of datasets. Beyond that, we analyze the obtained representations to find evidence that their spatial and chemical properties agree with chemical intuition.

K. T. Schütt
Machine Learning Group, Technische Universität Berlin, Berlin, Germany
e-mail: kristof.schuett@tu-berlin.de

A. Tkatchenko
Physics and Materials Science Research Unit, University of Luxembourg, Luxembourg, Luxembourg
e-mail: alexandre.tkatchenko@uni.lu

K.-R. Müller (✉)
Machine Learning Group, Technische Universität Berlin, Berlin, Germany

Max-Planck-Institut für Informatik, Saarbrücken, Germany

Department of Brain and Cognitive Engineering, Korea University, Seoul, Korea
e-mail: klaus-robert.mueller@tu-berlin.de

K. T. Schütt et al. (eds.), *Machine Learning Meets Quantum Physics*,
Lecture Notes in Physics 968, https://doi.org/10.1007/978-3-030-40245-7_11

11.1 Introduction

In recent years, machine learning has been successfully applied to the prediction of chemical properties for molecules and materials [1–14]. A significant part of the research has been dedicated to engineering features that characterize global molecular similarity [15–20] or local chemical environments [21–23] based on atomic positions. Then, a non-linear regression method—such as kernel ridge regression or a neural network—is used to correlate these features with the chemical property of interest.

A common approach to model atomistic systems is to decompose them into local environments, where a chemical property is expressed by a partitioning into latent atom-wise contributions. Based on these contributions, the original property is then reconstructed via a physically motivated aggregation layer. For example, Behler–Parrinello networks [21] or the SOAP kernel [22] decompose the total energy in terms of atomistic contributions

$$E = \sum_{i=1}^{n_{\text{atoms}}} E_i. \tag{11.1}$$

Atomic forces can be directly obtained as negative derivatives of the energy model. While this is often a suitable partitioning of extensive properties, intensive properties can be modeled as the mean

$$P = \frac{1}{n_{\text{atoms}}} \sum_{i=1}^{n_{\text{atoms}}} P_i. \tag{11.2}$$

However, this might still not be a sufficient solution for global molecular properties such as HOMO-LUMO gaps or excitation energies [24]. To obtain a better performance, output models that incorporate property-specific prior knowledge should be used. For example, the dipole moment can be written as

$$\boldsymbol{\mu} = \sum_{i}^{n_{\text{atoms}}} q_i \mathbf{r}_i \tag{11.3}$$

such that the atomistic neural network needs to predict atomic charges q_i [6,25–27].

The various atomistic models differ in how they obtain the energy contributions E_i, usually employing manually crafted atomistic descriptors. In contrast to such descriptor-based approaches, this chapter focuses on atomistic neural network architectures that learn efficient representations of molecules and materials *end-to-end*—i.e., directly from atom types Z_i and positions $\mathbf{r}_i$—while delivering accurate predictions of chemical properties across compositional and configurational space [11, 28, 29]. The presented models will encode important invariances, e.g., towards rotation and translation, directly into the deep learning architecture and obtain predicted property from physically motivated output layers. Finally, we will obtain spatially and chemically resolved insights from the learned representations regarding the inner workings of the neural network as well as the underlying data.

11.2 The Deep Tensor Neural Network Framework

In order to construct atom-centered representations $\mathbf{x}_i \in \mathbb{R}^F$, where i is the index of the center atom and F the number of feature dimension, a straight-forward approach is to expand the atomistic environment in terms of n-body interactions [30, 31], which can be written in general as

$$\mathbf{x}_i = f^{(1)}(Z_i) + \sum_{j \neq i} f^{(2)}((Z_i, \mathbf{r}_i), (Z_j, \mathbf{r}_j)) + \sum_{\substack{j,k \neq i \\ k \neq j}} f^{(3)}((Z_i, \mathbf{r}_i), (Z_j, \mathbf{r}_j), (Z_k, \mathbf{r}_k)) + \ldots \quad (11.4)$$

However, such an approach requires definition explicit n-body models $f^{(n)}$ (e.g., using neural networks) as well as computing of a large number of higher-order terms of the atom coordinates. At the same time, all many-body networks must respect the invariances w.r.t. rotation, translation, and the permutation of atoms.

An alternative approach is to incorporate higher-order interactions in a recursive fashion. Instead of explicitly modeling an n-body neural network, we design an interaction network $\mathbf{v} : \mathbb{R}^F \times \mathbb{R} \to \mathbb{R}^F$ that we use to model perturbations

$$\mathbf{x}_i^{(t+1)} = \mathbf{x}_i^{(t)} + \mathbf{v}^{(t)}(\mathbf{x}_1^{(t)}, \mathbf{r}_{i1}, \ldots, \mathbf{x}_n^{(t)}, \mathbf{r}_{in}), \quad (11.5)$$

of the chemical environment $\mathbf{x}_i^{(t)}$ by its neighboring environments $\mathbf{x}_j^{(t)}$ depending on their relative position $\mathbf{r}_{ij} = \mathbf{r}_j - \mathbf{r}_i$. On this basis, we define the *deep tensor neural network* (DTNN) framework [28]:

1. Use an embedding depending on the type of the center atom

$$\mathbf{x}_i^{(0)} = \mathbf{A}_{Z_i} \in \mathbb{R}^d$$

for the initial representation of local chemical environment i. This corresponds to the 1-body terms in Eq. 11.4.
2. Refine the embeddings repeatedly using the interaction networks from Eq. 11.5.
3. Arriving at the final embedding $\mathbf{x}_i^T$ after T interaction refinements, predict the desired chemical property using a property-specific output network (as described in Sect. 11.1).

The embedding matrix A as well as the parameters of the interaction networks $\mathbf{v}^{(t)}$ and the output network are learned during the training procedure. This framework allows for a family of atomistic neural network models—such as the deep tensor neural network [28] and SchNet [11, 29]—that differ in how the interactions $\mathbf{v}^{(t)}$ are modeled and the predictions are obtained from the atomistic representations $\mathbf{x}_i^T$.

11.3 SchNet

Building upon the principles of the previously described DTNN framework, we propose SchNet as a convolutional neural network architecture for learning representations for molecules and materials. Figure 11.1 depicts an overview of the SchNet architecture as well as how the interaction refinements are modeled by interaction blocks shown on the right. In the following, we will introduce the main component of SchNet—the continuous-filter convolutional layer—before describing how these are used to construct the interaction blocks.

11.3.1 Continuous-Filter Convolutional Layers

The commonly used convolutional layers [32] employ discrete filter tensors since they are usually applied to data that is sampled on a grid, such as digital images, video, or audio. However, such layers are not applicable for atomistic systems, since atoms can be located at arbitrary positions in space. For example, when predicting the potential energy, the output of a convolutional layer might change rapidly when an atom moves from one grid cell to the next. Especially when we aim to predict a smooth potential energy surface, a continuous and differentiable representation is

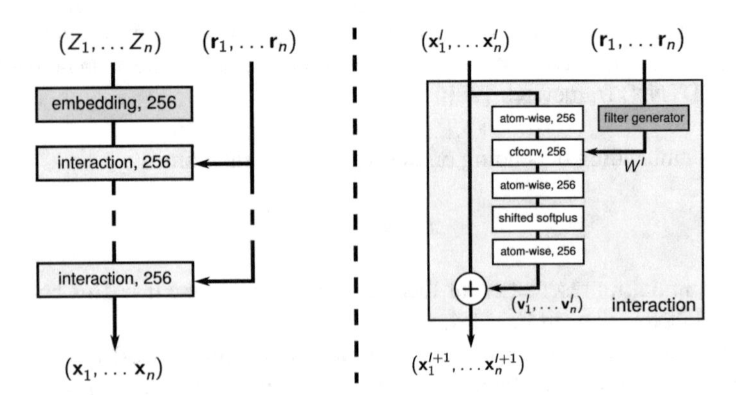

Fig. 11.1 The illustration shows an architectural overview of SchNet (left), the interaction block (right). The shifted softplus activation function is defined as $\text{ssp}(x) = \ln(0.5e^x + 0.5)$. The number of neurons used in the employed SchNet models is given for each parameterized layer

required. For this reason, we use a convolutional layer employing a continuous-filter function in order to model the interactions.

Given the representations $\mathbf{x}_i^l$ of the chemical environment of atom i at position $\mathbf{r}_i$ and layer l of the neural network, the atomistic system can be described by a function

$$\rho^l(\mathbf{r}) = \sum_{i=1}^{n_{\text{atoms}}} \delta(\mathbf{r} - \mathbf{r}_i)\mathbf{x}_i^l. \tag{11.6}$$

In order to include the interactions of the atom-centered environments, we convolve $\rho : \mathbb{R}^3 \rightarrow \mathbb{R}^F$ and a spatial filter $W : \mathbb{R}^3 \rightarrow \mathbb{R}^F$ as element-wise

$$(\rho * W)(\mathbf{r}) = \int\limits_{\mathbf{r}_a \in \mathbb{R}^3} \rho(\mathbf{r}_a) \circ W(\mathbf{r} - \mathbf{r}_a) d\mathbf{r}_a, \tag{11.7}$$

where "$\circ$" is the element-wise product. Here, the filter function W describes the interaction of a representation $\mathbf{x}_i$ with an atom at the relative position $\mathbf{r} - \mathbf{r}_i$. The filter functions can be modeled by a *filter-generating* neural network similar to those used in dynamic filter networks [33]. Considering the discrete location of atoms in Eq. 11.6, we obtain

$$\begin{aligned}(\rho^l * W)(\mathbf{r}) &= \sum_{j=1}^{n_{\text{atoms}}} \int\limits_{\mathbf{r}_a \in \mathbb{R}^3} \delta(\mathbf{r}_a - \mathbf{r}_j)\mathbf{x}_j^l \circ W(\mathbf{r} - \mathbf{r}_a) d\mathbf{r}_a \\ &= \sum_{j=1}^{n_{\text{atoms}}} \mathbf{x}_j^l \circ W(\mathbf{r} - \mathbf{r}_j). \end{aligned} \tag{11.8}$$

This yields a function representing how the atoms of the system act on another location in space. To obtain the rotationally invariant interactions between atoms,

$$\mathbf{x}_i^{l+1} = \left(\rho^l * W^l\right)(\mathbf{r}_i) = \sum_{j=1}^{n_{\text{atoms}}} \mathbf{x}_j^l \circ W(\mathbf{r}_j - \mathbf{r}_i), \tag{11.9}$$

i.e., we evaluate the convolution at discrete locations in space using continuous, radial filters.

11.3.2 Interaction Blocks

After introducing continuous-filter convolutional layers, we go on to construct the interaction blocks. Besides convolutions, we employ atom-wise, fully connected layers

$$\mathbf{x}_i^{(l+1)} = W^{(l)}\mathbf{x}_i^{(l)} + \mathbf{b}^{(l)} \tag{11.10}$$

that are applied separately to each atom i with tied weights $W^{(l)}$. Throughout the network, we use softplus non-linearities [34] that are shifted

$$f(x) = \ln\left(\frac{1}{2}e^x + \frac{1}{2}\right) \tag{11.11}$$

in order to conserve zero-activations: $f(0) = 0$. Figure 11.2 shows this activation function compared to exponential linear units (ELU) [35]:

$$f(x) = \begin{cases} e^x - 1 & \text{if } x < 0 \\ x & \text{otherwise} \end{cases} \tag{11.12}$$

The derivatives for ELU and softplus are shown in the middle and right panel of Fig. 11.2, respectively. A crucial difference is that the softplus is smooth while ELUs exhibit only first-order continuity. However, the higher-order differentiability of the model, and therefore also of the employed activation functions, is crucial for the prediction of atomic forces or vibrational frequencies.

Figure 11.1 (right) shows how the interaction block is assembled from these components. Since the continuous-filter convolutional layers are applied feature-wise, we achieve the mixing of feature maps by atom-wise layers before and after the convolution. This is analogous to depth-wise separable convolutional layers in Xception nets [36] which could outperform the architecturally similar InceptionV3 [37] on the ImageNet dataset [38] while having fewer parameters. Most importantly, feature-wise convolutional layers reduce the number of filters, which significantly reduces the computational cost. This is particularly important

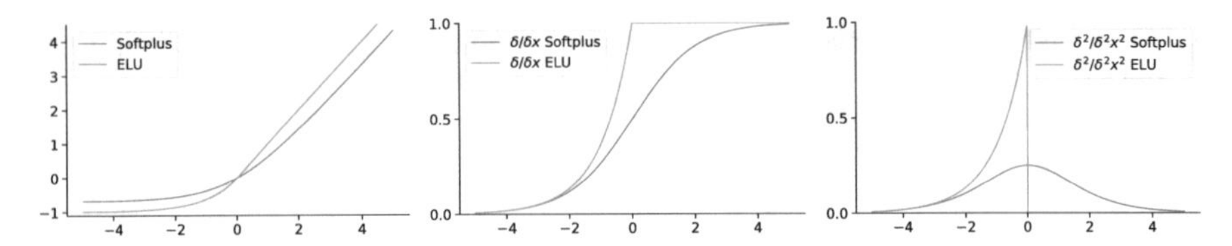

Fig. 11.2 Comparison of shifted softplus and ELU activation function. We show plots of the activation functions (left), and their first (middle) and second derivatives (right)

for continuous-filter convolutions, where each filter has to be computed by a filter-generating network.

11.3.3 Filter-Generating Networks

The architecture of the filter-generating network significantly influences the properties of the predicted filters and, consequently, the learned atomic interactions. Therefore, we can incorporate invariances or prior chemical knowledge into the filter. In the following, we describe the considerations that went into designing the SchNet filter-generating networks.

11.3.3.1 Self-Interaction

In an interatomic potential, we aim to avoid self-interaction of atoms, since this is fundamentally different than the interaction with other atoms. We can encode this in the filter network by constraining the filter network such that $W(\mathbf{r}_i - \mathbf{r}_j) = 0$ for $\mathbf{r}_i = \mathbf{r}_j$. Since two distinct atoms cannot be at the same position, this is an unambiguous condition to exclude self-interaction. This is equivalent to modifying Eq. 11.9 to exclude the center atom of the environment from the sum:

$$\mathbf{x}_i^{l+1} = \sum_{j \neq i} \mathbf{x}_j^l \circ W(\mathbf{r}_j - \mathbf{r}_i), \tag{11.13}$$

11.3.3.2 Rotational Invariance

As the input to the filter $W : \mathbb{R}^3 \rightarrow \mathbb{R}$ is only invariant to translations of the molecule, we additionally need to consider rotational invariance. We achieve this by using interatomic distances r_{ij} as input to the filter network, resulting in radial filters $W : \mathbb{R} \rightarrow \mathbb{R}^F$.

11.3.3.3 Local Distance Regimes

In the spirit of radial basis function (RBF) networks [39, 40], the filter-generating neural network $W(r_{ij})$ first expands the pair-wise distances

$$\hat{\mathbf{r}}_{ij} = \left[\exp(-\gamma(r_{ij} - k\Delta\mu)^2)\right]_{0 \leq k \leq r_{\mathrm{cut}}/\Delta\mu}, \tag{11.14}$$

with $\Delta\mu$ being the spacing of Gaussians with scale γ on a grid ranging from 0 to the distance cutoff r_{cut}. This helps to decouple the various regimes of atomic interactions and allow for an easier starting point for the training procedure. On top of the RBF expansion, we apply two fully connected layers with softplus activation functions.

As an illustrative example, Fig. 11.3 shows two linear models fitted to the potential energy surface of H_2. Using the distance as a feature directly, we obviously capture only a linear relationship. However, the expanded RBF feature space allows us to obtain a smooth and accurate fit of the potential.

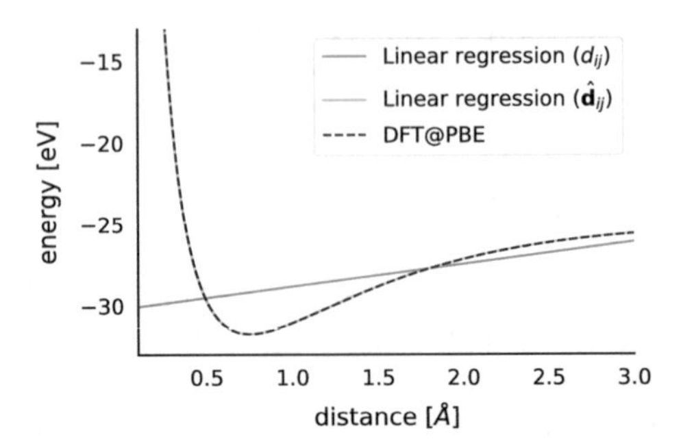

Fig. 11.3 Comparison of features for regression of bond stretching energies of H_2. We use scalar distances r_{ij} and distances in a radial basis $\hat{\mathbf{r}}_{ij}$ with $\Delta\mu = 0.1$ and $\gamma = 10$ as features, respectively. The energies were computed by Brockherde et al. [9] with DFT at the PBE level of theory

From an alternative viewpoint, if we initialize a neural network with the usual weight distributions and non-linearities, the resulting function is almost linear before training as the neuron activations are close to zero. Therefore, the filter values would be strongly correlated, leading to a plateauing cost function at the beginning of training. Radial basis functions solve this issue by decoupling the various distance regimes.

11.3.3.4 Cutoffs

While in principle the size of the filter in a continuous-filter convolutional layer can be infinite, there are natural limitations on how such a filter can be trained. The interatomic distances in a dataset of molecule have an upper bound determined by the size of the largest molecule. More importantly, we cannot consider interactions with an infinite number of atoms in case of atomistic systems with periodic boundary conditions. Therefore, it is often beneficial or even required to restrict the filter size using a distance cutoff.

While it is certainly possible to apply a hard cutoff, this may lead to rapidly changing energies in molecular dynamics simulations. Therefore, we apply a cosine cutoff function to the filter, to obtain a local filter

$$W_{\text{local}}(r_{ij}) = W(r_{ij}) f_{\text{cut}}(r_{ij}) \tag{11.15}$$

$$f_{\text{cut}}(r_{ij}) = \frac{1}{2}\cos\left(\frac{r_{ij}}{r_{\text{cut}}}\pi\right) + \frac{1}{2}. \tag{11.16}$$

In the following, we use a cutoff distance $r_{\text{cut}} = 5$Å for our models. Note that this constitutes only a cutoff for *direct interactions* between atoms. Since we perform multiple interaction passes, the effective perceptive field of SchNet is much larger.

11.3.3.5 Periodic Boundary Conditions (PBC)

For crystalline materials, we have to respect the PBCs when convolving with the interactions, i.e., we have to include interactions with periodic sites of neighboring unit cells: Due to the linearity of the convolution, we can move the sum over periodic images into the filter. Given atomistic representations $\mathbf{x}_i = \mathbf{x}_{ia} = \mathbf{x}_{ib}$ of site i for unit cells a and b, we obtain

$$\begin{aligned}\mathbf{x}_i^{l+1} = \mathbf{x}_{im}^{l+1} &= \sum_{j=0}^{n_{\text{atoms}}} \sum_{b=0}^{n_{\text{cells}}} \mathbf{x}_{jb}^{l} \circ \tilde{W}^l(\mathbf{r}_{jb} - \mathbf{r}_{ia}) \\ &= \sum_{j=0}^{n_{\text{atoms}}} \mathbf{x}_j^l \circ \underbrace{\left(\sum_{b=0}^{n_{\text{cells}}} \tilde{W}^l(\mathbf{r}_{jb} - \mathbf{r}_{ia}) \right)}_{W}. \end{aligned} \quad (11.17)$$

When using a hard cutoff, we have found that the filter needs to be normalized with respect to the number of neighboring atoms n_{nbh} for the training to converge:

$$W_{\text{normalized}}(r_{ij}) = \frac{1}{n_{\text{nbh}}} W(r_{ij}). \quad (11.18)$$

However, this is not necessary, when using a cosine cutoff function, as shown above.

11.4 Analysis of the Representation

Having introduced the SchNet architecture, we go on to analyze the representations that have been learned by training the models on QM9—a dataset of 130k small organic molecules with up to nine heavy atoms [41]—as well as a molecular dynamics trajectory of aspirin [20]. If not given otherwise, we use six interaction blocks and atomistic representations $\mathbf{x}_i \in \mathbb{R}^{256}$. The models have been trained using stochastic gradient descent with warm restarts [42] and the ADAM optimizer [43].

11.4.1 Locality of the Representation

As described above, atomistic models decompose the representation into local chemical environments. Since SchNet is able to learn a representation of such an environment, the locality of the representation may depend on whether a cutoff was used as well as the training data.

Table 11.1 shows the performance of SchNet models trained on various datasets with and without cutoff. We observe that the cutoff is beneficial for QM9 as well as the small aspirin training set with $N = 1000$ reference calculations. The cutoff function biases the model towards learning from local interactions, which helps with generalization since energy contributions from interactions at larger distances are

Table 11.1 Mean absolute (MAE) and root mean squared errors (RMSE) of SchNet with and without cosine cutoff for various datasets over three repetitions

Dataset	Property	Unit	r_{cut} [Å]	MAE	RMSE
QM9 ($N = 110$k)	U_0	kcal mol^{-1}	–	0.259	0.599
			5	**0.218**	**0.518**
	μ	Debye	–	0.019	0.037
			5	**0.017**	**0.033**
Aspirin ($N = 1$k)	Total energy	kcal mol^{-1}	–	0.438	0.592
			5	**0.402**	**0.537**
	Atomic forces	kcal mol^{-1}Å^{-1}	–	1.359	1.929
			5	**0.916**	**1.356**
Aspirin ($N = 50$k)	Total energy	kcal mol^{-1}	–	**0.088**	**0.113**
			5	0.102	0.130
	Atomic forces	kcal mol^{-1}Å^{-1}	–	**0.104**	**0.158**
			5	0.140	0.203
Materials project ($N = 62$k)	Formation energy	eV/atom	(hard[a]) 5	**0.037**	0.091
			5	0.039	**0.084**

[a]Due to the PBCs, there is no model without a cutoff for bulk crystals. The hard cutoff discards all atoms at distances $r_{ij} > r_{\text{cut}}$.
For the Materials Project data, we use a smaller model ($\mathbf{x}_i \in \mathbb{R}^{64}$) and compare the cosine cutoff to a normalized filter with hard cutoff. The number of reference calculations, N, is the size of the combined training and validation set.

much harder to disentangle. On the other hand, the SchNet model with trained on 50,000 aspirin reference calculation benefits from the large chemical environment when not applying a cutoff. This is because with such a large amount of training data, the model is now also able to infer more nonlocal interactions within the molecule. In the case of the Materials Project dataset, we observe that cosine cutoff and hard cutoff yield comparable results, where the cosine cutoff is slightly preferable since it obtains the lower root mean squared error (RMSE). Since the RMSE puts more emphasis on larger errors than MAE, this indicates that the cosine cutoff improves generalization. This may be due to the more local model which is obtained by focusing on smaller distances or by eliminating the discontinuities that are introduced by a hard cutoff.

Figure 11.4 shows the atomization energies of a carbon dimer as predicted by SchNet models trained on QM9 and the aspirin trajectory of the MD17 dataset. Since the models were trained on saturated molecules, this does not reflect the real energy or atomic forces of the dimer. The reason is that the energy contribution of carbon interactions in the context of equilibrium molecules or MD trajectories, respectively, includes the inferred contributions of other neighboring atoms. For instance, if we consider two carbon atoms at a distance of about 2.4 Å in aspirin, they are likely to be part of the aromatic ring with other carbon at a distance of 1.4 Å. We also observe a large offset for aspiring since the model was not trained on molecules with a varying number of atoms. If we wanted to eliminate these model

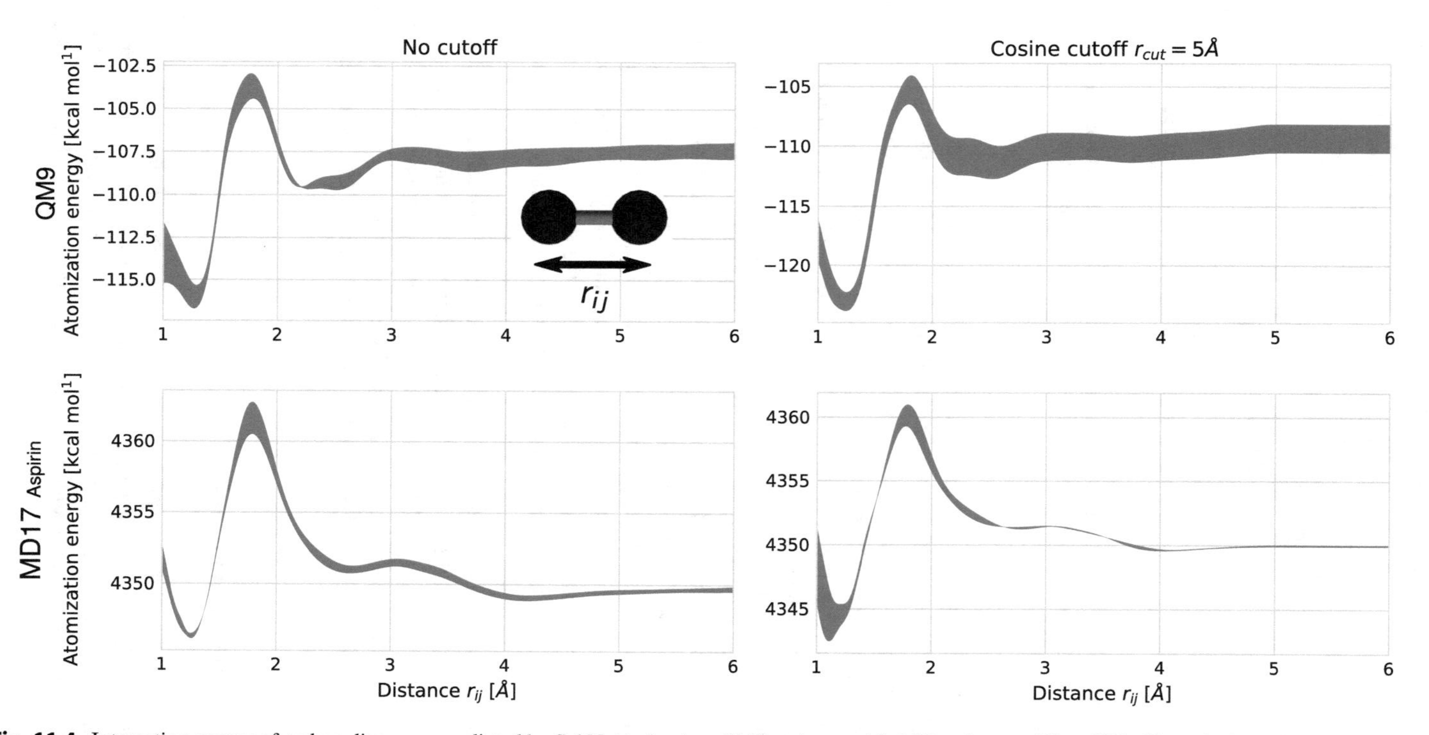

Fig. 11.4 Interaction energy of carbon dimers as predicted by SchNet trained on QM9 and an aspirin MD trajectory ($N = 50$k). Since the neural networks were not explicitly trained on carbon dimers, the energies assigned by SchNet are heavily influenced by the cutoff and inferred neighboring atoms of the respective training set. The width of the line represents the deviation of the energy over three models trained on different training splits

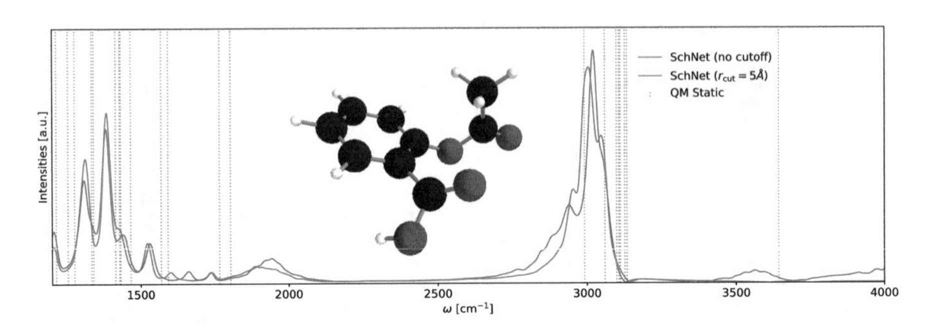

Fig. 11.5 Vibrational spectrum of aspirin as predicted by SchNet without cutoff on 50k reference calculations (DFT/PBE). The harmonic normal mode vibrations obtained with the electronic structure reference are shown in grey

biases, we needed to train the neural networks on more diverse datasets, e.g., by explicitly including dimers with large interatomic distances. While this is necessary to obtain a general model of quantum chemistry, it might even be detrimental for the prediction of a certain subset of molecules using a given amount of training data.

Considering the above, the analysis in Fig. 11.4 shows how the neural networks predict carbon–carbon interaction energies. Since there are no isolated carbon dimers in the training data, SchNet attributes molecular energies in the context of the data it was trained on, leading to large offsets in the assigned energies. Still, we observe that the general shape of the potential is consistent across all four models. Applying the cosine cutoff leads to constant energy contributions beyond $r_{\text{cut}} = 5$ Å, however, models without cutoff are nearly constant in this distance regime as well. SchNet correctly characterizes the carbon bond with its energy minimum between 1.2–1.3 Å and rising energy with larger distances. For the distance regime beyond about 1.8 Å, the inferred, larger environment dominates the attribution of interaction energies.

Given that the aspirin model trained on the larger dataset benefits from a larger attribution of interaction energies to larger distances, we analyze how the cutoff will affect the vibrational spectrum. Using the SchNet potentials, we have generated two molecular dynamics trajectories of 50 ps at 300 K using a Langevin thermostat with a time step of 0.5 fs. Figure 11.5 shows the vibrational spectra of the models with and without cosine cutoff.

11.4.2 Local Chemical Potentials

In order to further examine the spatial structure of the representation, we observe how SchNet models the influence of a molecule on a probe atom that is moved through space and acts as a test charge. This can be derived straight-forwardly from the definition of the continuous-filter convolutional layer in Eq. 11.8, which

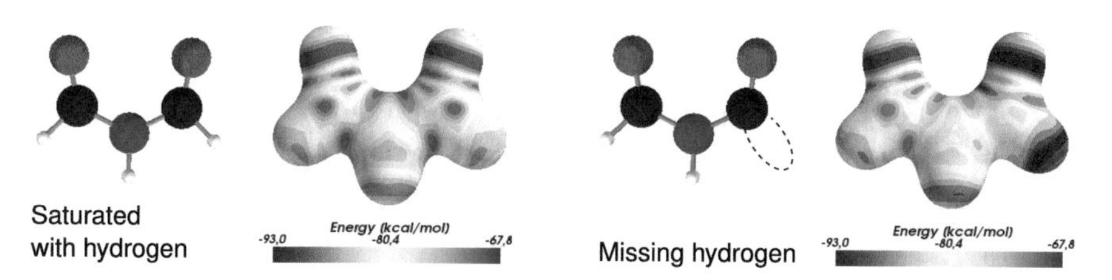

Fig. 11.6 Local chemical potentials of N-formylformamide generated by SchNet trained on QM9 using a hydrogen probe for the complete molecule (left) and after removing one of the hydrogens (right). The potentials are plotted on the $\sum_i \|\mathbf{r} - \mathbf{r}_i\| = 3.7$ Å isosurface of the saturated molecule

is defined for arbitrary positions in space:

$$\mathbf{x}_{\text{probe}} = \left(\rho^l * W\right)(\mathbf{r}_{\text{probe}}) = \sum_{j=1}^{n_{\text{atoms}}} \mathbf{x}_j^l \circ W(\mathbf{r}_{\text{probe}} - \mathbf{r}_j). \tag{11.19}$$

The remaining part of the model is left unchanged as those layers are only applied atom-wise. Finally, we visualize the predicted probe energy on a smooth isosurface around the molecule [11,27,28].

Figure 11.6 (left) shows this for N-formylformamide using a hydrogen probe. According to this, the probe is more likely to bond on the oxygens, as indicated by the lower probe energies. To further study this interpretation, we remove one of the hydrogens in Fig. 11.6 (right). In agreement with our analysis, this leads to even lower energies at the position of the missing hydrogen as well as the nearby oxygen due to the nearby unsaturated carbon.

11.4.3 Atom Embeddings

Having examined the spatial structure of SchNet representations, we go on to analyze what the model has learned about chemical elements included in the data. As described above, SchNet encodes atom types using embeddings $\mathbf{A}_Z \in \mathbb{R}^F$ that are learned during the training process. We visualize the two leading principal components of these embeddings to examine whether they resemble chemical intuition. Since QM9 only contains five atom types (H, C, N, O, F), we perform this analysis on the more diverse Materials Project dataset [44] as it includes 89 atom types ranging across the periodic table. Figure 11.7 shows the reduced embeddings of the main group elements of the periodic table. Atoms belonging to the same group tend to form clusters.

This is especially apparent for main groups 1–7, while group 8 appears to be more scattered. Beyond that, there are partial orderings of elements according to their period within some of the groups. We observe a partial order from light to heavier elements in some groups, e.g., in group 1 (left to right: H–Li–Na–[K, Rb, Cs]), group 2 (left to right: Be–Mg–Ca–Sr–Ba), and group 5 (left to right: P–As–

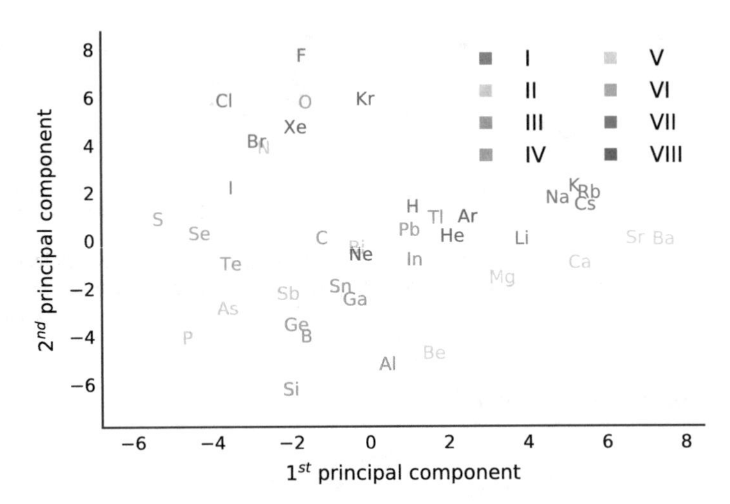

Fig. 11.7 Two leading principal components of the element embeddings learned by a SchNet model trained on 60k reference calculations of the Materials Project [44]

Sb–Bi). These results are consistent with those we obtained from previous SchNet models trained on earlier versions of the Materials Project repository [11, 27].

Note that these extracted chemical insights are not imposed by the SchNet architecture, but had to be inferred by the model based on the bulk systems and its energies in the training data.

11.5 Conclusions

We have presented the deep tensor neural network framework and its implementation SchNet, which obtains accurate predictions of chemical properties for molecules and materials. Representations of chemical environments are learned directly from atom types and position while filter-generating networks allow to incorporate invariance and prior knowledge.

In our analysis, we have found that atomic representations reflect an inferred chemical environment based on the bias of the training data. The obtained representations are dominated by local interactions and can be further localized using cosine cutoff functions that improve generalization. However, if a sufficient amount of data is available, interaction energies can be reliably attributed to larger interatomic distances, which will further improve the accuracy of the model. Moreover, we have defined local chemical potentials that allow for spatially resolved chemical insights and have shown that the models learn embeddings of chemical elements that show resemblance of the structure of the periodic table.

In conclusion, SchNet presents an end-to-end atomistic neural network that we expect to facilitate further developments towards interpretable deep learning architectures to assist chemistry research.

Acknowledgments The authors thank Michael Gastegger for valuable discussions and feedback. This work was supported by the Federal Ministry of Education and Research (BMBF) for the Berlin Big Data Center BBDC (01IS14013A) and the Berlin Center for Machine Learning (01IS18037A). Additional support was provided by the Institute for Information & Communications Technology Promotion and funded by the Korean government (MSIT) (No. 2017-0-00451, No. 2017-0-01779). A.T. acknowledges support from the European Research Council (ERC-CoG grant BeStMo).

References

1. K.T. Schütt, H. Glawe, F. Brockherde, A. Sanna, K.R. Müller, E. Gross, Phys. Rev. B **89**(20), 205118 (2014)
2. H. Huo, M. Rupp, (2017, preprint). arXiv:1704.06439
3. F.A. Faber, A.S. Christensen, B. Huang, O.A. von Lilienfeld, J. Chem. Phys. **148**(24), 241717 (2018)
4. S. De, A.P. Bartók, G. Csányi, M. Ceriotti, Phys. Chem. Chem. Phys. **18**(20), 13754 (2016)
5. T. Morawietz, A. Singraber, C. Dellago, J. Behler, Proc. Natl. Acad. Sci. **113**(30), 8368 (2016)
6. M. Gastegger, J. Behler, P. Marquetand, Chem. Sci. **8**(10), 6924 (2017)
7. F.A. Faber, L. Hutchison, B. Huang, J. Gilmer, S.S. Schoenholz, G.E. Dahl, O. Vinyals, S. Kearnes, P.F. Riley, O.A. von Lilienfeld, J. Chem. Theory Comput. **13**(11), 5255 (2017)
8. E.V. Podryabinkin, A.V. Shapeev, Comput. Mater. Sci. **140**, 171 (2017)
9. F. Brockherde, L. Vogt, L. Li, M.E. Tuckerman, K. Burke, K.R. Müller, Nat. Commun. **8**, 872 (2017)
10. A.P. Bartók, S. De, C. Poelking, N. Bernstein, J.R. Kermode, G. Csányi, M. Ceriotti, Sci. Adv. **3**(12), e1701816 (2017)
11. K.T. Schütt, H.E. Sauceda, P.J. Kindermans, A. Tkatchenko, K.R. Müller, J. Chem. Phys. **148**(24), 241722 (2018)
12. S. Chmiela, H.E. Sauceda, K.R. Müller, A. Tkatchenko, Towards exact molecular dynamics simulations with machine-learned force fields. Nat. Commun. **9**(1), 1–10 (2018)
13. A. Ziletti, D. Kumar, M. Scheffler, L.M. Ghiringhelli, Nat. Commun. **9**(1), 2775 (2018)
14. D. Dragoni, T.D. Daff, G. Csányi, N. Marzari, Phys. Rev. Mater. **2**(1), 013808 (2018)
15. A.P. Bartók, M.C. Payne, R. Kondor, G. Csányi, Phys. Rev. Lett. **104**(13), 136403 (2010)
16. M. Rupp, A. Tkatchenko, K.R. Müller, O.A. Von Lilienfeld, Phys. Rev. Lett. **108**(5), 058301 (2012)
17. G. Montavon, K. Hansen, S. Fazli, M. Rupp, F. Biegler, A. Ziehe, A. Tkatchenko, A.V. Lilienfeld, K.R. Müller, in *Advances in Neural Information Processing Systems 25*, ed. by F. Pereira, C.J.C. Burges, L. Bottou, K.Q. Weinberger (Curran Associates, Red Hook, 2012), pp. 440–448
18. K. Hansen, G. Montavon, F. Biegler, S. Fazli, M. Rupp, M. Scheffler, O.A. Von Lilienfeld, A. Tkatchenko, K.R. Müller, J. Chem. Theory Comput. **9**(8), 3404 (2013)
19. K. Hansen, F. Biegler, R. Ramakrishnan, W. Pronobis, O.A. von Lilienfeld, K.R. Müller, A. Tkatchenko, J. Phys. Chem. Lett. **6**, 2326 (2015)
20. S. Chmiela, A. Tkatchenko, H.E. Sauceda, I. Poltavsky, K.T. Schütt, K.R. Müller, Sci. Adv. **3**(5), e1603015 (2017)
21. J. Behler, M. Parrinello, Phys. Rev. Lett. **98**(14), 146401 (2007)
22. A.P. Bartók, R. Kondor, G. Csányi, Phys. Rev. B **87**(18), 184115 (2013)
23. M. Gastegger, L. Schwiedrzik, M. Bittermann, F. Berzsenyi, P. Marquetand, J. Chem. Phys. **148**(24), 241709 (2018)
24. W. Pronobis, K.T. Schütt, A. Tkatchenko, K.R. Müller, Eur. Phys. J. B **91**(8), 178 (2018)
25. A.E. Sifain, N. Lubbers, B.T. Nebgen, J.S. Smith, A.Y. Lokhov, O. Isayev, A.E. Roitberg, K. Barros, S. Tretiak, J. Phys. Chem. Lett. **9**(16), 4495 (2018)
26. K. Yao, J.E. Herr, D.W. Toth, R. Mckintyre, J. Parkhill, The TensorMol-0.1 model chemistry: a neural network augmented with long-range physics. Chem. Sci. **9**(8), 2261–2269 (2018)

27. K.T. Schütt, M. Gastegger, A. Tkatchenko, K.R. Müller, (2018, preprint). arXiv:1806.10349
28. K.T. Schütt, F. Arbabzadah, S. Chmiela, K.R. Müller, A. Tkatchenko, Nat. Commun. **8**, 13890 (2017)
29. K.T. Schütt, P.J. Kindermans, H.E. Sauceda, S. Chmiela, A. Tkatchenko, K.R. Müller, *Advances in Neural Information Processing Systems*, vol. 30 (Curran Associates, Red Hook, 2017), pp. 992–1002
30. A. Pukrittayakamee, M. Malshe, M. Hagan, L. Raff, R. Narulkar, S. Bukkapatnum, R. Komanduri, J. Chem. Phys. **130**(13), 134101 (2009)
31. M. Malshe, R. Narulkar, L.M. Raff, M. Hagan, S. Bukkapatnam, P.M. Agrawal, R. Komanduri, J. Chem. Phys. **130**(18), 184102 (2009)
32. Y. LeCun, B. Boser, J.S. Denker, D. Henderson, R.E. Howard, W. Hubbard, L.D. Jackel, Neural Comput. **1**(4), 541 (1989)
33. X. Jia, B. De Brabandere, T. Tuytelaars, L.V. Gool, in *Advances in Neural Information Processing Systems*, ed. by D.D. Lee, M. Sugiyama, U.V. Luxburg, I. Guyon, R. Garnett, vol. 29 (Curran Associates, Red Hook, 2016), pp. 667–675
34. C. Dugas, Y. Bengio, F. Bélisle, C. Nadeau, R. Garcia, *Advances in Neural Information Processing Systems* (Curran Associates, Red Hook, 2001), pp. 472–478
35. D.A. Clevert, T. Unterthiner, S. Hochreiter, (2015, preprint). arXiv:1511.07289
36. F. Chollet, *Proceedings of the IEEE Conference on Computer Vision and Pattern Recognition* (2017)
37. C. Szegedy, V. Vanhoucke, S. Ioffe, J. Shlens, Z. Wojna, *Proceedings of the IEEE Conference on Computer Vision and Pattern Recognition* (2016), pp. 2818–2826
38. J. Deng, W. Dong, R. Socher, L.J. Li, K. Li, L. Fei-Fei, *IEEE Conference on Computer Vision and Pattern Recognition, CVPR 2009* (IEEE, Piscataway, 2009), pp. 248–255
39. D. Broomhead, D. Lowe, Complex Syst. **2**, 321 (1988)
40. J. Moody, C.J. Darken, Neural Comput. **1**(2), 281 (1989)
41. R. Ramakrishnan, P.O. Dral, M. Rupp, O.A. von Lilienfeld, Sci. Data **1**, 140022 (2014)
42. I. Loshchilov, F. Hutter, (2016, preprint). arXiv:1608.03983
43. D.P. Kingma, J. Ba, (2014, preprint). arXiv:1412.6980
44. A. Jain, S.P. Ong, G. Hautier, W. Chen, W.D. Richards, S. Dacek, S. Cholia, D. Gunter, D. Skinner, G. Ceder, K.A. Persson, APL Mater. **1**(1), 011002 (2013). https://doi.org/10.1063/1.4812323

Part IV
Atomistic Simulations

Preface

An important problem of computational chemistry is to perform molecular dynamics simulations for molecules and materials. Therefore, one needs to obtain their potential energy surfaces (PESs) as accurately as possible. For large systems or long timescales, however, one has to trade off accuracy against computational cost. While *ab initio* electronic structure methods provide the most precise PESs, they are not affordable for large systems. On the other hand, (semi-)empirical force fields are fast to evaluate, but are not able to represent all aspects of the PES due to their constrained functional form and specific physical approximations. For these reasons, the modeling of PESs and corresponding atomic forces has become an important application of machine learning in quantum chemistry. In contrast to traditional force fields, machine learning potentials are more flexible, while at the same time a variety of proven techniques to avoid overfitting ensures generalization to unseen configurations [1].

Fully connected neural network was among the first machine learning models to be fitted to PESs [2, 3] before highly specialized descriptors and networks were developed. In Chap. 12, [4] give an overview about using high-dimensional neural network potentials (NNPs) [5] to learn PESs and drive atomistic simulations. The chapter nicely covers the whole procedure from generation of reference data sets over the construction of atomistic neural networks to performing molecular dynamics (MD) simulations and obtaining molecular spectra. Building on this, [6] go into detail about the future engineering aspects of NNPs using *atom-centered symmetry functions* and how atomic forces can be calculated from the NNP. Besides neural networks, notably Gaussian process (or kernel ridge regression) models have been successful in modeling molecules [7] and materials [8, 9].

Chapter 15 [10] details the important aspect of building a set of reference calculations using active learning and uncertainty estimation. This is explored for various applications (MD simulation, relaxation) and machine learning models (Gaussian processes, neural networks).

While the above chapters deal with the problem to efficiently *generate* an MD trajectory, Chap. 16 [11] introduces state-of-the-art methods to analyze MD trajectories on long timescales. Using these techniques, one can obtain low-dimensional models of the long-time dynamics of a given trajectory to gain insights about the metastable states of a system.

Berlin, Germany — Kristof T. Schütt
Berlin, Germany — Stefan Chmiela
Basel, Switzerland — O. Anatole von Lilienfeld
Luxembourg, Luxembourg — Alexandre Tkatchenko
Kashiwa, Japan — Koji Tsuda
Berlin, Germany — Klaus-Robert Müller
September 2019

References

1. K. Hansen, G. Montavon, F. Biegler, S. Fazli, M. Rupp, M. Scheffler, O.A. Von Lilienfeld, A. Tkatchenko, K.-R. Müller, J. Chem. Theory Comput. **9**(8), 3404 (2013)
2. T.B. Blank, S.D. Brown, A.W. Calhoun, D.J. Doren, J. Chem. Phys. **103**(10), 4129 (1995)
3. S. Manzhos, T. Carrington Jr., J. Chem. Phys. **125**(8), 084109 (2006)
4. M. Gastegger, P. Marquetand, in *Machine Learning for Quantum Simulations of Molecules and Materials*, ed. by K.T. Schütt, S. Chmiela, A. von Lilienfeld, A. Tkatchenko, K. Tsuda, K.-R. Müller. Lecture Notes Physics (Springer, Berlin, 2019)
5. J. Behler, M. Parrinello, Phys. Rev. Lett. **98**(14), 146401 (2007)
6. M. Hellström, J. Behler, in *Machine Learning for Quantum Simulations of Molecules and Materials*, ed. by K.T. Schütt, S. Chmiela, A. von Lilienfeld, A. Tkatchenko, K. Tsuda, K.-R. Müller. Lecture Notes Physics (Springer, Berlin, 2019)
7. S. Chmiela, A. Tkatchenko, H.E. Sauceda, I. Poltavsky, K.T. Schütt, K.-R. Müller, Sci. Adv. **3**(5), e1603015 (2017)
8. A.P. Bartók, M.C. Payne, R. Kondor, G. Csányi, Phys. Rev. Lett. **104**(13), 136403 (2010)
9. Z. Li, J.R. Kermode, A. De Vita, Phys. Rev. Lett. **114**(9), 096405 (2015)
10. A. Shapeev, K. Gubaev, E. Tsymbalov, E. Podryabinkin, in *Machine Learning for Quantum Simulations of Molecules and Materials*, ed. by K.T. Schütt, S. Chmiela, A. von Lilienfeld, A. Tkatchenko, K. Tsuda, K.-R. Müller. Lecture Notes Physics (Springer, Berlin, 2019)
11. F. Noé, in *Machine Learning for Quantum Simulations of Molecules and Materials*, ed. by K.T. Schütt, S. Chmiela, A. von Lilienfeld, A. Tkatchenko, K. Tsuda, K.-R. Müller. Lecture Notes Physics (Springer, Berlin, 2019)

Molecular Dynamics with Neural Network Potentials

12

Michael Gastegger and Philipp Marquetand

Abstract

Molecular dynamics simulations are an important tool for describing the evolution of a chemical system with time. However, these simulations are inherently held back either by the prohibitive cost of accurate electronic structure theory computations or the limited accuracy of classical empirical force fields. Machine learning techniques can help to overcome these limitations by providing access to potential energies, forces, and other molecular properties modeled directly after an accurate electronic structure reference at only a fraction of the original computational cost. The present text discusses several practical aspects of conducting machine learning driven molecular dynamics simulations. First, we study the efficient selection of reference data points on the basis of an active learning inspired adaptive sampling scheme. This is followed by the analysis of a machine learning based model for simulating molecular dipole moments in the framework of predicting infrared spectra via molecular dynamics simulations. Finally, we show that machine learning models can offer valuable aid in understanding chemical systems beyond a simple prediction of quantities.

M. Gastegger
Technical University of Berlin, Machine Learning Group, Berlin, Germany
e-mail: michael.gastegger@tu-berlin.de

P. Marquetand (✉)
University of Vienna, Faculty of Chemistry, Institute of Theoretical Chemistry, Vienna, Austria
e-mail: philipp.marquetand@univie.ac.at

K. T. Schütt et al. (eds.), *Machine Learning Meets Quantum Physics*, Lecture Notes in Physics 968, https://doi.org/10.1007/978-3-030-40245-7_12

12.1 Introduction

Chemistry is—in large part—concerned with the changes that matter undergoes. As such, chemistry is inherently time-dependent and if we want to model such chemical processes, then a time-dependent approach is most intuitive. The corresponding techniques can be summarized as dynamics simulations. In particular molecular dynamics (MD) simulations—defined usually as treating the nuclear motion with Newton's classical mechanics—are commonly used to "mimic what atoms do in real life" [1]. Such simulations have become indispensable not only in chemistry but also in adjacent fields like biology and material science [2, 3].

An important ingredient of MD simulations are molecular forces, which determine how the nuclei move. For the sake of accuracy, it is desirable to obtain these forces from quantum mechanical electronic structure calculations. However, such *ab initio* calculations are expensive from a computational perspective and hence only feasible for relatively small systems or short timescales. For larger systems (e.g., proteins) molecular forces are instead modeled by classical force fields, which are composed of analytic functions based on physical findings. As a consequence of these approximations, classical force fields are extremely efficient to compute but fail to reach the accuracy of electronic structure methods.

With the aim to obtain both a high accuracy and a fast evaluation, machine learning (ML) is employed to predict forces and an increasing number of research efforts are devoted to this idea. Since the forces are commonly evaluated as the derivative of the potential energy with respect to the coordinates, the learning of energies and forces are tightly connected and we often find the term "machine learning potential" in the literature.

Possibly the first work to use ML of potentials combined with dynamics simulations appeared in 1995 by Blank et al. [4], where diffusion of CO on a Ni surface and H_2 on Si was modeled. A comprehensive overview over the earlier work in this field, where mostly neural networks were employed, can be found in the reviews [5,6]. Later, also other ML methods like Gaussian approximation potentials [7] were utilized, diversifying the research landscape, which is reflected in various, more topical reviews, see, e.g., [8–11]. Today, the field has become so active that a review would be outdated by tomorrow. Here instead, we give an example of what can be achieved when combining ML and MD.

In this chapter, we describe simulations of one of the most fundamental experiments to detect moving atoms, namely infrared spectra. The simulations utilize MD based on potentials generated with high-dimensional neural networks, a special ML architecture. We show, in particular, how the training data can efficiently be gathered by an adaptive sampling scheme. Several practical aspects, tricks, and pitfalls are presented. Special emphasis is put also on the prediction of dipole moments and atomic charges, which are necessary ingredients besides potential energies and forces for the calculation of infrared spectra from MD.

12.2 Methods

12.2.1 High-Dimensional Neural Network Potentials

High-dimensional neural network potentials (NNPs) are a type of atomistic ML potentials [12]. Atomistic potentials model the properties of a system based on the contributions of individual atoms due to their local chemical environment (Fig. 12.1). In high-dimensional NNPs, atomic environments are represented via so-called atom-centered symmetry functions (ACSFs) [13]. Typically, ACSFs are radial and angular distribution functions, which account for rotational and translation invariances of the system. A radial symmetry function, for example, is a superposition of Gaussian densities:

$$G_i^{\mathrm{rad}} = \sum_{j \neq i}^{N} e^{-\eta (r_{ij} - r_0)^2} f_{\mathrm{cut}}(r_{ij}). \tag{12.1}$$

The sum includes all atoms j in vicinity of the central atom i. r_{ij} is the distance between i and j. η and r_0 are parameters which modulate the width and center of the Gaussian. A cutoff function f_{cut} ensures that only the local chemical environment contributes to the ACSF. For a more in-depth discussion on ACSFs and their features, we refer to References [11, 13–15].

Based on the ACSF representation of each atom, an atomistic neural network then predicts the contribution of this atom to the global molecular property. Finally, these contributions are recombined via an atomistic aggregation layer in order to recover the target property. For NNPs which model the potential energy E of a system, this

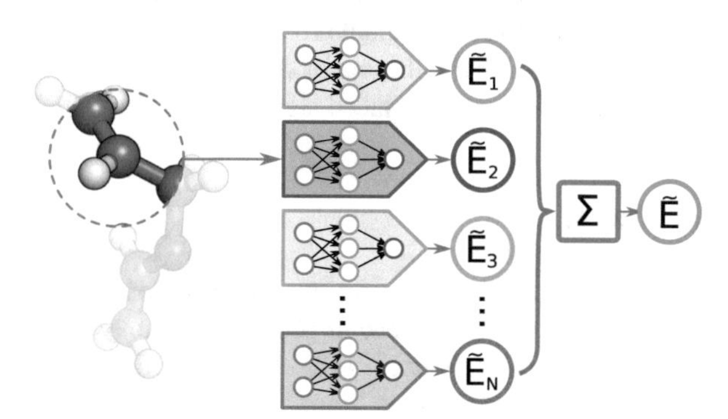

Fig. 12.1 In a high-dimensional neural network potential, the local chemical environment of each atom is first encoded in a structural descriptor. Based on these descriptors, neural networks predict atomistic energy contributions, where different networks are used for the different chemical elements. Finally, the atomic energies are summed in order to recover the total energy of the system

layer is usually chosen as a sum over the individual atomic energies E_i:

$$\tilde{E}^m = \sum_i^{N_m} \tilde{E}_i^m, \tag{12.2}$$

where N is the number of atoms in the molecule m. However, different aggregation layers can be formulated in order to model various properties, as will be discussed in the next section.

Due to the form of high-dimensional NNPs, expressions for analytic Cartesian derivatives of the model are readily available. Hence, NNPs provide access to energy conserving forces

$$\tilde{\mathbf{F}}_\alpha^m = -\sum_i^{N_m} \sum_d^{D_i^m} \frac{\partial \tilde{E}_i^m}{\partial G_d^m} \frac{\partial G_d^m}{\partial \mathbf{R}_\alpha^m}, \tag{12.3}$$

where $\tilde{\mathbf{F}}_\alpha^m$ and $\mathbf{R}_\alpha^m$ are the forces acting on atom α and its position, while D_i^m is the number of ACSFs centered on atom i. NNP forces can also be included into the training process by minimizing a loss function of the form

$$\mathcal{L} = \frac{1}{M} \sum_m^M \left(\tilde{E}^m - E^m\right)^2 + \frac{\vartheta}{M} \sum_m^M \frac{1}{3N_m} \sum_i^{N_m} \|\tilde{\mathbf{F}}_i^m - \mathbf{F}_i^m\|^2. \tag{12.4}$$

Here, M is the number of molecules present in the data set. ϑ controls the trade-off between fitting energies and forces, while $\mathbf{F}_i^m$ is the vector of Cartesian force components acting on atom i of molecule m. By using forces during training, $3N$ additional pieces of information are available for each molecule beside the potential energy. As a consequence, the overall number of reference computations required to construct an accurate NNP can be reduced significantly [16, 17].

12.2.2 Dipole Model

In addition to energies and forces, atomistic aggregation layers can also be formulated to model various other molecular properties. One example is the dipole moment model introduced in reference [18]. Here, the dipole moment $\boldsymbol{\mu}$ of a molecule is expressed as a system of atomic point charges, according to the relation

$$\tilde{\boldsymbol{\mu}} = \sum_i^N \tilde{q}_i \mathbf{r}_i. \tag{12.5}$$

$\tilde{q}_i$ is the charge located at atom i and $\mathbf{r}_i$ is the position vector of the atom relative to the molecules center of mass. The charges $\tilde{q}$ are modeled via atomistic networks and depend on the local chemical environment. However, these point charges are never learned directly, but instead represent latent variables inferred by the NNP dipole model during training, where the following loss function is minimized:

$$\mathcal{L} = \frac{1}{M}\sum_{m}^{M}\left(\tilde{Q}_m - Q_m\right)^2 + \frac{1}{3M}\sum_{m}^{M}\|\tilde{\boldsymbol{\mu}}_m - \boldsymbol{\mu}_m\|^2 + \dots. \tag{12.6}$$

$\tilde{\boldsymbol{\mu}}$ is the expression for the dipole moment given in Eq. 12.5 and $\boldsymbol{\mu}$ is the electronic structure reference, calculated as the expectation value of the dipole moment operator [19]. Note that the charges are not directly accessible from solving the Schrödinger equation but are usually obtained *a posteriori* from different *ad hoc* partitioning schemes [20]. The first term in the loss function introduces the additional constraint that the sum of latent charges $\tilde{Q} = \sum_i \tilde{q}_i$ should reproduce the total charge Q of the molecule. Formulated in this way, the machine learning model depends only on quantum mechanical observables in the form of a molecule's electrostatic moments (total charge and dipole moment), which are directly accessible by all electronic structure methods. Although the above formulation does not guarantee the conservation of total charge, it reduces the overall charge fluctuations to a minimum. The remaining deviations can then be corrected using simple rescaling or shifting schemes without loss of generality (see, e.g., references [21] and [22]).

Expression 12.6 can easily be extended to include higher moments such as the quadrupole moment $\boldsymbol{\Theta}$, as was suggested in reference [18]. In the context of the above model, $\boldsymbol{\Theta}$ takes the form

$$\tilde{\boldsymbol{\Theta}} = \sum_i \tilde{q}_i \left(3\mathbf{r}_i \otimes \mathbf{r}_i - \mathbf{1}\,\|\mathbf{r}_i\|\right), \tag{12.7}$$

where, $\mathbf{r}_i \otimes \mathbf{r}_i$ is the outer product of the Cartesian position vectors of atom i. However, it was found that the introduction of quadrupole moments offers no additional advantage, at least when modeling dipoles. Moreover, using an atomistic model for $\boldsymbol{\Theta}$ can be problematic for small molecules such as water, since the atom-centered point charges are not able to resolve features of the charge distribution arising from, e.g., the lone pair electrons of the oxygen.

It should be emphasized at this point that the atomistic aggregation layers presented here are not restricted to a single type of machine learning architecture. They can be coupled with any model in a modular fashion, as long as it provides access to atomic contributions. This was, for example, done recently with the SchNet architecture in order to model dipole moment magnitudes [22].

12.2.3 Adaptive Sampling Scheme

Before a NNP can be used for simulations, its free parameters need to be determined by training on a suitable set of reference data. Typically, a set of reference molecules is chosen in a two-step process. First, the PES is sampled to obtain a representative set of molecular configurations. Afterwards, the quantum chemical properties of these structures (e.g., energies, forces,...) are computed with an appropriate electronic structure method. The sampling can be performed in different ways, with molecular dynamics and normal mode sampling being only a few examples [14,23]. However, a feature shared by most sampling methods is that they either use approximate methods such as molecular force fields to guide the sampling or they perform all simulations at the final level of theory. Both approaches have drawbacks. In the first case, the PES regions explored with the lower level of theory need not correspond to regions relevant for the high-level model (e.g., different molecular geometries). In the second case, the unfortunate scaling of accurate electronic structure methods limits either the regions of the PES that can be covered or the overall accuracy of the reference computations.

A solution to these issues is to use an adaptive sampling scheme, where the ML model itself selects new data points to improve its accuracy [18]. This approach is inspired by active learning techniques and proceeds as follows (Fig. 12.2): First, a crude NNP model is used to explore an initial region of the PES. During simulation, an uncertainty measure for the NNP predictions is monitored. If this measure exceeds a threshold, the sampling is stopped and electronic structure computations are performed for the corresponding configuration. The resulting data is then added to the reference set and used to update the NNP. Sampling is continued with the improved model. This procedure is carried out in an iterative fashion until the desired level of accuracy is reached.

One advantage of this scheme is that the NNP model used to guide the sampling closely resembles the electronic structure reference method for large stretches of

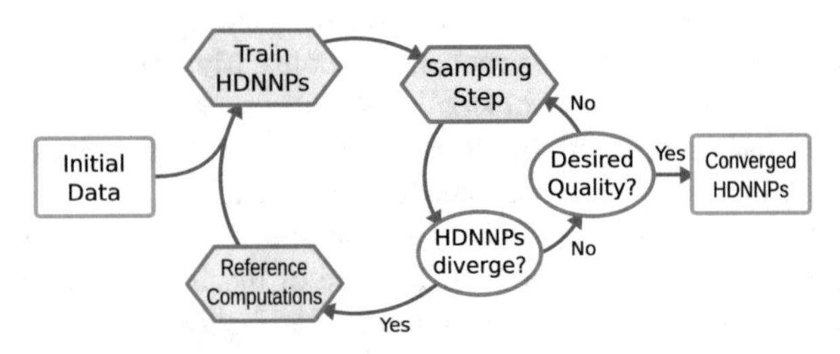

Fig. 12.2 The adaptive sampling scheme starts by training a preliminary ensemble of NNPs on a small set of reference computations. This ensemble is then used to sample new configurations via, e.g., molecular dynamics. For every sampled configuration, an uncertainty measure is computed. If this measure exceeds a predefined threshold, the simulation is stopped and a reference calculation is performed. The new data point is then added to the reference data and the next generation of NNPs is trained. This procedure is repeated in an iterative manner until a convergence threshold is reached

coordinate space. Hence, similar regions of the PES will be explored (e.g., bond lengths) as if the simulations were carried out with the reference method exclusively.

In addition, by using the model uncertainty to determine when additional reference computations should be performed, only a small number of expensive calculations are necessary. Due to the simplicity of this scheme, it can easily be combined with different sampling methods, such as those in molecular dynamics, metadynamics or Monte-Carlo based approaches [24, 25].

Perhaps the most important ingredient for the above scheme is an appropriate uncertainty measure. Here, it is possible to make use of a trait of NNs in general and NNPs in particular. Two NNPs trained on the same reference data will agree closely for regions of the PES described well by both models. However, in regions sampled insufficiently the predictions of both models will diverge quickly. Using the disagreement between different models to guide data selection is a popular approach in active learning called query by committee [26]. Based on the above behavior, one can formulate the following uncertainty measure for NNPs:

$$\sigma_E = \sqrt{\frac{1}{\mathfrak{N}-1}\sum_{\mathfrak{n}}^{\mathfrak{N}}\left(\tilde{E}_{\mathfrak{n}} - \overline{E}\right)^2}. \tag{12.8}$$

$\tilde{E}_{\mathfrak{n}}$ is the energy predicted by one of $\mathfrak{N}$ different NNPs and $\overline{E}$ is the average of all predictions. Hence, σ_E is the standard deviation of the different model predictions. Using an uncertainty measure of this form also has the following advantage: Since different NNPs are used to compute σ_E, they can be combined into an ensemble, where the prediction averages (e.g., $\overline{E}$) are used to guide PES exploration. The consequence is an improvement in the overall predictive accuracy of the ML approach at virtually no extra cost, due to error cancellation in the individual models.

12.3 Generation of Reference Data Sets

The following section discusses different practical aspects of the adaptive sampling scheme introduced above. After an investigation on the accuracy advantage offered by NNP ensembles, we study how frequently new reference computations are requested during a sampling run. Afterwards, the utility of different predicted properties as uncertainty measures for NNPs is analyzed. Finally, we introduce an extension to the standard sampling scheme, which improves overall sampling efficiency.

12.3.1 Accuracy of NNP Ensembles

In order to investigate the accuracy offered by ensembles of NNPs, we compare the predictions of ensembles containing up to five members to their respective electronic

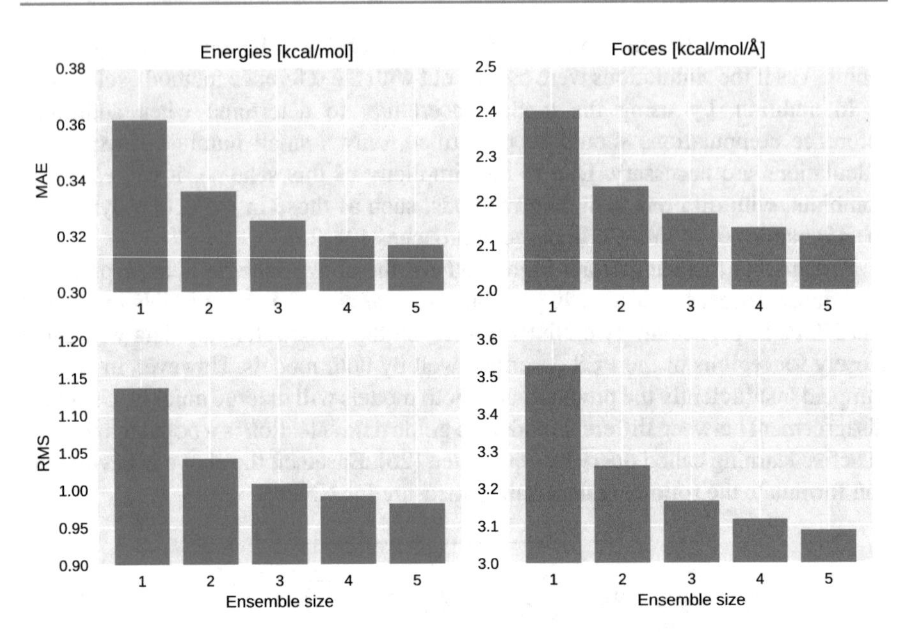

Fig. 12.3 Accuracy of ensemble predictions for molecular energies (blue) and forces (red) depending on the number of members. The computed error measures, MAE and RMSE, appear to decrease according to an $\frac{1}{\sqrt{\mathfrak{N}}}$ relation, where $\mathfrak{N}$ is the number of models in the ensemble. In all cases, the gain in accuracy is most pronounced when going from a single network to an ensemble of two

structure reference. This analysis is based on the protonated alanine tripeptide data set obtained in reference [18], which also serves as a basis for several other studies in this text. The data set contains 718 different peptide configurations at the BLYP/def2-DZVP level of theory sampled with the scheme described above. Figure 12.3 shows the energy and force mean absolute errors (MAEs) and root mean squared errors (RMSEs) for the different ensembles. Even the combination of only two different models already leads to a marked decrease in the prediction error. Since ensembles thrive on a cancellation of random error fluctuations, this gain in accuracy is particularly pronounced for the RMSEs. An interesting observation is that the forces profit to a greater extent than the energies, with a reduction in the error by approximately 0.3 $\text{kcal}\,\text{mol}^{-1}\,\text{Å}^{-1}$. This effect is expected to be of importance in the early stages of an adaptive sampling run, as the improved reliability of the model increases the likelihood that physically relevant configurations are sampled.

12.3.2 Choice of Uncertainty Measures

An important feature of atomistic NNPs is their ability to operate as fragmentation approaches, where they predict the energies of large molecules after being trained on only small fragments [27]. Hence, expensive reference computations never have to be performed for the whole system, but only for parts of it. This feature can be combined with the adaptive sampling scheme, as was, for example, done in reference [18]. In this setup, the uncertainty is not measured for the whole molecule, but instead for atom-centered fragments. Reference computations are only performed for those fragments where the uncertainty exceeds a predefined threshold, thus reducing the computational cost of constructing an accurate NNP even further. However, the deviation of ensemble energies (Eq. 12.8) can now no longer serve as the uncertainty measure.

Although substituting the total energies in Eq. 12.8 by their atom-wise counterparts $\tilde{E}_i$ would in theory be a straightforward choice for an atomistic uncertainty estimate, it is not feasible in practice. Due to the way NNPs are constructed (see Eq. 12.2), the partitioning of the total energy into latent atomic contributions is not unique. Hence, even if two NNPs yield almost identical predictions for the molecular energies, the distributions of atom-wise contributions can still differ significantly, as is shown for the alanine tripeptide in Fig. 12.4. If, e.g., the atomic energies of carbon atoms are used to compute the uncertainty, large deviations will be encountered for all regions of the PES, no matter how well the global predictions agree. As a consequence, reference computations will be performed for a

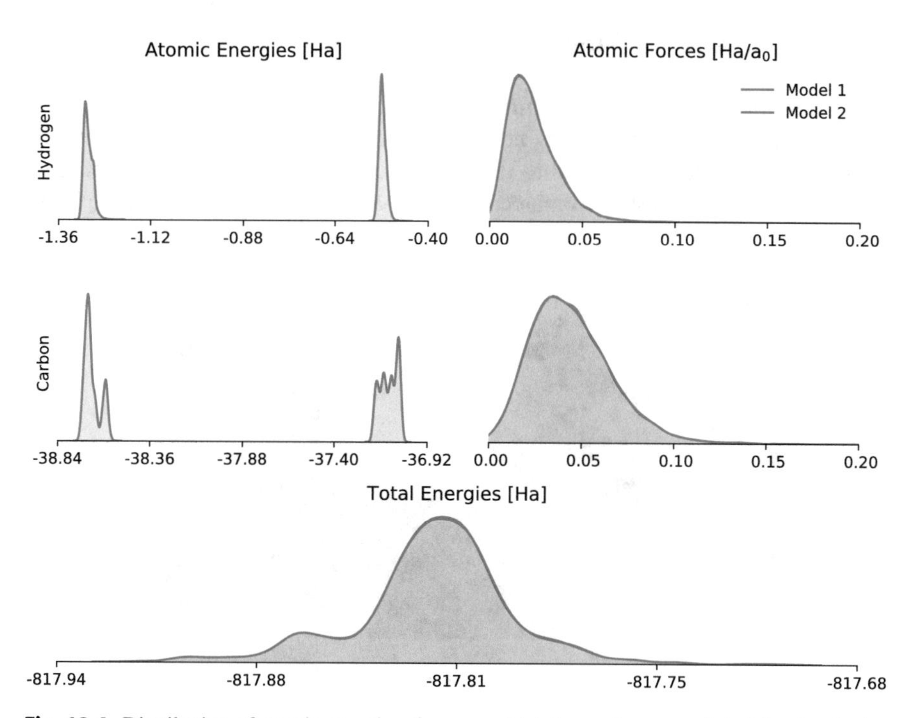

Fig. 12.4 Distribution of atomic energies, forces, and total energies as predicted for the alanine tripeptide by two NNP models (shown in red and blue). Although the NNP predictions agree well in the case of the total energies and atomic forces, the energy contributions of individual atoms vary dramatically between the models

large fraction of encountered configurations, thus effectively negating the advantage offered by the adaptive sampling scheme.

The better alternative is to reformulate the above measure to instead use the forces acting on each atom:

$$\sigma_F^{(\alpha)} = \sqrt{\frac{1}{\mathfrak{N}-1}\sum_{\mathfrak{n}}^{\mathfrak{N}} \left\| \tilde{\mathbf{F}}_{\mathfrak{n}}^{(\alpha)} - \overline{\mathbf{F}}^{(\alpha)} \right\|^2}, \tag{12.9}$$

where $\tilde{\mathbf{F}}_{\mathfrak{n}}^{(\alpha)}$ is the force acting on atom α as predicted by model $\mathfrak{n}$ of the ensemble. $\overline{\mathbf{F}}^{(\alpha)}$ is the average over all predictions. The measure $\sigma_F^{(\alpha)}$ has several advantages. Since it depends on the molecular forces it is purely atomistic. Moreover, due to how the forces are computed in NNPs (Eq. 12.3), they are insensitive to the learned partitioning in a similar manner as the total energy. This property can be observed in Fig. 12.4, where the distributions of forces acting on, e.g., hydrogen and carbon atoms show a similar agreement between models as do the molecular energies, but not the atomic energies.

12.3.3 Frequency of Reference Computations

An important aspect of the adaptive sampling scheme is how frequently new electronic structure computations need to be performed. Figure 12.5 depicts the number of configurations added to the reference data set versus the total number of sampling steps. The studied molecule is an n-alkane ($C_{69}H_{140}$, see Figure inset)

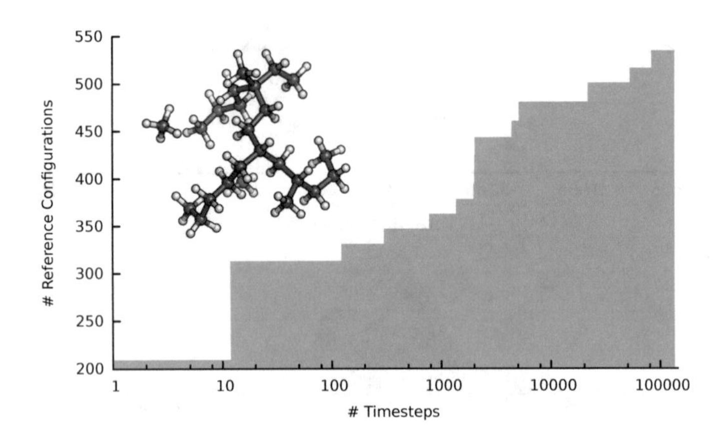

Fig. 12.5 Number of configurations accumulated during an adaptive sampling run for the $C_{69}H_{140}$ n-alkane plotted against the number of molecular dynamics steps. New samples are added frequently during the early stages of the sampling, while almost no configurations are collected during the later stages

and the sampling statistics were taken from the supporting information of Reference [18], obtained with a fragmentation procedure and the aforementioned force based uncertainty. As can be seen in the figure, there is a marked decrease in the number of electronic structure queries as the sampling progresses. Initially, new samples are added frequently, as the model explores the configuration space. More than half of the new samples are added within the first 2000 exploration steps. After this phase of determining a reliable first approximation of the electronic structure method, the sampling process is dominated by fine-tuning the NNP ensemble. Now only samples corresponding to insufficiently described regions of the PES are collected, reducing the requirement for expensive reference computations significantly. The efficiency of this simple approach is remarkable insofar, as only 534 configurations are needed to obtain an accurate model of the n-alkane sporting 621 degrees of freedom. The final model achieves RMSEs of 0.09 kcal mol^{-1} and 1.48 kcal mol^{-1} Å^{-1} compared to the reference energies and forces of the fragments.

12.3.4 Adaptive Sampling with Multiple Replicas

A potential problem of the adaptive sampling scheme is its serial nature. Currently, only one point of data is collected after each sampling period. Since the NNPs need to be retrained every time the reference data set is extended, the resulting procedure can become time consuming in its later stages, especially for large and flexible molecules (e.g., the tripeptide in reference [18]).

This problem can be overcome by introducing a parallel version of the adaptive scheme, as outlined in Fig. 12.6. Instead of simulating only a single system at a time, multiple sampling runs are performed in parallel, each using a copy of the NNP ensemble trained in the previous cycle. These independent simulations can be replicas of the system under various conditions (e.g., different temperatures), a range of conformations or even different molecules. Sampling is once again carried out until divergence is reached for all parallel simulations. The high uncertainty configurations are then computed with the reference method and added to the training data. This setup reduces the frequency with which NNPs need to be retrained, while at the same time improving PES exploration. A potential drawback of this scheme is that the collection of data points introduces periods of unproductivity, where some simulations are already finished while others are still running. However, this effect is negligible in praxis due to the high computational speed of the NNP models.

12.4 NNPs for Molecular Dynamics Simulations

Due to their combination of high accuracy and computational efficiency, NNPs are an excellent tool to accelerate MD simulations. A particularly interesting application is the computation of molecular spectra via the Fourier transform of different time autocorrelation functions [28]. Depending on the physical property

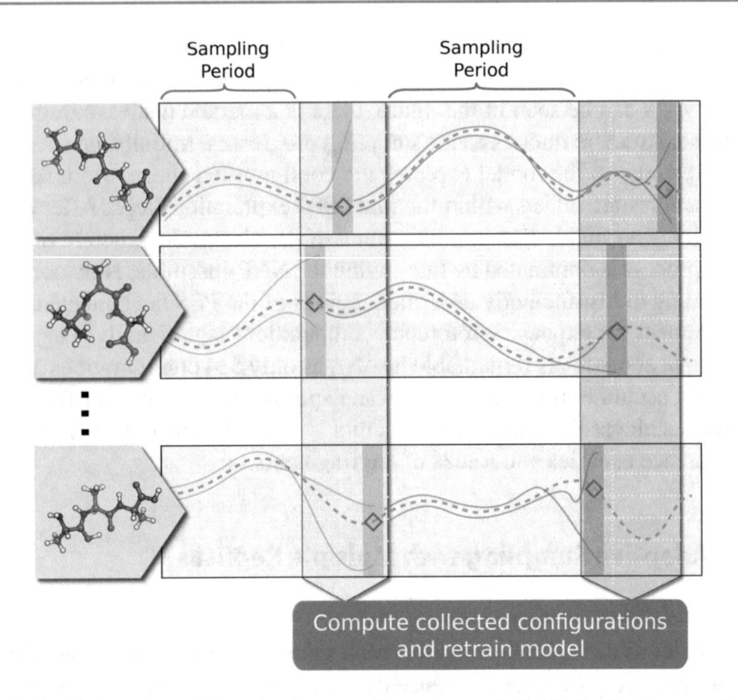

Fig. 12.6 Parallel version of the adaptive sampling scheme. Individual adaptive sampling runs are carried out for different replicas of the system (e.g., different configurations). For each replica, configurations with high uncertainty are identified. Once samples have been collected for all replicas, reference computations are carried out and the NNP ensemble model is retrained. Afterwards, the replica simulations are continued with the new model. In this manner, the NNPs have to be retrained less frequently and different regions of the PES can be explored more efficiently

underlying the autocorrelation function, different types of spectra can be obtained. One example are molecular infrared spectra, which can be modeled according to the relation:

$$I_{\mathrm{IR}} \propto \int_{-\infty}^{+\infty} \langle \dot{\boldsymbol{\mu}}(\tau)\dot{\boldsymbol{\mu}}(\tau + t)\rangle_{\tau}\, e^{-i\omega t}\mathrm{d}t, \tag{12.10}$$

where $\dot{\boldsymbol{\mu}}$ is the time derivative of the molecular dipole moment, τ is a time delay, ω is the vibrational frequency and t is the time.

The simulation of infrared spectra poses a particular challenge for machine learning techniques. Due to the dependence of Eq. 12.10 on $\dot{\boldsymbol{\mu}}$, a reliable model of the molecular dipole moment $\boldsymbol{\mu}$ is needed in addition to the PES description provided by conventional NNPs. In the next sections, we will explore various aspects and the potential pitfalls associated with such models.

12.4.1 Machine Learning for Molecular Dipole Moments

A straightforward way to model dipole moments in the context of NNPs is to train individual atomic networks to reproduce quantum chemical partial charges. The molecular dipoles can then be obtained via the point charge model given in Expression 12.5, where the $\tilde{q}_i$ are modeled by environment-dependent networks. A similar strategy was, e.g., used to model long-range electrostatic energies in Ref. [21].

However, such a model suffers from the inherent inhomogeneity of the various charge partitioning schemes available for electronic structure methods. The predicted partial charges can differ dramatically between schemes and some of them fail at reproducing molecular dipole moments entirely [20]. Even when considering only those methods which yield partial charges consistent with the molecular dipole moment, a strong dependence on the type of partitioning can still be observed. Hirshfeld charges [29], for example, appear to work well in the setup described above, as was demonstrated in reference [30]. Charges fit to the electrostatic potential (e.g., CHELPG [31]) on the other hand prove to be more problematic. To illustrate the issue at hand, Fig. 12.7 shows the MD IR spectrum of single methanol molecule computed with a partial charge model based on the CHELPG method in comparison to the electronic structure reference. The partial charge spectrum shows several marked differences from the reference. Small artificial peaks are introduced at 2100 and 3900 cm^{-1}, respectively. Moreover, the intensity of several peaks (e.g., at 1400 and 2800 cm^{-1}) is reproduced incorrectly.

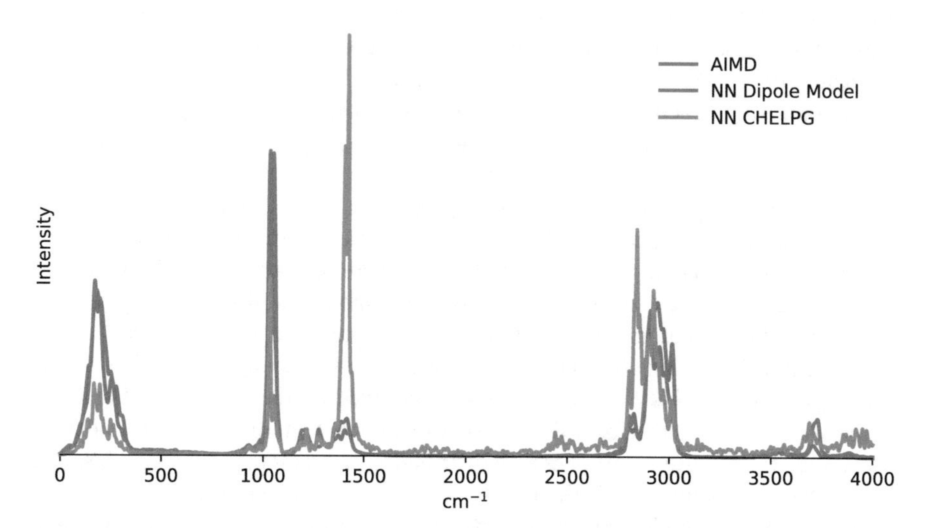

Fig. 12.7 Infrared spectra of a methanol molecule in the gas phase computed via *ab initio* molecular dynamics (blue), as well as machine-learned molecular dynamics using the dipole moment model introduced above (red) and a neural network model trained on CHELPG partial charges (gray). While the dipole moment model shows good agreement with the reference, the CHELPG model leads to erratic trends in peak positions and intensities

The most likely reason for these issues is the fitting procedure used to determine this particular type of reference charges. Since an independent least squares optimization is carried out for every molecular configuration, the obtained partial charges are not necessarily continuous with respect to incremental changes in the local environment of each atom. This makes it harder for the atomistic networks to learn a consistent charge assignment, leading to the erroneous behavior observed above.

A better approach is to incorporate the point charge model into the atomistic NNP architecture in the form of a dipole aggregation layer, as described in Sect. 12.2.2 and reference [18]. Instead of fitting to arbitrary partial charges, the model can now be trained directly on the molecular dipole moments, which are quantum mechanical observables. In this manner, the need for choosing an appropriate partitioning scheme is eliminated. The inherent advantage of such a model can be seen in Fig. 12.7, where it accurately reproduces the quantum chemical reference, although trained on the same set of configurations as the partial charge model.

12.4.2 Latent Partial Charges

A special feature of the above model is that it offers access to atomic partial charges. These charges are inferred by the NNP model based on the molecular electrostatic moments in a purely data driven fashion. Moreover, the charge models obtained with the above partitioning scheme depend on the local chemical environment of each atom. Hence, the charge distribution of the molecule can adapt to structural changes. Considering that partial charges are one of the most intuitive concepts in chemistry, the NNP latent charges represent an interesting analysis tool, e.g., for rationalizing reaction outcomes. In the following, we investigate how well the charges derived from the above dipole model agree with basic chemical insights.

One potential problem of atomistic properties (e.g., energies and charges) obtained via specialized aggregation layers is the extent with which the partitioning varies between different models. A good example are the atomic energies predicted by the tripeptide NNPs shown in Fig. 12.4. Although the total energies agree well, the partitioning into atomic contributions is highly non-unique. Such a behavior is detrimental if the latent contributions should serve as an analysis tool. In order to investigate whether this phenomenon is also observed for the latent partial charges, a similar analysis is performed for two dipole moment models of the alanine tripeptide. The resulting partial charge distributions are compared in Fig. 12.8. The latent charges obtained with the dipole model are significantly better conserved than the atomic energies and only small deviations are found between different NNPs.

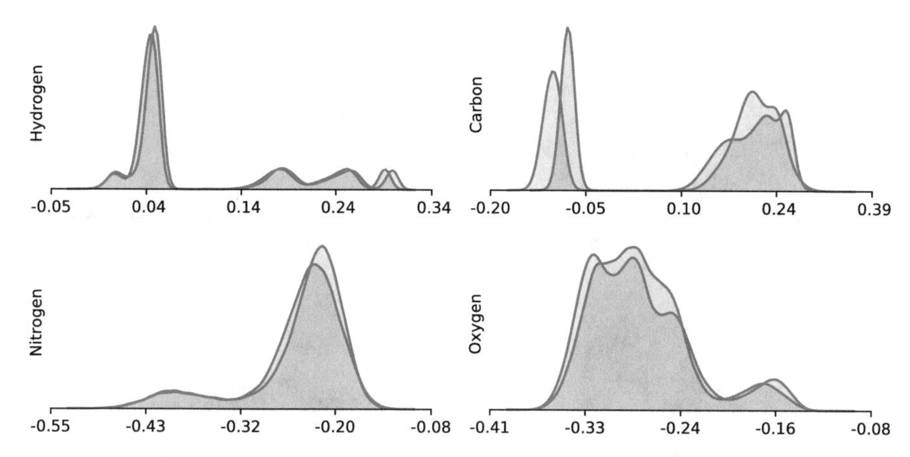

Fig. 12.8 Distribution of atomic partial charges predicted for the chemical elements present in the alanine tripeptide obtained with two dipole models (blue and red). Although differences between the two models are still present, the atomic charge distributions are better conserved than the atomic energies

This trend appears to hold in general, as can be seen by repeating the above experiment for the QM9 database [32] containing approximately 130,000 small organic molecules. Figure 12.9 shows the distributions of partial charges and atomic energies of oxygen predicted for all molecules in QM9. In each case, five different dipole and energy models were trained on growing subsets of the database using an adapted version of ACSFs, so-called weighted ACSFs [15] composed of 22 radial and 10 angular symmetry functions. In case of the dipole moment models, we replaced the dipole vector $\tilde{\boldsymbol{\mu}}$ in Eq. 12.6 by its magnitude $|\tilde{\boldsymbol{\mu}}| = \left\| \sum_i^N \tilde{q}_i \mathbf{r}_i \right\|$, as only the latter is available in QM9. All models used atomistic networks with three layers of 100 neurons each and shifted softplus nonlinearities and were trained using the ADAM algorithm [33] with a learning rate of 0.0001 (see Refs. [34] and [22] for details). Compared to the atomic energies, the distributions of atomic partial charges are not only better conserved between models, but also show systematic convergence upon inclusion of additional data. The reason for this behavior is the geometry-dependent term present in Eq. 12.5, which introduces an additional constraint into the partitioning procedure. This term encodes information on the spatial distribution of molecular charge and is different for every molecule. Moreover, due to the statistical nature of the training procedure, the latent charge model has to be consistent across a wide range of molecules and configurations. The combination of both properties strongly limits the number of valid latent charge assignments. These results are encouraging and demonstrate that the NNP partial charges are indeed capable to capture aspects of the chemistry underlying a system. However, care should be taken when using the latent charges as a direct replacement of their quantum chemical counterparts, as the resulting partitionings—although well behaved—are still not uniquely determined. This can lead to undesirable effects

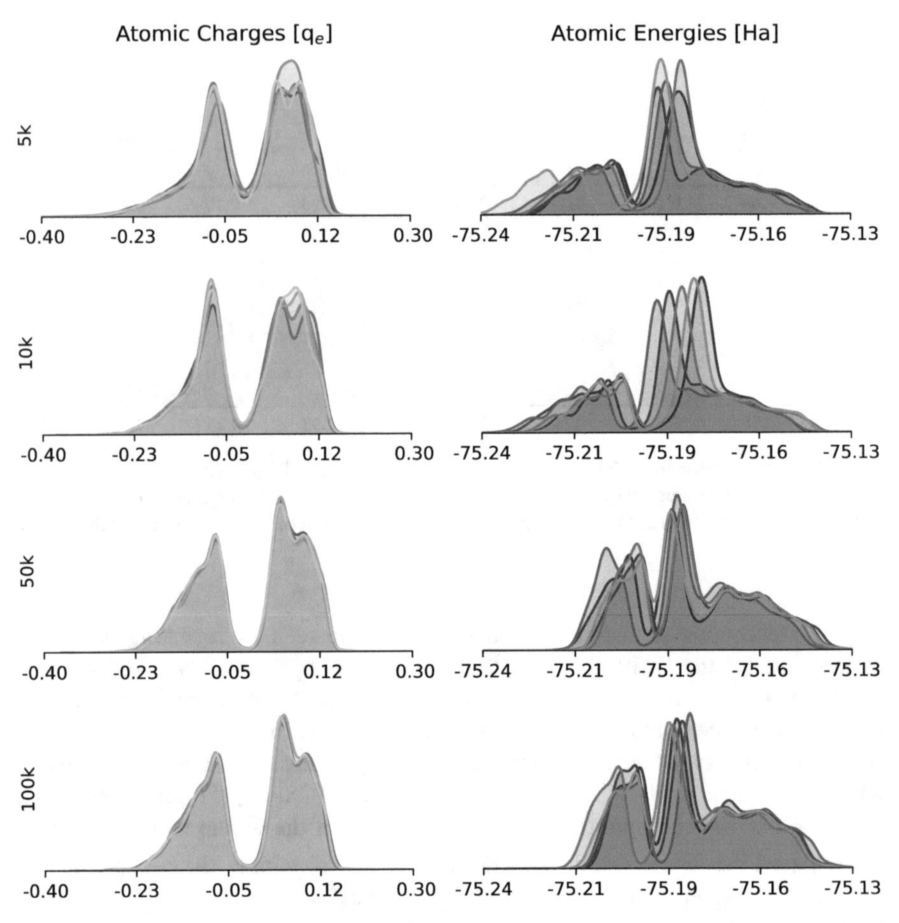

Fig. 12.9 Distribution of the atomic partial charges (blue) and energies (red) of oxygen predicted for all molecules in the QM9 database using different training set sizes (containing 5000, 10,000, 50,000 and 100,000 data points). In each case, five dipole and energy models were trained on the magnitude of the dipole moment and the total electronic energy each. In contrast to the atomic energies, the partial charge distributions converge to similar values upon increasing the training set size

when they are, e.g., used to model long-range electrostatic interactions without further processing, as it can introduce inconsistencies into the predicted model energies and forces [35].

A final point of interest is the extent of fluctuations of the total charge observed during a typical molecular dynamics simulation (see Sect. 12.2.2). Using the protonated alanine tripeptide as an example, the uncorrected total charge shows a standard deviation of 0.04 elementary charge units over a total of 150 ps of simulation time, while fluctuating around the expected average of 1.00. Hence, only minimal corrections are necessary to guarantee full charge conservation.

12.4.3 Electrostatic Potentials

Having ascertained the general reliability of the charge model, we now study how well the latent charge assignments agree with the predictions of electronic structure methods. In order to illustrate and compare different molecular charge distributions, we use partial charges to construct approximate electrostatic potentials (ESPs) of the form:

$$E(\mathbf{r}_0) = \sum_{i}^{N} \frac{q_i q_0}{||\mathbf{r}_i - \mathbf{r}_0||}, \tag{12.11}$$

where q_i and $\mathbf{r}_i$ are the partial charge and position vector of atom i. $\mathbf{r}_0$ is the position of a probe charge q_0, which was set to $q_0 = +1$ in all experiments.

Figure 12.10 shows the pseudo ESPs obtained with latent and Hirshfeld partial charges. The latter have been chosen for their general reliability and widespread use. To assess, how well the latent predictions of the dipole model capture the charge redistribution associated with changes in the molecular geometry, two configurations of the protonated alanine tripeptide are modeled, with a hydrogen transferred from the N-terminal NH_3^+ group to the neighboring carbonyl moiety.

In all cases, good agreement is found between the charge distributions predicted by the dipole moment model and the electronic structure equivalent. The latent charges are able to account for several important features, such as the localization of the positive charge of the molecule at the N-terminal NH_3^+ moiety in the first configuration, as well as its relocation towards the interior of the molecule

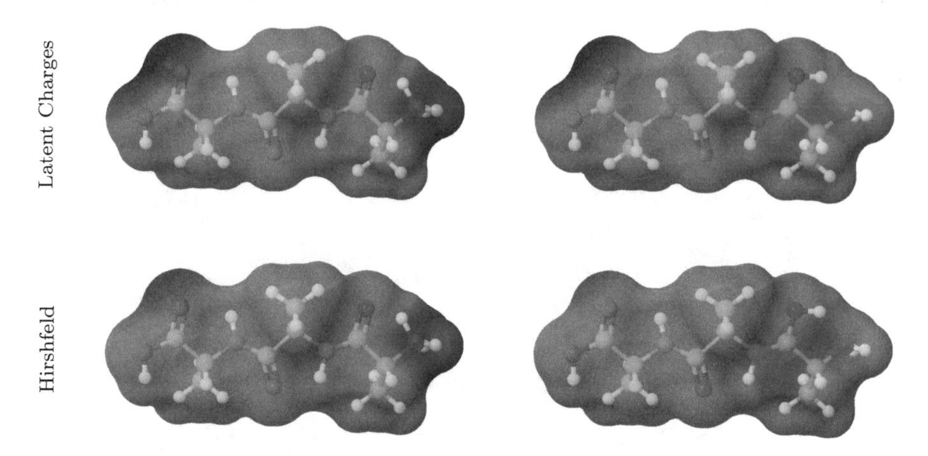

Fig. 12.10 Electrostatic potential surfaces of the alanine tripeptide based on Hirshfeld partial charges and latent partial charges yielded by the dipole model. The left-hand side shows a configuration protonated at the N-terminal NH_3^+ group, whereas the proton is situated on the adjacent carbonyl group in the right-hand side structure. Regions of negative charge are depicted in red, positively charged regions in blue

after proton transfer. The model also accounts for the regions of negative charge expected for the carbonyl and carboxylic acid groups. These findings are remarkable insofar, as all this information on the electronic structure of the molecule is inferred purely from the global dipole moments, demonstrating the power of the partitioning scheme.

12.4.4 Geometry Dependence of Latent Charges

A final analysis is dedicated to the behavior of the latent dipole model charges under changes in the local chemical environment. As an example, we study the evolution of the partial charge of the proton during the proton transfer event occurring in the alanine tripeptide. Figure 12.11 shows the NNP partial charge attributed to the proton plotted against the reaction coordinate. The curves for Hirshfeld, Mulliken [36] and CHELPG charges are included for comparison. Several interesting effects can be observed.

First, the dipole model curve exhibits a minimum close to the transition state of the transfer reaction. Since the latent charges can be seen as a proxy of the local electron density, this result can be interpreted as follows: At the initial and final stages of the transfer, the positive charge is located mainly at the proton itself. However, during the transfer and especially close to the transition regions, electron density is shared between the three participating atoms (O, N and proton). Hence, the positive charge is reduced for these configurations. This finding serves as an additional demonstration for the efficacy of the latent charge model. Although originally only conceived to model dipole moments, it is able to provide insights directly related to the electronic structure of the molecule at an atomistic resolution.

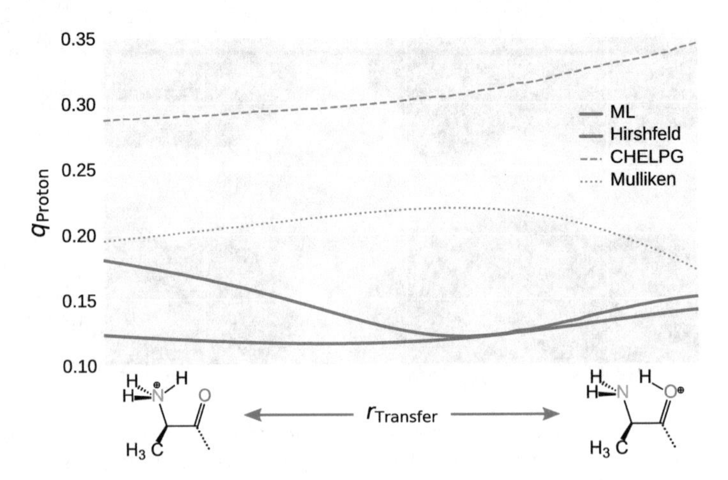

Fig. 12.11 Changes in the partial charge of the proton during different stages of the proton transfer event. Shown are charges computed via different conventional charge partitioning schemes (Hirshfeld, CHELPG and Mulliken), as well as the latent charges predicted by the ML model

Second, Fig. 12.11 illustrates the inherently different behavior found for various charge partitioning schemes. The Hirshfeld charges show a qualitatively similar curve to the machine-learned charge model and are well behaved in general, supporting the results reported in reference [30]. Mulliken and CHELPG charges on the other hand show completely different trends. The former are generally known for their unreliability, e.g., showing a strong dependence on the used basis set and large deviations when attempting to recover molecular dipole moments, hence the result is little surprising [20]. The counterintuitive behavior of the CHELPG charges serves as an additional confirmation for the effects observed in the methanol spectrum shown above (Fig. 12.7). Given this general discrepancy between various partitioning schemes, the charges derived via the dipole moment model constitute a viable alternative: They reproduce molecular dipole moments accurately, are derived directly from quantum mechanical observables and capture the influence of structural changes on the molecular charge distribution.

12.5 Conclusion

We have presented how molecular dynamics (MD) simulations can benefit from machine learning (ML) potentials and provided some background for the implementation of this ML-MD approach. The first challenge during such a task is to efficiently gather enough training data in order to create a converged potential. An adaptive sampling scheme can serve for this purpose and the efficiency can be improved when using (a) an ensemble of neural networks, (b) an adequate uncertainty measure as selection criterion, and (c) multiple replicas to parallelize the sampling.

As an ultimate test, experimental observables need to be calculated and compared to actual experimental results. In our case, infrared spectra are simulated, for which the neural networks do not only need to learn potentials and forces but also dipole moments and their atomistic counterparts, the atomic partial charges. If the latter are plotted in a geometry-dependent manner, e.g., along a reaction coordinate, these machine-learned charges provide insights directly related to the electronic structure of the molecule at an atomistic resolution. In this sense, machine learning can not only deliver potentials with supreme accuracy at compelling speed but also offer valuable insights beyond a simple prediction of quantities.

Despite this positive picture, many challenges remain, e.g., generalizing the machine learning models—ideally for all possible substances, different kinds of molecules and materials alike—or extending the range of properties to be learned. Possibly the biggest challenge is however to find a universally valid electronic structure method necessary for the generation of high-fidelity training data.

Acknowledgments M.G. was provided financial support by the European Union Horizon 2020 research and innovation program under the Marie Skłodowska-Curie grant agreement NO 792572. The computational results presented have been achieved in part using the Vienna Scientific Cluster (VSC). We thank J. Behler for providing the RuNNer code.

References

1. D. Marx, J. Hutter, *Ab Initio Molecular Dynamics: Basic Theory and Advanced Methods* (Cambridge University Press, Cambridge, 2009)
2. M.P. Allen, D.J. Tildesley, *Computer Simulation of Liquids* (Oxford University Press, Oxford, 1987)
3. D. Frenkel, B. Smit, *Understanding Molecular Simulation* (Academic, Cambridge, 2001)
4. T.B. Blank, S.D. Brown, A.W. Calhoun, D.J. Doren, J. Chem. Phys. **103**(10), 4129 (1995)
5. D.A.R.S. Latino, R.P.S. Fartaria, F.F.M. Freitas, J. Aires-De-Sousa, F.M.S. Silva Fernandes, Int. J. Quantum Chem. **110**(2), 432 (2010)
6. J. Behler, Phys. Chem. Chem. Phys. **13**, 17930 (2011)
7. A.P. Bartók, M.C. Payne, R. Kondor, G. Csányi, Phys. Rev. Lett. **104**, 136403 (2010)
8. B. Jiang, J. Li, H. Guo, Int. Rev. Phys. Chem. **35**(3), 479 (2016)
9. R. Ramakrishnan, O.A. von Lilienfeld, *Machine Learning, Quantum Chemistry, and Chemical Space*. Reviews in Computational Chemistry, chap. 5 (Wiley, Hoboken, 2017), pp. 225–256
10. V. Botu, R. Batra, J. Chapman, R. Ramprasad, J. Phys. Chem. C **121**(1), 511 (2017)
11. J. Behler, Angew. Chem. Int. Ed. **56**(42), 12828 (2017)
12. J. Behler, M. Parrinello, Phys. Rev. Lett. **98**(14), 146401 (2007)
13. J. Behler, J. Chem. Phys. **134**(7), 074106 (2011)
14. J. Behler, Int. J. Quantum Chem. **115**, 1032 (2015)
15. M. Gastegger, L. Schwiedrzik, M. Bittermann, F. Berzsenyi, P. Marquetand, J. Chem. Phys. **148**(24), 241709 (2018)
16. A. Pukrittayakamee, M. Malshe, M. Hagan, L.M. Raff, R. Narulkar, S. Bukkapatnum, R. Komanduri, J. Chem. Phys. **130**(13), 134101 (2009)
17. M. Gastegger, P. Marquetand, J. Chem. Theory Comput. **11**(5), 2187 (2015)
18. M. Gastegger, J. Behler, P. Marquetand, Chem. Sci. **8**, 6924 (2017)
19. F. Jensen, *Introduction to Computational Chemistry*, 2nd edn. (Wiley, Hoboken, 2007)
20. C.J. Cramer, *Essentials of Computational Chemistry*, 2nd edn. (Wiley, Hoboken, 2004)
21. T. Morawietz, V. Sharma, J. Behler, J. Chem. Phys. **136**(6), 064103 (2012)
22. K.T. Schütt, M. Gastegger, A. Tkatchenko, K.R. Müller, arXiv:1806.10349 [physics.comp-ph] (2018)
23. J.S. Smith, O. Isayev, A.E. Roitberg, Chem. Sci. **8**, 3192 (2017)
24. J.E. Herr, K. Yao, R. McIntyre, D.W. Toth, J. Parkhill, J. Chem. Phys. **148**(24), 241710 (2018)
25. J.S. Smith, B. Nebgen, N. Lubbers, O. Isayev, A.E. Roitberg, J. Chem. Phys. **148**(24), 241733 (2018)
26. H.S. Seung, M. Opper, H. Sompolinsky, *Proceedings of the Fifth Annual Workshop on Computational Learning Theory* (ACM, New York, 1992), pp. 287–294
27. M. Gastegger, C. Kauffmann, J. Behler, P. Marquetand, J. Chem. Phys. **144**(19), 194110 (2016)
28. M. Thomas, M. Brehm, R. Fligg, P. Vohringer, B. Kirchner, Phys. Chem. Chem. Phys. **15**, 6608 (2013)
29. F. Hirshfeld, Theor. Chim. Acta **44**(2), 129 (1977)
30. A.E. Sifain, N. Lubbers, B.T. Nebgen, J.S. Smith, A.Y. Lokhov, O. Isayev, A.E. Roitberg, K. Barros, S. Tretiak, J. Phys. Chem. Lett. **9**(16), 4495 (2018)
31. C.M. Breneman, K.B. Wiberg, J. Comput. Chem. **11**(3), 361 (1990)
32. R. Ramakrishnan, P.O. Dral, M. Rupp, O.A. Von Lilienfeld, Sci. Data **1**, 140022 (2014)
33. D.P. Kingma, J. Ba, arXiv preprint arXiv:1412.6980 (2014)
34. K. Schütt, P. Kessel, M. Gastegger, K. Nicoli, A. Tkatchenko, K.R. Müller, J. Chem. Theory Comput. **15**(1), 448 (2018)
35. K. Yao, J.E. Herr, D. Toth, R. Mckintyre, J. Parkhill, Chem. Sci. **9**, 2261 (2018)
36. R.S. Mulliken, J. Chem. Phys. **23**(10), 1833 (1955)

High-Dimensional Neural Network Potentials for Atomistic Simulations 13

Matti Hellström and Jörg Behler

Abstract

High-dimensional neural network potentials, proposed by Behler and Parrinello in 2007, have become an established method to calculate potential energy surfaces with first-principles accuracy at a fraction of the computational costs. The method is general and can describe all types of chemical interactions (e.g., covalent, metallic, hydrogen bonding, and dispersion) for the entire periodic table, including chemical reactions, in which bonds break or form. Typically, many-body atom-centered symmetry functions, which incorporate the translational, rotational, and permutational invariances of the potential energy surface exactly, are used as descriptors for the atomic environments. This chapter describes how such symmetry functions and high-dimensional neural network potentials are constructed and validated.

13.1 Introduction

Atomistic simulations in chemistry, physics, and materials science rely on calculations of the potential energy and forces for arbitrary nuclear configurations. This is typically done either with first-principles methods like density functional theory (DFT), which provide an approximate solution to the electronic Schrödinger equation, or by means of atomistic potentials, i.e., direct analytic relations between the structure and the potential energy.

M. Hellström · J. Behler (✉)
Universität Göttingen, Institut für Physikalische Chemie, Theoretische Chemie, Göttingen, Germany
e-mail: joerg.behler@uni-goettingen.de

K. T. Schütt et al. (eds.), *Machine Learning Meets Quantum Physics*, Lecture Notes in Physics 968, https://doi.org/10.1007/978-3-030-40245-7_13

A high-dimensional neural network potential [1], the topic of this chapter, is a type of atomistic potential based on machine learning [2]. In general, atomistic potentials are much faster to evaluate than first-principles methods, which enables them to be used in large-scale simulations with hundreds of thousands of atoms, for example, for the simulation of biomolecules in solution or of extended defects in crystalline materials, as well as in applications requiring extensive sampling of many configurations.

Most atomistic potentials are force fields. The equations for evaluating the energy in such force fields are based on approximations of known physical and chemical phenomena; for example, two noble gas atoms experience attractive van der Waals forces when they are far apart, but if the distance between the two atoms decreases, then at some point the interaction will invariably become repulsive due to the Pauli exclusion principle. Because this phenomenon is known and rather simple, it can be described approximately by an equation containing only few parameters. The most prominent example is the Lennard-Jones 12-6 potential [3], which contains only two parameters: one describing the strength of the attractive interaction, and one describing the distance at which the interaction becomes repulsive.

On the other hand, the class of atomistic potentials known as machine learning potentials [2] do not contain any physical approximations but instead utilize a very flexible potential energy expression. A variety of machine learning methods can be used in this context, for example, methods like Gaussian approximation potentials [4], kernel ridge regression [5], moment tensor potentials [6], and support vector machines [7]. This chapter focuses on artificial neural networks (NNs), which are used to construct high-dimensional neural network potentials (NNPs) [1, 8–10]. The flexible energy expression of an NNP typically contains many thousands of parameters that must be fitted (*learned*) to reproduce results from a typically very large training set of structures evaluated using first- principles methods.

High-dimensional NNPs evaluate the energy E of a structure by summing up individual atomic contributions [8]; each atomic contribution (*atomic energy*) is calculated using an individual neural network. The input to such a neural network is a fingerprint of the environment within some *cutoff sphere* around the atom. It is also possible to augment a high-dimensional neural network potential with the description of long-range interactions between atoms that are very far apart, e.g., electrostatic interactions [11–13].

For a high-dimensional neural network potential, the input features should be

- rotationally invariant,
- translationally invariant,
- permutationally invariant with respect to the order in which the atoms of a given element are provided in the input file,
- and sufficiently different for different atomic environments.

All of these features can be obtained by using the so-called *atom-centered symmetry function* (SFs) as input features [14].

High-dimensional NNPs have been applied to many different systems, e.g., in materials modelling [15] and aqueous chemistry [16, 17]. This chapter describes how high-dimensional NNPs can be used to evaluate the energy and forces acting on atoms in a system, how the neural networks can be fitted and validated, how to select suitable input features (symmetry functions), and how to construct a suitable training set.

13.2 Preliminaries

A high-dimensional neural network potential is parameterized to reproduce results from electronic structure (typically density functional theory) calculations. Thus, before the construction of a high-dimensional neural network potential can begin, a reference data set must be generated. This data set contains a set of atomic structures (periodic and/or non-periodic), together with the properties that are to be reproduced. In the context of high-dimensional neural network potentials, these properties are typically the total energies and the force vectors acting on the atoms. For the total energies, it is not necessary for the electronic structure reference calculation to output the "true" total energies (including, for example, interactions with core electrons in case of pseudopotential calculations); instead, the neural network will simply reproduce a total energy that is consistent with the used DFT code and the chosen settings like the exchange-correlation functional for a given system. Therefore, a NNP cannot be expected to produce more reliable results than a direct application of the reference method would, as any deviation is regarded as an error of the NNP and should be minimized. Consequently, a high-dimensional neural network potential can only be as good as the electronic structure method which is used to calculate the data in the reference set. Thus, the electronic structure method needs be chosen with care. At the same time, because high-dimensional neural network potentials often need a lot of training data (tens of thousands of structures), the electronic structure calculations should not be too costly.

Typically, only a part of the reference data set, the training set, is used to fit the neural network parameters. The rest of the reference data set, the test set, is used to evaluate the quality of the neural network potential on structures which were not present in the training set. This gives a measure of how well the NNP generalizes, at least as long as the available data spans the full configuration space that is relevant for the intended application.

13.3 Functional Form of a High-Dimensional Neural Network Potential

13.3.1 Energy Calculations

In a high-dimensional neural network potential, the potential energy of a system is a sum of the contributions of the individual atoms,

$$E = \sum_{i}^{N^{\mathrm{at}}} E_i, \tag{13.1}$$

where N^{at} is the number of atoms in the system and E_i is the contribution to the total energy from atom i. If the atom i is of some chemical element I, which we here write as $i \in I$, then the atomic energy

$$E_{i \in I} = \chi^I(\mathbf{G}^I(i)) \tag{13.2}$$

is calculated by means of an element-dependent neural network χ^I, which takes a representation of the chemical environment around the atom, $\mathbf{G}^I(i)$, as input.

Thus, the potential energy E can also be written as

$$E = \sum_{i}^{N^{\mathrm{at}}} E_i = \sum_{I} \sum_{i \in I} \chi^I(\mathbf{G}^I(i)), \tag{13.3}$$

where the outer sum is taken over all chemical elements in the system, and the inner sum over each atom of a particular chemical element I.

If there is more than one element in the system, there will be one type of neural network χ^I, χ^J, etc., per different elements I, J, etc. The different element-dependent neural networks can have different architectures, and will have different weight parameters defining the atomic energy output of the neural network. Similarly, the input vectors $\mathbf{G}^I(i)$, $\mathbf{G}^J(j)$, for two atoms $i \in I$ and $j \in J$ where $I \neq J$, can be computed in different ways, and can contain a different number of components. Of course, the number of components of the input vector $\mathbf{G}^I(i)$ must match the number of input features expected by the neural network χ^I. The input vectors $\mathbf{G}^I(i)$ are further discussed in Sect. 13.3.2.

The neural network χ^I is typically a standard fully-connected feed-forward neural network with continuously differentiable activation functions and a single output node, yielding the atomic energy E_i. The neural network is thus a function $\chi^I : \mathbb{R}^n \to \mathbb{R}$, where n denotes the number of input features.

Neural networks used in NNPs often contain two hidden layers, although it is also possible to use deeper neural networks. Some architectures that have been used in the literature include 30-25-25-1 for O atoms and 27-25-25-1 for H atoms in liquid water [16], 36-35-35-1 for Na atoms in NaOH(aq) solutions [17], 53-25-

25-1 for Cu atoms at Cu(s)/H_2O(l) interfaces [18], and 50-20-20-1 for Ru atoms at a Ru(s)/N_2(g) interface [19]. In all these examples the first number refers to the number of input SFs, the next two numbers to the number of nodes in the hidden layers, and the last number 1 specifies the single output node for the atomic energy E_i.

We denote the parameters of the neural network χ^I connecting layer k to layer $k+1$ as a matrix $\mathbf{A}^{I,[k]}$ and a vector $\mathbf{b}^{I,[k]}$ for the bias weights. These parameters must be fitted before the high-dimensional NNP can be used. The fitting process is described in more detail in Sect. 13.4.

Figure 13.1 shows a schematic representation of the calculation of the total energy for a system containing 100 atoms and two elements, I and J. The atomic energies of atoms of element I are calculated using the input features and NN weights colored in purple, while the atomic energies of atoms of element J are calculated using the input features and NN weights colored in green. As the

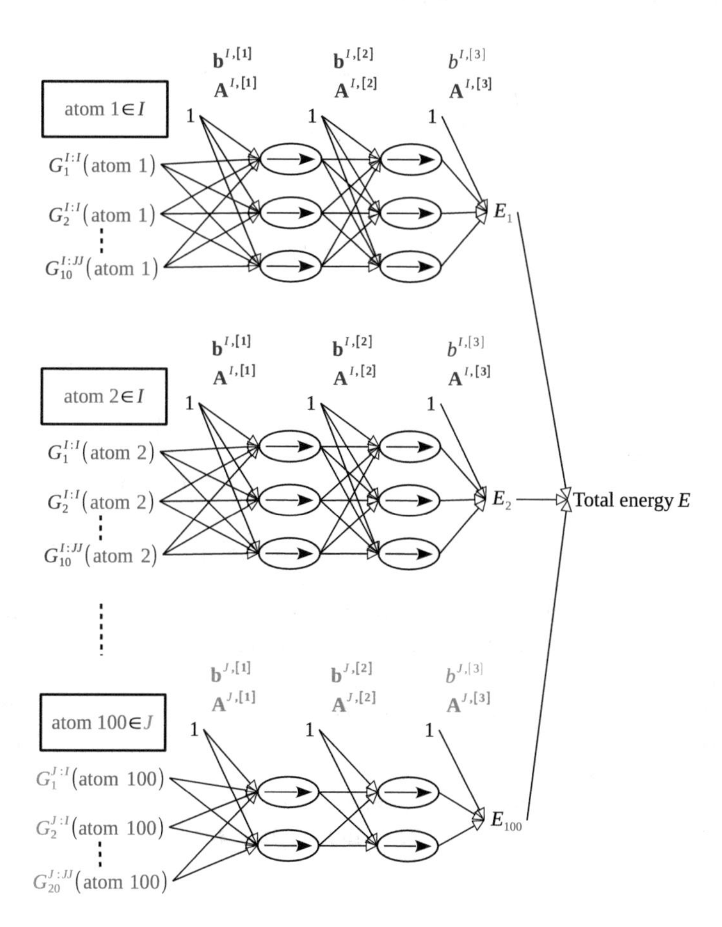

Fig. 13.1 A high-dimensional neural network potential for a system containing two elements I and J and in total 100 atoms

numerical values of the input vectors depend on the geometric environments of the atoms, they are different for each atom, while the definition of the features is the same within a given element. A particular naming scheme is used for the input features; it is further described in the next section.

13.3.2 Symmetry Functions

In the context of high-dimensional neural network potentials it is common to employ atom-centered symmetry functions as descriptors of the chemical environment around an atom [14]. Typically, only the local environment within a cutoff sphere of radius R_c is considered. This is achieved by means of a cutoff function $f_\mathrm{c}(R)$ that smoothly decays to 0 in value and slope at $R = R_\mathrm{c}$. Such functions are often used for constructing atomistic potentials in order to smoothly truncate the atomic interactions at some point. Some possible choices for $f_\mathrm{c}(R)$ are

$$f_\mathrm{c}(R) = \begin{cases} \frac{1}{2}\left[\cos\left(\frac{\pi R}{R_\mathrm{c}}\right) + 1\right] & R \leq R_\mathrm{c} \\ 0 & R > R_\mathrm{c} \end{cases} \tag{13.4}$$

or

$$f_\mathrm{c}(R) = \begin{cases} \frac{\tanh^3\left(1 - \frac{R}{R_\mathrm{c}}\right)}{\tanh^3(1)} & R \leq R_\mathrm{c} \\ 0 & R > R_\mathrm{c} \end{cases} \tag{13.5}$$

or

$$f_\mathrm{c}(R) = \begin{cases} \frac{20}{R_\mathrm{c}^7} R^7 - \frac{70}{R_\mathrm{c}^6} R^6 + \frac{84}{R_\mathrm{c}^5} R^5 - \frac{35}{R_\mathrm{c}^4} R^4 + 1 & R \leq R_\mathrm{c} \\ 0 & R > R_\mathrm{c}. \end{cases} \tag{13.6}$$

The first form (Eq. 13.4) is continuous in value and slope at $R = R_\mathrm{c}$; the second form (Eq. 13.5) also has a continuous second derivative; the third form (Eq. 13.6) additionally has a continuous third derivative. Figure 13.2 shows the three cutoff functions for $R_\mathrm{c} = 10$ Å, a typical cutoff value for high-dimensional neural network potentials.

We will use uppercase letters I, J, K, to indicate chemical elements, and lowercase letters i, j, k, to indicate individual atoms, and we use the notation $i \in I$ to indicate that the atom i is of element I.

The input to the neural network should be a descriptor of the atomic environment. This is done by means of a vector of symmetry function values $\mathbf{G}$ for each atom.

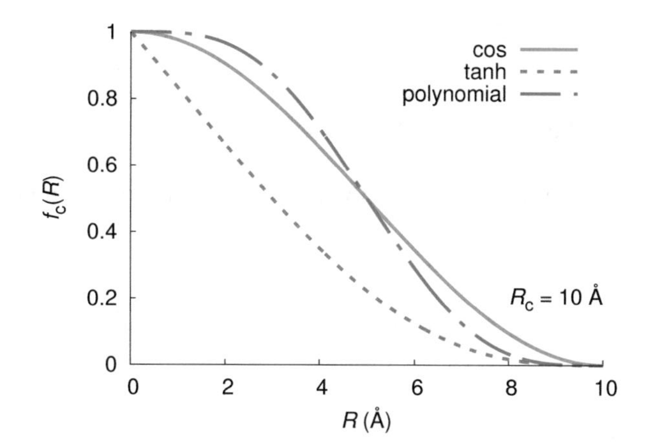

Fig. 13.2 Examples of cutoff functions for $R_\mathrm{c} = 10$ Å: cos (Eq. 13.4), tanh (Eq. 13.5), and polynomial (Eq. 13.6)

As an example, we consider a two-element system with elements I and J, and the vector of symmetry function values for an atom $i \in I$ is then given by

$$\mathbf{G}^I(i \in I) = \begin{pmatrix} G_1^{I:I}(i) \\ \vdots \\ G_{N^\mathrm{sym}(I:I)}^{I:I}(i) \\ G_1^{I:J}(i) \\ \vdots \\ G_{N^\mathrm{sym}(I:J)}^{I:J}(i) \\ G_1^{I:II}(i) \\ \vdots \\ G_{N^\mathrm{sym}(I:II)}^{I:II}(i) \\ G_1^{I:IJ}(i) \\ \vdots \\ G_{N^\mathrm{sym}(I:IJ)}^{I:IJ}(i) \\ G_1^{I:JJ}(i) \\ \vdots \\ G_{N^\mathrm{sym}(I:JJ)}^{I:JJ}(i) \end{pmatrix}. \tag{13.7}$$

In Eq. 13.7, the individual symmetry functions are denoted $G_1^{I:IJ}$, $G_2^{I:IJ}$, etc. The subscripts indicate the nth symmetry function of a particular kind (of which there are N^sym). A superscript of the form $^{I:JJ}$ denotes that the symmetry function is evaluated for the environment of an atom of element I (written before the colon), and that the value depends on all unique pairs of atoms of element J within the cutoff sphere around the atom for which the function is evaluated. As another example, the form $^{I:I}$ denotes that the symmetry function depends on all neighboring atoms of element I (see also the definitions of typical functions below).

The symmetry functions of the type $^{I:J}$ are often called *radial symmetry functions*. One commonly used functional form for a radial symmetry function is [14]

$$G_n^{I:J}(i \in I;\ \eta(n), R_\mathrm{s}(n), R_\mathrm{c}(n)) = \varphi_n^{I:J}\left(\sum_{\substack{j \in J \\ j \neq i}} e^{-\eta(R_{ij}-R_\mathrm{s})^2} \cdot f_\mathrm{c}(R_{ij})\right). \qquad (13.8)$$

Here, $\eta(n)$, $R_\mathrm{s}(n)$, and $R_\mathrm{c}(n)$ are the parameters that define the spatial shape and extension of the function. Each subscript n corresponds to different values of η and/or R_s, and potentially also to different cutoff radii R_c (although in practice R_c is often the same for all symmetry functions). In essence, the symmetry function in Eq. 13.8 is the summation of one-dimensional Gaussian functions of width η and centered at R_s evaluated for all distance R_{ij}, where j correspond to all neighboring atoms of element J (that can be the same as or different from I, cf. Eq. 13.7). Each contribution is weighted by the cutoff function $f_\mathrm{c}(R_{ij})$ in order to ensure that the contribution from atoms j for which $R_{ij} > R_\mathrm{c}$ smoothly vanishes. Finally, an optional scaling function $\varphi_n^{I:J}$ is applied that can scale the symmetry function values to some desired range (for example, between -1 and 1). The scaling function is used so that the different symmetry functions in $\mathbf{G}^I(i)$ (Eq. 13.7) have values in roughly the same range for different atoms and structures. This kind of feature scaling can be helpful for the optimization (fitting) algorithm, and also makes it possible to use dimensionality reduction techniques like principal component analysis.

Figure 13.3a shows the value of the summand in Eq. 13.8 for different distances, for a few selected values of η and R_s. Here, the cutoff distance is set to the typical

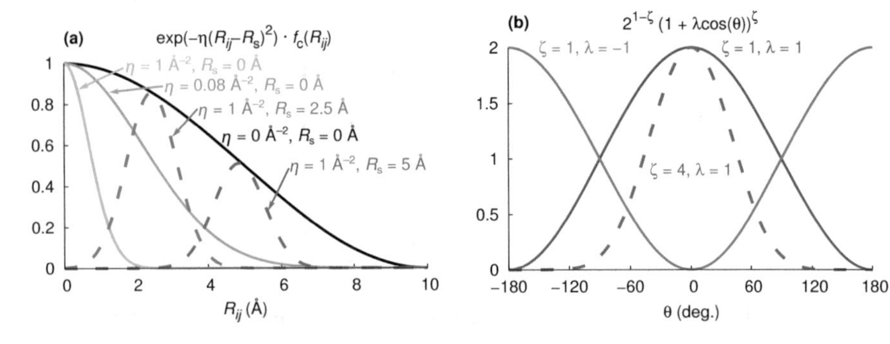

Fig. 13.3 (**a**) The summand of the radial symmetry function in Eq. 13.8 for some different values of η and R_s, using the cutoff function $f_\mathrm{c}(R_{ij})$ from Eq. 13.4. (**b**) The angular part of the summand of the symmetry function in Eq. 13.9, for some selected values of ζ and λ

value of $R_c = 10$ Å, and the cutoff function from Eq. 13.4 is used. The black line ($\eta = 0$ Å^{-2}) is equivalent to the plain cutoff function $f_c(R)$, as the Gaussian reduces to a factor 1.

For every combination of central element I and neighboring element J, several functions of the type in Eq. 13.8 with different values of η and/or R_s are used as input features to the NN. Using several such symmetry functions provides a significantly better fingerprint of the atomic environment than could be accomplished with only a single symmetry function. For example, if the single function with $\eta = 0$ Å^{-2} in Fig. 13.3 is used, then two neighbors around the atom i, both at a distance $R = 6.2$ Å, yield the same value of $G^I(i) = 0.632$ as a single neighbor at a distance $R = 4.15$ Å. By using several symmetry functions, all the relevant information about the atomic environment around an atom i can be encoded into the input vector $\mathbf{G}^I(i)$, which provides a structural fingerprint of the atomic environment as the input to the NN. However, the radial symmetry functions only depend on the interatomic distances R_{ij}. In order to incorporate angular dependencies, it is necessary to include a different type of symmetry function, an *angular symmetry function* [14]. For example, a common choice for an angular symmetry function is

$$G_n^{I:JK}(i \in I;\ \eta(n), \zeta(n), \lambda(n), R_c(n)) =$$

$$\varphi_n^{I:JK}\left(2^{1-\zeta} \sum_{\substack{j \in J,\, k \in K \\ j \neq i,\, k \neq i \\ k \neq j}} (1 + \lambda \cos\theta_{jik})^{\zeta} \cdot e^{-\eta(R_{ij}^2 + R_{ik}^2 + R_{jk}^2)} \cdot f_c(R_{ij}) \cdot f_c(R_{ik}) \cdot f_c(R_{jk})\right), \tag{13.9}$$

where the interatomic distances R_{ij}, R_{ik}, and R_{jk} and angle θ_{jik} between three atoms $i \in I$, $j \in J$, and $k \in K$ are used to compute the value of the symmetry function, for each of the possible unique combinations of neighbors j and k around the central atom i. Again, the elements J and K may be the same as or different from I, and $\varphi_n^{I:JK}$ is a scaling function. In Eq. 13.9, ζ determines the range of angles for which the angular term is approximately 0 (a larger value of ζ will make the peak less spread out), and λ takes on a value of either $+1$, for a maximum at $\theta_{jik} = 0°$, or -1, for a maximum at $\theta_{jik} = 180°$. The angular part [i.e., $(1 + \lambda \cos\theta_{jik})^{\zeta}$] of the symmetry function is shown in Fig. 13.3b, for a few different values of ζ and λ. The angular part is periodic with a period of 360° and symmetric around 0° and ±180°, which is mandatory to obtain the same symmetry function values for geometrically equivalent structures. It is important to note that the radial and angular types of symmetry functions in Eq. 13.7 do not correspond to two-body and three-body terms, but rather the summation of different two and three-body terms, making each symmetry function a many-body function.

One limitation of using symmetry functions as a description of the chemical environment around an atom is the use of the cutoff function $f_c(R_{ij})$, because as a consequence the symmetry functions only describe the chemical environment up to the cutoff distance R_c, which is usually taken to be a value in the range 6–10 Å. Thus, interactions that take place over longer distances, for example, electrostatic interactions, are effectively not captured. To overcome this limitation it is possible to combine a high-dimensional neural network potential with methods for computing long-range electrostatic interactions (see Sect. 13.5).

13.3.3 Choosing a Set of Symmetry Functions

The symmetry functions are rotationally and translationally invariant, and also invariant to permutations with respect to the order in which atoms of a given element appear in the input file. When applying a set of symmetry functions as input to a neural network (Eq. 13.7), one needs to decide which and how many symmetry functions to include for each combination of possible neighboring elements. Suitable values of the parameters η and R_s must be chosen for the radial symmetry functions, and η, λ, and ζ for the angular symmetry functions.

A good choice of symmetry functions will lead to a symmetry function vector $\mathbf{G}^I(i)$ for an atom $i \in I$ being sufficiently different when the atom i is in different atomic environments. This is what allows the neural network to distinguish between different chemical environments. Thus, there needs to be a sufficient number of different symmetry functions. At the same time, the number of symmetry functions should ideally be kept small, since a larger number of symmetry functions imply a larger number of parameters (neural network weights) that need to be fitted.

The range of values of a particular symmetry function should not be too small, when calculated, for example, over the different atoms in the training set. The range of values should also not be determined by a few outliers. It is possible to check for this by dividing the difference between the maximum and minimum value for a given symmetry function in the training set by its standard deviation; a value of about 10 is often good. Moreover, no two symmetry functions should be too strongly correlated. By calculating, e.g., the Pearson correlation coefficient, one can remove symmetry functions that are too strongly correlated (>0.9) [10, 20].

In choosing the parameter η for the radial symmetry functions, the largest value of η (the most quickly decaying Gaussian function) should be chosen so that the function starts to decay around the distance corresponding to the shortest possible meaningful bond between the two pertinent atom types.

If the magnitudes of the forces on two atoms are very different from each other, the atoms necessarily exist in different chemical environments. This means that at least one symmetry function should have a substantially different value for one atom as compared to the other. If this is not the case, the set of symmetry functions needs to be augmented [14].

Machine learning fitting algorithms usually work best if the different input features contain numbers of roughly the same order of magnitude. However, in, for example, the case of the radial symmetry functions in Eq. 13.8, symmetry functions with a larger value of the parameter η will always yield smaller values of G. For this reason, it is common to apply feature scaling (also called feature normalization), to the symmetry function values before training the neural network. This is a common preprocessing step for machine learning applications in general.

The scaling functions φ_n in Eqs. 13.8 and 13.9 provide such a scaling. One possible definition of φ_n, that scales the minimum and maximum values for the nth symmetry function in the training set to become -1 and 1, is

$$\varphi_n(x) = \frac{2(x - \min G_n^\circ)}{\max G_n^\circ - \min G_n^\circ} - 1, \tag{13.10}$$

where $\min G_n^\circ$ and $\max G_n^\circ$ are the smallest and largest values of the *non*-scaled (indicated by the circle superscript) nth symmetry function in the training set. One can also subtract the mean and divide by the standard deviation, to obtain a feature with mean 0 and unit variance:

$$\varphi_n(x) = \frac{x - \text{mean}\, G_n^\circ}{\text{std}\, G_n^\circ}. \tag{13.11}$$

13.3.4 Force Calculations

For many applications involving neural network potentials, for example, geometry optimizations and molecular dynamics simulations, it is of interest to calculate the forces acting on the atoms in the system. The forces also provide valuable local information about the PES, and can be used during the optimization (fitting) of the neural network parameters (see also Sect. 13.4).

The force with respect to some atomic coordinate α_i on an atom $i \in I$, which is conservative by construction, is equal to the negative gradient of the total energy with respect to this coordinate. Modern machine learning computational frameworks often include functionality for automatic differentiation, so that the gradients can be calculated automatically with a simple function call. For tutorial purposes, we will derive the actual equations used in the force calculations, which are needed for applications that do not include automatic differentiation.

The force can be expressed as

$$F_{\alpha_i} = -\frac{\partial E}{\partial \alpha_i} = -\sum_{j=1}^{N^{\text{at}}} \frac{\partial E_j}{\partial \alpha_i} = -\sum_J \sum_{j \in J} \sum_{n=1}^{N^{\text{sym}}(J)} \frac{\partial E_j}{\partial G_n^J(j)} \cdot \frac{\partial G_n^J(j)}{\partial \alpha_i}, \tag{13.12}$$

where the outermost sum runs over all chemical elements J in the system, and $G_n^J(j)$ is the nth symmetry function for the element J evaluated for the atom $j \in J$. Typically, α_i is one of the force components x_i, y_i, or z_i in a Cartesian coordinate system. The partial derivatives in Eq. 13.12 can be expressed analytically, if continuously differentiable activation functions are used and if symmetry functions such as those in Eq. 13.8 or 13.9 are used as descriptors for the chemical environment.

This analytical representation allows for a fast and accurate determination of the forces, for any possible atomic configuration.

The partial derivative $\frac{\partial E_j}{\partial G_n^J(j)}$ is the derivative of the neural network output for the atomic neural network on the atom j with respect to its nth input feature. This factor depends on the neural network architecture (number of input features, number of hidden layers, and the number of nodes per hidden layer), and can be evaluated in a standard fashion for common choices of activation functions. The Appendix illustrates how those terms are calculated for a simple high-dimensional neural network potential when using a logistic activation function. It should be noted that $\frac{\partial E_j}{\partial G_n^J(j)}$ depends on the value of *all* symmetry functions for the atom j, since it is different linear combinations of all of the symmetry functions that enter into the first hidden layer, where a nonlinear activation function is applied (the same is true also for the remaining hidden layers). Thus, the force acting on an atom $i \in I$ depends on the environment around the atom $j \in J$, as long as the second factor, $\frac{\partial G_n^J(j)}{\partial \alpha_i}$, is nonzero, which can happen only when the symmetry function $G_n^J(j)$ describes the environment of element I around the atom j (i.e., the symmetry function is of the type $G^{J:I}$ or $G^{J:IK}$) and i is inside the cutoff sphere of j.

The calculation of the gradient $\frac{\partial G_n^J(j)}{\partial \alpha_i}$ depends on the definition of the symmetry functions. For a radial symmetry function $G^{J:K}$ (with no scaling; cf. Eq. 13.8) defined as

$$G_n^{J:K}(j) = \sum_{\substack{k \in K \\ k \neq j}} e^{-\eta(R_{jk}-R_s)^2} \cdot f_c(R_{jk}), \tag{13.13}$$

where η and R_s are constants, the product rule of differentiation gives

$$\frac{\partial G_n^{J:K}(j)}{\partial \alpha_i} = \sum_{\substack{k \in K \\ k \neq j}} \mathrm{e}^{-\eta(R_{jk}-R_s)^2} \left(-2\eta(R_{jk} - R_s) f_c(R_{jk}) \frac{\partial R_{jk}}{\partial \alpha_i} + \frac{\partial f_c(R_{jk})}{\partial \alpha_i} \right). \tag{13.14}$$

The above expression requires the evaluation of the two partial derivatives $\frac{\partial R_{jk}}{\partial \alpha_i}$ and $\frac{\partial f_c(R_{jk})}{\partial \alpha_i}$. The distance R_{jk} between two atoms j and k is

$$R_{jk} = \sqrt{(x_k - x_j)^2 + (y_k - y_j)^2 + (z_k - z_j)^2}. \tag{13.15}$$

If, for example, $\alpha \equiv x$, then

$$\frac{\partial R_{jk}}{\partial \alpha_k} \equiv \frac{\partial R_{jk}}{\partial x_k} = \frac{2(x_k - x_j)}{2\sqrt{(x_k - x_j)^2 + (y_k - y_j)^2 + (z_k - z_j)^2}} = \frac{x_k - x_j}{R_{jk}}, \tag{13.16}$$

or, more generally, for any Cartesian component α_i,

$$\frac{\partial R_{jk}}{\partial \alpha_i} = \begin{cases} \dfrac{\alpha_j - \alpha_k}{R_{jk}} & i = j \\ \dfrac{\alpha_k - \alpha_j}{R_{jk}} & i = k \\ 0 & i \neq j \text{ and } i \neq k, \end{cases} \tag{13.17}$$

where the last equality follows from the fact that the distance between two atoms j and k does not depend on any other atom i.

If the cutoff function $f_c(R_{jk})$ is defined as in Eq. 13.4, then

$$\frac{\partial f_c(R_{jk})}{\partial \alpha_i} = \begin{cases} -\frac{1}{2}\sin\left(\frac{\pi R_{jk}}{R_c}\right)\frac{\pi}{R_c}\frac{\partial R_{jk}}{\partial \alpha_i} & R_{jk} \leq R_c \\ 0 & R_{jk} > R_c. \end{cases} \tag{13.18}$$

Thus, the terms in the sum in Eq. 13.14 all depend on $\frac{\partial R_{jk}}{\partial \alpha_i}$, which is naturally equal to zero if $i \neq j$ and $i \neq k$ (Eq. 13.17). As a consequence, the sum in Eq. 13.14 for a radial symmetry function $G^{J:K}$ reduces to just a single term if $i \neq j$ and $K = I$, and completely vanishes if $i \neq j$ and $K \neq I$:

$$\frac{\partial G_n^{J:K}(j)}{\partial \alpha_i} = \begin{cases} \displaystyle\sum_{\substack{k \in K \\ k \neq j}} e^{-\eta(R_{jk}-R_s)^2}\left(-2\eta(R_{jk}-R_s) f_c(R_{jk})\frac{\partial R_{jk}}{\partial \alpha_i} + \frac{\partial f_c(R_{jk})}{\partial \alpha_i}\right) & i = j \\ e^{-\eta(R_{ij}-R_s)^2}\left(-2\eta(R_{ij}-R_s) f_c(R_{ij})\frac{\partial R_{ij}}{\partial \alpha_i} + \frac{\partial f_c(R_{ij})}{\partial \alpha_i}\right) & i \neq j \text{ and } K = I \\ 0 & i \neq j \text{ and } K \neq I. \end{cases} \tag{13.19}$$

For an angular symmetry function defined as (cf. Eq. 13.9)

$$G^{J:KL}(j \in J) = 2^{1-\zeta} \sum_{\substack{k \in K,\, l \in L \\ k \neq j,\, l \neq j \\ k \neq l}} (1 + \lambda \cos\theta_{kjl})^{\zeta} e^{-\eta(R_{jk}^2 + R_{jl}^2 + R_{kl}^2)} \times f_c(R_{jk}) f_c(R_{jl}) f_c(R_{kl}) \tag{13.20}$$

with θ_{kjl} defined such that j is the central atom ($\cos\theta_{kjl} = \frac{\mathbf{R}_{jk}\cdot\mathbf{R}_{jl}}{R_{jk}R_{jl}}$), the partial derivative with respect to α_i becomes

$$\begin{aligned}
\frac{\partial G^{J:KL}(j)}{\alpha_i} &= 2^{1-\zeta} \sum_{\substack{k\in K\\ l\in L\\ j\neq k, j\neq l}} \mathrm{e}^{-\eta(R_{jk}^2+R_{jl}^2+R_{kl}^2)} \\
&\times \left[\zeta(1+\lambda\cos\theta_{kjl})^{\zeta-1}\lambda\frac{\partial\cos\theta_{kjl}}{\partial\alpha_i} f_\mathrm{c}(R_{jk})f_\mathrm{c}(R_{jl})f_\mathrm{c}(R_{kl})\right. \\
&- (1+\lambda\cos\theta_{kjl})^{\zeta}2\eta\left(R_{jk}\frac{\partial R_{jk}}{\partial\alpha_i} + R_{jl}\frac{\partial R_{jl}}{\alpha_i} + R_{kl}\frac{\partial R_{kl}}{\partial\alpha_i}\right) \\
&\times f_\mathrm{c}(R_{jk})f_\mathrm{c}(R_{jl})f_\mathrm{c}(R_{kl}) \\
&+ (1+\lambda\cos\theta_{kjl})^{\zeta}\frac{\partial f_\mathrm{c}(R_{jk})}{\partial\alpha_i} f_\mathrm{c}(R_{jl})f_\mathrm{c}(R_{kl}) \\
&+ (1+\lambda\cos\theta_{kjl})^{\zeta} f_\mathrm{c}(R_{jk})\frac{\partial f_\mathrm{c}(R_{jl})}{\partial\alpha_i} f_\mathrm{c}(R_{kl}) \\
&\left.+(1+\lambda\cos\theta_{kjl})^{\zeta} f_\mathrm{c}(R_{jk})f_\mathrm{c}(R_{jl})\frac{\partial f_\mathrm{c}(R_{kl})}{\partial\alpha_i}\right],
\end{aligned}$$

where

$$\frac{\partial\cos\theta_{kjl}}{\partial\alpha_i} = \begin{cases}
\dfrac{(\alpha_k-\alpha_j)(R_{jl}^2\cos\theta_{kjl}-R_{jk}R_{jl})+(\alpha_l-\alpha_j)(R_{jk}^2\cos\theta_{kjl}-R_{jk}R_{jl})}{R_{jk}^2R_{jl}^2} & i=j \\
\dfrac{(\alpha_l-\alpha_j)R_{jk}-(\alpha_k-\alpha_j)R_{jl}\cos\theta_{kjl}}{R_{jk}^2R_{jl}} & i=k \\
\dfrac{(\alpha_k-\alpha_j)R_{jl}-(\alpha_l-\alpha_j)R_{jk}\cos\theta_{kjl}}{R_{jl}^2R_{jk}} & i=l \\
0 & \text{otherwise.}
\end{cases} \tag{13.21}$$

An example, that illustrates how the force components are calculated on an atom, can be found in Appendix.

13.3.5 Other Types of Symmetry Functions

Designing good descriptors for machine learning potentials in general, and high-dimensional neural network potentials in particular, is currently a very active research field. Here, we give just a few recent examples.

Smith et al. [21] introduced a variation of the angular symmetry function in Eq. 13.9 (here given without any scaling function) as

$$G^{I:JK}(i \in I; \eta, \zeta, R_s, \theta_s) = 2^{1-\zeta} \sum_{\substack{j \in J,\, k \in K \\ j \neq i,\, k \neq i \\ k \neq j}} (1 + \cos(\theta_{jik} - \theta_s))^{\zeta} \times \exp\left[-\eta \left(\frac{R_{ij} + R_{ik}}{2} - R_s \right)^2 \right] \cdot f_c(R_{ij}) \cdot f_c(R_{ik}), \tag{13.22}$$

where the parameter θ_s provides an alternative control over the angular dependence of the symmetry function to the parameters λ and ζ in Eq. 13.9, and the parameter R_s is akin to the R_s-parameter from the radial symmetry functions in Eq. 13.8.

Gastegger et al. [22] introduced *weighted atom-centered symmetry functions*, wASCFs, where the element-dependent environment is captured not through a set of symmetry functions for every possible combination of elements, but instead by element-dependent prefactors. An example of a radial symmetry function would be

$$G^{I}(i \in I; \eta, R_s) = \sum_{J} \sum_{\substack{j \in J \\ j \neq i}} g(J) \mathrm{e}^{-\eta (R_{ij} - R_s)^2} f_c(R_{ij}), \tag{13.23}$$

where $g(J)$ is a function depending on the element J, for example, $g(J) = Z_J$, where Z_J is the atomic number. This symmetry function is thus calculated for *all* elements J, unlike the radial symmetry function $G^{I:J}$ in Eq. 13.8, that was calculated only for a *specific* element J. This type of approach could potentially decrease the number of needed symmetry functions for systems containing many elements. Another approach in a similar spirit has also been published by Artrith, Urban, and Ceder [23].

13.4 Construction of a High-Dimensional Neural Network Potential

A high-dimensional neural network potential is typically iteratively constructed using the following steps:

1. A reference electronic structure method (e.g., the employed DFT functional) is decided upon.

2. Initial training and test sets (structures calculated with the reference electronic structure method) are gathered.
3. Sets of descriptors (e.g., a set of symmetry functions) for each element are chosen.
4. A fitting algorithm and hyperparameters for the fitting algorithm (e.g., learning rate) are decided.
5. The neural network weights are fitted.
6. The quality of the fitted parameters is evaluated and structures which are not well described by the fitted potential are identified.
7. The training and test sets are augmented with more structures.
8. In case the potential is not yet satisfying, the process repeats from Step 3 using the new extended training and test sets.

The reference electronic structure method must describe the system(s) for which the potential is constructed with an acceptable accuracy. At the same time, the method should not be too costly, since typically many thousands of reference calculations are needed for constructing a large enough training set.

The initial training set can be obtained, by, for example, extracting snapshots from ab initio molecular dynamics simulations for the system under study. Another method is *normal mode sampling* [21], in which the atoms in optimized structures are displaced along the vibrational normal modes. A third possibility is to simply randomly displace the atoms in a structure by small amounts.

Some of the considerations that go into choosing a suitable set of symmetry functions were described in Sect. 13.3.3. Fortunately, it is often possible to reuse a set of symmetry functions for one system and apply it also to other systems (and elements). Moreover, automated unsupervised selection schemes, such as CUR decomposition, can be applied on a large library of symmetry functions, selecting the most suitable subset of symmetry functions based only on structural information, without the need for evaluating the neural network potential [20].

Many algorithms exist for fitting neural network weights. The algorithms work by minimizing a so-called cost function Γ (also called loss function), which is calculated by comparing the neural network output to the reference data given in the training set. The cost function is typically calculated as

$$\Gamma = \mathrm{MSE}(E) + \beta \mathrm{MSE}(F), \tag{13.24}$$

where MSE denotes the mean squared error and β is a coefficient for weighting the relative importance of errors, and accounting for the difference in units, in the energies and forces. The MSEs can also themselves be weighted, so that some energies or forces in the training set are assigned greater importance than others. Some of the algorithms for optimizing the neural network weights are, for example, gradient descent (*backpropagation*), the Levenberg–Marquardt algorithm [24, 25], and the global extended Kalman filter [26]. The description of those algorithms lies outside the scope of the current chapter.

Often, the quality of a set of NN weights is characterized by the root mean squared error, RMSE, for the energies E and forces F. RMSE(E) is often reported as a value normalized *per atom* to achieve a size-consistent error measure, because structures containing many atoms typically have larger absolute errors in the energy as compared to structures containing few atoms. In typical applications of NNPs, RMSE(E) is of the order of 1–2 meV per atom, and RMSE(F) of the order of about 0.1 eV/Å.

The RMSEs provide a single number as measure of the accuracy of the NNP. Although this is very useful, a single number can be quite deceiving, if the training set contains a great variety of structures, or if each individual structure contains atoms in different environments (e.g., bulk-like atoms and interfacial atoms). During the construction of the NNP, great care must be taken to ensure that the NNP does, in fact, describe all the pertinent parts of a system with the desired accuracy. This can be accomplished, for example, by visualizing the distribution of errors associated with different structures.

It is very likely that the initial training set used in the parameterization of the NNP does not contain enough information to fully map out the potential energy surface. Thus, once a NNP has been trained, it can be used to locate regions of the PES that are not well described. There are several ways of accomplishing this, in which the simplest is to simply apply the NNP to the kind of simulation that is targeted. If the resulting structures obtained from this simulation are sufficiently different from those in the training set, for example, if the value of a symmetry function on an atom lies outside the range of values of that symmetry function in the training set, then the structure is clearly different from those in the training set (the NNP is *extrapolating*). Structures obtained in that fashion can then be added to the training set.

Even if the NNP does not extrapolate, there may not be enough training data for some types of structures. Such structures can be found by fitting multiple NNPs (for example, using different random weight initializations) [10]. If different NNPs predict sufficiently different energies and forces on a single structure, then that type of structure is underrepresented in the training set and can be used to further expand the training set.

Finally, another way of gathering structures for the training set is to employ biased MD simulations, like metadynamics, in order to drive the system towards exploring new regions of the potential energy surface [27].

13.5 Long-Range Interactions

The use of a cutoff function in the atomic environment descriptors (Eq. 13.4) prevents an atom from interacting with atoms farther away than the cutoff radius. Some interactions decay only slowly with increasing distance, most notably dispersion and electrostatic interactions.

Dispersion interactions can be included in two ways: if the reference electronic structure method itself describes dispersion accurately, then the symmetry functions in the high-dimensional neural network potential only need a large enough distance cutoff. Alternatively, dispersion interactions can be added on-the-fly with some empirical dispersion correction, e.g., Grimme's D3 dispersion correction [28].

If some charge assignment scheme is used to calculate atomic charges, electrostatic interactions can be evaluated by an application of Coulomb's law or Ewald summation, and can then simply added to the total energy expression. In such cases, it is important to first subtract the long-range contribution to the energies and forces in the training set before the fitting of the NNP takes place, in order to avoid double-counting of the long-range contributions.

One type of charge assignment scheme that has been used in conjunction with high-dimensional NNPs is to use a second, different NNP, to predict atomic charges instead of atomic energies [11, 12]. The training set must then contain some approximation for the atomic charges, for example, the calculated Hirshfeld or Bader charges from the reference electronic structure calculations.

When including long-range interactions, the total energy becomes the sum of the short-range energy E^{short} (as described in Sect. 13.3.1), and the long-range energy E^{long}:

$$E = E^{\text{short}} + E^{\text{long}}. \tag{13.25}$$

The force with respect to some atomic coordinate α becomes

$$F_{\alpha_i} = F_{\alpha_i}^{\text{short}} + F_{\alpha_i}^{\text{long}} = -\frac{\partial E^{\text{short}}}{\partial \alpha_i} - \frac{\partial E^{\text{long}}}{\partial \alpha_i}, \tag{13.26}$$

where $F_{\alpha_i}^{\text{short}}$ is calculated as in Eq. 13.12. For a non-periodic system, $F_{\alpha_i}^{\text{long}}$ can be calculated as

$$\begin{aligned} F_{\alpha_i}^{\text{long}} &= -\frac{1}{2}\frac{\partial}{\partial \alpha_i}\sum_{j=1}^{N^{\text{at}}}\sum_{\substack{k=1\\k\neq j}}^{N^{\text{at}}}\frac{q_j q_k}{R_{jk}} \\ &= -\frac{1}{2}\sum_{j=1}^{N^{\text{at}}}\sum_{\substack{k=1\\j\neq i}}^{N^{\text{at}}}\frac{1}{R_{jk}^2}\left[\frac{\partial q_j}{\partial \alpha_i} q_k R_{jk} + q_j \frac{\partial q_k}{\partial \alpha_i} R_{jk} - q_j q_k \frac{\partial R_{jk}}{\partial \alpha_i}\right], \end{aligned} \tag{13.27}$$

where q_j is the charge on atom j. Note, that this expression contains partial derivatives of the charge on an atom j with respect to the position of the atom i, which is important since the atomic charges are environment-dependent. If a NNP with symmetry functions is used to determine the atomic charges, i.e., $q_j = \chi_q^J(\mathbf{G}^J(j \in J))$, then it can be shown that

$$F_{\alpha_i}^{\text{long}} = \sum_J \sum_{j\in J} \sum_{\substack{k=1\\k\neq j}}^{N^{\text{at}}} \frac{q_k}{R_{jk}} \cdot \left[\frac{1}{2}\frac{q_j}{R_{jk}}\frac{\partial R_{jk}}{\partial \alpha_i} - \sum_{n=1}^{N_q^{\text{sym}}(J)} \frac{\partial q_j}{\partial G_n^J(j)}\frac{\partial G_n^J(j)}{\partial \alpha_i}\right], \tag{13.28}$$

where $N_q^{\text{sym}}(J)$ refers to the number of symmetry functions for element J in the atomic neural network χ_q^J used to for the atomic charge determination.

For a periodic system, where the electrostatic energy is evaluated using, for example, Ewald summation, the expression for F_α^{long} becomes more complicated, although it can be derived in a similar fashion. Recently, also alternative approaches have been proposed, like the derivation of charges based on molecular dipole moments [13, 29].

13.6 Applications of High-Dimensional Neural Network Potentials

High-dimensional NNPs have been developed and applied to many different molecules and materials (for reviews see [1, 9]). Some examples include silicon [30], carbon [31], sodium [32], zinc oxide [11], germanium telluride [33], copper [34], Cu clusters on ZnO [35], Cu-Au nanoalloys [36], water–Cu interfaces [18], titanium dioxide [37], gold [38], copper-palladium-silver alloys [39], N_2 on Ru [19], water on ZnO [40, 41], aqueous NaOH solutions [17, 42], protonated water clusters [43, 44], and organic molecules [13, 21, 29, 45].

The above examples demonstrate the versatility of high-dimensional neural network potentials. In the next few years, many more applications of this method are likely to emerge, not least because it, as well as other machine learning potential methods, are under highly active development in the academic community.

13.7 Summary

High-dimensional neural network potentials are a type of machine learning potential and can be fitted to reproduce arbitrary potential energy surfaces from electronic structure calculations. In a high-dimensional neural network potential, the total energy is expressed as the sum of atomic contributions, which are evaluated by means of neural networks taking a representation of the atomic environment as input. Symmetry functions are examples of such representations, which are rotationally, translationally, and permutationally invariant. It is also possible to combine high-dimensional neural network potentials with methods for evaluating long-range contributions to the total energy, for example, electrostatic interactions.

A successful parameterization requires a large and diverse training set. The resulting errors from the fitting procedures can usually be made very small (about 1 meV per atom), making high-dimensional neural network potentials a very promising method for applications in materials modelling and computational chemistry.

Appendix: Calculating the Force Components on an Atom

The following example illustrates how the force components on the atom O1 in a three-atom system consisting of two O atoms, O1 and O2, and one H atom, H3, are calculated. The NN for O has the architecture 2-2-1 (one hidden layer containing two nodes; two input features described by two radial symmetry functions $G^{\mathrm{O:H}}$ and $G^{\mathrm{O:O}}$). Similarly, the NN for H also has the architecture 2-2-1 with two radial symmetry functions $G^{\mathrm{H:H}}$ and $G^{\mathrm{H:O}}$.

$$\mathbf{G}^{\mathrm{O}}(\mathrm{O1}) = \begin{pmatrix} G^{\mathrm{O:H}}(\mathrm{O1}) \\ G^{\mathrm{O:O}}(\mathrm{O1}) \end{pmatrix}, \quad \mathbf{G}^{\mathrm{O}}(\mathrm{O2}) = \begin{pmatrix} G^{\mathrm{O:H}}(\mathrm{O2}) \\ G^{\mathrm{O:O}}(\mathrm{O2}) \end{pmatrix}, \quad \mathbf{G}^{\mathrm{H}}(\mathrm{H3}) = \begin{pmatrix} G^{\mathrm{H:H}}(\mathrm{H3}) \\ G^{\mathrm{H:O}}(\mathrm{H3}) \end{pmatrix}$$

None of the symmetry functions are scaled, and $R_{\mathrm{s}} = 0\,\text{Å}$. The activation function in the hidden layer for both NNs is the logistic function, $f(x) = \frac{1}{1+\exp(-x)}$, for which the derivative can easily be expressed in terms of the function value: $f'(x) = f(x)(1 - f(x))$.

The total energy E is obtained as

$$E = E_{\mathrm{O1}} + E_{\mathrm{O2}} + E_{\mathrm{H3}}$$

and the force along the α-component of O1 (with α often being one of the Cartesian x, y, or z components) is calculated as

$$\begin{aligned}
-\frac{\partial E}{\partial \alpha_{\mathrm{O1}}} &= -\frac{\partial E_{\mathrm{O1}}}{\partial \alpha_{\mathrm{O1}}} - \frac{\partial E_{\mathrm{O2}}}{\partial \alpha_{\mathrm{O1}}} - \frac{\partial E_{\mathrm{H3}}}{\partial \alpha_{\mathrm{O1}}} \\
&= -\frac{\partial E_{\mathrm{O1}}}{\partial G^{\mathrm{O:H}}(\mathrm{O1})}\frac{\partial G^{\mathrm{O:H}}(\mathrm{O1})}{\partial \alpha_{\mathrm{O1}}} - \frac{\partial E_{\mathrm{O1}}}{\partial G^{\mathrm{O:O}}(\mathrm{O1})}\frac{\partial G^{\mathrm{O:O}}(\mathrm{O1})}{\partial \alpha_{\mathrm{O1}}} \\
&\quad - \frac{\partial E_{\mathrm{O2}}}{\partial G^{\mathrm{O:H}}(\mathrm{O2})}\frac{\partial G^{\mathrm{O:H}}(\mathrm{O2})}{\partial \alpha_{\mathrm{O1}}} - \frac{\partial E_{\mathrm{O2}}}{\partial G^{\mathrm{O:O}}(\mathrm{O2})}\frac{\partial G^{\mathrm{O:O}}(\mathrm{O2})}{\partial \alpha_{\mathrm{O1}}} \\
&\quad - \frac{\partial E_{\mathrm{H3}}}{\partial G^{\mathrm{H:H}}(\mathrm{H3})}\frac{\partial G^{\mathrm{H:H}}(\mathrm{H3})}{\partial \alpha_{\mathrm{O1}}} - \frac{\partial E_{\mathrm{H3}}}{\partial G^{\mathrm{H:O}}(\mathrm{H3})}\frac{\partial G^{\mathrm{H:O}}(\mathrm{H3})}{\partial \alpha_{\mathrm{O1}}}
\end{aligned}$$

The partial derivatives of the atomic energies with respect to the NN input features depend on the NN architecture (number of input features, number of hidden layers, number of nodes per hidden layer) as well as the activation function employed in the hidden units. Here, $y_m^{[1]}$ denotes the mth node in the first (and only) hidden layer. The NN weights are stored in matrices

$$\mathbf{A}^{[1],\mathrm{O}} = \begin{pmatrix} a_{11}^{[1],\mathrm{O}} & a_{12}^{[1],\mathrm{O}} \\ a_{21}^{[1],\mathrm{O}} & a_{22}^{[1],\mathrm{O}} \end{pmatrix}, \quad \mathbf{b}^{[1],\mathrm{O}} = \begin{pmatrix} b_1^{[1],\mathrm{O}} \\ b_2^{[1],\mathrm{O}} \end{pmatrix}, \quad \mathbf{A}^{[2],\mathrm{O}} = \begin{pmatrix} a_{11}^{[2],\mathrm{O}} \\ a_{21}^{[2],\mathrm{O}} \end{pmatrix}, \quad b^{[2],\mathrm{O}}$$

with a similar setup for the weights in the hydrogen NN.

$$
\begin{aligned}
y_1^{[1],\mathrm{O1}} &= f\left(a_{11}^{[1],\mathrm{O}} G^{\mathrm{O:H}}(\mathrm{O1}) + a_{21}^{[1],\mathrm{O}} G^{\mathrm{O:O}}(\mathrm{O1}) + b_1^{[1],\mathrm{O}}\right)\\
y_2^{[1],\mathrm{O1}} &= f\left(a_{12}^{[1],\mathrm{O}} G^{\mathrm{O:H}}(\mathrm{O1}) + a_{22}^{[1],\mathrm{O}} G^{\mathrm{O:O}}(\mathrm{O1}) + b_2^{[1],\mathrm{O}}\right)\\
E_{\mathrm{O1}} &= a_{11}^{[2],\mathrm{O}} y_1^{[1],\mathrm{O1}} + a_{21}^{[2],\mathrm{O}} y_2^{[1],\mathrm{O1}} + b^{[2],\mathrm{O}}\\
\frac{\partial E_{\mathrm{O1}}}{\partial G^{\mathrm{O:H}}(\mathrm{O1})} &= a_{11}^{[2],\mathrm{O}} \frac{\partial y_1^{[1],\mathrm{O1}}}{\partial G^{\mathrm{O:H}}(\mathrm{O1})} + a_{21}^{[2],\mathrm{O}} \frac{\partial y_2^{[1],\mathrm{O1}}}{\partial G^{\mathrm{O:H}}(\mathrm{O1})}\\
&= a_{11}^{[2],\mathrm{O}} a_{11}^{[1],\mathrm{O}} y_1^{[1],\mathrm{O1}}\left(1 - y_1^{[1],\mathrm{O1}}\right) + a_{21}^{[2],\mathrm{O}} a_{12}^{[1],\mathrm{O}} y_2^{[1],\mathrm{O1}}\left(1 - y_2^{[1],\mathrm{O1}}\right)\\
\frac{\partial E_{\mathrm{O1}}}{\partial G^{\mathrm{O:O}}(\mathrm{O1})} &= a_{11}^{[2],\mathrm{O}} a_{21}^{[1],\mathrm{O}} y_1^{[1],\mathrm{O1}}\left(1 - y_1^{[1],\mathrm{O1}}\right) + a_{21}^{[2],\mathrm{O}} a_{22}^{[1],\mathrm{O}} y_2^{[1],\mathrm{O1}}\left(1 - y_2^{[1],\mathrm{O1}}\right)\\
y_1^{[1],\mathrm{O2}} &= f\left(a_{11}^{[1],\mathrm{O}} G^{\mathrm{O:H}}(\mathrm{O2}) + a_{21}^{[1],\mathrm{O}} G^{\mathrm{O:O}}(\mathrm{O2}) + b_1^{[1],\mathrm{O}}\right)\\
y_2^{[1],\mathrm{O2}} &= f\left(a_{12}^{[1],\mathrm{O}} G^{\mathrm{O:H}}(\mathrm{O2}) + a_{22}^{[1],\mathrm{O}} G^{\mathrm{O:O}}(\mathrm{O2}) + b_2^{[1],\mathrm{O}}\right)\\
E_{\mathrm{O2}} &= a_{11}^{[2],\mathrm{O}} y_1^{[1],\mathrm{O2}} + a_{21}^{[2],\mathrm{O}} y_2^{[1],\mathrm{O2}} + b^{[2],\mathrm{O}}\\
\frac{\partial E_{\mathrm{O2}}}{\partial G^{\mathrm{O:H}}(\mathrm{O2})} &= a_{11}^{[2],\mathrm{O}} a_{11}^{[1],\mathrm{O}} y_1^{[1],\mathrm{O2}}\left(1 - y_1^{[1],\mathrm{O2}}\right) + a_{21}^{[2],\mathrm{O}} a_{12}^{[1],\mathrm{O}} y_2^{[1],\mathrm{O2}}\left(1 - y_2^{[1],\mathrm{O2}}\right)\\
\frac{\partial E_{\mathrm{O2}}}{\partial G^{\mathrm{O:O}}(\mathrm{O2})} &= a_{11}^{[2],\mathrm{O}} a_{21}^{[1],\mathrm{O}} y_1^{[1],\mathrm{O2}}\left(1 - y_1^{[1],\mathrm{O2}}\right) + a_{21}^{[2],\mathrm{O}} a_{22}^{[1],\mathrm{O}} y_2^{[1],\mathrm{O2}}\left(1 - y_2^{[1],\mathrm{O2}}\right)\\
y_1^{[1],\mathrm{H3}} &= f\left(a_{11}^{[1],\mathrm{H}} G^{\mathrm{H:H}}(\mathrm{H3}) + a_{21}^{[1],\mathrm{H}} G^{\mathrm{H:O}}(\mathrm{H3}) + b_1^{[1],\mathrm{H}}\right)\\
y_2^{[1],\mathrm{H3}} &= f\left(a_{12}^{[1],\mathrm{H}} G^{\mathrm{H:H}}(\mathrm{H3}) + a_{22}^{[1],\mathrm{H}} G^{\mathrm{H:O}}(\mathrm{H3}) + b_2^{[1],\mathrm{H}}\right)\\
E_{\mathrm{H3}} &= a_{11}^{[2],\mathrm{H}} y_1^{[1],\mathrm{H3}} + a_{21}^{[2],\mathrm{H}} y_2^{[1],\mathrm{H3}} + b^{[2],\mathrm{H}}\\
\frac{\partial E_{\mathrm{H3}}}{\partial G^{\mathrm{H:H}}(\mathrm{H3})} &= a_{11}^{[2],\mathrm{H}} a_{11}^{[1],\mathrm{H}} y_1^{[1],\mathrm{H3}}\left(1 - y_1^{[1],\mathrm{H3}}\right) + a_{21}^{[2],\mathrm{H}} a_{12}^{[1],\mathrm{H}} y_2^{[1],\mathrm{H3}}\left(1 - y_2^{[1],\mathrm{H3}}\right)\\
\frac{\partial E_{\mathrm{H3}}}{\partial G^{\mathrm{H:O}}(\mathrm{H3})} &= a_{11}^{[2],\mathrm{H}} a_{21}^{[1],\mathrm{H}} y_1^{[1],\mathrm{H3}}\left(1 - y_1^{[1],\mathrm{H3}}\right) + a_{21}^{[2],\mathrm{H}} a_{22}^{[1],\mathrm{H}} y_2^{[1],\mathrm{H3}}\left(1 - y_2^{[1],\mathrm{H3}}\right)
\end{aligned}
$$

The force component along α_{O1} thus depends on, for example, $\frac{\partial E_{\mathrm{O2}}}{\partial G^{\mathrm{O:O}}(\mathrm{O2})}$, which in turn depends on $y_1^{[1],\mathrm{O2}}$ and $y_2^{[1],\mathrm{O2}}$, which depend on $G^{\mathrm{O:H}}(\mathrm{O2})$ and $G^{\mathrm{O:O}}(\mathrm{O2})$. Thus all symmetry functions on the atom O2, and consequently the entire environment within the cutoff sphere around O2, contribute to the force acting on the atom O1.

The partial derivatives of the radial symmetry functions with respect to the coordinate α_{O1} become

$$\begin{aligned}
\frac{\partial G^{\mathrm{O:H}}(\mathrm{O1})}{\partial \alpha_{\mathrm{O1}}} &= \mathrm{e}^{-\eta R_{\mathrm{O1H3}}^2}\left(-2\eta R_{\mathrm{O1H3}} f_{\mathrm{c}}(R_{\mathrm{O1H3}})\frac{\partial R_{\mathrm{O1H3}}}{\partial \alpha_{\mathrm{O1}}} + \frac{\partial f_{\mathrm{c}}(R_{\mathrm{O1H3}})}{\partial \alpha_{\mathrm{O1}}}\right) \\
\frac{\partial G^{\mathrm{O:O}}(\mathrm{O1})}{\partial \alpha_{\mathrm{O1}}} &= \mathrm{e}^{-\eta R_{\mathrm{O1O2}}^2}\left(-2\eta R_{\mathrm{O1O2}} f_{\mathrm{c}}(R_{\mathrm{O1O2}})\frac{\partial R_{\mathrm{O1O2}}}{\partial \alpha_{\mathrm{O1}}} + \frac{\partial f_{\mathrm{c}}(R_{\mathrm{O1O2}})}{\partial \alpha_{\mathrm{O1}}}\right) \\
\frac{\partial G^{\mathrm{O:H}}(\mathrm{O2})}{\partial \alpha_{\mathrm{O1}}} &= \mathrm{e}^{-\eta R_{\mathrm{O2H3}}^2}\left(-2\eta R_{\mathrm{O2H3}} f_{\mathrm{c}}(R_{\mathrm{O2H3}})\frac{\partial R_{\mathrm{O2H3}}}{\partial \alpha_{\mathrm{O1}}} + \frac{\partial f_{\mathrm{c}}(R_{\mathrm{O2H3}})}{\partial \alpha_{\mathrm{O1}}}\right) \\
&= 0 \\
\frac{\partial G^{\mathrm{O:O}}(\mathrm{O2})}{\partial \alpha_{\mathrm{O1}}} &= \mathrm{e}^{-\eta R_{\mathrm{O1O2}}^2}\left(-2\eta R_{\mathrm{O1O2}} f_{\mathrm{c}}(R_{\mathrm{O1O2}})\frac{\partial R_{\mathrm{O1O2}}}{\partial \alpha_{\mathrm{O1}}} + \frac{\partial f_{\mathrm{c}}(R_{\mathrm{O1O2}})}{\partial \alpha_{\mathrm{O1}}}\right) \\
\frac{\partial G^{\mathrm{H:H}}(\mathrm{H3})}{\partial \alpha_{\mathrm{O1}}} &= 0 \\
\frac{\partial G^{\mathrm{H:O}}(\mathrm{H3})}{\partial \alpha_{\mathrm{O1}}} &= \mathrm{e}^{-\eta R_{\mathrm{O1H3}}^2}\left(-2\eta R_{\mathrm{O1H3}} f_{\mathrm{c}}(R_{\mathrm{O1H3}})\frac{\partial R_{\mathrm{O1H3}}}{\partial \alpha_{\mathrm{O1}}} + \frac{\partial f_{\mathrm{c}}(R_{\mathrm{O1H3}})}{\partial \alpha_{\mathrm{O1}}}\right) \\
&\quad + \mathrm{e}^{-\eta R_{\mathrm{O2H3}}^2}\left(-2\eta R_{\mathrm{O2H3}} f_{\mathrm{c}}(R_{\mathrm{O2H3}})\frac{\partial R_{\mathrm{O2H3}}}{\partial \alpha_{\mathrm{O1}}} + \frac{\partial f_{\mathrm{c}}(R_{\mathrm{O2H3}})}{\partial \alpha_{\mathrm{O1}}}\right) \\
&= \mathrm{e}^{-\eta R_{\mathrm{O1H3}}^2}\left(-2\eta R_{\mathrm{O1H3}} f_{\mathrm{c}}(R_{\mathrm{O1H3}})\frac{\partial R_{\mathrm{O1H3}}}{\partial \alpha_{\mathrm{O1}}} + \frac{\partial f_{\mathrm{c}}(R_{\mathrm{O1H3}})}{\partial \alpha_{\mathrm{O1}}}\right),
\end{aligned}$$

where η is the η-value of the pertinent symmetry function. All the above partial derivatives are calculated as sums over neighbors, but in this example, there are only 1 H and 1 O neighbor around each O atom. There are two O neighbors around H3, but one of the terms in the sum defining $\frac{\partial G^{\mathrm{H:O}}(\mathrm{H3})}{\partial \alpha_{\mathrm{O1}}}$ becomes 0, since the position of O1 does not affect the distance between O2 and H3.

References

1. J. Behler, Angew. Chem. Int. Ed. **56**(42), 12828 (2017)
2. J. Behler, J. Chem. Phys. **145**(17), 170901 (2016)
3. J.E. Jones, Proc. R. Soc. Lond. A **106**, 463 (1924)
4. A.P. Bartók, M.C. Payne, R. Kondor, G. Csányi, Phys. Rev. Lett. **104**, 136403 (2010)
5. M. Rupp, A. Tkatchenko, K.R. Müller, O.A. von Lilienfeld, Phys. Rev. Lett. **108**, 058301 (2012)
6. A.V. Shapeev, Multiscale Model. Simul. **14**, 1153 (2016)
7. R.M. Balabin, E.I. Lomakina, Phys. Chem. Chem. Phys. **13**, 11710 (2011)
8. J. Behler, M. Parrinello, Phys. Rev. Lett. **98**, 146401 (2007)
9. J. Behler, J. Phys. Condens. Matter **26**, 183001 (2014)
10. J. Behler, Int. J. Quantum Chem. **115**(16), 1032 (2015). https://doi.org/10.1002/qua.24890
11. N. Artrith, T. Morawietz, J. Behler, Phys. Rev. B **83**, 153101 (2011)

12. T. Morawietz, V. Sharma, J. Behler, J. Chem. Phys. **136**, 064103 (2012)
13. M. Gastegger, J. Behler, P. Marquetand, Chem. Sci. **8**, 6924 (2017)
14. J. Behler, J. Chem. Phys. **134**, 074106 (2011)
15. M. Hellström, J. Behler, in *Handbook of Materials Modeling: Methods: Theory and Modeling*, ed. by W. Andreoni, S. Yip (Springer International Publishing, Cham, 2018), pp. 1–20. https://doi.org/10.1007/978-3-319-42913-7_56-1
16. T. Morawietz, A. Singraber, C. Dellago, J. Behler, Proc. Natl. Acad. Sci. U.S.A. **113**(30), 8368 (2016). https://doi.org/10.1073/pnas.1602375113
17. M. Hellström, J. Behler, J. Phys. Chem. Lett. **7**, 3302 (2016). https://doi.org/10.1021/acs.jpclett.6b01448
18. S.K. Natarajan, J. Behler, Phys. Chem. Chem. Phys. **18**, 28704 (2016)
19. K. Shakouri, J. Behler, J. Meyer, G.J. Kroes, J. Phys. Chem. Lett. **8**, 2131 (2017)
20. G. Imbalzano, A. Anelli, D. Giofré, S. Klees, J. Behler, M. Ceriotti, J. Chem. Phys. **148**(24), 241730 (2018). https://doi.org/10.1063/1.5024611
21. J.S. Smith, O. Isayev, A.E. Roitberg, Chem. Sci. **8**, 3192 (2017). https://doi.org/10.1039/C6SC05720A
22. M. Gastegger, L. Schwiedrzik, M. Bittermann, F. Berzsenyi, P. Marquetand, J. Chem. Phys. **148**(24), 241709 (2018). https://doi.org/10.1063/1.5019667
23. N. Artrith, A. Urban, G. Ceder, Phys. Rev. B **96**, 014112 (2017)
24. K. Levenberg, Q. Appl. Math. **2**, 164 (1944)
25. D.W. Marquardt, SIAM J. Appl. Math. **11**, 431 (1963)
26. S. Haykin, *Kalman Filtering and Neural Networks* (Wiley, London, 2001)
27. J.E. Herr, K. Yao, R. McIntyre, D.W. Toth, J. Parkhill, J. Chem. Phys. **148**(24), 241710 (2018). https://doi.org/10.1063/1.5020067
28. S. Grimme, J. Antony, S. Ehrlich, H. Krieg, J. Chem. Phys. **132**(15), 154104 (2010). https://doi.org/10.1063/1.3382344
29. K. Yao, J.E. Herr, D. Toth, R. Mckintyre, J. Parkhill, Chem. Sci. **9**, 2261 (2018). https://doi.org/10.1039/C7SC04934J
30. J. Behler, R. Martoňák, D. Donadio, M. Parrinello, Phys. Rev. Lett. **100**, 185501 (2008)
31. R.Z. Khaliullin, H. Eshet, T.D. Kühne, J. Behler, M. Parrinello, Nat. Mater. **10**, 693 (2011)
32. H. Eshet, R.Z. Khaliullin, T.D. Kühne, J. Behler, M. Parrinello, Phys. Rev. Lett. **108**, 115701 (2012)
33. G.C. Sosso, G. Miceli, S. Caravati, J. Behler, M. Bernasconi, Phys. Rev. B **85**, 174103 (2012)
34. N. Artrith, J. Behler, Phys. Rev. B **85**, 045439 (2012)
35. N. Artrith, B. Hiller, J. Behler, Phys. Status Solidi B **250**, 1191 (2013)
36. N. Artrith, A.M. Kolpak, Comp. Mater. Sci. **110**, 20 (2015)
37. N. Artrith, A. Urban, Comp. Mater. Sci. **114**, 135 (2016)
38. J.R. Boes, M.C. Groenenboom, J.A. Keith, J.R. Kitchin, Int. J. Quantum Chem. **116**, 979 (2016)
39. S. Hajinazar, J. Shao, A.N. Kolmogorov, Phys. Rev. B **95**, 014114 (2017)
40. V. Quaranta, M. Hellström, J. Behler, J. Phys. Chem. Lett. **8**, 1476 (2017)
41. V. Quaranta, M. Hellström, J. Behler, J. Kullgren, P.D. Mitev, K. Hermansson, J. Chem. Phys. **148**(24), 241720 (2018). https://doi.org/10.1063/1.5012980
42. M. Hellström, J. Behler, J. Phys. Chem. B **121**(16), 4184 (2017). https://doi.org/10.1021/acs.jpcb.7b01490
43. S. Kondati Natarajan, T. Morawietz, J. Behler, Phys. Chem. Chem. Phys. **17**, 8356 (2015). https://doi.org/10.1039/C4CP04751F
44. C. Schran, F. Uhl, J. Behler, D. Marx, J. Chem. Phys. **148**(10), 102310 (2018). https://doi.org/10.1063/1.4996819
45. K.T. Schütt, M. Gastegger, A. Tkatchenko, K.R. Müller, arXiv:1806.10349 (2018)

Construction of Machine Learned Force Fields with Quantum Chemical Accuracy: Applications and Chemical Insights

14

Huziel E. Sauceda, Stefan Chmiela, Igor Poltavsky, Klaus-Robert Müller, and Alexandre Tkatchenko

Abstract

Highly accurate force fields are a mandatory requirement to generate predictive simulations. Here we present the path for the construction of machine learned molecular force fields by discussing the hierarchical pathway from generating the dataset of reference calculations to the construction of the machine learning model, and the validation of the physics generated by the model. We will use the symmetrized gradient-domain machine learning (sGDML) framework due to its ability to reconstruct complex high-dimensional potential energy surfaces (PES) with high precision even when using just a few hundreds of molecular conformations for training. The data efficiency of the sGDML model allows using reference atomic forces computed with high-level wave-function-based approaches, such as the *gold standard* coupled-cluster method with single,

H. E. Sauceda (✉)
Fritz-Haber-Institut der Max-Planck-Gesellschaft, Berlin, Germany
e-mail: sauceda@tu-berlin.de

S. Chmiela
Machine Learning Group, Technische Universität Berlin, Berlin, Germany

I. Poltavsky · A. Tkatchenko
Physics and Materials Science Research Unit, University of Luxembourg, Luxembourg, Luxembourg
e-mail: alexandre.tkatchenko@uni.lu

K.-R. Müller
Machine Learning Group, Technische Universität Berlin, Berlin, Germany

Max Planck Institute for Informatics, Stuhlsatzenhausweg, Saarbrücken, Germany

Department of Brain and Cognitive Engineering, Korea University, Anam-dong, Seongbuk-gu, Seoul, Korea
e-mail: klaus-robert.mueller@tu-berlin.de

K. T. Schütt et al. (eds.), *Machine Learning Meets Quantum Physics*, Lecture Notes in Physics 968, https://doi.org/10.1007/978-3-030-40245-7_14

double, and perturbative triple excitations (CCSD(T)). We demonstrate that the flexible nature of the sGDML framework captures local and non-local electronic interactions (e.g., H-bonding, lone pairs, steric repulsion, changes in hybridization states (e.g., $sp^2 \rightleftharpoons sp^3$), $n \rightarrow \pi^*$ interactions, and proton transfer) without imposing any restriction on the nature of interatomic potentials. The analysis of sGDML models trained for different molecular structures at different levels of theory (e.g., density functional theory and CCSD(T)) provides empirical evidence that a higher level of theory generates a smoother PES. Additionally, a careful analysis of molecular dynamics simulations yields new qualitative insights into dynamics and vibrational spectroscopy of small molecules close to spectroscopic accuracy.

14.1 Introduction

In silico studies of molecular systems and materials constitute one of the most important tools in physics, biology, materials science, and chemistry due to their great contributions in understanding systems ranging from small molecules (e.g., few atoms) up to large proteins and amorphous materials, guiding the exploration and the discovery of new materials and drugs. This requires the construction of physical models that faithfully describe interatomic interactions, and quantum mechanics (QM) is the pertinent methodology to engage such monumental task. Nevertheless, using the full machinery of QM (e.g., Dirac equation [1] and Quantum Electrodynamics [2]) would lead not far from simulations of diatomic molecules. To overcome this limitation, for most of the problems of interest, one can approximately describe a molecular system by the more tractable non-relativistic time-independent Schrödinger equation.

Additionally, one often decouples nuclear and electronic degrees of freedom by employing the Born–Oppenheimer (BO) approximation. This makes predictive simulations of molecular properties and thermodynamic functions possible by representing a N-atoms system by the global potential energy surface (PES) $V_{\mathrm{BO}}(\mathbf{x})$, where $\mathbf{x} = \{\mathbf{r}_1, \mathbf{r}_2, \ldots, \mathbf{r}_N\}$ and $\mathbf{r}_i$ the ith nuclear Cartesian coordinates. $V_{\mathrm{BO}}(\mathbf{x})$ is defined as the sum of the total electrostatic nuclear repulsion energy $\sum_{i,j>i} Z_i Z_j r_{ij}^{-1}$ and the electronic energy $\mathcal{E}_{\mathrm{elec}}$ solution of the electronic Schrödinger equation $\mathcal{H}_{\mathrm{elec}}\Psi = \mathcal{E}_{\mathrm{elec}}\Psi$ for a given set of nuclear coordinates $\mathbf{x}$. Therefore, V_{BO} contains all the information necessary to describe nuclear dynamics of the molecular system since all electronic quantum interactions are encoded in it via $\mathcal{E}_{\mathrm{elec}}$ within the BO approximation. A systematic partitioning of this energy could potentially help to gain further insights into the physics and chemistry of the system, nevertheless, in practice it is not known how to exactly expand the V_{BO} in different energetic contributions such as hydrogen bonding, electrostatics, dispersion interactions, or other electronic effects. Furthermore, any attempt in separating the PES in terms of known analytic forms or empirically derived interactions will always result in biasing the final model which limits its possible accuracy and may introduce non-physical artifacts. Therefore, the intricate form of V_{BO} resulting from an interplay between different quantum phenomena when solving the Schrödinger equation should be preserved.

In order to extract the dynamical properties and thermodynamics of molecular systems, the V_{BO} has to be sampled according to a thermodynamic ensemble (e.g., NVE, NVT, μVT, etc.) depending on the property being computed. The two most popular techniques are Monte Carlo sampling and molecular dynamics simulations (MD). In particular, MD constitutes the fundamental pillar of contemporary science by allowing remarkable advances and offering unprecedented insights into complex chemical and biological systems. However, sampling the V_{BO} using this technique in any of its flavors (e.g., Langevin or Verlet-velocity propagator) to obtain converged mechanical and thermodynamical properties often requires millions integration steps, meaning that the Schrödinger equation $\mathcal{H}_{\text{elec}}\Psi = \mathcal{E}_{\text{elec}}\Psi$ has to be solved and $-\mathbf{F} = \langle\Psi^*|\partial\mathcal{H}/\partial\mathbf{x}|\Psi\rangle$ evaluated a similar amount of times [3]. Such direct ab initio molecular dynamics (AIMD) simulations, where the quantum-mechanical energies and forces are computed on-the-fly for molecular configurations at every time step, are known to generate highly accurate but computationally very costly predictions. In practice, most of the works in AIMD use density functional theory (DFT) to approximate the solution of the Schrödinger equation for a system of electrons and nuclei. Unfortunately, in some cases different exchange-correlation functionals yield contrasting results for molecular properties [4] and it is not clear how to systematically improve their performance. Alternatively, wave-function based methods that account for electron correlation (e.g., post-Hartree–Fock methods) offer a systematically improvable framework but they are rarely used in AIMD simulations due to the steep increase in the required computational resources. For example, a nanosecond-long AIMD simulation for a single ethanol molecule using CCSD(T) method would demand approximately a million CPU years on modern hardware.

It is clear that AIMD is not an affordable route to pursue predictive simulations for most of the systems of interest. An alternative is to roughly approximate the V_{BO} by creating handcrafted interatomic and physically inspired potentials with parameters fitted to experimental data or quantum-mechanical calculations. This has been a common practice since the early works on molecular dynamics [5–8]. The complexity of creating reliable interatomic potentials using prior physical knowledge led to the development of dedicated force fields (FFs) for different chemical systems, a successful approach as highlighted by the 2013 Nobel Prize in Chemistry. Examples are the TIPnP FFs for water [9, 10], Tersoff potential for covalent materials [11], polarizable FFs [12], tight-binding potentials for semiconductors and metals [13]. This also includes a plethora of biomolecular FFs such as AMBER, MMFF, CHARMM, and GROMOS; FFs that often give reliable results for protein folding under ambient conditions [14–17]. The wide variety of available interatomic potentials highlights the fact that handcrafting a FF capable of describing different types of interactions (metallic bonding, covalent chemistry, hydrogen bonding, non-covalent interactions, etc.) in a unified and seamless fashion is a complex challenge. Furthermore, it is widely recognized that even dedicated molecular mechanic FFs cannot generate quantitative predictions from MD simulations due to their lack of accuracy. These increasingly pressing issues hinder truly predictive modeling, but at the same time encourage the development of more accurate and efficient methodologies.

One of the possible pathways is the employment of machine learning (ML) methods for the reconstruction of the PES function. Machine learned force fields (ML-FFs) exploit the correlation encoded in molecular datasets generated from AIMD trajectories (or any other sampling methodology) to reconstruct the underlying PES without imposing any particular explicit analytic form for the interatomic interactions. Furthermore, machine learning is based on rigorous statistical learning theory [18, 19], providing a powerful and general framework for FF learning. ML approaches can reconstruct complex high-dimensional objects with arbitrary precision given sufficient amount of data samples (e.g., molecular energies and atomic forces) for training. The accurate learning of V_{BO} is not a trivial task and it has driven a vast amount of work such as data sampling [20–24], molecular representations [25–41], neural networks architecture development [42–50], inference methods [51–66], and explanation methods [67–70]. A crucial contribution to the further development and understanding of the field is the releasing of ready-to-use software as well as molecular datasets which guaranties the reproducibility of published results [68, 71–73]. In terms of the performance, the computational cost of evaluating ML-FFs lies in between molecular mechanic FFs and ab initio calculations. In particular, the sGDML framework [64, 65] is 5–10 orders of magnitude faster than ab initio calculations and 2–3 orders of magnitude slower than molecular mechanic FFs.[1] A precise number depends on the molecular system under study. As a reference, the sGDML model can be up to 10^7 and 10^9 times faster than CCSD(T)/cc-pVTZ level of theory for a single-point calculation of malondialdehyde and aspirin, respectively [71], preserving the same accuracy. This allows the use of these ML-FFs for performing long-time MD simulations and exploring different molecular properties on the CCSD(T)-level of accuracy.

The PES reconstruction problem can be approached from two different but in principle equivalent ways,[2] by learning directly the scalar function V_{BO} or by first reconstructing the gradient field associated to the PES, ∇V_{BO}, and then recover the PES by analytic integration. These two types of ML models are called energy $\hat{f}_E$ and force $\hat{\mathbf{f}}_{\mathbf{F}}$ models, respectively.[3] The two most established methodologies to create such models are Neural Networks (NN) [42–45, 47, 48, 74–76] and kernel methods [20, 21, 23, 28, 29, 33, 36, 51, 53, 63–65]. An energy model, $\hat{f}_E$, can be based on NNs or kernel methods and trained on energies or using a combination

[1] It is important to notice that while the scaling of the performance in ML-FFs depends *only* on the number of atoms, while in the case of ab initio quantum chemical calculations their performance depends on the level of theory and on the size of the basis used to approximate the wave-function and the number of electrons.

[2] To the best knowledge of the authors up to this day these are the only two ways have been used in the PES reconstruction problem.

[3] The symbol $\hat{f}$ will be reserved to represent the predictor function of the machine learning model.

of energies and forces [21, 28, 36, 42, 45–48, 51, 53]. The associated FF to $\hat{f}_E$ is generated by analytic differentiation, $\hat{\mathbf{f}}_{\mathbf{F}\leftarrow E} = -\nabla \hat{f}_E$, which introduces some disadvantages to be discussed further in the text. In the case of force models $\hat{\mathbf{f}}_{\mathbf{F}}$, they could also be constructed using NNs but the problem is that to recover the underlying PES requires the analytic integration of the vector field predictor. This immensely limits their applicability since without an appropriate integration scheme they will not be able to recover the PES. A more common way to generate force models is using kernel methods [20,33,63,64] usually trained directly in the gradient domain. Contrary to the case of NNs based force models, kernels methods offer a much more flexible framework to conveniently define its analytic form, this is done by utilizing the robust framework of Gaussian processes which allows the incorporation of prior physical knowledge. Therefore, recovering the underlying PES $\hat{f}_{E\leftarrow\mathbf{F}}$ can be easily done by imposing that the mathematical formulation of $\hat{\mathbf{f}}_{\mathbf{F}}$ to be analytically integrable and consequently it will, by definition, encode the fundamental physical law of energy conservation [64].

In the limit of an infinite amount of data, energy and force models should converge to the same prediction error. Nevertheless, when dealing with finite or restricted amounts of data these two models do present very different performances. Some of the fundamental advantages of using force models instead of energy models are: (1) Learning in the gradient domain yield smoother PESs, (2) training exclusively on forces generates more accurate models than training using energies or a combination of both [47, 66, 71], (3) obtaining energies by analytical integration of force models tends give better behaved predictions as a result of the integral operator, this is in contrast with forces generated out of energy models by the gradient operator [66], and (4) force models are more data efficient [64, 65]. It is important to highlight that the data efficiency of force models arises not only because the greater amount of information in each force sample ($3N$ components, where N is the number of atoms), but also because each entry of the force vector is orthogonal to the rest,[4] therefore providing a complete linearized description of its immediate local neighborhood [77]. Continuing with the discussion of data efficiency, there is only a handful of models that fulfill this requirement. Even though formally both NN and kernel-based methods can achieve any desired accuracy, the realm of scarce data belongs to kernel models.[5] This is the case, for example, when the system under study requires to be described by a highly accurate reference method and it is only possible to compute a couple of hundreds of data points, as would be the case of some of the amino acids or large molecules. Such better reconstruction efficiency of kernel methods is due to their greater use of prior information, offering a unique and well-defined solution [64].

[4]The components of the force vector are orthogonal in $\mathbb{R}^{3N}$, space where the function is defined.

[5]The reason of such difference between NNs and kernel models is that, while kernels rely on feature engineering (i.e., handcrafted descriptors), NNs represent an end-to-end formalism to describe the data. This means that NNs require more data to infer the representation that optimally describes the system.

Here, we will focus on the *s*ymmetric ***G**radient **D**omain **M**achine **L***earning (sGDML) FF [65]. The sGDML is a kernel-based ML model which directly learns forces since it is trained explicitly in the gradient domain of V_{BO}. The principal feature of this model is that it was mathematically conceived as an analytically integrable curl-free framework. The energy conservation law is explicitly encoded into the model. Therefore, once the sGDML-FF $\hat{\mathbf{f}}_{\mathbf{F}}$ is trained, the potential energy function $\hat{f}_{E\leftarrow\mathbf{F}}$ is also available. It is worth highlighting that only forces are used for training given that there is empirical evidence that a loss function that combines energies and forces causes a degradation in the force prediction [47,71,78]. The second fundamental property of the model is that the complexity of the reconstruction process is reduced through the explicit incorporation of molecular symmetries (i.e., rigid and fluxional). These permutational symmetries are automatically extracted from the reference dataset via a multi-partite procedure [79]. Additionally, in this framework all atomic interactions are modeled globally, meaning that the learning problem is solved without any inherently non-unique atom-wise, pairwise, or any other many-body partitioning. Thus, the approach preserves the many-body nature of the quantum problem. These central properties contribute to the ability of the sGDML model to learn complex PES for molecules of intermediate size from limited amounts of reference calculations, an unachievable task for non-dedicated molecular FFs or even other ML methodologies. In particular, the sGDML model is able to reconstruct CCSD(T)-quality FFs from a limited amount (few hundreds) of reference molecular configurations [65].

In this chapter, we present an overview of the sGDML model from the construction of reliable datasets to the training and validation of the models to performing analysis of some relevant quantum effects captured by the model. The structure of the chapter is the following. In Sect. 14.2 we present the problem of imbalanced database and the idea behind the representative sampling. In Sect. 14.3 we introduce the idea of physically inspired ML-FFs and present the sGDML model as well as a comparative analysis of the differences between energy and force models. The evaluation of the performance of the model is presented in Sect. 14.4. Section 14.5 is dedicated to the analysis of smoothing of the PES by increasing the level of theory. In Sect. 14.6 the different types of interactions captured learned by the sGDML are highlighted. Finally, Sect. 14.7 we summarize and present the conclusions.

14.2 Data Generation and Sampling of the PES

The accurate reconstruction process of a high-dimensional surface via ML methods heavily relies on the available reference data. In the case of PES learning, a well-known approach is to construct the database by sampling the PES using molecular dynamics simulations. Of course the data generated with this methodology will depend on the temperature of MD simulation, therefore higher temperatures will explore higher energy regions (see Fig. 14.1). MD-generated database will be biased to lower energy regions of the PES, where the system spend most of the time. Consequently, this methodology is advisable only when the final application

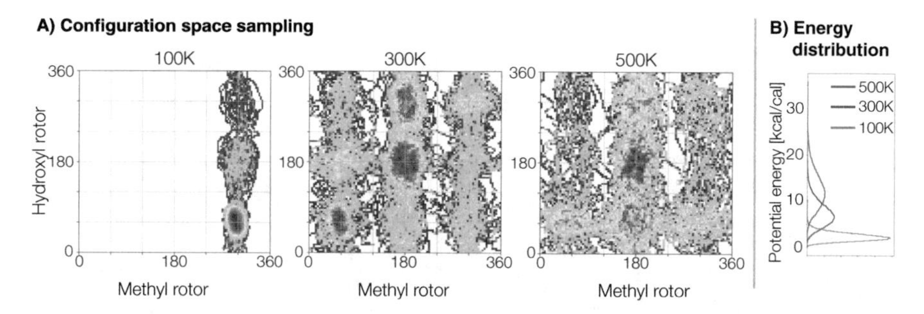

Fig. 14.1 (**a**) Sampling of ethanol's PES at 100, 300 and 500 K using AIMD at DFT/PBE+TS level of theory. (**b**) The potential energy profile is shown in for the different temperatures

involves MD simulation for equilibrium or close to equilibrium properties where rare events do not play a major role. Examples of this is the study of vibrational spectra, direct study of minima population, thermodynamic properties, etc. A general rule of thumb is to generate the database at a higher temperature compared to the intended use of the ML model trained on this data. For example, if we want to calculate the vibrational spectrum for ethanol at 300 K, generating the database at 500 K is a safe option since the subspace of configurations relevant at 300 K is contained in this database (see Fig. 14.1a).

The main databases used in this study were created by running AIMD (DFT) simulations at a temperature of 500 K using the FHI-aims package [80] at the generalized gradient approximation (GGA) level of theory with Perdew–Burke–Ernzerhof (PBE) [81] exchange-correlation functional and the Tkatchenko–Scheffler (TS) method [82] to account for van der Waals interactions using the light basis set. In the literature this is known as the MD17 dataset [64].

14.2.1 Imbalanced Sampling

From the ergodic hypothesis we know that the expected value of an observable A can be obtained from $\langle A\rangle_{\text{time}} = N_t^{-1}\sum_t^{N_t} A(\mathbf{x}_t)$, where $\mathbf{x}_t$ is the step t of the dynamics trajectory. This, of course, is valid only in the case in which the dynamics are long enough to visit all the possible configurations of the system under the given constraints. In practice, and in particular for AIMD this is not feasible due to its computational demands; therefore, in the context of databases generation this leads to biased databases. Figure 14.2 displays the sampling of the PESs for ethanol, keto form of malondialdehyde (keto-MDA), and Aspirin at 500 K using AIMD. It is easy to notice that even at high temperatures and more than 200 ps of simulation time, the sampling is biased and non-symmetric in the case of ethanol and Aspirin.

It is imperative to mention that when creating such databases and using them for generating ML models, many of the limitations of the database will be passed to the learned model. Then, the final user of ML model has to be aware of its

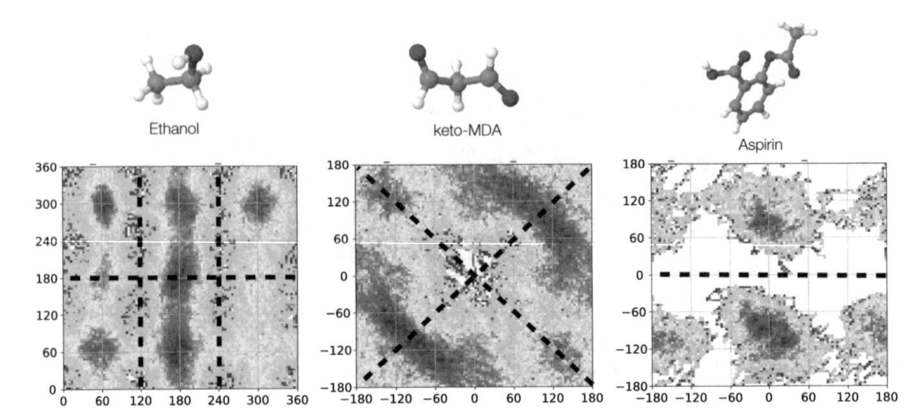

Fig. 14.2 Molecular dynamics' sampling of PESs for ethanol, keto-MDA and aspirin at 500 K using DFT/PBE+TS level of theory. The black dashed lines indicate the symmetries of the molecule

range of applicability. On the other hand, a robust ML framework would be able to remove some of the imperfections of the data by using prior information of the underlying nature of the data. As an example, if training a ML-FF using the ethanol's or Aspirin's data in Fig. 14.2 the ML model must be able to handle non-symmetric databases. Usually this is done by incorporating the indistinguishability between atoms of the same species.

14.2.2 Representative Sampling: From DFT to CCSD(T)

Constructing reliable molecular databases can be very complicated even for small molecules, since efficiently exploring the molecular PES not only depends on the size of the molecule but also on many other molecular features such as intramolecular interactions and fluxional groups. Generating $\sim 2 \times 10^5$ conformations from AIMD using a relatively affordable level of theory (e.g., PBE+TS with a small basis) can take from a couple of days to a couple of weeks. Higher levels of theory (e.g., PBE0+MBD) would require weeks or months of server time. Finally, whenever the system under study demands the use of highly accurate methodologies such as CCSD(T), generating an extensive database becomes computationally prohibitively expensive. To resolve this issue one can first sample the PES using a lower but representative level of theory in the AIMD simulations to generate trajectory $\{X_t^{\text{PBE+TS}}\}_{t=1}^{N_{\text{steps}}}$, and then sub-sample this database to generate a representative set of geometries. These geometries serve as an input for higher level of theory single-point calculations, e.g., $\{X_t^{\text{CCSD(T)}}\}_{t=1}^{N_{\text{sub-sample}}}$ (represented by red dots in Fig. 14.3), resulting in accurate and computationally affordable database.

From Fig. 14.3 we see that, in this 2D projection, the reference and the desired PESs look similar, which allows to use a PES@PBE+TS sampling as a good approx-

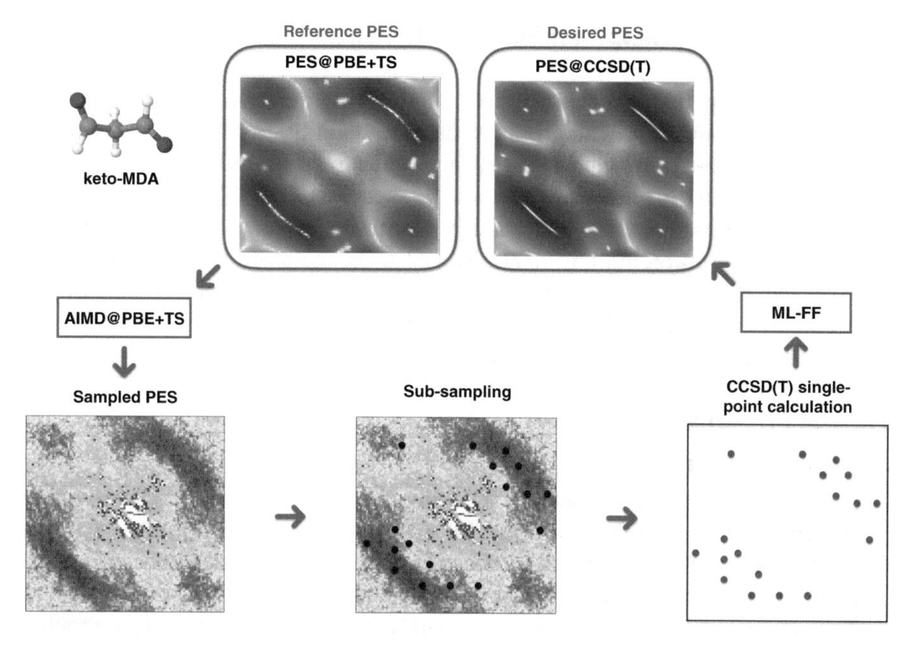

Fig. 14.3 Procedure followed to generate CCSD(T) database for the keto-MDA molecule. An AIMD simulation at 500 K using DFT/PBE+TS level of theory was used as a reference sampling of the molecular PES. Afterwards the obtained trajectory is sub-sampled (black dots) to generate a subset of representative geometries, then this is used to perform single-point calculations at a higher level of theory (red dots). In this case, CCSD(T) was the desired PES and the ML-FF used was sGDML

imation to the one that we would get by sampling PES@CCSD(T) directly [65]. This is a crucial concept that should be carefully used since even if the test error of the ML model is good, that does not mean that the predictions generated by the ML model will be physically valid. This would be the case in which the reference data comes from a PES that considerably differs from the desired PES, for example, the combination of HF and CCSD(T). Another example, when the reference data does not provide a reliable ML model, is the use of a database generated by an AIMD trajectory at 100 K for training a ML-FF, and then running MD simulations with this FF at higher temperatures. The problem is that the ML model will be generating predictions in the extrapolation regime, and therefore, there is no certainty that the results would be physically valid.

For building sGDML models, the CCSD and CCSD(T) databases were generated by using the subsampling scheme (Fig. 14.3) for some of the molecules from the MD17 database. In the case of keto-MDA, enol-MDA, and ethanol, the molecular configurations were recomputed using all-electron CCSD(T), while in the case of Aspirin all-electron CCSD were employed [83–85].

14.3 Physically Inspired Machine Learned Force Fields

Machine learning offers a wide variety of different universal approximators to reconstruct any function in the limit of data availability. In practice, the amount of accessible data is restricted, especially when reconstructing complex PESs from highly accurate reference calculations such as CCSD(T). Consequently, it is highly advantageous to mathematically constrain the space of solution of our approximator by enforcing universal physical laws, therefore naturally creating a data efficient model capable of delivering physically meaningful predictions. Below we summarize the desirable properties for a machine learned force field from the physics and computational point of view:

Physical Properties

- *Global model.* Building this property in the model will keep the many-body nature of the quantum interactions resulting from the solution of the Schrödinger equation $\mathcal{H}\Psi = E\Psi$ and from the evaluation of the Hellmann–Feynman forces $-\mathbf{F} = \langle\Psi^*|\partial\mathcal{H}/\partial\mathbf{x}|\Psi\rangle$. In practice, this means to avoid the non-unique partitioning of the total energy V_{BO} in atomic contributions.
- *Temporal symmetry.* This constraint demands that the ML generated Hamiltonian $\mathcal{H} = \mathcal{T} + \hat{f}_E$, with $\mathcal{T}$ and $\hat{f}_E$ the kinetic and potential energies, respectively, must be time-invariant, which means that the fundamental law of energy conservation has to be enforced in the ML model, $\hat{\mathbf{f}}_{\mathbf{F}} = -\nabla\hat{f}_E$.
- *Indistinguishability of atoms.* In quantum mechanics, two atoms of the same species cannot be distinguished.[6] This means that permuting two identical atoms in a molecule does not change the energy of the system: $V_{\text{BO}}(\ldots, \mathbf{x}_i, \ldots, \mathbf{x}_j, \ldots) = V_{\text{BO}}(\ldots, \mathbf{x}_j, \ldots, \mathbf{x}_i, \ldots)$. This spatial symmetry often represents a big challenge for ML global models, but it is trivially fulfilled by models that learn energy per atom.

Each one of the above mentioned physical properties of a quantum system constitute a constraint that narrows the space of solutions of the universal ML approximator down, contributing to a more efficient and accurate reconstruction of the original data generator.

Computational Requirements

- *Accuracy and data efficiency.* This is a highly desirable requirement in the reconstruction of PES from ab initio data since the generation of each data point constitute a considerable computational cost. As an example, a CCSD(T) single-point force calculation can take several days in a single processor for a medium sized molecule.

[6] Even though this is a fundamental property of quantum systems, the invariance of the energy to permutations of atoms of the same species is preserved even in classical mechanics. As will be the case in all the examples discussed in this chapter.

- *Robust and stable predictions.* To minimize the chance of artifacts in the reconstruction of the PES, the solution needs to be derived from a hypothesis space that satisfies the fundamental physical laws. Models that start from a general set of assumptions cannot be expected to generalize from small datasets.
- *Fast evaluation.* The main purpose of ML-FFs is their use in PES sampling techniques such as MD or Monte Carlo. This requires fast evaluations (few milliseconds per single-point energy/force calculations).

Whenever a ML model does not fulfill at least one of the properties or requirements mentioned above, it becomes either unreliable or inefficient for practical applications.

14.3.1 Symmetrized Gradient-Domain Machine Learning

Gradient-domain machine learning (GDML) is one of the approaches that fulfills all the properties discussed previously. The key idea is to use a Gaussian process (GP) to model the force field $\hat{\mathbf{f}}_{\mathbf{F}}$ as a transformation of an unknown potential energy surface $\hat{f}_E$ such that,

$$\hat{\mathbf{f}}_{\mathbf{F}} = -\nabla \hat{f}_E \sim \mathcal{GP}\left[-\nabla \mu_E(\mathbf{x}), \nabla_{\mathbf{x}} k_E\left(\mathbf{x}, \mathbf{x}'\right) \nabla_{\mathbf{x}'}^{\top}\right], \tag{14.1}$$

where μ_E and k_E are the mean and covariance of the energy GP, respectively [77]. Furthermore, the model is symmetrized (sGDML) to reflect the indistinguishability of atoms, while retaining the global nature of the interactions. With the inclusion of a descriptor $\mathbf{D} : \mathcal{X} \rightarrow \mathcal{D}$ as representation of the input, it takes the form

$$\hat{\mathbf{f}}_{\mathbf{F}}(\mathbf{x}) = \sum_{i}^{M} \sum_{q}^{S} \mathbf{P}_q \boldsymbol{\alpha}_i \mathbf{J}_{\mathbf{D}(\mathbf{x})} \mathbf{k}_{\mathbf{F}}\left(\mathbf{D}(\mathbf{x}), \mathbf{D}(\mathbf{P}_q \mathbf{x}_i)\right) \mathbf{J}_{\mathbf{D}(\mathbf{P}_q \mathbf{x}_i)}^{\top}, \tag{14.2}$$

where $\mathbf{J}_{\mathbf{D}(\mathbf{x})}$ is the Jacobian of the descriptor, M is the number or training data points, $\mathbf{P}_q$ is the qth permutation in the molecular permutational group, and S is the size of the group. The parameters $\boldsymbol{\alpha}_i$ are the ones to be learned during the training procedure. Due to linearity, the corresponding expression for the energy predictor can be simply obtained via (analytic) integration. It is generally assumed that overly smooth priors are detrimental to data efficiency, even if the prediction target is in fact indefinitely differentiable. For that reason, (s)GDML uses a Matérn kernel $k_E(\mathbf{x}, \mathbf{x}')$ with restricted differentiability to construct the force field kernel function,

$$\begin{aligned} \boldsymbol{k}_{F}(\mathbf{x}, \mathbf{x}') &= \nabla_{\mathbf{x}} k_E\left(\mathbf{x}, \mathbf{x}'\right) \nabla_{\mathbf{x}'}^{\top} \\ &= \left(5\left(\mathbf{x} - \mathbf{x}'\right)\left(\mathbf{x} - \mathbf{x}'\right)^{\top} - \mathbb{I}\sigma(\sigma + \sqrt{5}d)\right) \cdot \frac{5}{3\sigma^4} \exp\left(-\frac{\sqrt{5}d}{\sigma}\right). \end{aligned} \tag{14.3}$$

Instead of using directly the molecular coordinates as representations of the system, a descriptor is used to facilitate the learning procedure. In general, it is a non-linear transformation fulfilling a set of required invariances. Here, the geometry of the molecule is represented in terms of inverse distances between all atom pairs

$$D_{ij} = \begin{cases} \left\| \mathbf{r}_i - \mathbf{r}_j \right\|^{-1} & \text{for } i > j \\ 0 & \text{for } i \leq j \end{cases} \tag{14.4}$$

making the model invariant to roto-translations.

A full symmetrization of the model requires summing over all possible permutations of its inputs. To avoid the combinatorial challenge associated with summing over large symmetry groups, we restrict ourselves to the much smaller subset of physically plausible rigid space group and fluxional symmetries, $\{\mathbf{P}_q\}_{q=1}^S$. Extracting those symmetries usually requires chemical and physical intuition about the system under study, e.g., rotational barriers, which is impractical in a ML setting. To automate that step, we employ a multi-partite matching scheme that identifies and recovers the permutational transformations undergone by the system within the training dataset. This is achieved by finding the permutation operation τ that minimizes the cost function,

$$\arg\min_{\tau} \mathcal{L}(\tau) = \|\mathbf{P}(\tau)\mathbf{A}_G\mathbf{P}(\tau)^\top - \mathbf{A}_H\|^2, \tag{14.5}$$

between adjacency matrices $(\mathbf{A})_{ij} = \|\mathbf{r}_i - \mathbf{r}_j\|$ of all molecular graph pairs G and H in different energy states. A particular challenge is to find matchings that are consistent across the whole training set. The set of permutations $\{\mathbf{P}_q\}_{q=1}^S$ obtained by this method, also known as the *Higgins group*, omits unfeasible transformations that do not contribute any valuable information to the inference task and thus help in reducing the computational effort required to evaluate the model.

A sketch of the general training procedure is shown in Fig. 14.4, from sampling a molecular dynamics trajectory and extracting the Higgins group to solving the normal equation and generating the embedded PES in the data.

In Refs. [64–66, 71] it was demonstrated that the sGDML framework is highly data efficient being able to achieve state-of-the-art predictions even when trained on only a few hundred reference data points. As example, it is possible to reconstruct molecular PESs with a mean absolute error of less than 0.06 kcal mol^{-1} for small molecules (e.g., with up to 15 atoms) and 0.16 kcal mol^{-1} for more complex molecules (e.g., aspirin, paracetamol, and azobenzene) [65]. Such accuracy is achieved while following physical requirements and therefore resulting in robust learning models which are capable of decoding complex subtleties hidden in the reference data.

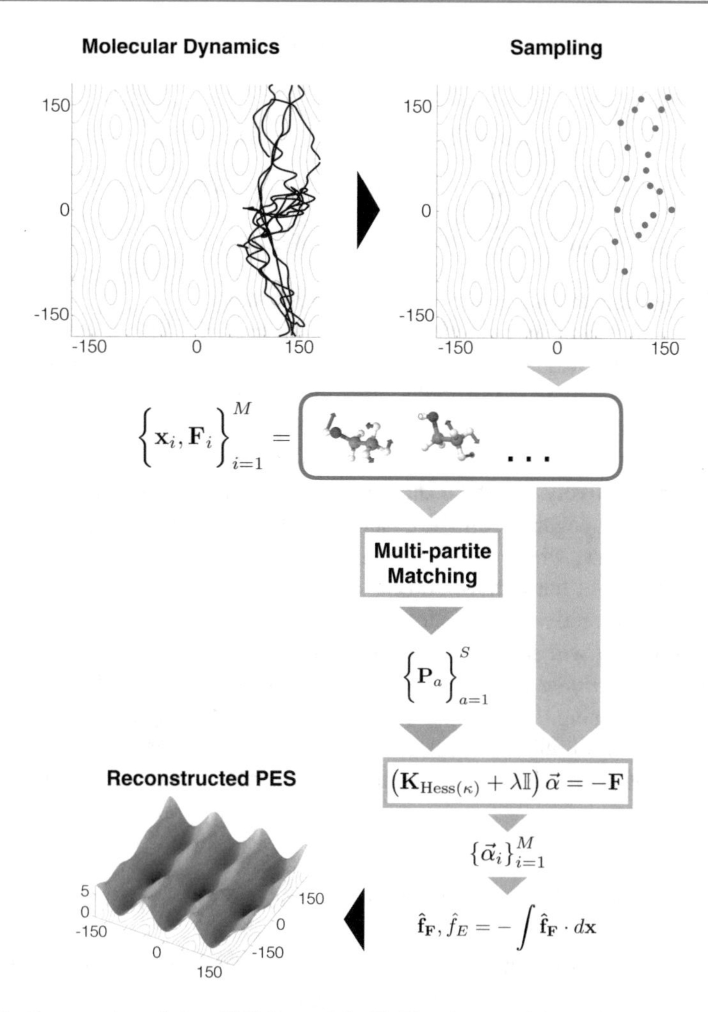

Fig. 14.4 Construction of the sGDML model. (1) The data used for training, $\{\mathbf{x}_i, \mathbf{F}_i\}_{i=1}^{M}$, is created by subsampling molecular dynamics trajectories (blue dots). The forces are represented by green arrows on top of each atom. (2) The set of molecular permutation symmetries, $\{\mathbf{P}_a\}_{a=1}^{S}$, are extracted from the training set by the multi-partite matching approach. This effectively enhances the size of the training set by a factor S and symmetrizes the PES. (3) The force field is trained by solving the linear system for $\{\alpha_j\}$. The reconstructed PES is obtained by analytical integration of the force predictor

14.3.2 Force vs. Energy ML Models

As stated in the previous section, the sGDML framework is constructed for being trained in the gradient domain of the energy function. This approach contrasts with conventional ML methodologies based on direct energy function learning (using energies and forces for training) in which the forces are computed via analytic differentiation [5–17, 28, 29, 33, 42–45, 47, 48, 51, 53, 74–76], as represented in the next diagram:

$$\begin{array}{llcl} & \text{Trained} & & \text{Derived} \\ \text{sGDML}: & \hat{\mathbf{f}}_{\mathbf{F}} & \longrightarrow & \hat{f}_{E\leftarrow\mathbf{F}} = -\int \hat{\mathbf{f}}_{\mathbf{F}} \cdot d\mathbf{x} + C \\ \text{E-ML}: & \hat{f}_E & \longrightarrow & \hat{\mathbf{f}}_{\mathbf{F}\leftarrow E} = -\nabla \hat{f}_E, \end{array}$$

where E-ML refers to energy machine learned models.

Any ML model has an associated learning uncertainty. This uncertainty is also present during the evaluation of the model. Given the nature of the operations in obtaining the derived quantities in the previous diagram, we can see that there is an advantage in learning the force field directly over the energy models. Let us consider the ensembles of models $\{\hat{\mathbf{f}}_{\mathbf{F}}\}$ and $\{\hat{f}_E\}$ with mean $\langle\hat{\mathbf{f}}_{\mathbf{F}}\rangle$ and $\langle\hat{f}_E\rangle$ and uncertainties γ_F and γ_E, respectively. It can be shown that, in the case of the sGDML model the uncertainty that propagates from the ensemble to the ensemble of energies $-\int \hat{\mathbf{f}}_{\mathbf{F}} \cdot d\mathbf{x}$ is given by $\sim\gamma_F \Delta x$, where Δx is a small number in the length scale. In the case of the uncertainty in the derived forces from E-ML, $-\nabla \hat{f}_E$ is given by $\sim\gamma_E/\Delta x$. From this simple analysis we conclude that: *the error attached to energies from the sGDML model will be attenuated while errors in predicted forces from E-ML models will be amplified* [66, 78]. Another intuitive proof of this effect was reported from signal processing theory point of view in the GDML original article [64]. This fundamental result highlights the irrefutable advantage of gradient-domain learning over energy-based learning, which evince the robustness and stability of such ML framework.

14.4 Gradient-Domain Learning and Its Performance

In this section we analyze the performance of the sGDML framework in reconstructing molecular force fields and their underlying potential energy surfaces. First, from the point of view of cross validation which judge its ability to predict unseen data, and second, perhaps a more physically relevant validation, a direct comparison with the reference method (e.g., DFT) of statistical properties computed from molecular dynamics simulation.

14.4.1 Static Validation

Table 14.1 shows the sGDML prediction results for six molecule datasets trained on 1000 geometries, sampled uniformly according to the MD@DFT trajectory energy distribution (see Fig. 14.4). It is easy to notice that for all the considered molecules the mean absolute error (MAE) in the energies is below 0.2 kcal mol^{-1}, and even lower than 0.1 kcal mol^{-1} for small molecules. Remarkable achievement considering that the model was trained using only 1000 training data points.

Table 14.1 Prediction accuracy for total energies and forces of the sGDML@DFT

Dataset		Energies		Forces					
						Magnitude		Angle	
Molecule	# ref.	MAE	RMSE	MAE	RMSE	MAE	RMSE	MAE	RMSE
keto-MDA	1000	0.10	0.13	0.41	0.62	0.39	0.56	0.0055	0.0087
Ethanol	1000	0.07	0.09	0.33	0.49	0.46	0.63	0.0051	0.0083
Salicylic acid	1000	0.12	0.15	0.28	0.44	0.32	0.45	0.0038	0.0064
enol-MDA	1000	0.07	0.09	0.13	0.22	–	–	–	–
Paracetamol	1000	0.15	0.20	0.49	0.70	0.60	0.84	0.0073	0.0118
Aspirin	1000	0.19	0.25	0.68	0.96	0.52	0.68	0.0094	0.0139

The mean absolute errors (MAE) and root mean squared error (RMSE) for the energy and forces are in kcal mol^{-1} and kcal mol^{-1}Å^{-1}, respectively. These results were originally published in Refs. [65] and [66]

This contrasts with pure energy-based models (e.g., other kernel models [64] or neural networks [86]) which require up to two orders of magnitude more samples to achieve a similar accuracy. As shown in the original GDML article [64], the superior performance of gradient based learning cannot be simply attributed to the greater information content of force samples (one energy value per $3N$ force components per sample). Let us consider a direct comparison of two kernel models, energy and gradient based, for energy learning with the same number of degrees of freedom (non-symmetrized versions for simplicity),

$$-\hat{f}_{E\leftarrow\mathbf{F}}(\mathbf{x}) = \sum_{i}^{M}\{\boldsymbol{\alpha}_i \cdot \nabla\}\kappa(\mathbf{x}, \mathbf{x}_i)\,. \tag{14.6}$$

$$-\hat{f}_E(\mathbf{x}) = \sum_{j}^{3N\times M} \beta_j\kappa(\mathbf{x}, \mathbf{x}_j)\,. \tag{14.7}$$

Then, each model has $3N \times M$ parameters with the difference that, in the energy model the $\{\beta_j\}_{j=1}^{3N\times M}$ parameters are correlation only by the learning procedure, while in the force model exist the additional correlation imposed in the triads $\{\alpha_i^x, \alpha_i^y, \alpha_i^z\}_{i=1}^{N\times M}$ by the gradient operator. Hence, this extra correlation between the parameters imposed by learning in the gradient domain reduces our space of solutions and therefore the model becomes more data efficient. Such fundamental characteristic positions the sGDML modes in a privileged place for learning force fields from highly accurate quantum chemical methodologies (e.g., CCSD(T)) in which data is very scarce, where even generating 100 data points is a monumental computational task. In the next section we will analyze this topic, but for now let us validate the sGDML models by direct comparison with MD simulations generated with the reference method.

14.4.2 Dynamic Validation

In the previous section, we saw that the prediction errors in sGDML learned models are very low. Nevertheless, a natural question to ask is if the molecular dynamics simulations using the learned models (i.e., MD@sGDML{DFT}) can actually replicate the statistical properties of the physical system as computed running MD simulations using the ab initio reference theory (i.e., MD@DFT). To address this issue, in this section we present MD simulations with sGDML and DFT forces for benzene, uracil, and aspirin molecules. All the simulations have been done within precisely the same conditions (temperature, integrator, integration step, software, etc.) using the i-PI molecular dynamics package [87].

14.4.2.1 Benzene and Uracil

In the case of benzene, we have performed MD simulations at 300 K using the same initial conditions for both MD@DFT and MD@sGDML{DFT}. Figure 14.5a shows the evolution of the potential energy in time and we can see a very good agreement. From this we can deduce that, at least in the first 10 ps of the trajectory, a MAE of 0.1 kcal mol^{-1}/0.06 kcal mol^{-1} $Å^{-1}$ in energies/forces for benzene's sGDML model does not generate significant deviations from the reference MD@DFT trajectory. In the case of uracil we repeated the same experiment but this time we started the simulations from different initial conditions and ran the simulations for 25 ps to collect more statistics. Figure 14.5b displays the evolution of the two potential energies, MD@DFT in red and MD@sGDML{DFT} in blue. It can be seen that both methods generate the same potential energy sampling (Fig. 14.5b-middle) and the same interatomic distance distribution (Fig. 14.5b-right). Therefore, the MAE of 0.11 kcal mol^{-1}/0.24 kcal mol^{-1} $Å^{-1}$ in energies/forces for uracil's model does not generate significant deviations from the exact reference data up to 25 ps of trajectory.

14.4.2.2 Aspirin

A more interesting case is aspirin, which is a much more complex molecule. In this case by running MD@GDML at 300 K, overall we observe a quantitative agreement in interatomic distance distribution between MD@DFT and MD@GDML simulations (Fig. 14.6-left). The small differences can be observed only in the distance range between 4.3 and 4.7 Å. This region mainly belongs to the distances between the two main functional groups in aspirin. Slightly higher energy barriers in the GDML model affect the collective motion of these groups, which results in a small difference in the interatomic distance distributions. These differences in the interatomic distance distributions vanish once the quantum nature of the nuclei is introduced via path integral molecular dynamics (PIMD) simulations (Fig. 14.6-right) [64]. Consequently, by running more realistic simulations we overcome the small imperfections in the reconstruction of the PES allowing to generate more accurate results.

By performing *static* and *dynamic* validations in sGDML learned models we have demonstrated the robustness and data efficiency (the models were trained only

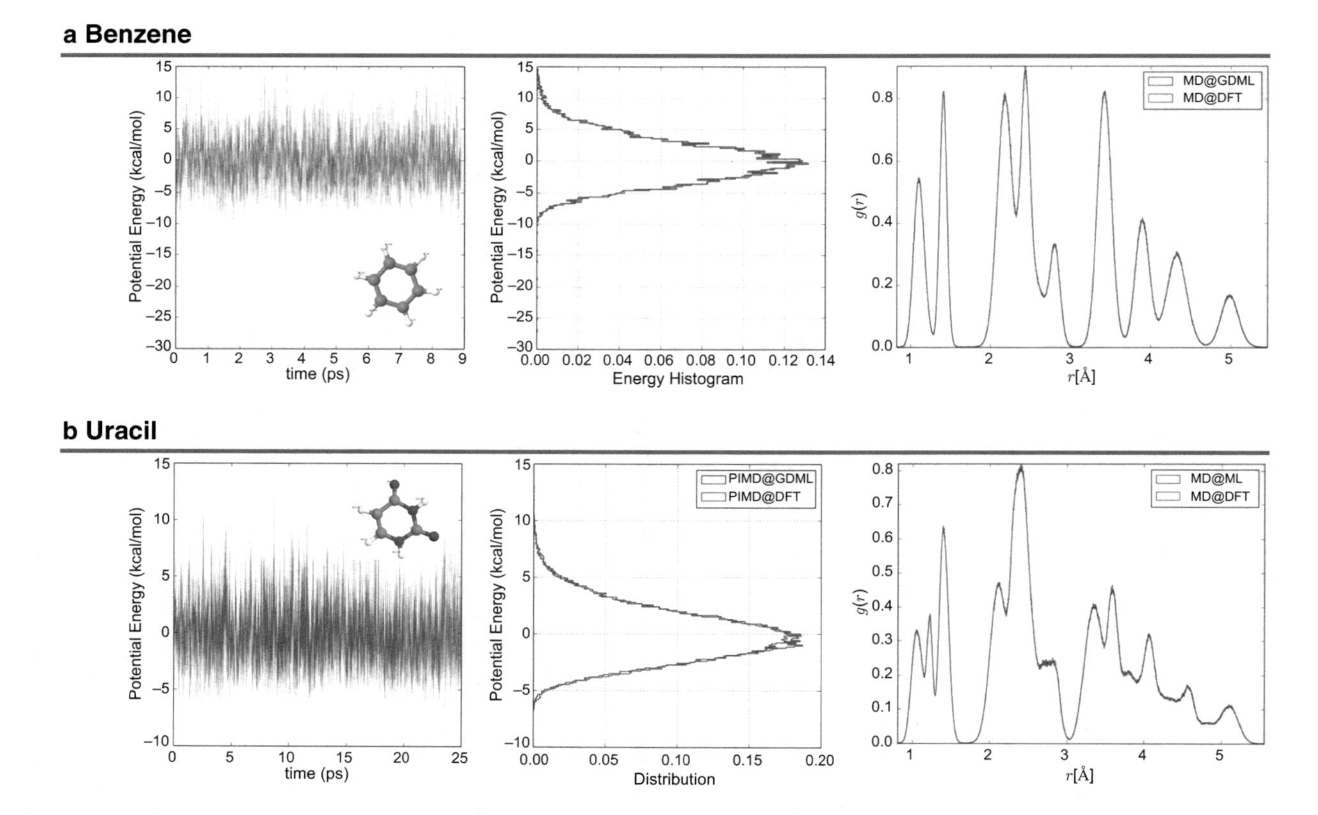

Fig. 14.5 Accuracy of potential energies (minus the mean value) sampling for sGDML@DFT (using PBE+TS functional) and sGDML@CCSD(T) models on various molecular dynamics datasets. Energy errors are in kcal mol^{-1}. These results, with the exception of enol-MDA, were originally published in Ref. [65]. All the models were trained using atomic forces for 1000 molecular conformations

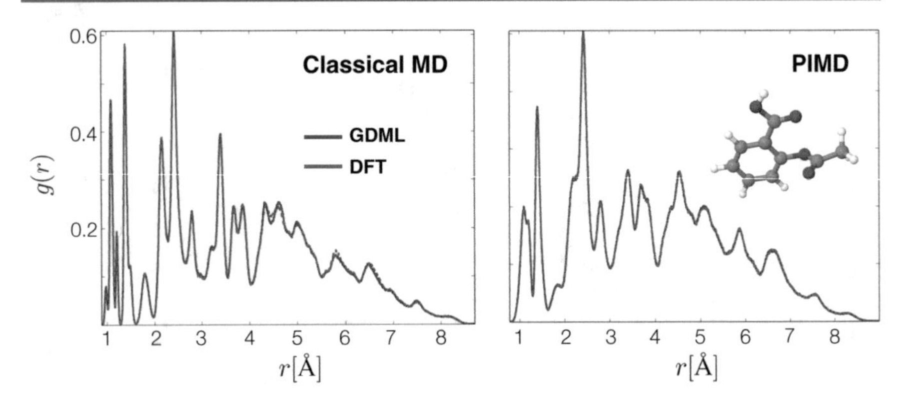

Fig. 14.6 Comparison of the interatomic distance distributions obtained from GDML (blue line) and DFT (dashed red line) with classical MD (left), and PIMD (right)

on 1000 data points) of the framework. In the next section, we briefly analyze interesting synergistic behavior between the data efficiency of the sGDML and using more accurate reference calculations.

14.5 Smoothness Hypothesis in Quantum Chemistry

Within the Born–Oppenheimer approximation, the potential energy surface $V_{\mathrm{BO}}(\mathbf{x})$ is the energy eigenvalue of the Schrödinger equation $\mathcal{H}\Psi = V_{\mathrm{BO}}\Psi$, which parametrically depends on a given set of nuclear coordinates $\mathbf{x}$, and the level of theory used to approximate its solution will of course define its accuracy. A very basic approximation is given by the Hartree–Fock theory (HF) in which the correlation between electrons of the same spin is treated as a mean field rather than as an instantaneous interaction and the correlation between electrons of opposite spins is omitted. To incorporate the missing *electron correlation*, other post-HF approximations were built on top of HF solutions, such as Møller–Plesset perturbation theory (e.g., MP2, MP3, and MP4), coupled cluster (e.g., CCSD, CCSD(T), and CCSDT), and configuration interaction (e.g., CISD and Full CI), etc. Unfortunately, moving to more accurate approximations is associated with a steep increase in the needed computational resources, making unfeasible to perform calculations, for example, using Full CI for molecules such as ethanol. In the case of density functional theory, which is less computationally demanding, it is not clear how to hierarchically increase electron correlation by going from one exchange-correlation functional to another one. Therefore, we focus only on post-HF methods.

The smoothness hypothesis states that *systematically increasing the amount of electron correlation will systematically smoothen the ground state potential energy surface* (see Fig. 14.7).

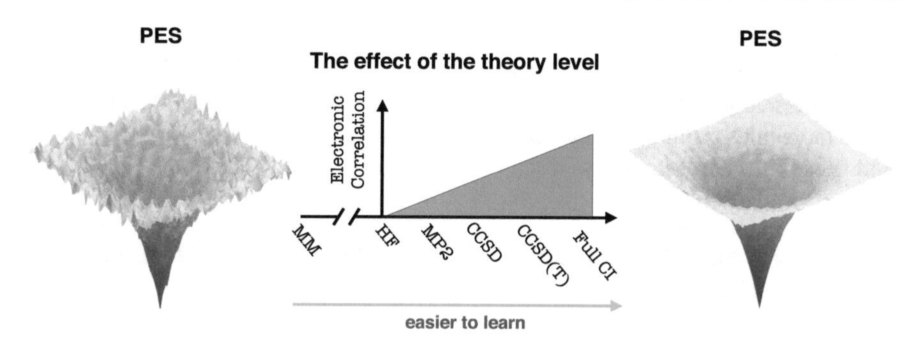

Fig. 14.7 Pictorial representation of the smoothness hypothesis in quantum chemistry

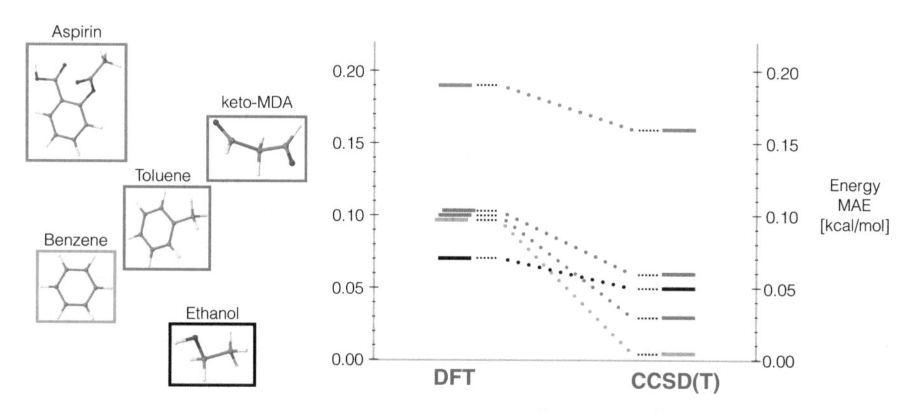

Fig. 14.8 Accuracy of total energies for sGDML@DFT (using PBE+TS functional) and sGDML@CCSD(T) models on various molecular dynamics datasets. Energy errors are in kcal mol^{-1}. These results, with the exception of enol-MDA, were originally published in Ref. [65]. All the models were trained using atomic forces for 1000 molecular conformations

As stated in the previous section, the sGDML framework is characterized for delivering state-of-the-art accuracies while using only a handful of training data points. This allows to construct compact sGDML models that faithfully reconstruct molecular force fields even from computationally costly ab initio methods such as the gold standard in quantum chemistry all-electron coupled cluster with single, double, and perturbative triple excitations (CCSD(T)). Now, by following the procedure described in Fig. 14.3 we trained a set of molecules using CCSD(T) reference data, giving very interesting results as displayed in Fig. 14.8. For all the molecules in this study, *the prediction energy error of the sGDML models dropped just by increasing the level of theory of the training data.* Furthermore, in the case of benzene the MAE drastically reduces to only few *cal mol*$^{-1}$!

From the signal reconstruction point of view, the smoother or the lower the complexity of the signal the easier to reconstruct. Meaning that less complex functions from the space of solutions can be used to capture the intrinsic features encoded in the reference data. Hence, given that increasing the electron correlation (going from DFT to CCSD(T)) makes the problem easier to learn (see Fig. 14.8) and because of the above given argument, we can say that for the studied molecules our results support the smoothness hypothesis (Fig. 14.7) [66]. An explanation why some molecules profit more than others by increasing the level of theory is not clear and needs further research.

14.6 Learning Molecular PES: What Type of Interactions Can Be Captured?

In this section, we exemplify the insights obtained with sGDML model for ubiquitous and challenging features of general interest in chemical physics: intramolecular hydrogen bonds, electron lone pairs, electrostatic interactions, proton transfer effect, and other electronic effects (e.g., bonding–antibonding orbital interaction and change in the bond nature).

The PES, V_{BO}, contains all the information necessary to describe the dynamics of a molecular system. Its intricate functional form results from the interplay between different quantum interactions, characteristic that should be preserved during the learning process. Consequently, it is not known how to expand the V_{BO} in different energetic contributions (e.g., hydrogen bonding, electrostatics, dispersion interactions, or other electronic effects) to make it more interpretable. Nevertheless, by accurately learning the V_{BO} at a high level of theory using the sGDML framework, we can perform careful analysis on the learned models and its results from applications (e.g., MD simulations) to decode many of the complex features contained in the quantum chemical data.

In practice, these features or intramolecular interactions (e.g., van der Waals interactions, energy barriers or H-bond interactions) are subtle variations in the energy surface of less than $0.1\,\text{kcal mol}^{-1}$, one order of magnitude lower than so-called chemical accuracy. An particular example is the ethanol molecule. The relative stability of its *trans* and $gauche^{(l,r)}$ conformers is within $0.1\,\text{kcal mol}^{-1}$. Furthermore, the energetic barriers $trans \rightleftharpoons gauche^{(l,r)}$ and $gauche^{(l)} \rightleftharpoons gauche^{(r)}$ differ only by $\sim 0.1\,\text{kcal mol}^{-1}$ too. Any machine learning model with an expected error above those stringent accuracies risk misrepresenting the molecular system or even inverting this subtle energy difference, which will lead to incorrect configuration probabilities and hence quantitatively wrong dynamical properties. The robust sGDML framework has been shown to satisfy such stringent demands, obtaining MAEs of 0.1–$0.2\,\text{kcal mol}^{-1}$ for molecules with up to 15 atoms [65].[7] Moreover, as shown in Fig. 14.8, the prediction error can be even lower by training on coupled-cluster reference data. With the certainty that we are working with very accurate ML models, we can confidently analyze and interpret their results.

[7]Even though the MAE is in the same order as the required accuracy, we have to mention that this error is computed in the whole dataset. This means that the error in the highly sampled regions (e.g., local minima) will be lower than the reported MAE.

14.6.1 Electrostatic Interactions and Electron Lone Pairs

First, we focus our attention on electrostatic interactions, in particular lone-pair–atom interaction. The concept of electron lone pairs plays a central role in chemistry, these are ubiquitous atomic features responsible for a wide variety of phenomena. A simple way to define lone pairs is as atomic valence electrons that are not shared with any other atom in a molecule, i.e. they are not involved in bond formation. They are often present as lone pairs of nitrogen and oxygen atoms in a molecule.

14.6.1.1 Electron Lone Pairs in Ethanol

A very illustrative case used along this chapter is ethanol molecule: (1) it has two rotors—hydroxyl and methyl groups—as main degrees of freedom making very easy to visualize its PES, (2) due to its complex electronic structure it requires at least CCSD(T) to correctly describe its PES, (3) despite its simple appearance it is not trivial to reconstruct its force field, and (4) it presents a rich variety of intramolecular interactions such as the strong effects of electron lone pairs on its dynamics.

By analyzing its PES, we find a subtle quasi-linear coupling between the methyl and hydroxyl rotors in the *trans* configuration (highlighted by the gray arrow in Fig. 14.9a). This dihedral dependence between the two functional groups is due to the electrostatic attraction between the lone pairs (negative charge) in the oxygen atom and the partially positively charged hydrogen atoms in the methyl rotor as shown in the inset in Fig. 14.9a. Such coupling becomes clear when analyzing configurational sampling obtained from molecular dynamics simulations (Fig. 14.9b), where the dynamical implications of the coupling between the two rotors at finite temperature is evident. Accurately capturing such interaction is crucial to correctly

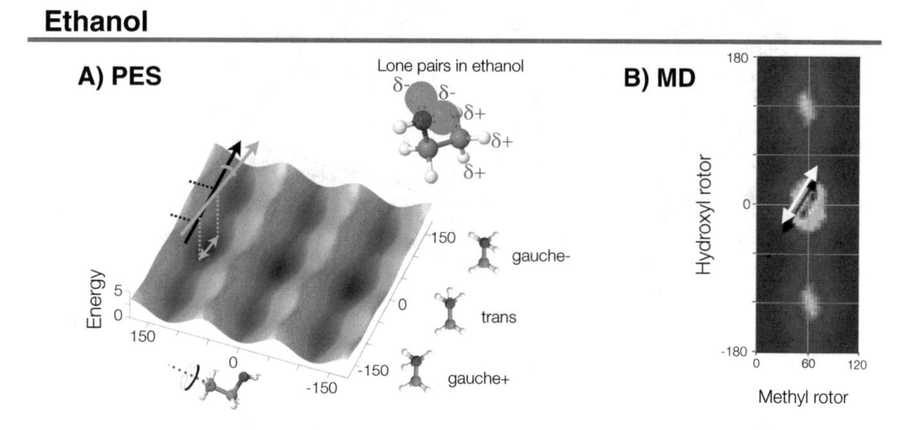

Fig. 14.9 (**a**) Ethanol's PES (sGDML@CCSD(T) model). The molecule shows the effect of oxygen's lone pair and the partial positive charges in methyl's hydrogen atoms and their coupling is represented by a gray arrow in the PES of ethanol. (**b**) PES sampling generated by MD@sGDML simulations at 300 K using the NVE ensemble

replicate and explain experimental measurements such as population analysis and vibrational spectra [65].

14.6.1.2 Oxygen–Oxygen Atom Repulsion in Keto-MDA

From the subtle interaction described in the previous section we move to a stronger electrostatic repulsion in the keto-MDA molecule as shown in Fig. 14.10. In a similar way as the ethanol molecule, keto-MDA can be taken as benchmark system in learning, given the complexity of its PES despite its small size (see also Fig. 14.3). The PES of keto-MDA molecule contains flat regions corresponding to global minimum (dark blue region in Fig. 14.10) represented by the molecular structures in Fig. 14.10d and convoluted pathways to move from minimum to minimum (see Fig. 14.10F1, F2, F3). Additionally, one can notice the sudden increase in the molecular energy when the two oxygen atoms are in the closest configuration as illustrated in Fig. 14.10a. From Fig. 14.10 we can see that even though the molecule

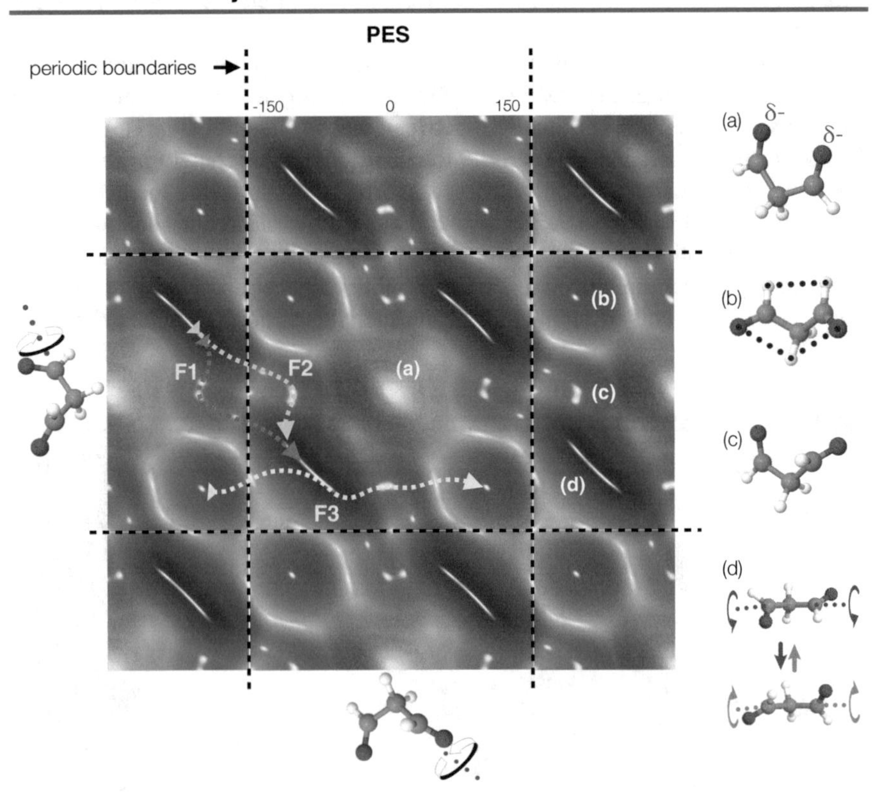

Fig. 14.10 PES for keto-MDA with periodic boundary conditions. The structure (**a**) leads to a steep increase in energy due to the close distance between two negatively charged oxygen atoms. (**b**) and (**c**) represent local minima and (**d**) display the dynamics of the global minimum. By analyzing the dynamics, the trajectories F1, F2, and F3 were found to be the most frequent transition paths

only has two main degrees of freedom (the two rotors) it has a rough PES as a result of many complex interactions. By considering the electron lone pairs in each oxygen atom and their closeness in configuration Fig. 14.10a, it suggests that the steep increase in the energy can be primarily attributed to the electrostatic repulsion between the lone pairs in each atom. Additionally, it could be that steric effects caused by electron cloud overlap could play also an important role.

14.6.2 Intramolecular H-Bond and Proton Transfer

One of the most important phenomenon in biology and materials science is hydrogen bonding (H-bond), which is responsible of a plethora of chemical and physical effects [88–92]. Molecular mechanic force fields fail in representing this interaction due to the simple fact that we do not have an appropriate analytical model for it. Therefore, ML is a very promising framework to attack this problem as recently shown by the low errors accomplished by the sGDML model. This includes good performances in describing two different types of H-bonds: standard donor–acceptor H-bond and the symmetric H-bond. A pictorial representation of their PES and two examples of molecules containing such interaction, salicylic acid, and the enol form of malondialdehyde (enol-MDA) are shown in Fig. 14.11.

(A) Regular Hydrogen bond

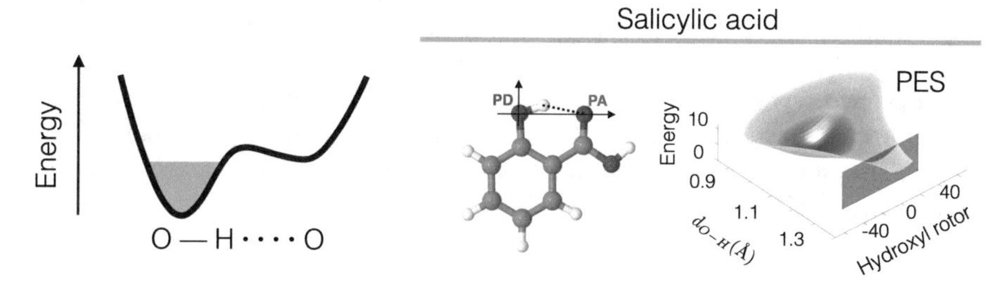

(B) Symmetric Hydrogen bond

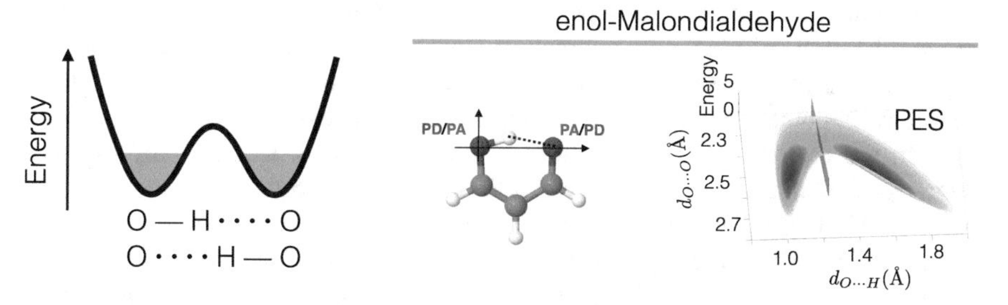

Fig. 14.11 Intramolecular hydrogen bond of (**a**) conventional type with salicylic acid as example and (**b**) symmetric type exemplified by enol-Malondialdehyde molecule

In the particular case of regular asymmetric H-bond, as salicylic acid molecule, the interaction is a standard donor–acceptor kind of H-bond between the hydroxyl and carboxylic acid groups. The main characteristic of this kind of interaction consists in allowing the proton to stretch from the PD oxygen towards the PA oxygen (Fig. 14.11a), which results in the well-known red-shift in the stretching frequency of O–H in the participating hydroxyl group [89–91]. Additionally, the H-bond also generates a blue shift in normal modes perpendicular to the H-bond, which is directly related to a O–H$\cdots$O interaction [66]. These effects can be measured experimentally via IR and Raman spectroscopy.

In the case of the symmetric H-bond, we observe a symmetric double-well PES as schematically represented in Fig. 14.11b. The energetic barrier separating the two minima will be determined by the nature of the molecule under study and on the participating functional groups in the H-bond. In this case the symmetrized nature of the sGDML approach is crucial to consistently describe such interaction. When the energy barrier is low, as in the case of enol-MDA: $\sim$4 kcal mol^{-1}, proton transfer between the two oxygen atoms is allowed even at room temperature and it is enhanced when considering nuclear quantum effects. Something to highlight here is that the energy barrier can depend strongly on the level of theory used to generate the reference data. This is due to the intricate and subtle quantum nature of this interaction, which requires high-level quantum chemistry methods. It has been found that by systematically increasing the amount of electron correlation energy in the case of enol-MDA, the energy barrier decreases as $\sim 13 \to \sim 5 \to \sim 4$ kcal mol^{-1} for HF $\to$ CCSD $\to$ CCSD(T), respectively [66]. Results that demonstrate the importance of the correlation energy in such complex phenomena as the H-bond interaction and their potential effects in proton transfer.

These two types of intramolecular H-bond are ubiquitous in nature and their presence can drastically change the chemistry and physical properties of any molecular system. Therefore the accuracy achieved by the sGDML model for the description of this interaction is particularly important.

14.6.3 Hybridization and Electronic Delocalization

Previously in this section we have shown that sGDML can learn interactions such as electrostatics and H-bonds. Here we analyze much more subtle interactions: change of hybridization and electronic interaction. Contrary to electrostatic interactions and H-bonds, that can often be approximated and implemented in empirical FFs, purely quantum phenomena (i.e., no classical analogue) are always missing in conventional FFs. In the case of flexible and fully data-driven model, learning any quantum interaction coming from $-\mathbf{F} = \langle \Psi^* | \partial \mathcal{H} / \partial \mathbf{x} | \Psi \rangle$, is a trivial task accomplished without relying on prior knowledge of the phenomena or its connection to any classical electrodynamic or mechanical concepts. Two examples are capturing the configuration and energetic features associated to changes in hybridization states and $n \to \pi^*$ interactions.

Hybridization change

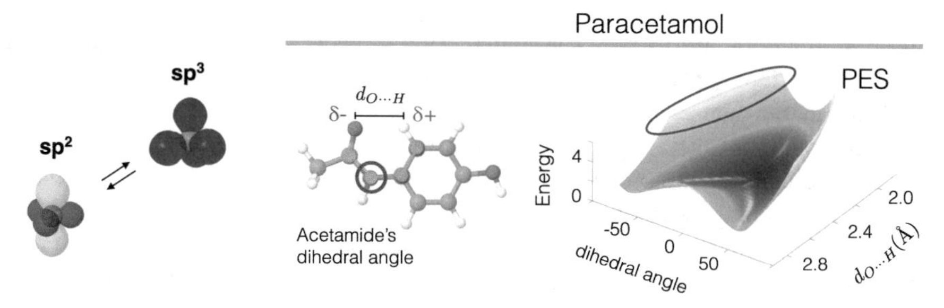

Fig. 14.12 Hybridization change. The hybridization change of the nitrogen atom is illustrated for the paracetamol molecule

In general, changes in the molecular electronic state are related to rearrangements of the electronic configuration to minimize the energy for a given molecular conformation. This could be, for example, a transition from a singlet to a triplet state, as in the case of some metallic clusters or molecules [93], or locally modifying an atomic hybridization state. The paracetamol molecule, for example, is a system that presents a $sp^2 \leftrightarrow sp^3$ hybridization change in the nitrogen atom for configurations in which the dihedral angle of the acetamide group is increasing while keeping the interatomic distance $d_{O\cdots H}$ constant (see Fig. 14.12). This generates a steep energy increase as illustrated by yellow regions in the PES in Fig. 14.12. In that region of configuration space the conjugated state in the molecule breaks, given that the nitrogen atom changes its hybridization state from $sp^2 \rightarrow sp^3$.

Another important, but less studied electronic interaction, is the overlap between occupied (lone pair n) and antibonding (π^*) orbitals: $n \rightarrow \pi^*$ interactions. The analysis of this interaction is beyond the scope of the current book chapter but it is worth to mention a couple of things. The $n \rightarrow \pi^*$ interactions is a ubiquitous interaction in biological and other molecular systems but only recently it was found the importance of such weak interaction [94]. In particular, it plays a very decisive role in the dynamics of the aspirin molecule. The $n \rightarrow \pi^*$ attraction interaction is responsible for the binding between the ester and carbonyl groups, defining the structure of the global minimum even at room temperature [65].

There are many other electronic effects (e.g., hyperconjugation, configuration dependent charge densities, Jahn–Teller effect, π–hole interactions, etc. [95]) for which we do not have analytical approximations. Therefore, they cannot be incorporated in conventional FFs limiting their reliability and predictive power. The rigorous requirement of accurately capturing such effects in ML models is justified by the increasing precision in state-of-the-art spectroscopic experimental results [95–101] which demand computationally inexpensive and highly accurate PESs to interpret and obtain further insights.

In summary, in this section we have analyzed a wide variety of interatomic interaction via high fidelity energy landscapes learning with the sGDML model. In particular, we described hydrogen bonds, electrostatic and electronic interactions. But as a final comment in this chapter, it is fair to ask the question: How relevant these interactions are in larger systems? This is because up to this moment, we have shown the importance of these phenomena in small molecules where it is understandable that such interactions play a major role. The answer is yes, all these interactions together play a major role in protein folding as recently suggested by Deepak and Sankararamakrishnan [94]. The edge where some proteins fold is a result of a complex interplay between many of the interactions analyzed here. Consequently, one of the main challenges in the route to model bigger systems is to preserve the reliability of the sGDML framework on describing such interactions.

14.7 Conclusions

In this book chapter we have presented the construction of molecular force fields using the symmetrized gradient-domain machine learning model. In particular, we have introduced what are the desirable requirements of machine learning force fields from the point of view of physics and computation efficiency. In this context, the sGDML framework is able to reconstructs high-dimensional manifolds embedded in the training data even from a few 100s of samples. Achievement that allows the use of highly accurate ab initio reference data such as the "gold standard" CCSD(T) method. The flexibility of such universal approximator comes from its fully data-driven nature, characteristic that grants the adaptability to describe any quantum interaction coming from $-\mathbf{F} = \langle\Psi^*|\partial\mathcal{H}/\partial\mathbf{x}|\Psi\rangle$. Here we have also described a simple way to systematically increase the level of theory from DFT to CCSD(T) by the subsampling—and—recomputing method, keeping in mind that the DFT's PES is already close to the CCSD(T) one.

The main advantages over other machine learning methods are: (1) highly data efficient originated by being trained in the gradient domain, (2) its robustness acquired by modeling all atomic interactions globally without any inherent non-unique partitioning of the energy or force, (3) it encodes the fundamental physical law of energy conservation as well as (4) atomic indistinguishability as a prior, correctly representing spatial and temporal symmetries.

Some challenges remain to be solved within the sGDML framework, mainly consisting in how to extend its applicability to larger systems. Many of the advantages of the model are related to its global nature, unfortunately this also imposes limits on the maximum size of the molecules that can be considered as well as the training set size. Solving this fundamental problem requires careful and well-reasoned fragmentation schemes to divide the problem into smaller independent subproblems without compromising its robustness. A possible direction to go can be a data-driven approach in a way that is tailored to preserving the intricate phenomena and quantum interactions studied in this chapter. The existence of such approach would benefit from the explicit knowledge of fluxional symmetries within the system and well-defined functional groups. In its current formulation, the sGDML framework captures different types of interaction as well as interaction scales, with no need to separating them. Nevertheless, an explicit decoupling of long-range interactions, e.g., van der Waals forces, could be a new avenue to further increase its applicability to increasingly larger and complex molecules.

Acknowledgments S.C., A.T., and K.-R.M. thank the Deutsche Forschungsgemeinschaft, Germany (projects MU 987/20-1 and EXC 2046/1 [ID: 390685689]) for funding this work. A.T. is funded by the European Research Council with ERC-CoG grant BeStMo. This work was supported by the German Ministry for Education and Research as Berlin Big Data Centre (01IS14013A) and Berlin Center for Machine Learning (01IS18037I). This work was also supported by the Information & Communications Technology Planning & Evaluation (IITP) grant funded by the Korea government (No. 2017-0-00451). This publication only reflects the authors views. Funding agencies are not liable for any use that may be made of the information contained herein. Part of this research was performed while the authors were visiting the Institute for Pure and Applied Mathematics, which is supported by the National Science Foundation, United States.

References

1. T. Saue, L. Visscher, H.J. Aa. Jensen, R. Bast, with contributions from V. Bakken, K.G. Dyall, S. Dubillard, U. Ekström, E. Eliav, T. Enevoldsen, E. Faßhauer, T. Fleig, O. Fossgaard, A.S.P. Gomes, E.D. Hedegård, T. Helgaker, J. Henriksson, M. Iliaš, Ch.R. Jacob, S. Knecht, S. Komorovský, O. Kullie, J.K. Lærdahl, C.V. Larsen, Y.S. Lee, H.S. Nataraj, M.K. Nayak, P. Norman, G. Olejniczak, J. Olsen, J.M.H. Olsen, Y.C. Park, J.K. Pedersen, M. Pernpointner, R. di Remigio, K. Ruud, P. Sałek, B. Schimmelpfennig, A. Shee, J. Sikkema, A.J. Thorvaldsen, J. Thyssen, J. van Stralen, S. Villaume, O. Visser, T. Winther, S. Yamamoto, DIRAC, a relativistic ab initio electronic structure program, Release DIRAC18 (2018). Available at https://doi.org/10.5281/zenodo.2253986, see also http://www.diracprogram.org
2. L.F. Pašteka, E. Eliav, A. Borschevsky, U. Kaldor, P. Schwerdtfeger, Phys. Rev. Lett. **118**(2), 023002 (2017)
3. M. Tuckerman, *Statistical Mechanics: Theory and Molecular Simulation* (Oxford University Press, Oxford, 2010)
4. W. Koch, M.C. Holthausen, *A Chemist's Guide to Density Functional Theory* (Wiley, Weinheim, 2015)
5. B.J. Alder, T.E. Wainwright, J. Chem. Phys. **31**(2), 459 (1959). https://doi.org/10.1063/1.1730376
6. A. Rahman, Phys. Rev. **136**, A405 (1964). https://doi.org/10.1103/PhysRev.136.A405
7. L. Verlet, Phys. Rev. **159**, 98 (1967). https://doi.org/10.1103/PhysRev.159.98
8. A. Rahman, F.H. Stillinger, J. Chem. Phys. **55**(7), 3336 (1971). https://doi.org/10.1063/1.1676585
9. W.L. Jorgensen, J. Chandrasekhar, J.D. Madura, R.W. Impey, M.L. Klein, J. Chem. Phys. **79**(2), 926 (1983). https://doi.org/10.1063/1.445869
10. M.W. Mahoney, W.L. Jorgensen, J. Chem. Phys. **112**(20), 8910 (2000). https://doi.org/10.1063/1.481505
11. J. Tersoff, Phys. Rev. B **37**, 6991 (1988). https://doi.org/10.1103/PhysRevB.37.6991
12. A. Warshel, P.K. Sharma, M. Kato, W.W. Parson, Biochim. Biophys. Acta Proteins Proteomics **1764**(11), 1647 (2006). https://doi.org/10.1016/j.bbapap.2006.08.007
13. M.S. Daw, M.I. Baskes, Phys. Rev. B **29**, 6443 (1984). https://doi.org/10.1103/PhysRevB.29.6443
14. P.K. Weiner, P.A. Kollman, J. Comput. Chem. **2**(3), 287 (1981). https://doi.org/10.1002/jcc.540020311
15. B.R. Brooks, R.E. Bruccoleri, B.D. Olafson, D.J. States, S. Swaminathan, M. Karplus, J. Comput. Chem. **4**(2), 187 (1983). https://doi.org/10.1002/jcc.540040211

16. T.A. Halgren, J. Comput. Chem. **17**(5–6), 490 (1996). https://doi.org/10.1002/(SICI)1096-987X(199604)17:5/6<490::AID-JCC1>3.0.CO;2-P
17. T.A. Soares, P.H. Hünenberger, M.A. Kastenholz, V. Kräutler, T. Lenz, R.D. Lins, C. Oostenbrink, W.F. van Gunsteren, J. Comput. Chem. **26**(7), 725 (2005). https://doi.org/10.1002/jcc.20193
18. J. Friedman, T. Hastie, R. Tibshirani, *The Elements of Statistical Learning*, vol. 1 (Springer Series in Statistics, New York, 2001)
19. V.N. Vapnik, *The Nature of Statistical Learning Theory* (Springer, Berlin, 1995)
20. Z. Li, J.R. Kermode, A. De Vita, Phys. Rev. Lett. **114**, 096405 (2015). https://doi.org/10.1103/PhysRevLett.114.096405
21. E.V. Podryabinkin, A.V. Shapeev, Comput. Mater. Sci. **140**, 171 (2017). https://doi.org/10.1016/j.commatsci.2017.08.031
22. P.O. Dral, A. Owens, S.N. Yurchenko, W. Thiel, J. Chem. Phys. **146**(24), 244108 (2017). https://doi.org/10.1063/1.4989536
23. A. Mardt, L. Pasquali, H. Wu, F. Noé, Nat. Commun. **9**(1), 5 (2018). https://doi.org/10.1038/s41467-017-02388-1
24. F. Noé, S. Olsson, J. Köhler, H. Wu, Boltzmann generators: Sampling equilibrium states of many-body systems with deep learning. Science, **365**(6457), eaaw1147 (2019)
25. M. Rupp, A. Tkatchenko, K.R. Müller, O.A. von Lilienfeld, Phys. Rev. Lett. **108**(5), 58301 (2012). https://doi.org/10.1103/PhysRevLett.108.058301
26. A.P. Bartók, M.C. Payne, R. Kondor, G. Csányi, Phys. Rev. Lett. **104**(13), 136403 (2010). https://doi.org/10.1103/PhysRevLett.104.136403
27. K. Hansen, G. Montavon, F. Biegler, S. Fazli, M. Rupp, M. Scheffler, O.A. von Lilienfeld, A. Tkatchenko, K.R. Müller, J. Chem. Theory Comput. **9**(8), 3404 (2013). https://doi.org/10.1021/ct400195d
28. A.P. Bartók, G. Csányi, Int. J. Quantum Chem. **115**(16), 1051 (2015). https://doi.org/10.1002/qua.24927
29. M. Rupp, R. Ramakrishnan, O.A. von Lilienfeld, J. Phys. Chem. Lett. **6**(16), 3309 (2015). https://doi.org/10.1021/acs.jpclett.5b01456
30. S. De, A.P. Bartok, G. Csányi, M. Ceriotti, Phys. Chem. Chem. Phys. **18**, 13754 (2016). https://doi.org/10.1039/C6CP00415F
31. N. Artrith, A. Urban, G. Ceder, Phys. Rev. B **96**(1), 014112 (2017). https://doi.org/10.1103/PhysRevB.96.014112
32. A.P. Bartók, S. De, C. Poelking, N. Bernstein, J.R. Kermode, G. Csányi, M. Ceriotti, Sci. Adv. **3**(12), e1701816 (2017). https://doi.org/10.1126/sciadv.1701816
33. A. Glielmo, P. Sollich, A. De Vita, Phys. Rev. B **95**, 214302 (2017). https://doi.org/10.1103/PhysRevB.95.214302
34. K. Yao, J.E. Herr, J. Parkhill, J. Chem. Phys. **146**(1), 014106 (2017). https://doi.org/10.1063/1.4973380
35. F.A. Faber, L. Hutchison, B. Huang, J. Gilmer, S.S. Schoenholz, G.E. Dahl, O. Vinyals, S. Kearnes, P.F. Riley, O.A. von Lilienfeld, J. Chem. Theory Comput. **13**(11), 5255 (2017). https://doi.org/10.1021/acs.jctc.7b00577
36. M. Eickenberg, G. Exarchakis, M. Hirn, S. Mallat, L. Thiry, J. Chem. Phys. **148**(24), 241732 (2018). https://doi.org/10.1063/1.5023798
37. A. Glielmo, C. Zeni, A. De Vita, Phys. Rev. B **97**(18), 184307 (2018). https://doi.org/10.1103/PhysRevB.97.184307
38. A. Grisafi, D.M. Wilkins, G. Csányi, M. Ceriotti, Phys. Rev. Lett. **120**, 036002 (2018). https://doi.org/10.1103/PhysRevLett.120.036002
39. Y.H. Tang, D. Zhang, G.E. Karniadakis, J. Chem. Phys. **148**(3), 034101 (2018). https://doi.org/10.1063/1.5008630
40. W. Pronobis, A. Tkatchenko, K.R. Müller, J. Chem. Theory Comput. **14**(6), 2991 (2018). https://doi.org/10.1021/acs.jctc.8b00110
41. F.A. Faber, A.S. Christensen, B. Huang, O.A. von Lilienfeld, J. Chem. Phys. **148**(24), 241717 (2018). https://doi.org/10.1063/1.5020710

42. J. Behler, M. Parrinello, Phys. Rev. Lett. **98**(14), 146401 (2007). https://doi.org/10.1103/PhysRevLett.98.146401
43. K.V.J. Jose, N. Artrith, J. Behler, J. Chem. Phys. **136**(19), 194111 (2012). https://doi.org/10.1063/1.4712397
44. J. Behler, J. Chem. Phys. **145**(17), 170901 (2016). https://doi.org/10.1063/1.4966192
45. M. Gastegger, J. Behler, P. Marquetand, Chem. Sci. **8**, 6924 (2017). https://doi.org/10.1039/C7SC02267K
46. K.T. Schütt, F. Arbabzadah, S. Chmiela, K.R. Müller, A. Tkatchenko, Nat. Commun. **8**, 13890 (2017). https://doi.org/10.1038/ncomms13890
47. K.T. Schütt, H.E. Sauceda, P.J. Kindermans, A. Tkatchenko, K.R. Müller, J. Chem. Phys. **148**(24), 241722 (2018). https://doi.org/10.1063/1.5019779
48. K.T. Schütt, P.J. Kindermans, H.E. Sauceda, S. Chmiela, A. Tkatchenko, K.R. Müller, in *Advances in Neural Information Processing Systems 30* (Curran Associates, New York, 2017), pp. 991–1001
49. K. Ryczko, K. Mills, I. Luchak, C. Homenick, I. Tamblyn, Comput. Mater. Sci. **149**, 134 (2018). https://doi.org/10.1016/j.commatsci.2018.03.005
50. L. Zhang, J. Han, H. Wang, R. Car, E. Weinan, Phys. Rev. Lett. **120**(14), 143001 (2018). https://doi.org/10.1103/PhysRevLett.120.143001
51. A.P. Bartók, R. Kondor, G. Csányi, Phys. Rev. B **87**(18), 184115 (2013). https://doi.org/10.1103/PhysRevB.87.184115
52. G. Montavon, M. Rupp, V. Gobre, A. Vazquez-Mayagoitia, K. Hansen, A. Tkatchenko, K.R. Müller, O.A. von Lilienfeld, New J. Phys. **15**(9), 95003 (2013)
53. V. Botu, R. Ramprasad, Phys. Rev. B **92**, 094306 (2015). https://doi.org/10.1103/PhysRevB.92.094306
54. F. Brockherde, L. Vogt, L. Li, M.E. Tuckerman, K. Burke, K.R. Müller, Nat. Commun. **8**(1), 872 (2017). https://doi.org/10.1038/s41467-017-00839-3
55. T.D. Huan, R. Batra, J. Chapman, S. Krishnan, L. Chen, R. Ramprasad, NPJ Comput. Mater. **3**(1), 37 (2017). https://doi.org/10.1038/s41524-017-0042-y
56. T. Bereau, R.A. DiStasio Jr., A. Tkatchenko, O.A. Von Lilienfeld, J. Chem. Phys. **148**(24), 241706 (2018). https://doi.org/10.1063/1.5009502
57. N. Lubbers, J.S. Smith, K. Barros, J. Chem. Phys. **148**(24), 241715 (2018). https://doi.org/10.1063/1.5011181
58. K. Kanamori, K. Toyoura, J. Honda, K. Hattori, A. Seko, M. Karasuyama, K. Shitara, M. Shiga, A. Kuwabara, I. Takeuchi, Phys. Rev. B **97**(12), 125124 (2018). https://doi.org/10.1103/PhysRevB.97.125124
59. T.S. Hy, S. Trivedi, H. Pan, B.M. Anderson, R. Kondor, J. Chem. Phys. **148**(24), 241745 (2018). https://doi.org/10.1063/1.5024797
60. J.S. Smith, O. Isayev, A.E. Roitberg, Chem. Sci. **8**, 3192 (2017). https://doi.org/10.1039/C6SC05720A
61. J. Wang, S. Olsson, C. Wehmeyer, A. Perez, N.E. Charron, G. de Fabritiis, F. Noé, C. Clementi, Machine learning of coarse-grained molecular dynamics force fields. ACS Cent. Sci. **5**(5), 755–767 (2019)
62. R. Winter, F. Montanari, F. Noé, D.A. Clevert, Chem. Sci. **10**, 1692 (2019). https://doi.org/10.1039/C8SC04175J
63. A.S. Christensen, F.A. Faber, O.A. von Lilienfeld, J. Chem. Phys. **150**(6), 064105 (2019). https://doi.org/10.1063/1.5053562
64. S. Chmiela, A. Tkatchenko, H.E. Sauceda, I. Poltavsky, K.T. Schütt, K.R. Müller, Sci. Adv. **3**(5), e1603015 (2017). https://doi.org/10.1126/sciadv.1603015
65. S. Chmiela, H.E. Sauceda, K.R. Müller, A. Tkatchenko, Nat. Commun. **9**(1), 3887 (2018). https://doi.org/10.1038/s41467-018-06169-2
66. H.E. Sauceda, S. Chmiela, I. Poltavsky, K.R. Müller, A. Tkatchenko, J. Chem. Phys. **150**(11), 114102 (2019)
67. M. Alber, S. Lapuschkin, P. Seegerer, M. Hägele, K.T. Schütt, G. Montavon, W. Samek, K.R. Müller, S. Dähne, P.J. Kindermans, iNNvestigate neural networks. J. Mach. Learn. Res.

20(93), 1–8 (2019)
68. M. Meila, S. Koelle, H. Zhang, A regression approach for explaining manifold embedding coordinates. Preprint. (2018). arXiv:1811.11891
69. S. Lapuschkin, S. Wäldchen, A. Binder, G. Montavon, W. Samek, K.R. Müller, Nat. Commun. **10**(1), 1096 (2019)
70. W. Samek, G. Montavon, A. Vedaldi, L.K. Hansen, K.R. Muller (eds.), *Explainable AI: Interpreting, Explaining and Visualizing Deep Learning*. LNCS, vol. 11700 (Springer, 2019). https://doi.org/10.1007/978-3-030-28954-6
71. S. Chmiela, H.E. Sauceda, I. Poltavsky, K.R. Müller, A. Tkatchenko, Comput. Phys. Commun. (2019). https://doi.org/10.1016/j.cpc.2019.02.007
72. K. Yao, J.E. Herr, D.W. Toth, R. Mckintyre, J. Parkhill, Chem. Sci. **9**(8), 2261 (2018). https://doi.org/10.1039/C7SC04934J
73. K.T. Schütt, P. Kessel, M. Gastegger, K.A. Nicoli, A. Tkatchenko, K.R. Müller, J. Chem. Theory Comput. **15**(1), 448 (2019). https://doi.org/10.1021/acs.jctc.8b00908
74. J. Behler, S. Lorenz, K. Reuter, J. Chem. Phys. **127**(1), 014705 (2007). https://doi.org/10.1063/1.2746232
75. J. Behler, J. Chem. Phys. **134**(7), 074106 (2011). https://doi.org/10.1063/1.3553717
76. J. Behler, Phys. Chem. Chem. Phys. **13**, 17930 (2011). https://doi.org/10.1039/C1CP21668F
77. E. Solak, R. Murray-smith, W.E. Leithead, D.J. Leith, C.E. Rasmussen, in *Advances in Neural Information Processing Systems 15* (MIT Press, Cambridge, 2003), pp. 1057–1064
78. S. Chmiela, Towards exact molecular dynamics simulations with invariant machine-learned models. Ph.D. thesis, Technische Universität Berlin, 2019. https://doi.org/10.14279/depositonce-8635
79. D. Pachauri, R. Kondor, V. Singh, in *Advances in Neural Information Processing Systems* (2013), pp. 1860–1868
80. V. Blum, R. Gehrke, F. Hanke, P. Havu, V. Havu, X. Ren, K. Reuter, M. Scheffler, Comput. Phys. Commun. **180**(11), 2175 (2009). https://doi.org/10.1016/j.cpc.2009.06.022
81. J.P. Perdew, K. Burke, M. Ernzerhof, Phys. Rev. Lett. **77**, 3865 (1996). https://doi.org/10.1103/PhysRevLett.77.3865
82. A. Tkatchenko, M. Scheffler, Phys. Rev. Lett. **102**, 073005 (2009). https://doi.org/10.1103/PhysRevLett.102.073005
83. J.M. Turney, A.C. Simmonett, R.M. Parrish, E.G. Hohenstein, F.A. Evangelista, J.T. Fermann, B.J. Mintz, L.A. Burns, J.J. Wilke, M.L. Abrams, N.J. Russ, M.L. Leininger, C.L. Janssen, E.T. Seidl, W.D. Allen, H.F. Schaefer, R.A. King, E.F. Valeev, C.D. Sherrill, T.D. Crawford, WIREs Comput. Mol. Sci. **2**(4), 556 (2012). https://doi.org/10.1002/wcms.93
84. R.M. Parrish, L.A. Burns, D.G.A. Smith, A.C. Simmonett, A.E. DePrince, E.G. Hohenstein, U. Bozkaya, A.Y. Sokolov, R. Di Remigio, R.M. Richard, J.F. Gonthier, A.M. James, H.R. McAlexander, A. Kumar, M. Saitow, X. Wang, B.P. Pritchard, P. Verma, H.F. Schaefer, K. Patkowski, R.A. King, E.F. Valeev, F.A. Evangelista, J.M. Turney, T.D. Crawford, C.D. Sherrill, J. Chem. Theory Comput. **13**(7), 3185 (2017). https://doi.org/10.1021/acs.jctc.7b00174
85. D.G.A. Smith, L.A. Burns, D.A. Sirianni, D.R. Nascimento, A. Kumar, A.M. James, J.B. Schriber, T. Zhang, B. Zhang, A.S. Abbott, E.J. Berquist, M.H. Lechner, L.A. Cunha, A.G. Heide, J.M. Waldrop, T.Y. Takeshita, A. Alenaizan, D. Neuhauser, R.A. King, A.C. Simmonett, J.M. Turney, H.F. Schaefer, F.A. Evangelista, A.E. DePrince, T.D. Crawford, K. Patkowski, C.D. Sherrill, J. Chem. Theory Comput. **14**(7), 3504 (2018). https://doi.org/10.1021/acs.jctc.8b00286
86. B. Anderson, T.S. Hy, R. Kondor (2019). Preprint. arXiv:1906.04015
87. M. Ceriotti, J. More, D.E. Manolopoulos, Comput. Phys. Commun. **185**(3), 1019 (2014). https://doi.org/10.1016/j.cpc.2013.10.027
88. S. Scheiner, Molecules **22**(9), 1521 (2017). https://doi.org/10.3390/molecules22091521
89. P. Hobza, Int. J. Quantum Chem. **90**(3), 1071 (2002). https://doi.org/10.1002/qua.10313
90. A. Karpfen, E.S. Kryachko, J. Phys. Chem. A **113**(17), 5217 (2009). https://doi.org/10.1021/jp9005923

91. C. Wang, D. Danovich, S. Shaik, Y. Mo, J. Chem. Theory Comput. **13**(4), 1626 (2017). https://doi.org/10.1021/acs.jctc.6b01133
92. B. Kuhn, P. Mohr, M. Stahl, J. Med. Chem. **53**(6), 2601 (2010). https://doi.org/10.1021/jm100087s
93. A. Cembran, F. Bernardi, M. Garavelli, L. Gagliardi, G. Orlandi, J. Am. Chem. Soc. **126**(10), 3234 (2004)
94. R. Deepak, R. Sankararamakrishnan, Biophys. J. **110**(9), 1967 (2016). https://doi.org/10.1016/j.bpj.2016.03.034
95. R. Sarkar, S.R. Reddy, S. Mahapatra, H. Köppel, Chem. Phys. **482**, 39 (2017). https://doi.org/10.1016/j.chemphys.2016.09.011
96. P. Gruene, D.M. Rayner, B. Redlich, A.F.G. van der Meer, J.T. Lyon, G. Meijer, A. Fielicke, Science **321**(5889), 674 (2008). https://doi.org/10.1126/science.1161166
97. C. Romanescu, D.J. Harding, A. Fielicke, L.S. Wang, J. Chem. Phys. **137**(1), 014317 (2012). https://doi.org/10.1063/1.4732308
98. R.M. Balabin, Phys. Chem. Chem. Phys. **12**, 5980 (2010). https://doi.org/10.1039/b924029b
99. J.A. Ruiz-Santoyo, J. Wilke, M. Wilke, J.T. Yi, D.W. Pratt, M. Schmitt, L. Álvarez Valtierra, J. Chem. Phys. **144**(4), 044303 (2016). https://doi.org/10.1063/1.4939796
100. J.A. Davies, L.E. Whalley, K.L. Reid, Phys. Chem. Chem. Phys. **19**, 5051 (2017). https://doi.org/10.1039/C6CP08132K
101. F. Gmerek, B. Stuhlmann, E. Pehlivanovic, M. Schmitt, J. Mol. Struct. **1143**, 265 (2017). https://doi.org/10.1016/j.molstruc.2017.04.092

Active Learning and Uncertainty Estimation 15

Alexander Shapeev, Konstantin Gubaev, Evgenii Tsymbalov, and Evgeny Podryabinkin

Abstract

Active learning refers to collections of algorithms of systematically constructing the training dataset. It is closely related to uncertainty estimation—we, generally, do not need to train our model on samples on which our prediction already has low uncertainty. This chapter reviews active learning algorithms in the context of molecular modeling and illustrates their applications on practical problems.

15.1 Introduction

Active learning refers, generally, to a class of machine learning algorithms for automatic assembling of the training set from a list of samples or a statistical distribution. Their goal is to reduce the error compared to a simple (for instance,

A. Shapeev (✉) · E. Podryabinkin
Skolkovo Institute of Science and Technology, Center for Energy Science and Technology, Moscow, Russia
e-mail: a.shapeev@skoltech.ru

K. Gubaev
Skolkovo Institute of Science and Technology, Center for Energy Science and Technology, Moscow, Russia

Present Address: Materials Science and Engineering, Delft University of Technology, Delft, The Netherlands

E. Tsymbalov
Skolkovo Institute of Science and Technology, Center for Energy Science and Technology, Moscow, Russia

Skolkovo Institute of Science and Technology, Center for Computational and Data-Intensive Science and Engineering, Moscow, Russia

K. T. Schütt et al. (eds.), *Machine Learning Meets Quantum Physics*, Lecture Notes in Physics 968, https://doi.org/10.1007/978-3-030-40245-7_15

random) sampling. It is achieved by, roughly speaking, avoiding similar samples in the training set—adding a similar configuration to the training set gives little extra information to the model. Such algorithms are introduced and reviewed in Sect. 15.2. In molecular modeling, however, active learning has found another important application than merely reducing the error. In some applications, like sampling molecular reactions paths, finding the relevant molecular configurations *is* a part of the problem—in order to sample representative configurations we must have an accurate approximation to the potential energy surface obtaining which requires the set of representative configurations. Such a vicious circle can be broken by active learning as will be explained in Sect. 15.3.

The scope of this chapter is limited to regression problems, although, traditionally, active learning is applied to classification problems more often. For the ease of exposition, we will often limit generality; e.g., we will assume a least-square loss functional in the regression problem without emphasizing that this is not the only option.

15.2 Active Selection from Given Samples: Uncertainty Estimation

Suppose we have a way of finding a property $\bar{f}(x)$ of a given molecular configuration x, which we may refer to simply as molecule. For instance, x could be the geometry of the molecular configuration, expressed as a collection of the types and coordinates of its atoms and $\bar{f}(x)$ could be its energy. When finding such a property is a computationally or experimentally hard problem, machine learning can be applied in order to reduce the related costs.

Let us introduce the notations. Let our machine-learning model $f = f(x;\theta)$ be a family of functions parametrized by a collection of parameters θ among which we look for an approximant of $\bar{f}(x)$. The configurations for which we need to predict $\bar{f}(x)$ are encoded by a probability distribution X from which possible x are sampled. Our goal is to find θ that minimizes the mean-square error $\mathbb{E}\big|f(\theta;x) - \bar{f}(x)\big|^2$, where the expectation is taken with respect to $x \sim X$. The machine-learning, or data-driven, approach to that problem is to generate a training set $\{x_1,\ldots,x_N\}$, compute the labels $\bar{f}(x_1),\ldots,\bar{f}(x_N)$, and find $\bar{\theta}$ minimizing the mean-square training error

$$\frac{1}{N}\sum_{i=1}^{N}\big|f(\theta;x_i) - \bar{f}(x_i)\big|^2. \tag{15.1}$$

It is often a bad idea to take the training error (15.1) as an approximation of the true error $\mathbb{E}\big|f(\theta;x) - \bar{f}(x)\big|^2$, often called *generalization error*. Indeed, by adding parameters one can often make the training error as small as one wishes, but the generalization error will be large—this is known as overfitting. To reliably estimate the generalization error, one needs to choose another set $x_1',\ldots,x_N'$, called test set, on which the error is measured. If x'_j are chosen from the distribution X independently from each other and the training set, then the *test error*,

$$\frac{1}{N'}\sum_{j=1}^{N'}\left|f(\theta; x'_j) - \bar{f}(x'_j)\right|^2, \tag{15.2}$$

is a good (unbiased) estimation of the generalization error.

Suppose that we already have a training set $x_1, \ldots, x_N$ and a trained model $f(\bar{\theta}; x)$, but we are additionally allowed to make evaluation of $\bar{f}(x_i)$ for a small number[1] of molecules x_i. We can, of course, sample molecules randomly from our distribution X and add them to the training set. Instead, we may decide to add molecules systematically—the latter is called active learning. While deciding if x_i should be added we, of course, are not allowed to evaluate $\bar{f}(x_i)$, but we can use the structure of our machine-learning model $f(\bar{\theta}; x)$.

The problem of active learning can be reduced to the following problem: given a new molecular structure x^* sampled from X, decide whether to add x^* to the training set. This, in turn, can be reduced to the problem of *uncertainty estimation* of the model. An uncertainty of (the prediction on x^* of) the model $f(\theta; x)$ is a function $\gamma(x)$ that is expected to correlate with the generalization error, the larger the error is, the higher $\gamma(x)$ should be. In active learning one uses $\gamma(x)$ to rank molecular structures from the pool for the query strategy. The *query strategy* is an algorithm that decides whether x^* should be added to the training set. A query strategy could simply be to add x^* to the training set if $\gamma(x^*) > \gamma_{\mathrm{tsh}}$ for a selected threshold γ_{tsh} or to add a certain number of configurations with the highest value of γ. More complicated strategies can be employed, e.g., accounting for the differences in the cost of obtaining $\bar{f}(x^*)$.

It should be emphasized that designing a reliable uncertainty measure $\gamma(x)$ is often hard; one reason being that if we have a model $f(x^*; \theta)$ and be able to reliably predict the error $f(x^*; \theta) - \bar{f}(x^*)$, then we could immediately create a better model by subtracting the error from $f(x^*; \theta)$. For this reason, we can only predict an absolute value of the error $|f(x^*; \theta) - \bar{f}(x^*)|$, but not its sign.

One should be aware that actively selected training samples $x_{N+1}, x_{N+2}, \ldots$ statistically differ from the distribution X. This could lead, for example, to an increase in mean absolute generalization error while significantly decreasing the maximal absolute generalization error, see, e.g., [10, 28]. This disturbing feature of active learning could be desirable in some applications. For example, a critical superconductivity temperature T_{c} of an "average" material is zero, however, when searching for high-temperature superconductors we want accurate predictions for those rare cases when T_{c} is large, rather than competing for a milli-Kelvin accuracy of predicting T_{c} for an "average" material. Even if we were lucky to find those rare cases, we would not know them in advance and therefore we cannot encode them into the distribution X upon designing the search space. Below we briefly review several common ways to defining the model uncertainly.

[1] Small because evaluation of $\bar{f}(x_i)$ is expensive.

15.2.1 Predictive Variance for Gaussian Process Regression

This uncertainty estimate is defined in the context of Gaussian processes. In Gaussian process regression it is assumed that the predicted function is Gaussian-distributed and its variance given by some function $k(x, x')$ called *kernel*:

$$\mathrm{cov}(f(x), f(x')) = k(x, x').$$

The kernel $k(x, x')$ defines how similar the molecules are, the larger $k(x, x')$ is the more similar x and x' are (up to physical symmetries). The examples of kernels include SOAP (Smooth Overlap of Atomic Positions) [2], MBTR [16], BoB [12], and the one used within the symmetrized gradient-domain machine learning (sGDML) approach [4].

Suppose that for a certain molecule x^* we need to predict the variance of observation of $f(x^*)$ given training data. To that end, we denote by $\boldsymbol{x} := (x_1, \ldots, x_N)$ the collection of training data, by $k(\boldsymbol{x}, \boldsymbol{x})$ the matrix whose (i, j)-th element is $k(x_i, x_j)$, $k(x^*, \boldsymbol{x}) := \big(k(x^*, x_1), \ldots, k(x^*, x_N)\big)$, $\boldsymbol{y} := (y_1, \ldots, y_N)$.

One can then derive that the most probable guess for the underlying function $\bar{f}(x)$, in the sense of mathematical expectation, is[2]

$$f(\theta, x) := k(x^*, \boldsymbol{x})k(\boldsymbol{x}, \boldsymbol{x})^{-1}\boldsymbol{y} \tag{15.3}$$

and its variance is

$$\gamma_{\mathrm{pv}}(x) = k(x^*, x^*) - k(x^*, \boldsymbol{x})\, k(\boldsymbol{x}, \boldsymbol{x})^{-1} k(\boldsymbol{x}, x^*), \tag{15.4}$$

which is called the *predictive variance*. Typically, if a kernel represents well the underlying law (distribution) that the data follow, the predictive variance is a good estimate of the actual generalization error. The corresponding query strategy can assume that the higher $\gamma_{\mathrm{pv}}(x^*)$ is, the more useful x^* is. A detailed exposition of Gaussian processes can be found in [31].

15.2.2 Query by Committee

In the context of the previous subsection, if we sampled independent functions $f_1, \ldots, f_M$ with mean (15.3) and variance (15.4), then

$$\gamma_{\mathrm{qc}}(x) = \frac{1}{M-1} \sum_{m=1}^{M} \left(f_m(x) - \frac{1}{M} \sum_{\ell=1}^{M} f_\ell(x) \right)^2 \tag{15.5}$$

[2]Here we have implicitly assumed that the distribution of $f(\theta, x)$ has zero mean.

is an unbiased estimate of the predictive variance (15.4). This fact is often used in practice: instead of fitting one single model f to the training data, an ensemble of models $f_1, \ldots, f_M$ is fitted. One then uses (15.5) as the uncertainty estimate.

In neural networks one can start with different realizations of random initial parameters and do the fitting. The fitting procedure would stop (if early stopping is used) near a random local minimum of the loss function. There are many works aimed at further diversification of the ensemble $f_1, \ldots, f_M$—models are made different from each other structurally (using different architectures [27] and models [20]), use different data subsets (bagging) [41]. This approach is being actively used for molecular modeling [1, 36] due to the simplicity of both the method and its implementation. The main disadvantage of this approach is an increased training time: using M model trained independently increases the time required for training M times and also requires M times more memory to store weights. This leads to a significant drop of the performance for the modern large neural network architectures. One of the options for speeding up this approach in case of neural networks is to use weights from the previous training steps (snapshot ensemble [15]) but this reduces the diversification of models and imposes additional costs on the model weights storage.

15.2.3 D-Optimality

The next uncertainty estimator considered in this chapter is D-optimality [34]. Here we present it in a way it was introduced in molecular modeling [28], which is slightly different from the conventional way, as will be discussed below.

D-optimality is easiest to understand in the context of linear regression,

$$f(x;\theta) = \sum_{\ell=1}^{L} \theta_\ell B_\ell(x),$$

where $B_\ell(x)$ is a set of basis functions. Suppose that each training sample y_i is given with some normally distributed noise ϵ_i:

$$y_i = \bar{f}(x_i) + \epsilon_i.$$

Assuming the same number of training data points as the number of basis functions, we can solve $f(x_i;\theta) = y_i$ for θ and express $f(x;\theta)$ through $\bar{f}(x_i) + \epsilon_i$. The noise of the prediction of $f(x)$ will hence be

$$f(x;\theta) - \bar{f}(x) = \big(B_1(x) \ldots B_L(x)\big) \underbrace{\begin{pmatrix} B_1(x_1) & \ldots & B_L(x_1) \\ \vdots & \ddots & \vdots \\ B_1(x_L) & \ldots & B_L(x_L) \end{pmatrix}^{-1}}_{=:\mathsf{B}^{-1}} \begin{pmatrix} \epsilon_1 \\ \ldots \\ \epsilon_L \end{pmatrix}.$$

Assuming ϵ_i are independent and identically distributed with variance 1 then the variance of $f(x;\theta) - \bar{f}(x)$ would be

$$\sum_{i=1}^{L} \gamma_i^2,$$

where γ_i is the ith element of the following vector:

$$\boldsymbol{\gamma} = \big(B_1(x) \ \dots \ B_L(x)\big)\, \mathsf{B}^{-1}. \qquad (15.6)$$

In practice, instead of this expression, the uncertainty estimate is defined as

$$\gamma_{\text{do}} := \max_i |\gamma_i|.$$

The query strategy based on γ_{do} is known as a D-optimality criteria. The reason for calling it D-optimality is that γ_{do} is the maximal value by which the determinant det(B) can increase (by absolute value) if we try changing one of its rows by $\big(B_1(x) \ \dots \ B_L(x)\big)$.

In this implementation of D-optimality, we have an active set of exactly L configurations—those that correspond to the L rows of the square matrix B. The training set could be larger and consist of all the configurations that ever were in the active set. This is slightly different from the conventional way of implementing D-optimality which allows for non-square matrices B by considering det $\big(\mathsf{B}^T\mathsf{B}\big)$.

In our learning-on-the-fly molecular simulations, Sect. 15.3, our query strategy is to add a configuration x^* to the active set if $|\gamma_{\text{do}}(x^*)| > \gamma_{\text{tsh}}$ for some $\gamma_{\text{tsh}} = 1 + \delta_{\text{tsh}} > 1$ and remove the configuration corresponding to the maximal γ_i from (15.6) from the active set. As mentioned, when following this strategy, det(B) always increases by at least $1 + \delta_{\text{tsh}}$ when a new configuration is added, thus guaranteeing that only a limited number of configuration will be added in practice to the training set.[3]

To further motivate D-optimality, we note that we can express $\theta = \mathsf{B}^{-1}(y_1, \dots, y_L)^T$ and hence for any configuration x we have

$$f(x;\theta) = \big(B_1(x) \ \dots \ B_L(x)\big) \cdot \big(\theta_1, \dots, \theta_L\big) = \sum_i \gamma_i y_i.$$

Hence not adding x to the training set implies that $f(x;\theta)$ is predicted as a linear combination of the labels y_i with coefficients that are less than $\gamma_{\text{tsh}} = 1 + \delta_{\text{tsh}}$ by absolute value. This is a way to mathematically formalize the statement that if x

[3]To be precise, it could be mathematically proved that only a limited number configurations will be added to the training set if the configurations are sampled from a distribution with a compact support.

is not added to the training set then its prediction is an interpolation, or at most an extrapolation by δ_{tsh}.

Finally, to give an information-theoretic justification to D-optimality, let us consider the vector $\big(B_1(x) \ \ldots \ B_L(x)\big)$ as a descriptor of a configuration x. Then one can analytically compute the L-dimensional volume of the region formed by configurations x for which the prediction of $f(x;\theta)$ involves only interpolation. This volume is equal to $2^L|\det(\mathsf{B})|$, which justifies us introducing the information the training set has for making predictions for other configurations as $\log(|\det(\mathsf{B})|)$. (The training set-independent factor of 2^L is safely ignored.) Hence another way to interpret D-optimality is to say that it adds those configurations to the training set that increase the information contained in the training set by at least $\log(1+\delta_{\mathrm{tsh}})$.

15.2.4 Bayesian Methods for Neural Networks

In the context of neural network-based models, a typical approach to uncertainty estimation is often associated with Bayesian neural networks [3, 18]. In Bayesian neural networks, weights are represented as random variables with an explicitly defined distribution (prior), and the output consists of both predictive mean and predictive variance, as in Gaussian processes. In some cases, it is possible to explicitly express the variance of the model output [32] thus producing an uncertainty estimate γ in the spirit of Sect. 15.2.1. Alternatively, this framework allows one to rapidly generate an ensemble of models on-the-fly, producing an uncertainty estimate in the spirit of Sect. 15.2.2. Unfortunately, the straightforward use of the Bayesian approach for the neural networks is very costly in terms of computational resources due to a large number of neural network parameters and large amounts of data often accompanying the use of this type of models. Most of the works in this field are aimed at reducing this complexity and theoretically justifying the existing heuristic methods for training models from the Bayesian point of view [22, 24, 25, 38].

To the best of our knowledge, pure Bayesian neural networks have not been applied to the problems of computational materials science. Implicitly, however, researchers may use dropout [14, 37]—random omission of the weights in fully connected layers of the neural network. First introduced as an empirical method to fight the correlation of weights of a neural network, dropout found its theoretical justification as stochastic averaging over an ensemble of models [37] or realization of a Bayesian neural network with Bernoulli weights distribution [8]. The idea of the use of dropout not only at the training stage but also at the inference stage was a breakthrough; this approach appeared recently in the works of Gal [7], where uncertainty estimates based on the Bayesian approach and the dropout were proposed and analyzed. This approach is attractive due to the simplicity of its implementation and the possibility of application to the already trained models and established architectures. Dropout-based uncertainty estimates are often less computationally expensive compared to the Bayesian analogues. However, most of the works in this direction focus on the classification task [9, 17] rather the regression.

15.2.4.1 Example of Active Learning with SchNet

To demonstrate the power of a neural network-based approach to active learning, we conducted a series of numerical experiments with a state-of-the-art neural network architecture in the field of chemoinformatics—SchNet [33]. This network takes information about an organic molecule as input, and, after preprocessing and intricate training procedure, outputs a property of the molecule. Despite its complex structure, SchNet contains two consecutive fully connected layers, therefore it is possible to use a dropout layer between them.

We tested our approach on the problem of predicting the molecular internal energy at 0 K from the QM9 dataset [30]. We used the TensorFlow implementation of SchNet with the same architecture as in the original paper [33] except for an increased size of hidden layers (from 64 and 32 units to 256 and 128 units, respectively) and a dropout layer placed between them. It is expected that wider hidden layers lead to a better uncertainty estimation, as the infinite-width layer will theoretically result in an unbiased estimate of both mean and variance for the corresponding Bayesian neural network [7]. This layer was turned on during inference only.

We compare the basic dropout-based approach for uncertainty estimation (MCDUE, as in [39]), its improved version that uses Gaussian process (NN+GP, as in [40]), and random sampling. In our experiment, we separate the whole dataset of 133,885 molecules into the initial set of 10,000 molecules, the test set of 5000 molecules, and the rest of the data allocated as the pool. On each active learning iteration, we perform 100,000 training epochs and then calculate the uncertainty estimates using either MCDUE or NN+GP approach. We then select 2000 molecules with the highest uncertainty from the pool, add them to the training set, and perform another active learning iteration.

The results are shown in Fig. 15.1. Both MCDUE and NN+GP demonstrate steady 15% accuracy increase in terms of mean absolute test error. This might not

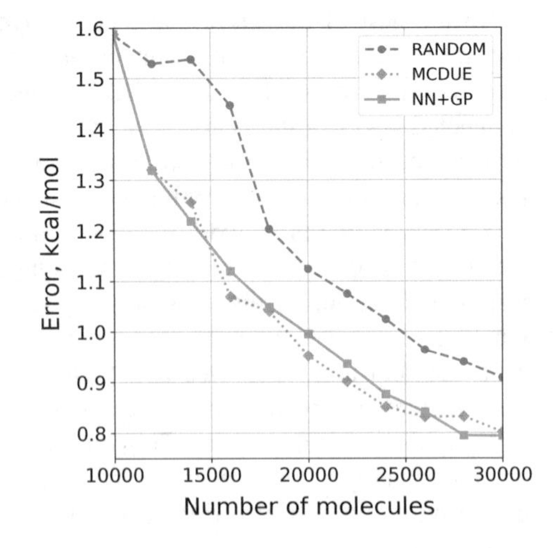

Fig. 15.1 The training curve for the active learning scenario: starting from 10,000 random molecules we pick 2000 based on the uncertainty estimate. Dropout-based algorithms result in the 15% decrease in mean absolute test error compared to random sampling. The results are averaged over three independent runs with random choices of the starting 10,000 molecules

seem substantial, however, it may be significant in terms of the time savings for the computationally expensive quantum-mechanical calculations. For example, to reach the error of 1 kcal/mol starting from the SchNet trained on 10,000 molecules, one need to additionally sample 15,000 molecules in case of random sampling or just 10,000 molecules using the NN+GP uncertainty estimation procedure.

15.3 Learning-On-the-Fly

In molecular simulations active learning approaches are used to solve a more important problem than merely reducing prediction errors. Active learning can ensure reliability of a molecular simulation, as will be illustrated in this section. In the majority of such applications the molecular configurations to be labeled are sampled by a generating algorithm which typically takes the labels of the previously generated configurations as an input. The typical examples are molecular dynamics (MD), crystal structure prediction, Monte-Carlo methods, structure relaxation, etc. In these cases one deals with the stream of molecular configurations $x_1, \ldots, x_i$ rather than a pool of them and the next configurations $x_{i+1}, x_{i+2}, \ldots$ are typically unknown at the ith step. At some step i of such process the configuration x_i can significantly differ from those in the training set. Since machine-learning methods are typically inaccurate outside the training domain, calculation of the forces may have a poor accuracy due to extrapolation. At the same time, incorrect values of the forces can lead to failure of the simulation process, because they are used as an input for generation of the next configuration x_{i+1}. This issue makes the simulation process non-reliable.

Thus, to train a reliable machine-learning model one needs to have a set of relevant configurations sampled from the stream $x_{i+1}, x_{i+2}, \ldots$ by a generating algorithm. On the other hand, to generate a stream of configurations, the generating algorithm requires a well-trained machine-learning model. Breaking this vicious circle can be done by *learning-on-the-fly*. Here we present an adaptation of active learning for learning-on-the-fly. All the examples involve the moment tensor potentials [35]—a particular form of machine-learning potentials, and the D-optimality approach [11, 28].

15.3.1 Active Learning in Molecular Dynamics

Algorithmically, an MD simulation is numerical integration of Newton's equation of motion, which is done in two alternating steps:

1. for a given configuration, forces acting on atoms are calculated;
2. for given forces, the equation of motion is numerically integrated and new atomic positions and velocities are calculated, thereby the next configuration is generated.

To calculate forces for a given configuration one needs an interatomic interaction model. Traditionally, either an empirical interatomic potential or a quantum-mechanical model is used for these purposes. Empirical potentials are typically computationally efficient but often have insufficient accuracy for a quantitative prediction; quantum-mechanical models, on the contrary, can make a quantitative prediction but are computationally very expensive. Machine-learning potentials come to rescue—they combine the advantages of the two aforementioned models: the accuracy and efficiency.

Machine-learning potentials are fitted to quantum-mechanical energy and forces—the derivatives of the energy with respect to atomic positions—on the training set of configurations. To avoid extrapolation during MD, these configurations should be sampled from a region in the configurational space accessible to the MD trajectories. But to sample an MD trajectory, a reliable potential is required for an MD algorithm.

The simplest learning-on-the-fly method for sampling training configurations while exploring a potential energy surface as proposed in [19] (see also [5, 6]) is to add configurations to the training set regularly, with a certain frequency. For instance, during MD for crystalline Si [19], all the configurations from the first 1000 MD steps were added to the training set and then one configuration was added every 30 time steps. The speed-up of this approach, in the limit of long molecular-dynamics trajectories is a factor of 30. This strategy is somewhat analogous to random sampling: in an ideal situation, the configurations in the training set and the configurations in molecular dynamics are samples drawn from the same (Boltzmann) distribution. This learning strategy does not guarantee that extrapolation is completely avoided—30 time steps may be enough for a trajectory to completely escape the region spanned by the training set, e.g., where the distances between atoms become unphysically small which a potential was not trained for. This could be mitigated by learning more often which would, however, reduce computational efficiency.

As it was shown in [28] the use of active learning offers an elegant solution to the problem of extrapolation. The active learning algorithm detects and learns only the extrapolative configurations—see the workflow diagram in Fig. 15.2. Namely, the configuration x is considered extrapolative if the prediction uncertainty exceeds some threshold value: $\gamma(x) > \gamma_{\rm tsh}$. At each MD time step, prior to calculation of the forces, the configuration is tested and if it appears extrapolative then the corresponding quantum-mechanical energy is calculated and this configuration is learned by the machine-learning potential.

Thus, the training domain is expanded automatically by adding only relevant configurations sampled from an MD trajectory. This ensures reliable prediction of the forces and keeps the MD trajectory within the relevant domain of the configurational space, as illustrated by the following example. In [28] we studied the reliability of an MD process with three potentials trained in a different manner. For testing we take an NVT ensemble of 128 atoms of BCC-lithium at the temperature $T = 300\,\mathrm{K}$. We ran a quantum-mechanical MD and added the first 1000 configurations to the training set of all the three potentials. The first potential

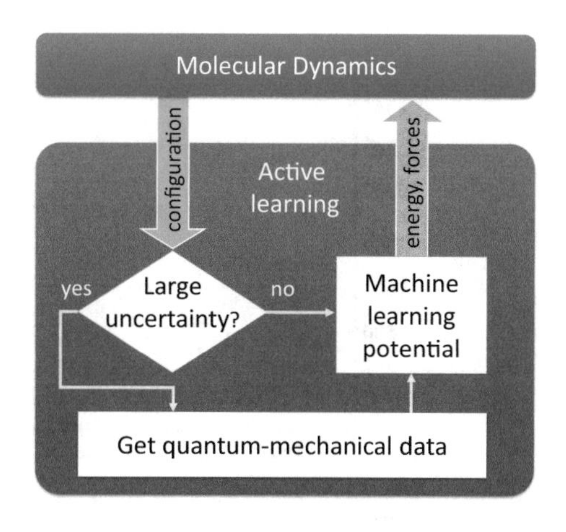

Fig. 15.2 Workflow in actively training a potential on-the-fly. An active-learning scheme gets a molecular configuration as an input and returns as output its energy, forces, and stresses, by possibly retraining the interatomic potential

was trained passively on these 1000 configurations. The second one was trained passively on-the-fly by adding one configuration every 100 time steps. The third potential was trained actively on-the-fly from scratch. In the course of an MD simulation we measured the time until the minimal interatomic distance becomes smaller than 1.5 Å—this was an indicator of MD failure. The distance of 1.5 Å was chosen as a safe threshold to avoid false positives in determining failure—this is about 1.5 times smaller than the typical minimal distance seen in a simulation at $T = 300\,\mathrm{K}$ [28]. Figure 15.3 illustrates the average time until failure and the number of quantum-mechanical calculations made during an MD. In the test the third potential, the one trained on-the-fly, was completely robust in contrast to the other two training strategies. This illustrates how active learning makes an MD with a machine-learning potential reliable.

In addition to ensuring reliability of an MD simulation, the active learning-on-the-fly algorithm is very computationally efficient. Indeed, the overhead of evaluating the uncertainty $\gamma(x)$ in the above example is less than the complexity of forces computation, and the main computational efforts are spent on quantum-mechanical calculations. The frequency of the latter decreases as the MD simulation time advances. This is seen well in another computational experiment in which the actively learning potential starts learning from an empty training set, the results are shown in Fig. 15.4.

As one can see from Fig. 15.4, most of quantum-mechanical simulations take place at the initial stage of MD. When the region spanned by the MD trajectory in the configurational space is well-explored, the extrapolative configurations appear very rarely. Therefore a practical way to reduce the amount of quantum-mechanical calculations is to pre-explore the relevant region in configurational space with another model of interatomic interaction. Such model can be from the class of empirical potentials or a quantum-mechanical model with low accuracy but computationally cheap. The purpose of this model is to sample unlabeled configurations to a pool for initial pre-training of the potential using the active learning approach. The energies and forces for the pre-training set are still computed with a

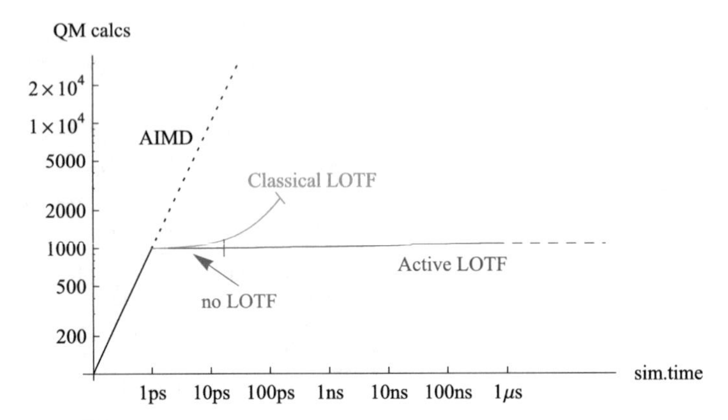

Fig. 15.3 Comparison of ab initio molecular dynamics (AIMD) with no-learning MD, classical learning-on-the-fly (LOTF) inspired by [19], and active LOTF. The no-learning and classical LOTF MD are not completely reliable: on average every 15 ps the no-learning MD fails, i.e., escapes into an unphysical region in the phase space. The classical LOTF makes this ten times more reliable (failure time of 150 ps) at the expense of extra 1500 quantum-mechanical calculations. In contrast, the active LOTF makes MD completely reliable (i.e., failures are not observed) at the cost of only 50 quantum-mechanical calculations as measured over the first 0.5 μs of simulation time. Reprinted from [28], with permission from Elsevier

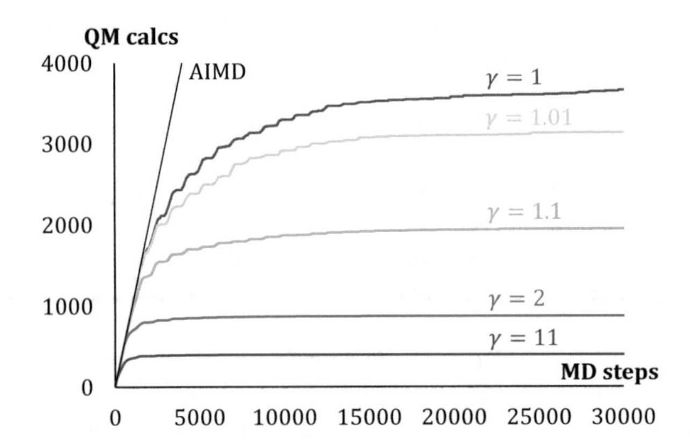

Fig. 15.4 Number of quantum-mechanical calculations in a learning-on-the-fly MD as a function of the MD time step with different thresholds γ_{tsh} (first 30 ps). Reprinted from [28], with permission from Elsevier

quantum-mechanical model, therefore the cheap model is only required to produce geometrically reasonable configurations. After pre-training, active learning-on-the-fly will still add configurations to the training set, however, the total number of configurations will be less—see Fig. 15.5.

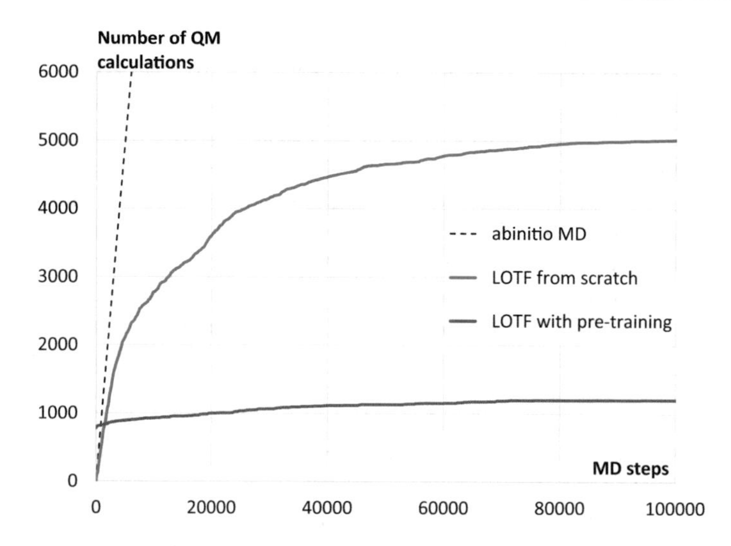

Fig. 15.5 Number of quantum-mechanical calculations in a learning-on-the-fly MD as a function of the MD time step. The blue curve corresponds to learning "from scratch," the red curve corresponds to learning from the pre-trained potential

The machine-learning potential learning-on-the-fly can be considered, effectively, as an independent model of interatomic interaction from an algorithmic point of view, as seen on the diagram in Fig. 15.2. Indeed, it takes a molecular configuration as input and returns energy and forces as the output—just like a quantum-mechanical model, for example. It should be noted that the model itself slightly changes during an MD run—it is undesired if one needs, for instance, the exact energy conservation, but it does not appear critical for an MD with a thermostat. It may cause more trouble for structure relaxation: even a slight change in potential energy may trick a minimization algorithm that it found a local minimum. Therefore for structure relaxation the procedure needs to be slightly modified as explained in the next section.

15.3.2 Active Learning in Crystal Structure Prediction

Crystal structure prediction is aimed at searching for the most stable structures for a given atomic composition. It is typically done at zero Kelvin temperature, at which the problem is reduced to finding of the global minimum, or several deepest minima, on the potential energy surface. Methods for crystal structure prediction involve two main components: an algorithm for sampling the configurational space and the interatomic interaction model. The first one is used for exploration and efficient sampling of the potential energy surface. This algorithm uses the interatomic interaction model to calculate the energy and also to locate the local minima. The examples of sampling algorithms include evolutionary algorithms, particle swarm optimization, metadynamics, and minima hopping; a detailed survey on crystal structure prediction methods can be found in [26]. For the interatomic interaction model, the density functional theory (DFT) is typically used, since it is able to provide the sufficient quantitative accuracy for predicting the correct structures.

Practical crystal structure prediction involves evaluation of energies (sometimes with forces and stresses) for extremely large amount of different structures that typically takes more than 99.99% of the total computational time. This fact motivates application of machine-learning to crystal structure prediction.

Applying machine-learning potentials to crystal structure prediction yields similar problems as in MD: a potential should predict properties of configurations not present in a training set, with a chance of extrapolation resulting in an inaccurate prediction for a given structure. Similarly to the case of MD, using active learning makes it possible to perform a simulation in which evaluation of new structures and their sampling for the training set is done simultaneously, i.e., a machine-learning model is learned on-the-fly.

Active learning in crystal structure prediction is used similarly as in MD (see Fig. 15.2), with the molecular dynamics "driver" changed to the crystal structure generation and relaxation "driver." We thus replace computationally expensive DFT with a learning-on-the-fly interatomic potential, whereas the sampling algorithm is kept as is. At the same time, in crystal structure prediction additional difficulties arise, not typical for MD. First of all, the structures are searched for over a much larger domain of configuration space than is explored during an MD. Indeed, running an MD at some temperature without phase transitions would result in configurations of the same composition and the same topology, only perturbed by temperature fluctuations. On the contrary, in the present context we need to calculate the properties of many structures that differ in both composition and geometry. Therefore, to efficiently exploit the active learning approach, a scheme different from the one described in Sect. 15.3.1 is used, as explained below.

Our active learning algorithm is tested together with two sampling algorithms: in [29] with the evolutionary method USPEX [21], and in [11] with simply generating a pool of structures to reproduce a diverse set of geometries and chemical compositions typical for most of alloys [13, 23]. For the ease of exposition of our active learning algorithm, we will consider only the latter structure sampling strategy. Here relaxation produces a sequence of structures with stresses and forces decreasing in absolute value and ending in a structure with zero forces and stresses (equilibrium structure). The process of relaxation of a configuration x (also called equilibration or structure optimization), essentially, consists of repeating the following two steps:

1. for a given configuration x_i ($x_0 \equiv x$) the energy gradients (forces acting on atoms and stresses acting on the lattice) are calculated;
2. based on the gradients, a new configuration x_{i+1} with lower energy is derived from the configuration x_i by slight displacements of atoms and a slight lattice deformation.

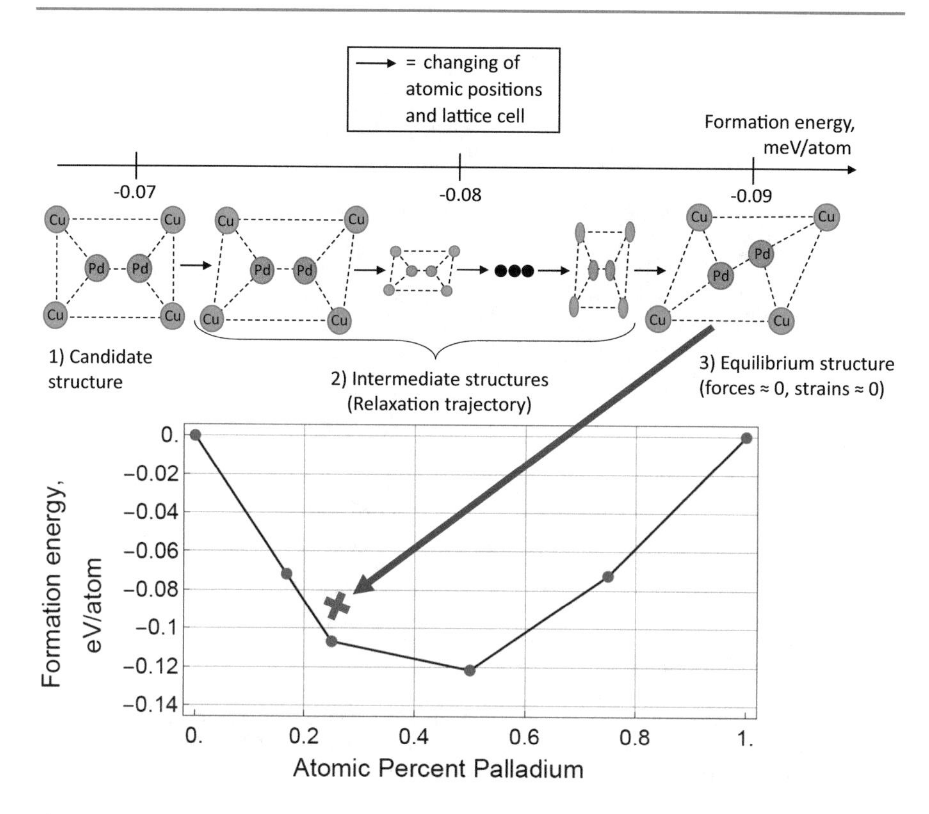

Fig. 15.6 Illustration of the relaxation trajectory concept: the lattice vectors and positions of the atoms are gradually changed to minimize the formation energy of the structure. The equilibrium structures are then added to a formation energy vs concentration graph based on which the convex hull is constructed. Reprinted from [10], with permission from Elsevier

The initial structures (before relaxation) are sometimes called the *candidate structures*. Similarly to the MD case, while performing relaxation of some candidate structure, we encounter new configurations (see Fig. 15.6) forming a *relaxation trajectory* ending in some equilibrated structure.

We demonstrate our active learning approach based on an example of application of active learning to the search for stable ternary Al-Ni-Ti alloys [11]. We choose a set of candidate structures containing 377,000 structures with different types of lattices and populated with different types of atoms (up to 12 atoms in the unit cell). To pre-train the potential we randomly sampled from candidate structures few hundreds of configurations for the training set. Once the potential is trained, we perform relaxation of each candidate structure with an active learning approach, where each relaxation can have two outcomes: it can either be finished successfully resulting in an equilibrated structure, or it can be terminated if an extrapolative configuration occurs (see Fig. 15.7). We do not immediately train on each extrapolative configuration occurred in some particular relaxation trajectory:

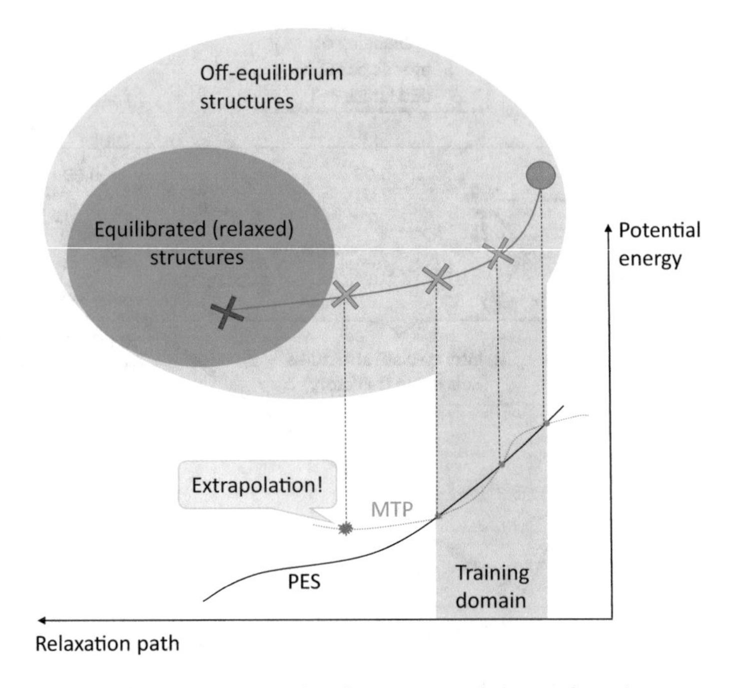

Fig. 15.7 While performing structure relaxation, an extrapolative configuration occurs when the potential, MTP [11], fails to approximate the potential energy surface (PES) outside the training domain. Reprinted from [10], with permission from Elsevier

there can be hundreds of thousands of relaxation trajectories, and selecting even one configuration from each trajectory would be computationally infeasible. Thus, we use an algorithm of selecting an optimal subset for training from a pool of configurations, exactly in the framework introduced in Sect. 15.2, with a size of such a subset being up to few hundred. Once it has been calculated with DFT, the selected subset is added to the training set. After retraining on the new data, the machine-learning potential can successfully relax more structures, as its training set is larger and more diverse (see Fig. 15.8). We then perform the next iteration of the algorithm: relax the candidate structures with the next-generation potential. Some relaxation trajectories will still contain extrapolative configurations, from which we again shortlist the configurations to enter the next-generation training set. After several iterations of:

- relaxing the candidate structures,
- selecting the optimal subset of extrapolative structures to be calculated with DFT and added to the training set, and
- retraining of the potential,

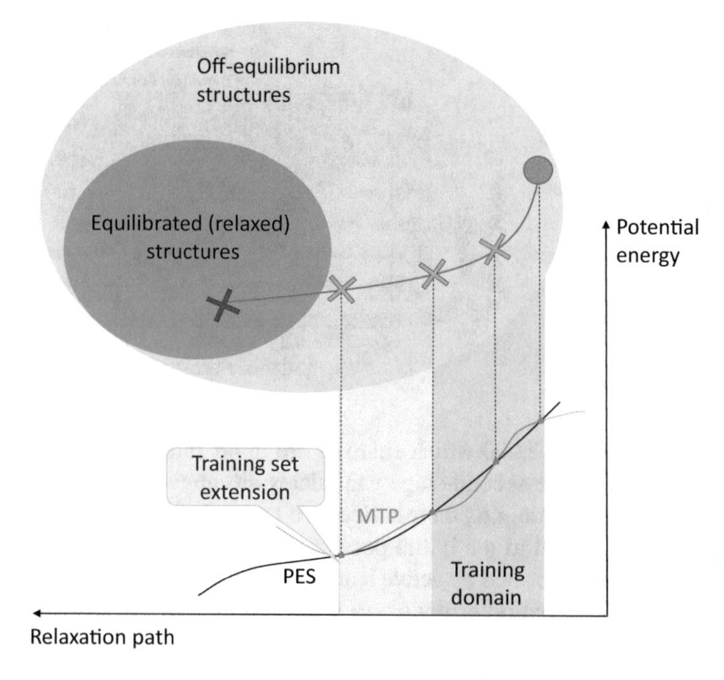

Fig. 15.8 After retraining on extrapolative configuration (extension of the training domain) the potential can provide a reliable prediction. Reprinted from [10], with permission from Elsevier

we obtain a training set of 2400 configurations formed by the active learning. The root-mean-square error on this training set is $\sigma := 27$ meV/atom. Once trained on this training set, the potential is able to relax all the candidate structures without exceeding the threshold γ_{tsh} for extrapolation grade. Once all the structures are relaxed, we calculate their energies and select the most stable structures, which are put on a *convex hull*, sometimes informally called a zero-temperature phase diagram. If we had the exact values of energies, the task would have already been solved at this stage, as we need nothing more. However, as any potential has finite accuracy, the obtained convex hull is not accurate—it may contain structures with non-minimal energy and, conversely, some minimal-energy structures may be missing. We therefore perform an additional procedure to make the convex hull more accurate.

After constructing a convex hull based on the approximate energies, we assume that all the stable structures are contained within a 4σ interval from the lowest energy level (convex hull level). Following this assumption, we throw out all the structures not falling into this range—this gives us the new region for the search for the stable structures. We call this process *screening* (see Fig. 15.9 for a graphical illustration of screening for a binary alloy case). Thus, after the first screening, 62,000 structures remained. We then start the next stage of convex hull constructing: we repeat the procedure from scratch, but with smaller amount of candidate

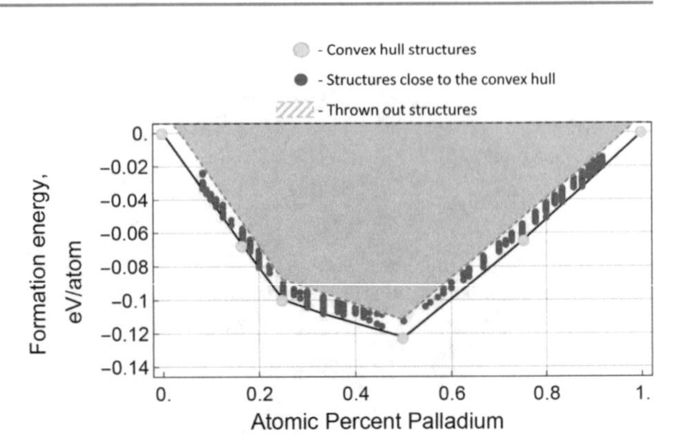

Fig. 15.9 Illustration of the screening concept on a different, binary Ag-Pd system: only the structures with formation energies lying within $4\sigma = 16$ meV/atom from the convex hull level are selected, the others are thrown out

structures—only those 62,000 which energies are in 4σ interval from the previous convex hull level. These second-stage candidates are already equilibrated during the first stage of relaxation, i.e., their lattices are less deformed and their atoms are less displaced, compared to the initial pool of candidate structures. Consequently, once the training set is formed by active learning (976 configurations), the potential exhibits lower approximation errors ($\sigma = 9$ meV/atom root-mean-square training error), compared to the first stage of relaxation, as the same functional form incorporated in the potential is applied to approximate the data (energies, forces, and stresses) in a smaller and less diverse domain of configurations space. Once the second generation of candidate structures is relaxed, we again construct a convex hull with the approximate energies, narrowing down the list of structures to a $4\sigma = 36$ meV/atom interval from the convex hull level.

After the second screening only 7000 structures remained, which were then relaxed with DFT to obtain the exact energies. This way, applying a machine-learning potential allowed us to relax 377,000 structures doing only 7000 actual DFT relaxations (starting from already pre-equilibrated structures, which is much faster), providing time savings of two orders of magnitude. If we were interested in constructing a convex hull with the 9 meV/atom error, we could have done only about 3400 single-point VASP calculations (time savings of three orders of magnitude), which would consume 90% of the computational time, while the rest 10% would be spent on fitting and relaxation with a machine-learning potential. The details of the proposed methodology are provided in [11].

The scheme of using active learning, described in this section, can be applied not only to the search of stable metallic alloys, but to the search of other materials, e.g., non-periodic organic molecules. The benefit it provides comes from the fact that the majority of costly quantum-mechanical calculations are replaced by much faster calculations, which in turn provide sufficient accuracy for successful screening of candidate structures. The evaluation of relatively small amount of shortlisted structures can be done with a quantum-mechanical model to obtain the exact values of the energies.

15.4 Conclusion

One of the major problems in molecular modeling solved by machine learning is the reduction of the amount of expensive quantum-mechanical calculations. When the molecules of interest, or more generally, atomistic configurations, are known in advance, active learning offers a way to reduce the number of quantum-mechanical calculations by picking the most informative molecules to be evaluated with quantum mechanics. In this chapter we have described four active learning methods: predictive variance in the framework of Gaussian processes, query by committee, D-optimality, and Bayesian methods for neural networks.

In some applications, for instance, discovery of stable crystal structures, exploration and learning of the potential energy surface has to be done simultaneously. Active learning-on-the-fly becomes indeed indispensable for this purpose as was demonstrated on the practical examples of molecular dynamics and crystal structure prediction.

Acknowledgments The work was supported by the Skoltech NGP Program No. 2016-7/NGP (a Skoltech-MIT joint project). The authors acknowledge the usage of the Skoltech CEST cluster (Magnus) from Prof. Shapeev's group for obtaining the results presented in this work.

References

1. N. Artrith, J. Behler, High-dimensional neural network potentials for metal surfaces: a prototype study for copper. Phys. Rev. B **85**(4), 045439 (2012)
2. A.P. Bartók, R. Kondor, G. Csányi, On representing chemical environments. Phys. Rev. B **87**(18), 184115 (2013)
3. C.M. Bishop, Bayesian neural networks. J. Braz. Comput. Soc. **4**(1), 61–68 (1997)
4. S. Chmiela, H.E. Sauceda, K.-R. Muller, A. Tkatchenko, Towards exact molecular dynamics simulations with machine-learned force fields. Nat. Commun. **9**(1), 1–10 (2018)
5. G. Csányi, T. Albaret, M. Payne, A. De Vita, Learn on the fly: a hybrid classical and quantum-mechanical molecular dynamics simulation. Phys. Rev. Lett. **93**(17), 175503 (2004)
6. A. De Vita, R. Car, A novel scheme for accurate MD simulations of large systems, in *MRS Proceedings*, vol. 491 (Cambridge University Press, Cambridge, 1997), p. 473
7. Y. Gal, Uncertainty in deep learning. PhD thesis, University of Cambridge, 2016
8. Y. Gal, Z. Ghahramani, Dropout as a Bayesian approximation: representing model uncertainty in deep learning, in *International Conference on Machine Learning* (2016), pp. 1050–1059
9. Y. Gal, R. Islam, Z. Ghahramani, Deep Bayesian active learning with image data. in *Proceedings of the 34th International Conference on Machine Learning*, vol. 70 (2017), pp. 1183–1192. www.JMLR.org
10. K. Gubaev, E.V. Podryabinkin, A.V. Shapeev, Machine learning of molecular properties: locality and active learning. J. Chem. Phys. **148**(24), 241727 (2018)
11. K. Gubaev, E.V. Podryabinkin, G.L. Hart, A.V. Shapeev, Accelerating high-throughput searches for new alloys with active learning of interatomic potentials. Comput. Mater. Sci. **156**, 148–156 (2019)
12. K. Hansen, F. Biegler, R. Ramakrishnan, W. Pronobis, O.A. Von Lilienfeld, K.-R. Muller, A. Tkatchenko, Machine learning predictions of molecular properties: accurate many-body potentials and nonlocality in chemical space. J. Phys. Chem. Lett. **6**(12), 2326–2331 (2015)

13. G.L.W. Hart, L.J. Nelson, R.W. Forcade, Generating derivative structures at a fixed concentration. Comput. Mater. Sci. **59**, 101–107 (2012)
14. G.E. Hinton, N. Srivastava, A. Krizhevsky, I. Sutskever, R.R. Salakhutdinov, Improving neural networks by preventing co-adaptation of feature detectors (2012). Preprint. arXiv:1207.0580
15. G. Huang, Y. Li, G. Pleiss, Z. Liu, J.E. Hopcroft, K.Q. Weinberger, Snapshot ensembles: Train 1, get M for free. Paper presented at the 5th International Conference on Learning Representations, ICLR 2017, Toulon, France, 24–26 April 2017. Conference Track Proceedings, 2017. https://openreview.net
16. H. Huo, M. Rupp, Unified representation for machine learning of molecules and crystals for machine learning (2017). Preprint. arXiv:1704.06439
17. M. Kampffmeyer, A.-B. Salberg, R. Jenssen, Semantic segmentation of small objects and modeling of uncertainty in urban remote sensing images using deep convolutional neural networks, in *Proceedings of the IEEE Conference on Computer Vision and Pattern Recognition Workshops* (2016), pp. 1–9
18. I. Kononenko, Bayesian neural networks. Biol. Cybern. **61**(5), 361–370 (1989)
19. Z. Li, J.R. Kermode, A. De Vita, Molecular dynamics with on-the-fly machine learning of quantum-mechanical forces. Phys. Rev. Lett. **114**, 096405 (2015)
20. Z. Lu, J. Bongard, Exploiting multiple classifier types with active learning, in *Proceedings of the 11th Annual Conference on Genetic and Evolutionary Computation* (ACM, New York, 2009), pp. 1905–1906
21. A.O. Lyakhov, A.R. Oganov, H.T. Stokes, Q. Zhu, New developments in evolutionary structure prediction algorithm USPEX. Comput. Phys. Commun. **184**(4), 1172–1182 (2013)
22. A.G. de G. Matthews, J. Hron, M. Rowland, R.E. Turner, Z. Ghahra-mani, Gaussian process behaviour in wide deep neural networks. Paper presented at the *6th International Conference on Learning Representations, ICLR 2018*, Vancouver, Canada, 30 April–3 May 2018. Conference Track Proceedings, 2018. https://openreview.net
23. M.J. Mehl, D. Hicks, C. Toher, O. Levy, R.M. Hanson, G.L.W. Hart, S. Curtarolo, The AFLOW library of crystallographic prototypes: part 1. Comput. Mater. Sci. **136**:S1–S828 (2017)
24. D. Molchanov, A. Ashukha, D. Vetrov, Variational dropout sparsifies deep neural networks, in *Proceedings of the 34th International Conference on Machine Learning*, vol. 70 (2017), pp. 2498–2507. www.JMLR.org
25. K. Neklyudov, D. Molchanov, A. Ashukha, D.P. Vetrov, Structured Bayesian pruning via log-normal multiplicative noise, in *Advances in Neural Information Processing Systems* (2017), pp. 6775–6784
26. A. Oganov (ed.), *Modern Methods of Crystal Structure Prediction* (Wiley-VCH, Weinheim, 2010)
27. D.W. Opitz, J.W. Shavlik, Generating accurate and diverse members of a neural-network ensemble, in *Advances in Neural Information Processing Systems* (1996), pp. 535–541
28. E.V. Podryabinkin, A.V. Shapeev, Active learning of linearly parametrized interatomic potentials. Comput. Mater. Sci. **140**, 171–180 (2017)
29. E. Podryabinkin, E. Tikhonov, A. Shapeev, A. Oganov, Accelerating crystal structure prediction by machine-learning interatomic potentials with active learning. Phys. Rev. B **99**(6), 064114 (2019)
30. R. Ramakrishnan, P.O. Dral, M. Rupp, O.A. Von Lilienfeld, Quantum chemistry structures and properties of 134 kilo molecules. Sci. Data **1**, 140022 (2014)
31. C.E. Rasmussen, Gaussian processes in machine learning, in *Advanced Lectures on Machine Learning* (Springer, Berlin, 2004), pp. 63–71
32. M.D. Richard, R.P. Lippmann, Neural network classifiers estimate Bayesian a posteriori probabilities. Neural Comput. **3**(4), 461–483 (1991)
33. K. Schütt, P.-J. Kindermans, H.E.S. Felix, S. Chmiela, A. Tkatchenko, K.-R. Müller, SchNet: a continuous-filter convolutional neural network for modeling quantum interactions, in *Advances in Neural Information Processing Systems* (2017), pp. 991–1001
34. B. Settles, Active learning. Synth. Lect. Artif. Intell. Mach. Learn. **6**(1), 1–114 (2012)

35. A.V. Shapeev, Moment tensor potentials: a class of systematically improvable interatomic potentials. Multiscale Model. Simul. **14**(3), 1153–1173 (2016)
36. J.S. Smith, B. Nebgen, N. Lubbers, O. Isayev, A.E. Roitberg, Less is more: sampling chemical space with active learning. J. Chem. Phys. **148**(24), 241733 (2018)
37. N. Srivastava, G. Hinton, A. Krizhevsky, I. Sutskever, R. Salakhutdinov, Dropout: a simple way to prevent neural networks from overfitting. J. Mach. Learn. Res. **15**(1), 1929–1958 (2014)
38. M. Teye, H. Azizpour, K. Smith, Bayesian uncertainty estimation for batch normalized deep networks, ed. by J.G. Dy, A. Krause, in *Proceedings of the 35th International Conference on Machine Learning, ICML2018*, Stockholm, Sweden, 10–15 July 2018, vol. 80. *Proceedings of Machine Learning Research PMLR (2018)*, pp. 4914–4923
39. E. Tsymbalov, M. Panov, A. Shapeev, Dropout-based active learning for regression, in *International Conference on Analysis of Images, Social Networks and Texts* (Springer, Cham, 2018), pp. 247–258
40. E. Tsymbalov, S. Makarychev, A. Shapeev, M. Panov, Deeper connections between neural networks and Gaussian processes speed-up active learning, in *Proceedings of the Twenty-Eighth International Joint Conference on Artificial Intelligence*. Main track (2019), pp. 3599–3605
41. Z.-H. Zhou, J. Wu, W. Tang, Ensembling neural networks: many could be better than all. Artif. Intell. **137**(1–2), 239–263 (2002)

Machine Learning for Molecular Dynamics on Long Timescales

16

Frank Noé

Abstract

Molecular dynamics (MD) simulation is widely used to analyze the properties of molecules and materials. Most practical applications, such as comparison with experimental measurements, designing drug molecules, or optimizing materials, rely on statistical quantities, which may be prohibitively expensive to compute from direct long-time MD simulations. Classical machine learning (ML) techniques have already had a profound impact on the field, especially for learning low-dimensional models of the long-time dynamics and for devising more efficient sampling schemes for computing long-time statistics. Novel ML methods have the potential to revolutionize long timescale MD and to obtain interpretable models. ML concepts such as statistical estimator theory, end-to-end learning, representation learning, and active learning are highly interesting for the MD researcher and will help to develop new solutions to hard MD problems. With the aim of better connecting the MD and ML research areas and spawning new research on this interface, we define the learning problems in long timescale MD, present successful approaches, and outline some of the unsolved ML problems in this application field.

16.1 Introduction

Molecular dynamics (MD) simulation is a widely used method of computational physics and chemistry to compute properties of molecules and materials. Examples include to simulate how a drug molecule binds to and inhibits a protein, or how

F. Noé (✉)
Freie Universität Berlin, Berlin, Germany
e-mail: frank.noe@fu-berlin.de

K. T. Schütt et al. (eds.), *Machine Learning Meets Quantum Physics*, Lecture Notes in Physics 968, https://doi.org/10.1007/978-3-030-40245-7_16

a battery material conducts ions. Despite its high computational cost, researchers use MD in order to get a principled understanding of how the composition and the microscopic structure of a molecular system translate into such macroscopic properties. In addition to scientific knowledge, this understanding can be used for designing molecular systems with better properties, such as drug molecules or enhanced materials.

MD has many practical problems, but at least four of them can be considered to be fundamental, in the sense that none of them is trivial for a practically relevant MD simulation, and there is extensive research on all of them. We refer to these four fundamental MD problems as SAME (Sampling, Analysis, Model, Experiment):

1. **S**ampling: To compute expectation values via MD simulations the simulation time needs to significantly exceed the slowest equilibration process in the molecular system. For most nontrivial molecules and materials, the presence of rare events and the sheer cost per MD time step make sufficient direct sampling unfeasible.
2. **A**nalysis: If enough statistics can be collected, we face huge amounts of simulation data (e.g., millions of time steps, each having 100,000s of dimensions). How can we analyze such data and obtain comprehensive and comprehensible models of the most relevant states, structures, and events sampled in the data?
3. **M**odel: MD simulations employ an empirical model of the molecular system studied. As the simulation computes forces from an energy model, this model is often referred to a MD force field. MD energy models are built from molecular components fitted to quantum-mechanical and experimental data. The accuracy of such a model is limited by the accuracy of the data used and the errors involved in transferring the training data usually obtained for small molecules to the often larger molecules simulated.
4. **E**xperiment: Experiments and simulations cannot access the same observables. While in MD simulation, the positions and velocities of all particles are available at all times, experiments usually probe complex functions of the positions and velocities, such as emission or absorption spectra of certain types of radiation. Computing these functions from first principles often requires the solution of a quantum-mechanical calculation with an accuracy that is unfeasible for a large molecular system. The last problem thus consists of finding good approximations to compute how an experiment would "see" a given MD state.

Machine learning (ML) has the potential to tackle these problems, and has already had profound impact on alleviating them. Here I will focus on the analysis problem and its direct connections to the sampling problem specifically for the case of long-time MD where these problems are most difficult and interesting. I believe that the solution of these problems lies on the interface between chemical physics and ML, and will therefore describe these problems in a language that should be understandable to audiences from both fields.

Let me briefly link to MD problems and associated ML approaches not covered by this chapter. The present description focuses on low-dimensional models of long-time MD and these can directly be employed to attack the sampling problem. The direct effect of these models is that short MD simulations that are individually not sampling all metastable states can be integrated, and thus an effective sampling that is much longer than the individual trajectory length, and on the order of the total simulation time can be reached [84]. The sampling efficiency can be further improved by adaptively selecting the starting points of MD simulations based on the long-time MD model, and iterating this process [21, 22, 36, 81, 82, 125, 137]. This approach is called "adaptive sampling" in the MD community, which is an active learning approach in ML language. Using this approach, timescales beyond seconds have been reached and protein–protein association and dissociation has recently been sampled for the first time with atomistic resolution [81].

A well-established approach to speed up rare events in MD is to employ the so-called enhanced sampling methods that change the thermodynamic conditions (temperature, adding bias potentials, etc.) [26, 30, 31, 49, 113], and to subsequently reweight to the unbiased target ensemble [3,4,24,27,59,106]. Recently, ML methods have been used to adaptively learn optimal biasing functions in such approaches [86, 118]. A conceptually different approach to sampling is the Boltzmann generator [71], a directed generative network to directly draw statistically independent samples from equilibrium distributions. While these approaches are usually limited to compute stationary properties, ML-based MD analysis models have recently been integrated with enhanced sampling methods in order to also compute unbiased dynamical properties [89, 126, 128, 129]. These methods can now also access all-atom protein dynamics beyond seconds timescales [77].

ML methods that use MD trajectory data to obtain a low-dimensional models of the long-time dynamics are extensively discussed here. Not discussed are manifold learning methods that purely use the data distribution, such as kernel PCA [98], isomap [18, 111], or diffusion maps [17, 88]. Likewise, there is extensive research on geometric clustering methods—both on the ML and the MD application side—which only plays a minor role in the present discussion.

Learning an accurate MD model—the so-called force-field problem—is one of the basic and most important problems of MD simulation. While this approach has traditionally been addressed by relatively ad hoc parametrization methods it is now becoming more and more a well-defined ML problem where universal function approximators (neural networks or kernel machines) are trained to reproduce quantum-mechanical potential energy surfaces with high accuracy [5–7, 13, 14, 91, 99, 100]. See other chapters in this book for more details. A related approach to scale to the next-higher length-scale is the learning of coarse-grained MD models from all-atom MD data [119, 120, 134]. These approaches have demonstrated that they can reach high accuracy, but employing the kernel machine or neural network to run MD simulations is still orders of magnitude slower than simulating a highly optimized MD code with an explicitly coded model. Achieving high accuracy while approximately matching the computational performance of commonly used MD codes is an important future aim.

Much less ML work has been done on the interpretation and integration of experimental data. MD models are typically parametrized by combining the matching of energies and forces from quantum-mechanical simulations with the matching of thermodynamic quantities measured by experiments, such as solvation free energies of small molecules. As yet, there is no rigorous ML method which learns MD models following this approach. Several ML methods have been proposed to integrate simulation data on the level of a model learned from MD simulation data (e.g., a Markov state model), typically by using information-theoretic principles such as maximum entropy or maximum caliber [20, 37, 74]. Finally, there is an emerging field of ML methods that predict experimental quantities, such as spectra, from chemical or molecular structures, which is an essential task that needs to be solved to perform data integration between simulation and experiment. An important step-stone for improving our ability to predict experimental properties is the availability of training datasets where chemical structures, geometric structures, and experimental measurements under well-defined conditions are linked [85].

16.2 Learning Problems for Long-Time Molecular Dynamics

16.2.1 What Would We Like to Compute?

The most basic quantitative aim of MD is to compute equilibrium expectations. When $\mathbf{x}$ is state of a molecular system, such as coordinates and velocities of the atoms in a protein system in a periodic solvent box, the average value of an observable A is given by:

$$\mathbb{E}[A] = \int A(\mathbf{x})\, \mu(\mathbf{x})\, \mathrm{d}\mathbf{x} \tag{16.1}$$

where $\mu(\mathbf{x})$ is the equilibrium distribution, i.e. the probability to find a molecule in state $\mathbf{x}$ at equilibrium conditions. A common choice is the Boltzmann distribution in the canonical ensemble at temperature T:

$$\mu(\mathbf{x}) \propto \mathrm{e}^{-\frac{U(\mathbf{x})}{k_B T}} \tag{16.2}$$

where $U(\mathbf{x})$ is a potential energy and the input constant $k_B T$ is the mean thermal energy per degree of freedom. The observable A can be chosen to compute, e.g., the probability of a protein to be folded at a certain temperature, or the probability for a protein and a drug molecule to be bound at a certain drug concentration, which relates to how much the drug inhibits the protein's activity. Other equilibrium expectations, such as spectroscopic properties, do not directly translate to molecular function, but are useful to validate and calibrate simulation models.

Molecules are not static but change their state $\mathbf{x}$ over time. Under equilibrium conditions, these dynamical changes are due to thermal fluctuations, leading to trajectories that are stochastic. For Markovian dynamics (e.g., classical MD), given configuration $\mathbf{x}_t$ at time t, the probability of finding the molecule in configuration

$\mathbf{x}_{t+\tau}$ at a later time can be expressed by the transition density p_τ:

$$\mathbf{x}_{t+\tau} \sim p_\tau(\mathbf{x}_{t+\tau} \mid \mathbf{x}_t). \tag{16.3}$$

Thus, a second class of relevant quantities is that of dynamical expectations:

$$\mathbb{E}[G; \tau] = \int\int \mu(\mathbf{x}_t)\, p_\tau(\mathbf{x}_{t+\tau} \mid \mathbf{x}_t)\, G(\mathbf{x}_t, \mathbf{x}_{t+\tau})\, \mathrm{d}\mathbf{x}_t\, \mathrm{d}\mathbf{x}_{t+\tau}. \tag{16.4}$$

As above, the observable G determines which dynamical property we are interested in. With an appropriate choice we can measure the average time a protein takes to fold or unfold, or dynamical spectroscopic expectations such as fluorescence correlations or dynamical scattering spectra.

16.2.2 What Is Molecular Dynamics?

MD simulation mimics the natural dynamics of molecules by time-propagating the state of a molecular system, such as coordinates and velocities of the atoms in a protein system in a periodic solvent box. MD is a Markov process involving deterministic components such as the gradient of a model potential $U(\mathbf{x})$ and stochastic components, e.g., from a thermostat. The specific choice of these components determines the transition density (16.3). Independent of these choices, a reasonable MD algorithm should be constructed such that it samples from $\mu(\mathbf{x})$ in the long run:

$$\lim_{\tau \to \infty} p_\tau(\mathbf{x}_{t+\tau} \mid \mathbf{x}_t) = \mu(\mathbf{x}) \propto \mathrm{e}^{-U(\mathbf{x})/k_B T}. \tag{16.5}$$

Thus, if a long enough MD trajectory can be generated, the expectation values (16.1) and (16.4) can be computed as direct averages. Unfortunately, this idea can only be implemented directly for very small and simple molecular systems. Most of the interesting molecular systems involve rare events, and as a result generating MD trajectories that are long enough to compute the expectation values (16.1) and (16.4) by direct averaging becomes unfeasible. For example, the currently fastest special-purpose supercomputer for MD, Anton II, can generate simulations on the order of 50 μs per day for a protein system [104]. The time for two strongly binding proteins to spontaneously dissociate can take over an hour, corresponding to a simulation time of a century for a single event [81].

16.2.3 Learning Problems for Long-Time MD

Repeated sampling from $p_\tau(\mathbf{x}_{t+\tau} \mid \mathbf{x}_t)$ "simulates" the MD system in time steps of length τ and will, due to (16.5), result in configurations sampled from $\mu(\mathbf{x}_t)$. Hence, knowing $p_\tau(\mathbf{x}_{t+\tau} \mid \mathbf{x}_t)$ is sufficient to compute any stationary or dynamical expectation ((16.1), (16.4)). The primary ML problem for long-time MD is thus to learn a model of the probability distribution $p_\tau(\mathbf{x}_{t+\tau} \mid \mathbf{x}_t)$ from simulation data pairs $(\mathbf{x}_t, \mathbf{x}_{t+\tau})$ which allows $\mathbf{x}_{t+\tau} \sim p_\tau(\mathbf{x}_{t+\tau} \mid \mathbf{x}_t)$ to be efficiently sampled. However, this problem is almost never addressed directly, because it is unnecessarily difficult. Configurations $\mathbf{x}$ live in a very high-dimensional space (typically 10^3 to 10^6 dimensions); the probability distributions $p_\tau(\mathbf{x}_{t+\tau} \mid \mathbf{x}_t)$ and $\mu(\mathbf{x})$ are multimodal and complex such that direct sampling is not tractable, and because of the exponential relationship between energies and probabilities (16.2), small mistakes in sampling $\mathbf{x}$ will lead to completely unrealistic molecular structures.

Because of these difficulties, ML methods for long-time MD usually take the detour of finding a low-dimensional *representation*, often called latent space representation, $\mathbf{y} = E(\mathbf{x})$, using the encoder E, and learning the dynamics in that space

$$\begin{array}{ccc} \mathbf{x}_t & \xrightarrow{E} & \mathbf{y}_t \\ \text{MD} \downarrow & & \downarrow \mathbf{P} \\ \mathbf{x}_{t+\tau} & \xleftarrow{D/G} & \mathbf{y}_{t+\tau} \end{array}$$

A relatively recent but fundamental insight is that for many MD systems there exists a natural low-dimensional representation in which the stationary and dynamical properties can be represented exactly if we give up time resolution by choosing a large lag time τ [60, 83, 92]. Thus, for long-time MD the intractable problem to learn $p_\tau(\mathbf{x}_{t+\tau} \mid \mathbf{x}_t)$ can be broken down into three learning problems (LPs) out of which two are much less difficult, and the third one does not need to be solved in order to compute stationary or dynamical expectations ((16.1), (16.4)), and that will be treated in the remainder of the article:

1. **LP1: Learn propagator P in representation y**. The simplest problem is to learn a model to propagate the latent state $\mathbf{y}_t$ in time for a given encoding $E(\mathbf{x}_t)$. This model is often linear using the propagator matrix $\mathbf{P}$, and hence shallow learning methods such as regression are used. In addition to obtaining an accurate model, it is desirable for $\mathbf{P}$ to be compact and easily interpretable/readable for a human specialist.
2. **LP2: Learn encoding *E* to representation y**. Learning the generally nonlinear encoding $\mathbf{y} = E(\mathbf{x})$ is a harder problem. Both shallow methods (regression in kernel and feature spaces, clustering, and likelihood maximization) and deep methods (neural networks) are used. LP1 and LP2 can be coupled to an end-to-end learning problem for $p_\tau(E(\mathbf{x}_{t+\tau}) \mid E(\mathbf{x}_t))$. LP2 has only become a well-defined ML problem recently with the introduction of a variational approach that defines a meaningful loss function for LP2 [66, 127].
3. **LP3: Learn decoding *D*/*G* to configuration space**. The most difficult problem is to decode the latent representation $\mathbf{y}$ back to configuration space. Because configuration space is much higher dimensional than latent space, this is an inverse problem where each $\mathbf{y}$ corresponds to many $\mathbf{x}$ configurations. The most

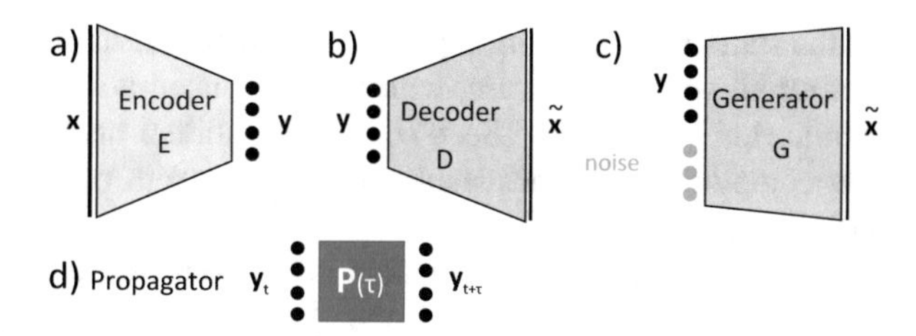

Fig. 16.1 Overview of network structures for learning Markovian dynamical models

faithful solution is to learn a generator G, representing a conditional probability distribution, $\mathbf{x} \sim G(\mathbf{y})$. This problem contains the hardest parts of the full learning problem for $p_\tau(\mathbf{x}_{t+\tau} \mid \mathbf{x}_t)$ and addressing it is still in its infancy.

These learning problems lead to different building blocks that can be implemented by neural networks or linear methods and can be combined towards different architectures (Fig. 16.1).

16.3 LP1: Learning Propagator in Feature Space

The simplest and most established learning problem is to learn a propagator, **P**, for a given, fixed encoding E. Therefore we discuss this learning problem first before defining what a "good" encoding E is and how to find it. As will be discussed below, for most MD systems of interest, there exists an encoding E to a *spectral representation* in which the dynamics is linear and low-dimensional. Although this spectral representation can often not be found exactly, it can usually be well enough approximated such that a linear dynamic model

$$\mathbb{E}\left[\mathbf{y}_{t+\tau}\right] = \mathbf{P}^\top \mathbb{E}\left[\mathbf{y}_t\right] \tag{16.6}$$

is an excellent approximation as well. $\mathbb{E}$ denotes an expectation value over time that accounts for stochasticity in the dynamics, and can be omitted for deterministic dynamical systems. For example, if $\mathbf{y}_t$ indicates which state the system is in at time t, $\mathbb{E}\left[\mathbf{y}_t\right]$ corresponds to a probability distribution over states.

Finding a linear model **P** is a shallow, unsupervised learning problem that in many cases has an algebraic expression for the optimum. Having a linear propagator also has great advantages for the analysis of the dynamical system. The analyses that can be done depend on the type of the representation and the mathematical properties of **P**. If E performs a one-hot-encoding that indicates which "state" the system is in, then the pair $(E, \mathbf{P})$ is called Markov state model (MSM [12, 16, 67, 84, 101, 108]), and **P** is the transition matrix of a Markov chain whose elements p_{ij} are nonnegative and can be interpreted as the conditional probabilities to be in a state j at time $t + \tau$ given that the system was in a state i at time t (Sects. 16.3.2 and 16.3.3). For MSMs, the whole arsenal of Markov chains analysis algorithms is available, e.g., for computing limiting distributions, first passage times, or the statistics of transition pathways [58,68]. If the transition matrix additionally has a real-valued spectrum, which is associated with dynamics at thermodynamic equilibrium conditions (Sect. 16.3.3), additional analyses are applicable, such as the computation of metastable (long-lived) sets of states by spectral clustering [19,67,101].

A broader class of propagators arise from encodings E that are partitions of unity, i.e. where $y_i(\mathbf{x}) > 0$ and $\sum_i y_i(\mathbf{x}) = 1$ for all $\mathbf{x}$ [33, 48, 55]. Such encodings correspond to a "soft clustering," where every configuration $\mathbf{x}$ can still be assigned to a state, but the assignment is no longer unique. The resulting propagators $\mathbf{P}$ are typically no longer transition matrices whose elements can be guaranteed to be nonnegative, but they can still be used to propagate probability densities by means of Eq. (16.6), and if they have a unique eigenvalue of 1, the corresponding eigenvector $\boldsymbol{\pi} = [\pi_i]$ still corresponds to the unique equilibrium distribution:

$$\boldsymbol{\pi} = \mathbf{P}^\top \boldsymbol{\pi}. \tag{16.7}$$

For arbitrary functions E, we can still use $\mathbf{P}$ to propagate state vectors according to Eq. (16.6), although these state vectors do no longer have a probabilistic interpretation, but are simply coefficients that model the configuration in the representation's basis. Owing to the Markovianity of the model, we can test how well the time-propagation of the model in time coincides with an estimation of the model at longer times, by means of the Chapman–Kolmogorov equation:

$$\mathbf{P}^n(\tau) \approx \mathbf{P}(n\tau) \tag{16.8}$$

In order to implement this equation, one has to decide which matrix norm should be used to compare the left and right hand side. A common choice is to compare the leading eigenvalues $\lambda_i(\tau)$. As these decay exponentially with time in a Markov process, it is common to transform them to relaxation rates or timescales by means of:

$$t_i(\tau) = -\frac{\tau}{\log|\lambda_i(\tau)|}. \tag{16.9}$$

A consequence of the Chapman–Kolmogorov equality is that these relaxation timescales are independent of the lag time τ at which $\mathbf{P}$ is estimated [108]. For real-valued eigenvalues, t_i corresponds to an ordinary relaxation time of the corresponding dynamical process. If $\mathbf{P}$ has complex-valued eigenvalues, t_i is the decay time of the envelope of an oscillating process whose oscillation frequency depends on the phase of λ_i.

16.3.1 Loss Function and Basis Statistics

Given one or many MD simulation trajectories $\{\mathbf{x}_t\}$, we apply E in order to map them to the representation $\{\mathbf{y}_t\}$, defining the input to LP1. The basic learning problem is the parameter estimation problem which consists of obtaining the optimal estimator $\hat{\mathbf{P}}$ as follows:

1. Define a loss function $\mathcal{L}(\mathbf{P}; \{\mathbf{y}_t\})$
2. Obtain the optimal estimator as $\hat{\mathbf{P}} = \arg\min_{\mathbf{P}} \mathcal{L}(\mathbf{P}; \{\mathbf{y}_t\})$

As most texts about molecular kinetics do not use the concept of a loss function, I would like to highlight the importance of a loss (or score) function from a ML point of view. The difference between fitting a training dataset $\{\mathbf{y}_t\}$ and ML is that ML aims at finding the estimator that performs best on an independent test dataset. To this end we need to not only optimize the parameters (such as the matrix elements of $\mathbf{P}$), but also hyper-parameters (such as the size of $\mathbf{P}$), which requires the concept of a loss function. Another important learning problem is to estimate the uncertainties of the estimator $\hat{\mathbf{P}}$.

To express the loss function and the optimal estimator of linear propagators $\mathbf{P}$, we do not actually need the full trajectory $\{\mathbf{y}_t\}$, but only certain sufficient statistics that are usually more compact than $\{\mathbf{y}_t\}$ and thus may require less storage space and lead to faster algorithms. The most prominent statistics are the empirical means and covariance matrices:

$$\boldsymbol{\mu}_0 = \frac{1}{T}\sum_{t=1}^{T-\tau} \mathbf{y}_t \tag{16.10}$$

$$\boldsymbol{\mu}_\tau = \frac{1}{T}\sum_{t=1}^{T-\tau} \mathbf{y}_{t+\tau} \tag{16.11}$$

$$\mathbf{C}_{00} = \frac{1}{T}\sum_{t=1}^{T-\tau} \mathbf{y}_t\mathbf{y}_t^\top \tag{16.12}$$

$$\mathbf{C}_{0\tau} = \frac{1}{T}\sum_{t=1}^{T-\tau} \mathbf{y}_t\mathbf{y}_{t+\tau}^\top \tag{16.13}$$

$$\mathbf{C}_{\tau\tau} = \frac{1}{T}\sum_{t=1}^{T-\tau} \mathbf{y}_{t+\tau}\mathbf{y}_{t+\tau}^\top. \tag{16.14}$$

A common modification to (16.12), (16.14) is the so-called shrinkage estimator that is used in ridge or Tikhonov regularization [50,93]. Since many algorithms involve the inversion of (16.12), (16.14) which might be rank-deficient, these estimators are often modified by adding a second matrix which ensures full rank, e.g.:

$$\tilde{\mathbf{C}}_{00} = \mathbf{C}_{00} + \lambda \mathbf{I} \tag{16.15}$$

$$\tilde{\mathbf{C}}_{\tau\tau} = \mathbf{C}_{\tau\tau} + \lambda \mathbf{I}, \tag{16.16}$$

where the small number λ is a regularization hyper-parameter.

16.3.2 Maximum Likelihood and Markov State Models

The concepts of maximum likelihood estimators and Markov state models (MSMs) are naturally obtained by defining an encoding which is a set characteristic function/indicator function:

$$y_{t,i} = \begin{cases} 1 & \mathbf{x}_t \in S_i \\ 0 & \text{else}, \end{cases} \tag{16.17}$$

where $S_1, \ldots, S_n$ is a partition of configuration space into n discrete states, i.e. each point $\mathbf{x}$ is assigned to exactly one state S_i, indicated by the position of the 1 in the encoding vector. In ML, (16.17) is called one-hot encoding. A consequence of (16.17) is that the covariance matrix (16.13) becomes:

$$c_{0\tau,ij} = N_{ij},$$

where N_{ij} counts the total number of transitions observed from i to j. The covariance matrix (16.12) is a diagonal matrix with diagonal elements

$$c_{00,ii} = N_i = \sum_j N_{ij},$$

where we use N_i to count the total number of transitions starting in state i. With this encoding, a natural definition for the propagator $\mathbf{P}$ is a transition matrix whose elements indicate the transition probability from any state i to any state j in a time step τ:

$$p_{ij} = \mathbb{P}\left[y_{t+\tau,j} = 1 \mid y_{t,i} = 1\right]$$

A natural optimality principle is then the maximum likelihood estimator (MLE): find the transition matrix $\hat{\mathbf{P}}$ that has the highest probability to produce the observation $\{\mathbf{y}_t\}$. The likelihood is given by:

$$L \propto \prod_{i,j} p_{ij}^{N_{ij}}. \tag{16.18}$$

Where the last term collects equal transition events along the trajectory and discards the proportionality factor. Maximizing L is equivalent to minimizing $-L$. However, as common in likelihood formulations we instead use $-\log L$ as a loss, which is minimal at the same $\hat{\mathbf{P}}$ but avoids the product:

$$\mathcal{L}_{\mathrm{ML}}(\mathbf{P};\{\mathbf{y}_t\}) = -\log L = -\sum_{i,j} N_{ij} \log p_{ij}. \tag{16.19}$$

The MLE $\hat{\mathbf{P}}$ can be easily found by minimizing (16.19) with the constraint $\sum_j p_{ij} = 1$ using the method of Lagrange multipliers. The result is intuitive: the maximum likelihood transition probability equals the corresponding fraction of transitions observed out of each state:

$$p_{ij} = \frac{N_{ij}}{N_i}.$$

In matrix form we can express this estimator as

$$\mathbf{P} = \mathbf{C}_{00}^{-1}\mathbf{C}_{0\tau}, \tag{16.20}$$

an expression that we will find also for other optimization principles. As we have a likelihood (16.18), we can also define priors and construct a full Bayesian estimator that not only provides the maximum likelihood result (16.20), but also posterior means and variances for estimating uncertainties. Efficient samplers are known that allow us to sample transition matrices directly from the distribution (16.18), and these samples can be used to compute uncertainties on quantities derived from $\mathbf{P}$ [36, 107].

An important property of a transition matrix is its stationary distribution $\boldsymbol{\pi}$ (which we will assume to exist and be unique here) with

$$\pi_i = \int_{\mathbf{x}\in S_i} \mu(\mathbf{x})\,\mathrm{d}\mathbf{x},$$

$\boldsymbol{\pi}$ that can be computed by solving the eigenvalue problem (16.7).

16.3.3 MSMs with Detailed Balance

In thermodynamic equilibrium, i.e. when a molecular system is evolving purely as a result of thermal energy at a given thermodynamic condition and no external force is applied, the absolute probability of paths between any two end-points is symmetric. As a consequence of this, there exists no cycle in state space which contains net flux in either direction, and no network can be extracted from the system, consistently with the second law of thermodynamics. We call this condition *detailed balance* and write it as:

$$\mu(\mathbf{x})\, p_\tau(\mathbf{y} \mid \mathbf{x}) = \mu(\mathbf{y})\, p_\tau(\mathbf{x} \mid \mathbf{y})\ \forall \mathbf{x}, \mathbf{y}, \tau > 0. \tag{16.21}$$

Integrating $\mathbf{x}$ and $\mathbf{y}$ over the sets S_i and S_j in this equation leads to detailed balance for MSMs:

$$\pi_i p_{ij} = \pi_j p_{ji}. \tag{16.22}$$

When the molecular system is simulated such that Eq. (16.21) hold, we also want to ensure that the estimator $\hat{\mathbf{P}}$ fulfills the constraint (16.22). Enforcing (16.21) in the estimator reduces the number of free parameters and thus improves the statistics. More importantly, propagators that fulfill (16.21) or (16.22) have a real-valued spectrum for which additional analyses can be made (see beginning of Sect. 16.3).

The trivial estimator (16.20) does not fulfill (16.22), unless N_{ij} is, by chance, a symmetric matrix. Maximum likelihood estimation with (16.22) as a constraint can be achieved by an iterative algorithm first developed in [9] and reformulated as in Algorithm 1 in [116]. Enforcing (16.22) is only meaningful if there is a unique stationary distribution, which, requires the transition matrix to define a fully connected graph. For this reason, graph algorithms are commonly used to find the largest connected set of states before estimating an MSM with detailed balance [9, 84, 94].

Algorithm 1 Detailed balance $\pi_i p_{ij} = \pi_j p_{ji}$ with unknown π [9, 116]

1. Initialize: $\pi_i^{(0)} = \frac{\sum_{j=1}^n c_{ij}}{\sum_{i,j=1}^n c_{ij}}$
2. Iterate until convergence: $\pi_i^{(k+1)} = \sum_{j=1}^n \frac{c_{ij}+c_{ji}}{c_i/\pi_i^{(k)} + c_j/\pi_j^{(k)}}$
3. $p_{ij} = \frac{(c_{ij}+c_{ji})\pi_j}{c_i\pi_j + c_j\pi_i}$

When the equilibrium distribution $\boldsymbol{\pi}$ is known a priori or obtained from another estimator as in [115, 128, 129], the maximum likelihood estimator can be obtained by the iterative Algorithm 2 developed in [116]:

Algorithm 2 Detailed balance $\pi_i p_{ij} = \pi_j p_{ji}$ with known π [116]

1. Initialize Lagrange parameters: $\lambda_i^{(0)} = \frac{1}{2}\sum_j (c_{ij} + c_{ji})$
2. Iterate until convergence: $\lambda_i^{(k+1)} = \sum_{j, c_{ij}+c_{ji}>0}^n \frac{(c_{ij}+c_{ji})\lambda_i^{(k)}\pi_j}{\lambda_j^{(k)}\pi_i + \lambda_i^{(k)}\pi_j}$
3. $p_{ij} = \frac{(c_{ij}+c_{ji})\pi_j}{\lambda_i\pi_j + \lambda_j\pi_i}$

As for MSMs without detailed balance, methods have been developed to perform a full Bayesian analysis of MSMs with detailed balance. No method is known to sample independent transition matrices from the likelihood (16.18) subject to the detailed balance constraints (16.22); however, efficient Markov Chain Monte Carlo methods have been developed and implemented to this end [2, 15, 57, 63, 94, 114, 116].

16.3.4 Minimal Regression Error

We can understand Eq. (16.6) as a regression from $\mathbf{y}_t$ onto $\mathbf{y}_{t+\tau}$ where $\mathbf{P}$ contains the unknown coefficients. The regression loss is then directly minimizing the error in Eq. (16.6):

$$\min \mathbb{E}_t \left[\left\| \mathbf{y}_{t+\tau} - \mathbf{P}^\top \mathbf{y}_t \right\|^2 \right]$$

and for a given dataset $\{\mathbf{y}_t\}$ we can define matrices $\mathbf{Y}_0 = (\mathbf{y}_0, \ldots, \mathbf{y}_{T-\tau})^\top$ and $\mathbf{Y}_\tau = (\mathbf{y}_\tau, \ldots, \mathbf{y}_T)^\top$ resulting in the loss function:

$$\mathcal{L}_{\mathrm{LSQ}}(\mathbf{P}; \{\mathbf{y}_t\}) = \|\mathbf{Y}_0 - \mathbf{Y}_\tau \mathbf{P}\|_F^2 , \tag{16.23}$$

where F indicates the Frobenius norm, i.e. the sum over all squares. The direct solution of the least squares regression problem in (16.23) is identical with the trivial MSM estimator (16.20). Thus, the estimator (16.20) is more general than for MSMs—it can be applied for to any representation $\mathbf{y}_t$. Dynamic mode decomposition (DMD) [90,96,97,117] and extended dynamic mode decomposition (EDMD) [124] are also using the minimal regression error, although they usually consider low-rank approximations of $\mathbf{P}$.

In general, the individual dimensions of the encoding E may not be orthogonal, and if not, the matrix $\mathbf{C}_{00}$ is not diagonal, but contains off-diagonal elements quantifying the correlation between different dimensions. When there is too much correlation between them, $\mathbf{C}_{00}$ may have some vanishing eigenvalues, i.e. not full rank, causing it not to be invertible or only invertible with large numerical errors. A standard approach in least squares regression is to then apply the ridge regularization (Eq. (16.15)). Using (16.15) in the estimator (16.20) is called ridge regression.

16.3.5 Variational Approach for Dynamics with Detailed Balance (VAC)

Instead of using an optimality principle to estimate $\mathbf{P}$ directly, we will now derive a variational principle for the eigenvalues and eigenvectors of $\mathbf{P}$, from which we can then easily assemble $\mathbf{P}$ itself. At first, this approach seems to be a complication compared to the likelihood or least squares approach, but this approach is key in making progress on LP2 because the variational principle for

$\mathbf{P}$ has a fundamental relation to the spectral properties of the transition dynamics in configuration space (16.3). It also turns out that the variational approach leads to a natural representation of configurations that we can optimize in end-to-end learning frameworks. We first define the balanced propagator:

$$\tilde{\mathbf{P}} = \mathbf{C}_{00}^{-\frac{1}{2}} \mathbf{C}_{0\tau} \mathbf{C}_{\tau\tau}^{-\frac{1}{2}}. \tag{16.24}$$

In this section, we will assume that detailed balance holds with a unique stationary distribution, Eq. (16.21). In the statistical limit this means that $\mathbf{C}_{00} = \mathbf{C}_{\tau\tau}$ holds and $\mathbf{C}_{0\tau}$ is a symmetric matrix. Using these constraints, we find the stationary balanced propagator:

$$\tilde{\mathbf{P}} = \mathbf{C}_{00}^{-\frac{1}{2}} \mathbf{C}_{01} \mathbf{C}_{00}^{-\frac{1}{2}} = \mathbf{C}_{00}^{\frac{1}{2}} \mathbf{P} \mathbf{C}_{00}^{-\frac{1}{2}}. \tag{16.25}$$

Where we have used Eq. (16.6). Due to the symmetry of $\mathbf{C}_{0\tau}$, $\tilde{\mathbf{P}}$ is also symmetric and we have the symmetric eigenvalue decomposition (EVD):

$$\tilde{\mathbf{P}} = \tilde{\mathbf{U}} \mathbf{\Lambda} \tilde{\mathbf{U}}^\top \tag{16.26}$$

with eigenvector matrix $\tilde{\mathbf{U}} = \left[\tilde{\mathbf{u}}_1, \ldots, \tilde{\mathbf{u}}_n\right]$ and eigenvalue matrix $\mathbf{\Lambda} = \mathrm{diag}(\lambda_1, \ldots, \lambda_n)$ ordered as $\lambda_1 \geq \lambda_2 \geq \cdots \geq \lambda_n$. This EVD is related to the EVD of $\mathbf{P}$ via a basis transformation:

$$\mathbf{P} = \mathbf{C}_{00}^{-\frac{1}{2}} \tilde{\mathbf{U}} \mathbf{\Lambda} \left(\tilde{\mathbf{U}} \mathbf{C}_{00}^{-\frac{1}{2}} \right)^\top = \mathbf{U} \mathbf{\Lambda} \mathbf{U}^{-1} \tag{16.27}$$

such that $\mathbf{U} = \mathbf{C}_{00}^{-\frac{1}{2}} \tilde{\mathbf{U}}$ are the eigenvectors of $\mathbf{P}$, their inverse is given by $\mathbf{U}^{-1} = \mathbf{C}_{00}^{-\frac{1}{2}} \tilde{\mathbf{U}}^\top$, and both propagators share the same eigenvalues. The above construction is simply a change of viewpoint: instead of optimizing the propagator $\mathbf{P}$, we might as well optimize its eigenvalues and eigenvectors, and then assemble $\mathbf{P}$ via Eq. (16.27).

Now we seek an optimality principle for eigenvectors and eigenvalues. For symmetric eigenvalue problems such as (16.26), we have the following variational principle: The dominant k eigenfunctions $\tilde{\mathbf{r}}_1, \ldots, \tilde{\mathbf{r}}_k$ are the solution of the maximization problem:

$$\begin{aligned} \sum_{i=1}^{k} \lambda_i &= \max_{\tilde{\mathbf{f}}_1, \ldots, \tilde{\mathbf{f}}_k} \sum_{i=1}^{k} \frac{\tilde{\mathbf{f}}_i^\top \tilde{\mathbf{P}} \tilde{\mathbf{f}}_i}{\left(\tilde{\mathbf{f}}_i^\top \tilde{\mathbf{f}}_i\right)^{\frac{1}{2}} \left(\tilde{\mathbf{f}}_i^\top \tilde{\mathbf{f}}_i\right)^{\frac{1}{2}}} = \max_{\mathbf{f}_1, \ldots, \mathbf{f}_k} \sum_{i=1}^{k} \frac{\mathbf{f}_i^\top \mathbf{C}_{0\tau} \mathbf{f}_i}{\left(\mathbf{f}_i^\top \mathbf{C}_{00} \mathbf{f}_i\right)^{\frac{1}{2}} \left(\mathbf{f}_i^\top \mathbf{C}_{00} \mathbf{f}_i\right)^{\frac{1}{2}}} \\ &= \sum_{i=1}^{k} \frac{\mathbf{u}_i^\top \mathbf{C}_{0\tau} \mathbf{u}_i}{\left(\mathbf{u}_i^\top \mathbf{C}_{00} \mathbf{u}_i\right)^{\frac{1}{2}} \left(\mathbf{u}_i^\top \mathbf{C}_{00} \mathbf{u}_i\right)^{\frac{1}{2}}} = \left(\mathbf{U}^\top \mathbf{C}_{00} \mathbf{U}\right)^{-\frac{1}{2}} \mathbf{U}^\top \mathbf{C}_{0\tau} \mathbf{U} \left(\mathbf{U}^\top \mathbf{C}_{00} \mathbf{U}\right)^{-\frac{1}{2}}. \end{aligned} \tag{16.28}$$

This means: we vary a set of vectors $\mathbf{f}_i = \mathbf{C}_{00}^{-\frac{1}{2}}\tilde{\mathbf{f}}_i$, and when the so-called Rayleigh quotients on the right hand side are maximized, we have found the eigenvectors. In this limit, the argument of the Rayleigh quotient equals the sum of eigenvalues. As the argument above can be made for every value of k starting from $k = 1$, we have found each single eigenvalue and eigenvector at the end of the procedure (assuming no degeneracy). This variational principle becomes especially useful for LP2, because using the variational approach of conformation dynamics (VAC [66, 72]), it can also be shown that the eigenvalues of $\mathbf{P}$ are lower bounds to the true eigenvalues of the Markov dynamics in configurations $\mathbf{x}$ (Sect. 16.4.2).

Now we notice that this variational principle can also be understood as a direct correlation function of the data representation. We define the *spectral representation* as:

$$\mathbf{y}_t^s = \left(\mathbf{y}_t^\top \mathbf{u}_1, \ldots, \mathbf{y}_t^\top \mathbf{u}_n\right) \tag{16.29}$$

inserting the estimators for $\mathbf{C}_{00}$ and $\mathbf{C}_{0\tau}$ (Eqs. (16.12), (16.13)) into Eq. (16.28), we have:

$$\sum_{i=1}^{k} \lambda_i = \frac{\sum_{t=1}^{T-\tau} \mathbf{y}_t^s \mathbf{y}_{t+\tau}^{s\top}}{\sum_{t=1}^{T-\tau} \mathbf{y}_t^s \mathbf{y}_t^{s\top}} = \left(\mathbf{C}_{00}^s\right)^{-\frac{1}{2}} \mathbf{C}_{0\tau}^s \left(\mathbf{C}_{00}^s\right)^{-\frac{1}{2}},$$

where the superscript s denotes the covariance matrices computed in the spectral representation.

The same calculation as above can be performed with powers of the eigenvalues, e.g., $\sum_{i=1}^{k} \lambda_i^2$. We therefore get a whole family of VAC-optimization principles, but two choices are especially interesting: we define the VAC-1 loss, that is equivalent to the generalized matrix Rayleigh quotient employed in [56], as:

$$\mathcal{L}_{\mathrm{VAC-1}}(\mathbf{U}; \{\mathbf{y}_t\}) = -\mathrm{trace}\left[\left(\mathbf{U}^\top \mathbf{C}_{00} \mathbf{U}\right)^{-\frac{1}{2}} \mathbf{U}^\top \mathbf{C}_{0\tau} \mathbf{U} \left(\mathbf{U}^\top \mathbf{C}_{00} \mathbf{U}\right)^{-\frac{1}{2}}\right] \tag{16.30}$$

$$\mathcal{L}_{\mathrm{VAC-1}}(\{\mathbf{y}_t^s\}) = -\mathrm{trace}\left[\left(\mathbf{C}_{00}^s\right)^{-\frac{1}{2}} \mathbf{C}_{0\tau}^s \left(\mathbf{C}_{00}^s\right)^{-\frac{1}{2}}\right]. \tag{16.31}$$

The VAC-2 loss is the Frobenius norm, i.e. the sum of squared elements of the matrix:

$$\mathcal{L}_{\mathrm{VAC-2}}(\mathbf{U}; \{\mathbf{y}_t\}) = -\left\|\left(\mathbf{U}^\top \mathbf{C}_{00} \mathbf{U}\right)^{-\frac{1}{2}} \mathbf{U}^\top \mathbf{C}_{0\tau} \mathbf{U} \left(\mathbf{U}^\top \mathbf{C}_{00} \mathbf{U}\right)^{-\frac{1}{2}}\right\|_F^2 \tag{16.32}$$

$$\mathcal{L}_{\mathrm{VAC-2}}(\{\mathbf{y}_t^s\}) = -\left\|\left(\mathbf{C}_{00}^s\right)^{-\frac{1}{2}} \mathbf{C}_{0\tau}^s \left(\mathbf{C}_{00}^s\right)^{-\frac{1}{2}}\right\|_F^2. \tag{16.33}$$

This loss induces a natural spectral embedding where the variance along each dimension equals the squared eigenvalue and geometric distances in this space are related to kinetic distances [64].

16.3.6 General Variational Approach (VAMP)

The variational approach for Markov processes (VAMP) [127] generalizes the above VAC approach to dynamics that do not obey detailed balance and may not even have an equilibrium distribution. We use the balanced propagator (16.24) that is now no longer symmetric. Without symmetry we cannot use the variational principle for eigenvalues, but there is a similar variational principle for singular values. We therefore use the singular value decomposition (SVD) of the balanced propagator:

$$\tilde{\mathbf{P}} = \tilde{\mathbf{U}}\mathbf{\Sigma}\tilde{\mathbf{V}}^{\top}. \tag{16.34}$$

Again, this SVD is related to the SVD of $\mathbf{P}$ via a basis transformation:

$$\mathbf{P} = \mathbf{C}_{00}^{-\frac{1}{2}}\tilde{\mathbf{U}}\mathbf{\Sigma}\left(\mathbf{C}_{\tau\tau}^{-\frac{1}{2}}\tilde{\mathbf{V}}\right)^{\top} = \mathbf{U}\mathbf{\Sigma}\mathbf{V}^{\top} \tag{16.35}$$

with $\mathbf{U} = \mathbf{C}_{00}^{-\frac{1}{2}}\tilde{\mathbf{U}}$ and $\mathbf{V} = \mathbf{C}_{\tau\tau}^{-\frac{1}{2}}\tilde{\mathbf{V}}$. Using two sets of search vectors $\mathbf{f}_i = \mathbf{C}_{00}^{-\frac{1}{2}}\tilde{\mathbf{f}}_i$ and $\mathbf{g}_i = \mathbf{C}_{\tau\tau}^{-\frac{1}{2}}\tilde{\mathbf{g}}_i$, we can follow the same line of derivation as above and obtain:

$$\begin{aligned}
\sum_{i=1}^{k}\sigma_i &= \max_{\tilde{\mathbf{f}}_1,\ldots,\tilde{\mathbf{f}}_k,\tilde{\mathbf{g}}_1,\ldots,\tilde{\mathbf{g}}_k}\sum_{i=1}^{k}\frac{\tilde{\mathbf{f}}_i^{\top}\tilde{\mathbf{P}}\tilde{\mathbf{g}}_i}{\left(\tilde{\mathbf{f}}_i^{\top}\tilde{\mathbf{f}}_i\right)^{\frac{1}{2}}\left(\tilde{\mathbf{g}}_i^{\top}\tilde{\mathbf{g}}_i\right)^{\frac{1}{2}}} \\
&= \left(\mathbf{U}^{\top}\mathbf{C}_{00}\mathbf{U}\right)^{-\frac{1}{2}}\mathbf{U}^{\top}\mathbf{C}_{0\tau}\mathbf{V}\left(\mathbf{V}^{\top}\mathbf{C}_{\tau\tau}\mathbf{V}\right)^{-\frac{1}{2}}.
\end{aligned}$$

Now we define again a spectral representation. If we set $\mathbf{C}_{00} = \mathbf{C}_{\tau\tau}$ (equilibrium case) as above, we can define a single spectral representation, otherwise we need two sets of spectral coordinates:

$$\mathbf{y}_t^{s,0} = \left(\mathbf{y}_t^{\top}\mathbf{u}_1,\ldots,\mathbf{y}_t^{\top}\mathbf{u}_n\right) \tag{16.36}$$

$$\mathbf{y}_t^{s,\tau} = \left(\mathbf{y}_t^{\top}\mathbf{v}_1,\ldots,\mathbf{y}_t^{\top}\mathbf{v}_n\right). \tag{16.37}$$

As in the above procedure, we can define a family of VAMP scores, where the VAMP-1 and VAMP-2 scores are of special interest:

$$\mathcal{L}_{\text{VAMP-1}}(\mathbf{U},\mathbf{V};\{\mathbf{y}_t\}) = -\text{trace}\left[\left(\mathbf{U}^\top \mathbf{C}_{00}\mathbf{U}\right)^{-\frac{1}{2}} \mathbf{U}^\top \mathbf{C}_{0\tau}\mathbf{V}\left(\mathbf{V}^\top \mathbf{C}_{\tau\tau}\mathbf{V}\right)^{-\frac{1}{2}}\right] \tag{16.38}$$

$$\mathcal{L}_{\text{VAMP-1}}(\{\mathbf{y}_t^{s,0},\mathbf{y}_t^{s,\tau}\}) = -\text{trace}\left[\left(\mathbf{C}_{00}^s\right)^{-\frac{1}{2}} \mathbf{C}_{0\tau}^s \left(\mathbf{C}_{\tau\tau}^s\right)^{-\frac{1}{2}}\right]. \tag{16.39}$$

The VAMP-2 score is again related to an embedding where geometric distance corresponds to kinetic distance [78]:

$$\mathcal{L}_{\text{VAMP-2}}(\mathbf{U},\mathbf{V};\{\mathbf{y}_t\}) = -\left\|\left(\mathbf{U}^\top \mathbf{C}_{00}\mathbf{U}\right)^{-\frac{1}{2}} \mathbf{U}^\top \mathbf{C}_{0\tau}\mathbf{V}\left(\mathbf{V}^\top \mathbf{C}_{\tau\tau}\mathbf{V}\right)^{-\frac{1}{2}}\right\|_F^2 \tag{16.40}$$

$$\mathcal{L}_{\text{VAMP-2}}(\{\mathbf{y}_t^{s,0},\mathbf{y}_t^{s,\tau}\}) = -\left\|\left(\mathbf{C}_{00}^s\right)^{-\frac{1}{2}} \mathbf{C}_{0\tau}^s \left(\mathbf{C}_{\tau\tau}^s\right)^{-\frac{1}{2}}\right\|_F^2. \tag{16.41}$$

16.4 Spectral Representation and Variational Approach

Before turning to LP2, we will relate the spectral decompositions in the VAC and VAMP approaches described above to spectral representations of the transition density of the underlying Markov dynamics in $\mathbf{x}_t$. These two representations are connected by variational principles. Exploiting this principle leads to the result that a meaningful and feasible formulation of the long-time MD learning problem is to seek a spectral representation of the dynamics. This representation may be thought of as a set of collective variables (CVs) pertaining to the long-time MD, or slow CVs [65].

16.4.1 Spectral Theory

We can express the transition density (16.3) as the action of the Markov propagator in continuous-space, and by its spectral decomposition [92, 127]:

$$p(\mathbf{x}_{t+\tau}) = \int p(\mathbf{x}_{t+\tau} \mid \mathbf{x}_t;\tau)\,p(\mathbf{x}_t)\,\mathrm{d}\mathbf{x}_t \tag{16.42}$$

$$\approx \sum_{k=1}^{n} \sigma_k^* \langle p(\mathbf{x}_t) \mid \phi(\mathbf{x}_t)\rangle \psi(\mathbf{x}_{t+\tau}). \tag{16.43}$$

The spectral decomposition can be read as follows: The evolution of the probability density can be approximated as the superposition of basis functions ψ. A second set of functions, ϕ is required in order to compute the amplitudes of these functions.

In general, Eq. (16.43) is a singular value decomposition with left and right singular functions ϕ_k, ψ_k and true singular values σ_k^* [127]. The approximation then is a low-rank decomposition in which the small singular values are discarded. For the special case that dynamics are in equilibrium and satisfy detailed balance (16.21), Eq. (16.43) is an eigenvalue decomposition with the choices:

$$\sigma_k^* = \lambda_k^*(\tau) = \mathrm{e}^{-\tau\kappa_k} \in \mathbb{R}$$
$$\phi_k(\mathbf{x}) = \psi_k(\mathbf{x})\mu(\mathbf{x}).$$

Hence Eq. (16.43) simplifies: we only need one set of functions, the eigenfunctions ψ_k. The true eigenvalues λ_k^* are real-valued and decay exponentially with the time step τ (hence Eq. (16.9)). The characteristic decay rates κ_k are directly linked to experimental observables probing the processes associated with the corresponding eigenfunctions [12, 69]. The approximation in Eq. (16.43) is due to truncating all terms with decay rates faster than κ_n. This approximation improves exponentially with increasing τ.

Spectral theory makes it clear why learning long-time MD via LP1-3 is significantly simpler than trying to model $p(\mathbf{x}_{t+\tau} \mid \mathbf{x}_t; \tau)$ directly: For long time steps τ, $p(\mathbf{x}_{t+\tau} \mid \mathbf{x}_t; \tau)$ becomes intrinsically low-dimensional, and the problem is thus significantly simplified by learning to approximate the low-dimensional representation $(\psi_1, \ldots, \psi_n)$ for a given τ.

16.4.2 Variational Principles

The spectral decomposition of the exact dynamics, Eq. (16.43), is the basis for the usefulness of the variational approaches described in Sects. 16.3.5 and 16.3.6. The missing connection is filled by the following two variational principles that are analogous to the variational principle for energy levels in quantum mechanics. The VAC variational principle [66] is that for dynamics obeying detailed balance (16.21), the eigenvalues λ_k of a propagator matrix $\mathbf{P}$ via any encoding $\mathbf{y} = E(\mathbf{x})$ are, in the statistical limit, lower bounds of the true λ_k^*. The VAMP variational principle is more general, as it does not require detailed balance (16.21), and applies to the singular values:

$$\lambda_k \le \lambda_k^* \quad \text{(with DB)}$$
$$\sigma_k \le \sigma_k^* \quad \text{(no DB)}.$$

Equality is only achieved for $E(\mathbf{x}) = \mathrm{span}(\psi_1, \ldots, \psi_n)$ when detailed balance holds, and for $E(\mathbf{x}) = \mathrm{span}(\psi_1, \ldots, \psi_n, \phi_1, \ldots, \phi_n)$ when detailed balance does not hold. Specifically, the eigenvectors or the singular vectors of the propagator then approximate the individual eigenfunctions or singular functions (assuming no degeneracy):

$$\lambda_k = \lambda_k^* \longrightarrow \mathbf{u}_k^\top E(\mathbf{x}) = \psi(\mathbf{x})$$

$$\sigma_k = \sigma_k^* \longrightarrow \begin{cases} \mathbf{u}_k^\top E(\mathbf{x}) = \psi(\mathbf{x}) \\ \mathbf{v}_k^\top E(\mathbf{x}) = \phi(\mathbf{x}). \end{cases}$$

As direct consequence of the variational principles above, the loss function associated with a given embedding E is, in the statistical limit, also an upper bound to the sum of true eigenvalues:

$$\mathcal{L}_{VAC-r} \geq -\sum_{k=1}^{n} \left(\lambda_k^*\right)^r$$

$$\mathcal{L}_{VAMP-r} \geq -\sum_{k=1}^{n} \left(\sigma_k^*\right)^r$$

and for the minimum possible loss, E has identified the dominant eigenspace or singular space.

16.4.3 Spectral Representation Learning

We have seen in Sect. 16.3 (LP1) that a propagator $\mathbf{P}$ can be equivalently represented by its eigenspectrum or singular spectrum. We can thus define a spectral encoding that attempts to directly learn the encoding to the spectral representation:

$$\mathbf{y}_t^s = E^s(\mathbf{x}_t)$$

with the choices (16.29) or (16.36), (16.37), depending on whether the dynamics obey detailed balance or not. In these representations, the dynamics are linear. After encoding to this representation, the eigenvalues or singular values can be directly estimated from:

$$\mathbf{\Lambda} = \left(\mathbf{R}^\top \mathbf{C}_{00}^s \mathbf{R}\right)^{-1} \mathbf{R}^\top \mathbf{C}_{0\tau}^s \mathbf{R} \tag{16.44}$$

$$\mathbf{\Sigma} = \left(\mathbf{U}^\top \mathbf{C}_{00}^s \mathbf{U}\right)^{-\frac{1}{2}} \mathbf{U}^\top \mathbf{C}_{0\tau}^s \mathbf{V} \left(\mathbf{V}^\top \mathbf{C}_{\tau\tau}^s \mathbf{V}\right)^{-\frac{1}{2}}. \tag{16.45}$$

Based on these results, we can formulate the learning of the spectral representation, or variants of it, as the key approach to solve LP2.

16.5 LP2: Learning Features and Representation

Above we have denoted the full MD system configuration $\mathbf{x}$ and $\mathbf{y}$ the latent space representation in which linear propagators are used. We have seen that there is a special representation $\mathbf{y}^s$. In general there may be a whole pipeline of transformations, e.g.,

$$\mathbf{x} \rightarrow \mathbf{x}^f \rightarrow \mathbf{y} \rightarrow \mathbf{y}^s,$$

where the first step is a featurization from full configurations $\mathbf{x}$ to features, e.g., the selection of solute coordinates or the transformation to internal coordinates such as distances or angles. On the latent space side $\mathbf{y}$ we may have a handcrafted or a learned spectral representation. Instead of considering these transformations individually, we may construct a direct end-to-end learning framework that performs multiple transformation steps.

To simplify notation, we commit to the following notation: $\mathbf{x}$ coordinates are the input to the learning algorithm, whether these are full Cartesian coordinates of the MD system or already transformed by some featurization. $\mathbf{y}$ are coordinates in the latent space representations that are the output of LP2, $\mathbf{y} = E(\mathbf{x})$. We only explicitly distinguish between different stages within configuration or latent space (e.g., $\mathbf{y}$ vs $\mathbf{y}^s$) when this distinction is explicitly needed.

16.5.1 Suitable and Unsuitable Loss Functions

We first ask: What is the correct formulation for LP2? More specifically: which of the loss functions introduced in LP1 above are compatible with LP2? Looking at the sequence of learning problems:

$$\mathbf{x} \overset{LP2}{\rightarrow} \mathbf{y} \overset{LP1}{\rightarrow} \mathbf{P}.$$

It is tempting to concatenate them to an end-to-end learning problem and try to solve it by minimizing any of the three losses defined for learning of $\mathbf{P}$ in Sect. 16.3. However, if we make the encoding $\mathbf{y} = E(\mathbf{x})$ sufficiently flexible, we find that only one of the loss functions remains as being suitable for end-to-end learning, while two others must be discarded as they have trivial and useless minima:

Likelihood Loss The theoretical minimum of the likelihood loss (16.19) is equal to 0 and is achieved if all $p_{ij} \equiv 1$ for the transitions observed in the dataset. However, this maximum can be trivially achieved by learning a representation that assigns all microstates to a single state, e.g., the first state:

$$\arg\max_{E,P} \mathcal{L}_{\mathrm{ML}}(\mathbf{P}; \{E(\mathbf{x}_t)\}) = \begin{pmatrix} E(\mathbf{x}) \equiv & 1 \\ \mathbf{P} = & \begin{pmatrix} 1 & 0 & \cdots & 0 \\ n/a & \cdots & \cdots & n/a \\ \vdots & & & \vdots \end{pmatrix} \end{pmatrix}.$$

Maximizing the transition matrix likelihood while varying the encoding E is therefore meaningless.

Regression Loss A similar problem is encountered with the regression loss. The theoretical minimum of (16.23) is equal to 0 and is achieved when $\mathbf{y}_{t+\tau} \equiv \mathbf{P}^{\top}\mathbf{y}_t$ for all t. This, can be trivially achieved by learning the uninformative representation:

$$\arg\max_{E,P} \mathcal{L}_{\mathrm{LSQ}}(\mathbf{P}; \{E(\mathbf{x}_t)\}) = \begin{pmatrix} E(\mathbf{x}) \equiv 1 \\ \mathbf{P} \quad = \mathbf{Id} \end{pmatrix}.$$

Minimizing the propagator least squares error while varying the encoding E is therefore meaningless. See also discussion in [76].

Variational Loss The variational loss (VAC or VAMP) does not have trivial minima. The reason is that, according to the variational principles [66, 127], the variational optimum coincides with the approximation of the dynamical components. A trivial encoding such as $E(\mathbf{x}) \equiv 1$ only identifies a single component and is therefore variationally suboptimal. The variational loss is thus the only choice among the losses described in LP1 that can be used to learn both $\mathbf{y}$ and $\mathbf{P}$ in an end-to-end fashion.

16.5.2 Feature Selection

We next address the problem of learning optimal feature vectors $\mathbf{x}^f$. We can view this problem as a feature selection problem, i.e. we consider a large potential set of features and ask which of them leads to an optimal model of the long-time MD. In this view, learning the featurization is a model selection problem that can be solved by minimizing the validation loss.

We can solve this problem by employing the variational losses as follows:

1. Compute the eigenvectors $\mathbf{U}$ (Eq. (16.27)) or the singular vectors $\mathbf{U}, \mathbf{V}$ (Eq. (16.35)) directly from the training set $\mathbf{X}^{\mathrm{train}} = \left(\mathbf{x}_0^f, \ldots, \mathbf{x}_T^f\right)^{\top}$.
2. Using $\mathbf{U}$ or $\mathbf{U}, \mathbf{V}$, transform both training and test trajectories into spectral representation (Eq. (16.29) or (16.36)–(16.37)).

3. Compute VAC or VAMP validation scores as $\mathcal{L}_{\text{VAC}}(\mathbf{U}^{\text{train}}; \{\mathbf{y}_t^{\text{test}}\})$ (Eqs. (16.30), (16.32)) or $\mathcal{L}_{\text{VAMP}}(\mathbf{U}^{\text{train}}, \mathbf{V}^{\text{train}}; \{\mathbf{y}_t^{\text{test}}\})$ (Eqs. (16.38), (16.40)).

In [95] we follow this feature selection procedure using VAMP-2 validation for describing protein folding. We find that a combination of torsion backbone angles and $\exp(-d_{ij})$ with d_{ij} being the minimum distances between amino acids performs best among a large set of candidate features.

16.5.3 Blind Source Separation and TICA

For a given featurization, a widely used linear learning method to obtain the spectral representation is an algorithm first introduced in [61] as a method for blind source separation that later became known as time-lagged independent component analysis (TICA) method [35, 62, 80, 102], sketched in Algorithm 3. In [80], it was shown that the TICA algorithm directly follows from the minimization of the VAC variational loss ((16.31), (16.33)) to best approximate the Markov operator eigenfunctions by a linear combination of input features. As a consequence, TICA approximates the eigenvalues and eigenfunctions of Markov operators that obey detailed balance (16.21), and therefore approximates the slowest relaxation processes of the dynamics.

Algorithm 3 performs a symmetrized estimation of covariance matrices in order to guarantee that the eigenvalue spectrum is real. In most early formulations, one usually symmetrizes only $\mathbf{C}_{0\tau}$ while computing $\mathbf{C}_{00}$ by (16.12), which is automatically symmetric. However these formulations might lead to eigenvalues larger than 1, which do not correspond to any meaningful relaxation timescale in the present context—this problem is avoided by the step 1 in Algorithm 3 [130]. Note that symmetrization of $\mathbf{C}_{0\tau}$ introduces an estimation bias if the data is non-stationary, e.g., because short MD trajectories are used that have not been started from the equilibrium distribution. To avoid this problem, please refer to Ref. [130] which introduces the Koopman reweighting procedure to estimate symmetric covariance matrices without this bias, although at the price of an increased estimator variance.

Furthermore, the covariance matrices in step 1 of Algorithm 3 are computed after removing the mean. Removing the mean has the effect of removing the eigenvalue 1 and the corresponding stationary eigenvector, hence all components return by Algorithm 3 approximate dynamical relaxation processes with finite relaxation timescales estimates according to Eq. (16.9).

The TICA propagator can be directly computed as $\bar{\mathbf{P}} = \bar{\mathbf{C}}_{00}^{-1}\bar{\mathbf{C}}_{0\tau}$, and is a least squares result in the sense of Sect. 16.3.4. Various extensions of the TICA algorithm were developed: Kernel formulations of TICA were first presented in machine learning [32] and later in other fields [103, 123]. An efficient way to solve TICA for multiple lag times simultaneously was introduced as TDSEP [135, 136]. Efficient computation of TICA for very large feature sets can be performed with a hierarchical decomposition [79], or a compressed sensing approach [53]. TICA is closely related to the dynamic mode decomposition (DMD) [90, 96, 97, 117] and the subsequently developed extended dynamic mode decomposition (EDMD) algorithms [124]. DMD approximates the left eigenvectors ("modes") instead of the Markov operator eigenfunctions described here. EDMD is algorithmically identical to VAC/TICA, but is in practice also used for dynamics that do not fulfill detailed balance (16.21), although this leads to complex-valued eigenfunctions that are more difficult to interpret and work with.

Algorithm 3 TICA($\{\mathbf{x}_t\}, \tau, n$)

1. Compute symmetrized mean free covariance matrices

$$\bar{\mathbf{C}}_{00} = \lambda\mathbf{I} + \sum_{t=1}^{T-\tau}(\mathbf{x}_t - \boldsymbol{\mu}_0)(\mathbf{x}_t - \boldsymbol{\mu}_0)^\top + (\mathbf{x}_{t+\tau} - \boldsymbol{\mu}_\tau)(\mathbf{x}_{t+\tau} - \boldsymbol{\mu}_\tau)^\top$$

$$\bar{\mathbf{C}}_{0\tau} = \sum_{t=1}^{T-\tau}(\mathbf{x}_t - \boldsymbol{\mu}_0)(\mathbf{x}_{t+\tau} - \boldsymbol{\mu}_\tau)^\top + (\mathbf{x}_{t+\tau} - \boldsymbol{\mu}_\tau)(\mathbf{x}_t - \boldsymbol{\mu}_0)^\top$$

   with means $\boldsymbol{\mu}_0, \boldsymbol{\mu}_\tau$ defined analogously as in (16.10)–(16.11), where λ is an optional ridge parameter.
2. Compute the largest n Eigenvalues and Eigenvectors of:

$$\bar{\mathbf{C}}_{0\tau}\mathbf{u}_i = \lambda_i \bar{\mathbf{C}}_{00}\mathbf{u}_i$$

3. Project to spectral representation: $\mathbf{y}_t = \left(\mathbf{x}_t^\top \mathbf{u}_1, \ldots, \mathbf{x}_t^\top \mathbf{u}_n\right)$ for all t
4. Return $\{\mathbf{y}_t\}$

16.5.4 TCCA/VAMP

When the dynamics do not satisfy detailed balance (16.21), e.g., because they are driven by an external force or field, the TICA algorithm is not meaningful, as it will not even in the limit of infinite data approximate the true spectral representation. If detailed balance holds for the dynamical equations, but the data is non-stationary, i.e. because short simulation trajectories started from a non-equilibrium distribution are used, the symmetrized covariance estimation in Algorithm 3 introduces a potentially large bias.

These problems can be avoided by going from TICA to the time-lagged or temporal canonical correlation analysis (TCCA, Algorithm 4) [8], which can be viewed as a direct implementation of the VAMP approach [127], i.e. it results from minimizing the VAMP variational loss ((16.39),(16.41)), when approximating the Markov operator singular functions with a linear combination of features. The TCCA algorithm performs a canonical correlation analysis (CCA) applied to time series. The price of using TCCA instead of TICA is that the interpretation of the embedding is less straightforward: in the detailed balance case, the Markov operator eigenvalues and their variational approximation are directly related to relaxation timescales, and thus one finds a "slow" embedding. The singular values maximized in TCCA have no direct relation to a relaxation timescale, TCCA rather approximates the Markov operator at time step τ with minimal error [127].

TCCA returns two sets of features approximating the left and right singular functions of the Markov operator and that can be interpreted as the optimal spectral representation to characterize state of the system "before" and "after" the transition with time step τ. For non-stationary dynamical systems, these representations are valid for particular points in time, t and $t+\tau$ [47].

VAMP/TCCA as a method to obtain a low-dimensional spectral representation of the long-time MD is discussed in detail in [78], where the algorithm is used to identify low-dimensional embeddings of driven dynamical systems, such as an ion channel in an external electrostatic potential.

Algorithm 4 TCCA($\{\mathbf{y}_t\}, \tau, n$)

1. Compute covariance matrices $\mathbf{C}_{00}, \mathbf{C}_{0\tau}, \mathbf{C}_{\tau\tau}$ from $\{\mathbf{x}_t\}$, as in Eqs. ((16.12)–(16.14)) or Eqs. ((16.15), (16.16)).
2. Perform the truncated SVD:

$$\tilde{\mathbf{P}} = \mathbf{C}_{00}^{-\frac{1}{2}} \mathbf{C}_{0\tau} \mathbf{C}_{\tau\tau}^{-\frac{1}{2}} \approx \mathbf{U}'\mathbf{S}\mathbf{V}'^{\top}$$

   where $\tilde{\mathbf{P}}$ is the propagator for the representations $\mathbf{C}_{00}^{-\frac{1}{2}}\mathbf{x}_t$ and $\mathbf{C}_{\tau\tau}^{-\frac{1}{2}}\mathbf{x}_{t+\tau}$, $\mathbf{S}$=diag$(s_1, \ldots, s_k)$ is a diagonal matrix of the first k singular values that approximate the true singular values $\sigma_1, \ldots, \sigma_k$, and $\mathbf{U}'$ and $\mathbf{V}'$ consist of the k corresponding left and right singular vectors respectively.
3. Compute $\mathbf{U} = \mathbf{C}_{00}^{-\frac{1}{2}}\mathbf{U}'$, $\mathbf{V} = \mathbf{C}_{\tau\tau}^{-\frac{1}{2}}\mathbf{V}'$
4. Project to spectral representation: $\mathbf{y}_t^0 = \left(\mathbf{x}_t^\top \mathbf{u}_1, \ldots, \mathbf{x}_t^\top \mathbf{u}_n\right)$ and $\mathbf{y}_t^\tau = \left(\mathbf{x}_t^\top \mathbf{v}_1, \ldots, \mathbf{x}_t^\top \mathbf{v}_n\right)$ for all t
5. Return $\{(\mathbf{y}_t^0, \mathbf{y}_t^\tau)\}$

16.5.5 MSMs Based on Geometric Clustering

For the spectral representations found by TICA and TCCA, a propagator $\mathbf{P}(\tau)$ can be computed by means of Eq. (16.6); however, this propagator is harder to interpret than a MSM propagator whose elements correspond to transition probabilities between states. For this reason, TICA, TCCA, and other dimension reduction algorithms are frequently used as a first step towards building an MSM [79,80,102]. Before TICA and TCCA were introduced into the MD field, MSMs were directly built upon manually constructed features such as distances, torsions, or in other metric spaces that define features only indirectly, such as the pairwise distance of aligned molecules [43,112]—see Ref. [40] for an extensive discussion.

In this approach, the trajectories in feature space, $\{\mathbf{x}_t^f\}$, or in the representation $\{\mathbf{y}_t\}$, must be further transformed into a one-hot encoding (16.17) before the MSM can be estimated via one of the methods described in Sect. 16.3. In other words, the configuration space must be divided into n sets that are associated with the n MSM states. Typically, clustering methods group simulation data by means of geometric similarity. When MSMs were built on manually constructed feature spaces, research on suitable clustering methods was very active [1, 9, 11, 12, 16, 39, 42, 44, 67, 94, 105, 109, 132]. Since the introduction of TICA and TCCA that identify a spectral representation that already approximates the leading eigenfunctions, the choice of the clustering method has become less critical, and simple methods such as k-means^{++} lead to robust results [39,41,94]. The final step towards an easily interpretable MSM is coarse-graining of $\mathbf{P}$ down to a few states [23,28,38,48,70,75,133].

The geometric clustering step introduces a different learning problem and objective whose relationship to the original problem of approximating long-term MD is not clear. Therefore, geometric clustering must be at the moment regarded as a pragmatic approach to construct an MSM from a given embedding, but this approach departs from the avenue of a well-defined machine learning problem.

16.5.6 VAMPnets

VAMPnets [55] were introduced to replace the complicated and error-prone approach of constructing MSMs by (1) searching for optimal features $\mathbf{x}^f$, (2) combining them to a representation $\mathbf{y}$, e.g., via TICA, (3) clustering it, (4) estimating the transition matrix $\mathbf{P}$, and (5) coarse-graining it, by a single end-to-end learning approach in which all of these steps are replaced by a deep neural network. This is possible because with the VAC and VAMP variational principles, loss functions are available that are suitable to train the sequence of learning problems 1 and 2 simultaneously. A similar architecture is used by EDMD with dictionary learning [51], which avoids the problem of the regression error to collapse to trivial encodings E (Sect. 16.5.1) by fixing some features that are not learnable.

VAMPnets contain two network lobes that transform the molecular configurations found at a time delay τ along the simulation trajectories (Fig. 16.2a). VAMPnets can be minimized with any VAC or VAMP variational loss. In Ref. [55], the VAMP-2 loss (16.41) was used, which is meaningful for both dynamics with and without detailed balance. When detailed balance (16.22) is enforced in the propagator obtained by (16.6), the loss function automatically becomes VAC-2. VAMPnets may either use two distinct network lobes to encode the spectral representation of the left and right singular functions (which is important for non-stationary dynamics [46, 47]), whereas for MD with a stationary distribution we generally use parameter sharing and have two identical lobes. For dynamics with detailed balance, the VAMPnet output then encodes the space of the dominant Markov operator eigenfunctions (Fig. 16.3b).

In order to obtain a propagator that can be interpreted as an MSM, [55] chose to use a Softmax layer as an output layer, thus transforming the spectral representation to a soft indicator function similar to spectral clustering methods such as PCCA+ [19, 87]. As a result, the propagator computed by Eq. (16.6) is *almost* a transition

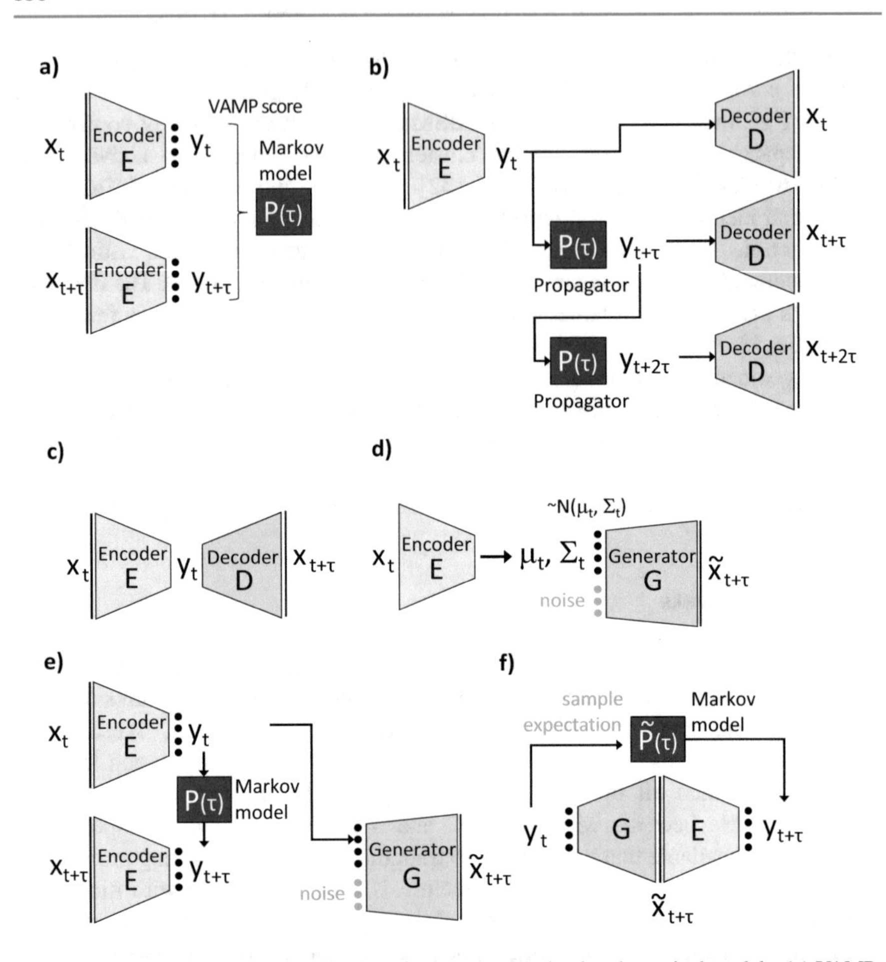

Fig. 16.2 Overview of network structures for learning Markovian dynamical models. (**a**) VAMPnets [55]. (**b**) Time-autoencoder with propagator [54, 76]. (**c**) Time-autoencoder (TAE) [121]. (**d**) Variational time-encoder [34]. (**e**) Deep Generative Markov State Models [131]. (**f**) The rewiring trick to compute the propagator **P** for a deep generative MSM

matrix. It is guaranteed to be a true transition matrix in the limit where the output layer performs a hard clustering, i.e. one-hot encoding (16.17). Since this is not true in general, the VAMPnet propagator may still have negative elements, but these are usually very close to zero. The propagator is still valid for transporting probability distributions in time and can therefore be interpreted as an MSM between metastable states (Fig. 16.4d).

The results described in [55] (see, e.g., Figs. 16.3, and 16.4) were competitive with and sometimes surpassed the state-of-the-art handcrafted MSM analysis pipeline. Given the rapid improvements of training efficiency and accuracy of deep neural networks seen in a broad range of disciplines, we expect end-to-end learning approaches such as VAMPnets to dominate the field eventually.

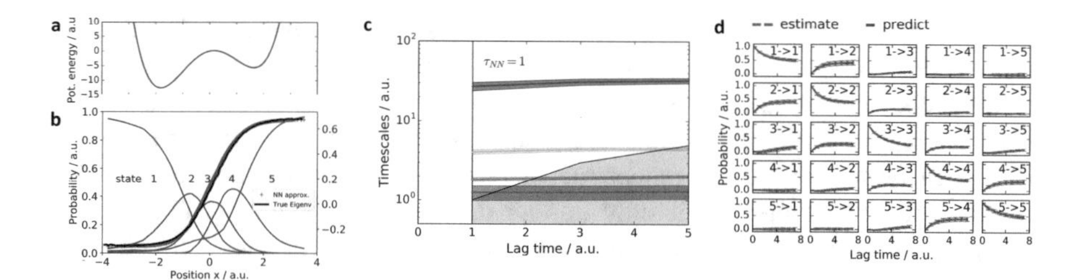

Fig. 16.3 Figure adapted from [55]: Approximation of the slow transition in a bistable potential by a VAMPnet with one input node (x) and five output nodes. (**a**) Potential energy function $U(x) = x^4 - 6x^2 + 2x$. (**b**) Eigenvector of the slowest process calculated by direct numerical approximation (black) and approximated by a VAMPnet with five output nodes (red). Activation of the five Softmax output nodes defines the state membership probabilities (blue). (**c**) Relaxation timescales computed from the Koopman model using the VAMPnet transformation. (**d**) Chapman–Kolmogorov test comparing long-time predictions of the Koopman model estimated at $\tau = 1$ and estimates at longer lag times. Panels (**c**) and (**d**) report 95% confidence interval error bars over 100 training runs

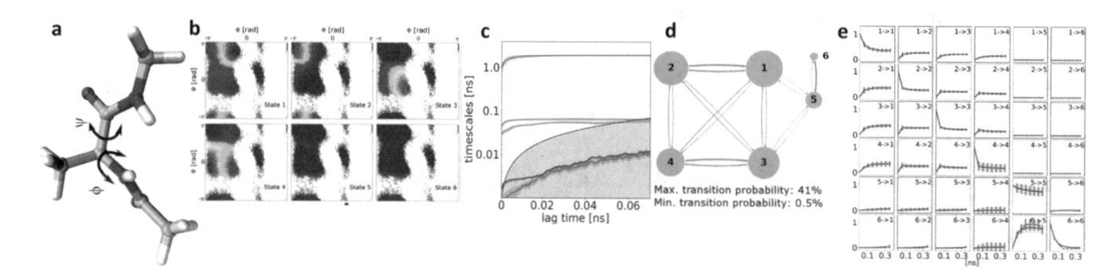

Fig. 16.4 Figure adapted from [55]: Kinetic model of alanine dipeptide obtained by a VAMPnet with 30 input nodes (x, y, z Cartesian coordinates of heavy atoms) and six output nodes. (**a**) Structure of alanine dipeptide. The main coordinates describing the slow transitions are the backbone torsion angles ϕ and ψ; however, the neural network inputs are only the Cartesian coordinates of heavy atoms. (**b**) Assignment of all simulated molecular coordinates, plotted as a function of ϕ and ψ, to the six Softmax output states. Color corresponds to activation of the respective output neuron, indicating the membership probability to the associated metastable state. (**c**) Relaxation timescales computed from the Koopman model using the neural network transformation. (**d**) Representation of the transition probabilities matrix of the Koopman model; transitions with a probability lower than 0.5% have been omitted. (**e**) Chapman–Kolmogorov test comparing long-time predictions of the Koopman model estimated at $\tau = 50$ ps and estimates at longer lag times. Panels (**c**) and (**e**) report 95% confidence interval error bars over 100 training runs excluding failed runs

16.6 LP3 Light: Learn Representation and Decoder

As discussed in Sect. 16.5.1, end-to-end learning combining LP1 and LP2 are limited in their choice of losses applied to the propagator resulting from LP2: Variational losses can be used, leading to the methods described in Sect. 16.5, while using the likelihood and regression losses are prone to collapse to a trivial representation that does not resolve the long-time dynamical processes.

One approach to "rescue" these approaches is to add other loss functions to prevent this collapse to a trivial, uninformative representation from happening. An obvious choice is to add a decoder that is trained with some form of reconstruction loss: the representation $\mathbf{r}$ should still contain enough information that the input ($\mathbf{x}$ or $\mathbf{y}$) can be approximately reconstructed. We discuss several approaches based on this principle. Note that if only finding the spectral embedding and learning the propagator $\mathbf{P}$ is the objective, VAMPnets solve this problem directly and employing a reconstruction loss unnecessarily adds the difficult inverse problem of reconstructing a high-dimensional variable from a low-dimensional one. However, approximate reconstruction of inputs may be desired in some applications, and is the basis for LP3.

16.6.1 Time-Autoencoder

The time-autoencoder [121] shortcuts LP2 and constructs a direct learning problem between $\mathbf{x}_t$ and $\mathbf{x}_{t+\tau}$ (Fig. 16.2c).

$$\mathbf{x}_t \overset{E}{\longrightarrow} \mathbf{y}_? \overset{D}{\longrightarrow} \mathbf{x}_{t+\tau}. \tag{16.46}$$

The time-autoencoder is trained by reconstruction loss:

$$\mathcal{L}_{\mathrm{TAE}}(E, D; \{\mathbf{x}_t\}) = \sum_{t=0}^{T-\tau} \|\mathbf{x}_{t+k\tau} - D\left(E(\mathbf{x}_t)\right)\|, \tag{16.47}$$

where $\|\cdot\|$ is a suitable norm, e.g., the squared 2-norm.

The TAE has an interesting interpretation: If E and D are linear transformations, i.e. encoder and decoder matrices $\mathbf{E} \in \mathbb{R}^{N\times n}$, $\mathbf{D} \in \mathbb{R}^{n\times N}$, the minimum of (16.47) is found by VAMP/TCCA, and for data that is in equilibrium and obeys detailed balance by VAC/TICA [121]. The reverse interpretation is not true: the solution found by minimizing (16.47) does not lead to TICA/TCCA modes, as there is no constraint in the time-autoencoder for the components $\mathbf{r}_t$—they only span the same space. Within this interpretation, the time-autoencoder can be thought of a nonlinear version of TCCA/TICA in the sense of being able to find a slow but nonlinear spectral representation.

Time-autoencoders have several limitations compared to VAMPnets: (1) Adding the decoder network makes the learning problem more difficult. (2) As indicated in scheme (16.46), it is not clear what the time step pertaining to the spectral representation $\mathbf{y}$ is (t, $t+\tau$, or something in between), as the time stepping is done throughout the entire network. (3) Since the decoding problem from any given $\mathbf{y}$ to $\mathbf{x}_{t+\tau}$ is underdetermined but the decoder network D is deterministic, it will only be able to decode to a "mean" $\mathbf{x}$ for all $\mathbf{x}$ mapping to the same $\mathbf{y}$. Thus, time-autoencoders cannot be used to sample the transition density (16.3) to generated sequences $\mathbf{x}_t \rightarrow \mathbf{x}_{t+\tau}$.

16.6.2 Time-Autoencoder with Propagator

Both [54, 76] have introduced time-autoencoders that additionally learn the propagator in the spectral representation, and thus fix problem (2) of time-autoencoders, while problems (1) and (3) still remain.

Instead of scheme (16.46), time-autoencoders with propagator introduce a time-propagation step that makes the time step explicit for every step:

$$\mathbf{x}_t \xrightarrow{E} \mathbf{y}_t \xrightarrow{\mathbf{P}} \mathbf{y}_{t+\tau} \xrightarrow{D} \mathbf{x}_{t+\tau}, \tag{16.48}$$

where $\mathbf{P}$ is the matrix defined by a $n \times n$ linear layer. Training this network exclusively with the standard autoencoder loss would not impose the correct internal structure—in particular, it would not be possible to control that E learns only the representation and $\mathbf{P}$ performs the time step. Lusch et al. [54] and Otto and Rowley [76] enforce the dynamical consistency by training several lag times simultaneously with variants of the following type of loss:

$$\mathcal{L}_{\text{TAE-P}} = \sum_{t=0}^{T-k\tau} \left(\sum_{k=0}^{K} \alpha_k \left\| \mathbf{x}_{t+k\tau} - D\left(\mathbf{P}^k E(\mathbf{x}_t)\right) \right\| + \sum_{k=1}^{K} \beta_k \left\| E(\mathbf{x}_{t+k\tau}) - \mathbf{P}^k E(\mathbf{x}_t) \right\| \right), \tag{16.49}$$

where α_k, β_k are coefficients, the first term correspond to a autoencoder reconstruction loss and the second term trains the correct time-propagation of $\mathbf{P}$ in latent space. The number of lag times, K, to be considered is a user-defined choice. Note that it is not a typical hyper-parameter as matching the dynamics at more lag times makes the learning problem harder, and thus the cross-validation score of (16.49) cannot be used to select K. Unrolling the network for $K = 2$ results in Fig. 16.2b. This approach works excellently in deterministic (but highly nonlinear) dynamical systems with short time steps [54, 76].

In stochastic systems such as MD, it appears more difficult to learn $\mathbf{r}_t$ and $\mathbf{P}$ such that they span the spectral components of the underlying propagator and recover its largest eigenvalues. While this observation needs more study, potential explanations are that in long-time MD we need large time steps τ, in order to make the spectral representation learning problem low-dimension (see Sect. 16.4.1), and that the stochastic fluctuations are large which makes learning a decoder D difficult.

16.6.3 Variational (Time-)Autoencoders

Several recent approaches employ variational autoencoders (VAEs) for the long-time MD or related learning problems. Variational autoencoders [45] learn to sample a probability distribution that approximates the distribution underlying observation data. To this end, VAEs employ variational Bayesian inference [25] in order to approximately minimize the KL divergence between the generated and the observed distribution. VAEs have a similar structure as usual autoencoders, with an inference network mapping from a high-dimensional variable $\mathbf{x}$ to a typically lower-dimensional latent variable $\mathbf{r}$, and attempting to reconstruct $\mathbf{x}$ in a decoder network. The main difference is that every latent point $\mathbf{r}$ encodes the moments of a distribution which are used to sample $\mathbf{x}$ such that the distributions become similar.

VAEs have been used in RAVE [86] for enhancing the sampling by identifying a space of "reaction coordinates" in which MD sampling can be efficiently driven, and in Autograin [119] to find a way to coarse-grain a molecule into effective beads. Both methods use VAEs without an inference network that employs a time step τ, and therefore they address learning problems that are conceptually different from long-time MD learning problem as treated here.

A much more closely related work are variational time-encoders [34] (Fig. 16.2d), which employ a VAE between time steps $\mathbf{x}_t$ at the input and $\mathbf{x}_{t+\tau}$ at the output:

$$\begin{array}{c} \mathbf{x}_t \xrightarrow{E} \mu(\mathbf{x}_t) \rightarrow \mathbf{y}_t \rightarrow \oplus \rightarrow \mathbf{y}_{t+\tau} \xrightarrow{D} \mathbf{x}_{t+\tau} \\ \uparrow \\ \mathcal{N}(0,1) \end{array}$$

As [34] notes, this approach does not achieve the sampling of the $\mathbf{x}_{t+\tau}$ distribution (the variational theory underlying VAEs requires that the same type of variable is used at input and output) and hence does not act as a propagator $\mathbf{x}_t \rightarrow \mathbf{x}_{t+\tau}$, but succeeds in learning a spectral representation of the system. For this reason, the variational time-encoder is listed in this section rather than in LP3.

16.7 LP3 Heavy: Learn Generative Models

The full solution of LP3 involves learning to generate samples $\mathbf{x}_{t+\tau}$ from the lower-dimensional feature embedding or spectral representation. This is a very important goal as its solution would yield an ability to sample the MD propagator $\mathbf{x}_t \rightarrow \mathbf{x}_{t+\tau}$ at long time steps τ, which would yield a very efficient simulator. However, because of the high dimensionality of configuration space and the complexity of distributions there, this aim is extremely difficult and still in its infancy.

Clearly standard tools for learning directed generative networks, such as Variational Autoencoders [45] and generative adversarial nets [29] are "usual suspects" for the solution of this problem. However, existing applications of VAEs and GANs on the long-time MD problem have focused on learning a latent representation that is suitable to encode the long-time processes or a coarse-graining, and the decoder has been mostly used to regularize the problem (Sect. 16.6.3). The first approach to actually reconstruct molecular structures in configuration space, so as to achieve long time step sampling, was made in [131], which will be analyzed in some detail below.

16.7.1 Deep Generative MSMs

The deep generative MSMs described [131] (Fig. 16.2e), we propose to address LP1-3 in the following manner. We first formulate a machine learning problem to learn the following two functions:

- A probabilistic encoding of the input configuration to a low-dimensional latent space, $\mathbf{x}_t \rightarrow E(\mathbf{x}_t)$. Similar to VAMPnets with a probabilistic output (Sect. 16.5.6), E has n elements, and each element represents the probability of configuration $\mathbf{x}$ to be in a metastable (long-lived) state i:

$$E_i(\mathbf{x}) = \mathbb{P}(\mathbf{x}_t \in \text{state } i \mid \mathbf{x}_t = \mathbf{x}).$$

  Consequently, these functions are nonnegative ($E_i(x) \geq 0 \;\forall x$) and sum up to one ($\sum_i E_i(x) = 1 \;\forall x$). The functions $E(x)$ can, e.g., be represented by a neural network mapping from $\mathbb{R}^d$ to $\mathbb{R}^m$ with a Softmax output layer.
- An n-element probability distribution $\mathbf{q}(\mathbf{x};\tau) = (q_1(\mathbf{x};\tau), \ldots, q_n(\mathbf{x};\tau))$, which assigns to each configuration $\mathbf{x}$ a probability density that a configuration that was in metastable state i at time t will "land" in $\mathbf{x}$ at time $t+\tau$:

$$q_i(\mathbf{x};\tau) = \mathbb{P}(\mathbf{x}_{t+\tau} = \mathbf{x} | \mathbf{x}_t \in \text{state } i).$$

  We thus briefly call these densities "landing densities."

  Schematically, deep generative MSMs treat LP1-3 in the way:

$$\begin{array}{ccc} \mathbf{x}_t & \overset{E}{\rightarrow} & \mathbf{y}_t \\ & \swarrow & \\ & q & \\ \mathbf{x}_{t+\tau} & & \end{array}$$

Deep generative MSMs represent the transition density (16.3) in the following form (Fig. 16.2e):

$$p_\tau(\mathbf{x}_{t+\tau}|\mathbf{x}_t) = E(\mathbf{x}_{t+\tau})^\top \mathbf{q}(\mathbf{y};\tau) = \sum_{i=1}^{m} E_i(\mathbf{x}_t) q_i(\mathbf{x}_{t+\tau};\tau). \tag{16.50}$$

To work with this approach we finally need a generator G, which is a structure that samples from the density $\mathbf{q}$:

$$G(i, \epsilon; \tau) = \mathbf{y} \sim q_i(\mathbf{y};\tau) \tag{16.51}$$

It appears that deep generative MSMs do not learn the propagator explicitly. However, the propagator can be obtained from E and $\mathbf{q}$ by means the "rewiring" trick (Fig. 16.2f): By exchanging the order in which E and G are applied and then computing the propagator $\mathbf{P}$ as a sample average over $\mathbf{q}$, obtained from repeatedly applying the generator:

$$p_{ij}(\tau) = \mathbb{E}_G \left[E_j \left(G(i, \epsilon; \tau) \right) \right]. \tag{16.52}$$

In contrast to VAMPnets (Sect. 16.5.6), it is guaranteed that the propagator (16.52) is a true transition matrix with nonnegative elements.

16.7.2 Deep Resampling MSMs

We first describe a very simple generator that generates no new (unseen) configurations, but only learns a function $\mathbf{q}$ that can be used to resample known configurations [131]. While this approach is clearly limited, it has two advantages: it will not generate any illegal configuration, and it can be trained with maximum likelihood. For this approach, we model the landing densities by

$$q_i(\mathbf{x}_{t+\tau}) = \frac{w(\mathbf{x}_{t+\tau})\gamma_i(\mathbf{x}_{t+\tau})}{\sum_{s=0}^{T-\tau} w(\mathbf{x}_{s+\tau})\gamma_i(\mathbf{x}_{s+\tau})}. \tag{16.53}$$

Where $\gamma_i(\mathbf{x}_{t+\tau})$ is a trainable, unnormalized density function and w is an additional weight function which may be employed to change the weights of configurations, but is usually identical to 1. In [131], $\gamma_i(\mathbf{y})$ is a deep neural network that receives $\mathbf{y}$ as an input as well as the condition i by means of a one-hot-encoding with n input units, and has a single output node encoding the probability weight. The normalized density $\mathbf{q}$ is computed by evaluating the γ-network for all configurations at time points $\tau, \ldots, T$ and then normalizing over all time points.

Deep resampling MSMs can be trained by maximizing the likelihood based on expression (16.50), resulting in the following loss function:

$$\begin{aligned}
\mathcal{L}_{\text{DeepResampleMSM}} &= \sum_{t=1}^{T-\tau} p_\tau(\mathbf{x}_{t+\tau}|\mathbf{x}_t) \\
&= \sum_{t=1}^{T-\tau} \sum_{i=1}^{m} E_i(\mathbf{x}_t) q_i(\mathbf{x}_{t+\tau}; \tau)
\end{aligned}$$

where q_i is evaluated by (16.53). Alternatively, we can optimize χ_i and γ_i using the Variational Approach for Markov Processes (VAMP) [127]. However, we found the ML approach to perform significantly better in [131].

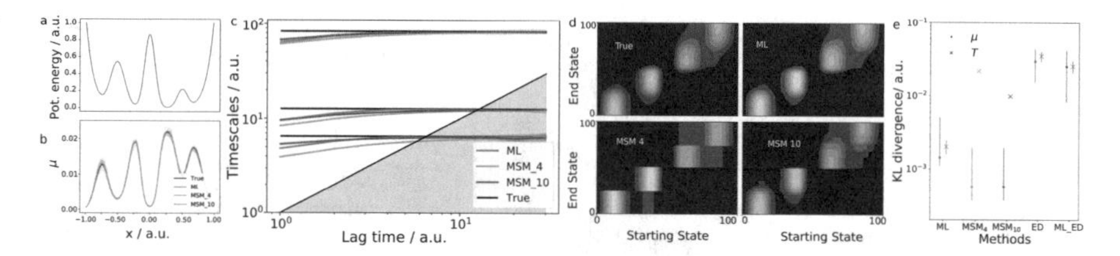

Fig. 16.5 Reproduced from [131]: Performance of deep versus standard MSMs for diffusion in the Prinz potential. (**a**) Potential energy as a function of position x. (**b**) Stationary distribution estimates of all methods with the exact distribution (black). (**c**) Implied timescales of the Prinz potential compared to the real ones (black line). (**d**) True transition density and approximations using maximum likelihood (ML) DeepResampleMSM, four and ten state MSMs. (**e**) KL divergence of the stationary and transition distributions with respect to the true ones for all presented methods (also DeepResampleMSM)

In deep resample MSMs, the propagator according to (16.52) becomes simply:

$$\mathbf{P} = \frac{1}{T-\tau} \sum_{t=1}^{T-\tau} \mathbf{q}(\mathbf{x}_{t+\tau}) E(\mathbf{x}_{t+\tau})^{\top}. \tag{16.54}$$

Deep resample MSMs were found to accurately reproduce the eigenfunctions and dominant relaxation timescales of benchmark examples [131], and learn to represent the transition density in configuration space (Fig. 16.5).

16.7.3 Deep Generative MSMs with Energy Distance Loss

In contrast to resampling MSMs, we now want to develop generative MSMs, which can produce genuinely new configurations. This makes the method promising for performing active learning in MD [10, 81], and to predict the future evolution of the system in other contexts. To this end, we train a directed generative network to represent (16.51). Such a generator can be trained with various principles, e.g. by means of a variational autoencoder or with adversarial training [29, 45]. In [131], we found that a third principle works well: training the generator G by minimizing the conditional energy distance (ED). The standard ED, introduced in [110], is a metric between the distributions of random vectors, defined as

$$D_E\left(\mathbb{P}(\mathbf{x}), \mathbb{P}(\mathbf{y})\right) = \mathbb{E}\left[2\left\|\mathbf{x}-\mathbf{y}\right\| - \left\|\mathbf{x}-\mathbf{x}'\right\| - \left\|\mathbf{y}-\mathbf{y}'\right\|\right] \tag{16.55}$$

for two real-valued random vectors $\mathbf{x}$ and $\mathbf{y}$. $\mathbf{x}', \mathbf{y}'$ are independently distributed according to the distributions of $\mathbf{x}, \mathbf{y}$. Based on this metric, we introduce the conditional energy distance between the transition density of the system and that

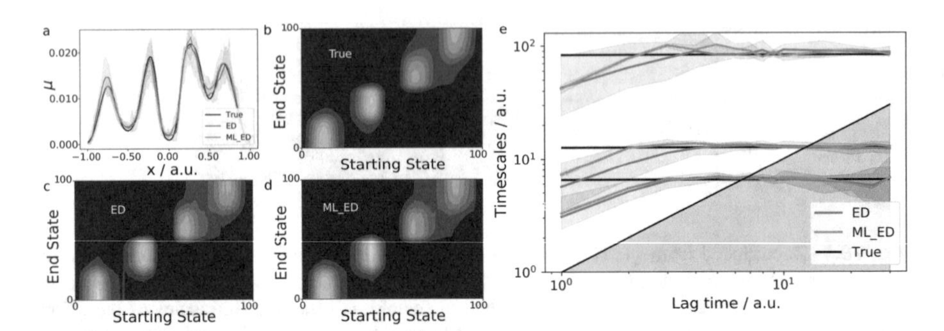

Fig. 16.6 Reproduced from [131]: Performance of deep generative MSMs for diffusion in the Prinz Potential. Comparison between exact reference (black), deep generative MSMs estimated using only energy distance (ED) or combined ML-ED training. (**a**) Stationary distribution. (**b–d**) Transition densities. (**e**) Relaxation timescales

of the generative model:

$$D \triangleq \mathbb{E}\left[D_E\left(\mathbb{P}(\mathbf{x}_{t+\tau} \mid \mathbf{x}_t), \mathbb{P}(\hat{\mathbf{x}}_{t+\tau} \mid \mathbf{x}_t)\right) \mid \mathbf{x}_t\right] \tag{16.56}$$
$$= \mathbb{E}\left[2\left\|\hat{\mathbf{x}}_{t+\tau} - \mathbf{x}_{t+\tau}\right\| - \left\|\hat{\mathbf{x}}_{t+\tau} - \hat{\mathbf{x}}'_{t+\tau}\right\| - \left\|\mathbf{x}_{t+\tau} - \mathbf{x}'_{t+\tau}\right\|\right].$$

Here $\mathbf{x}_{t+\tau}$ and $\mathbf{x}'_{t+\tau}$ are distributed according to the transition density for given $\mathbf{x}_t$ and $\hat{\mathbf{x}}_{t+\tau}, \hat{\mathbf{x}}'_{t+\tau}$ are independent outputs of the generative model. Implementing the expectation value with an empirical average results in an estimate for D that is unbiased, up to an additive constant. We train G to minimize D, and subsequently estimate $\mathbf{P}$ by using the rewiring trick and sampling (16.52).

Deep generative MSMs trained with the energy distance were also found to accurately reproduce the eigenfunctions and dominant relaxation timescales of benchmark examples [131], and learn to represent the transition density in configuration space (Fig. 16.6). In contrast to resampling MSMs described in the previous section, they can also be used to generalize to sampling new, previously unseen, configurations, and are therefore a first approach to sample the long-time propagator $\mathbf{x}_t \rightarrow \mathbf{x}_{t+\tau}$ in configuration space (Fig. 16.7).

16.8 Data and Software

Many of the algorithms described above are implemented in the PyEMMA [94] software—see [122] and www.pyemma.org for a tutorials and installation instructions. Some of the deep learning algorithms can be found on Github.[1]

The field is still lacking good resources with public datasets, partially because long-time MD data of nontrivial systems is typically extremely large (giga- to terabytes), and due to the unsupervised nature of the learning problems, the role of a

[1] https://github.com/markovmodel/deeptime.

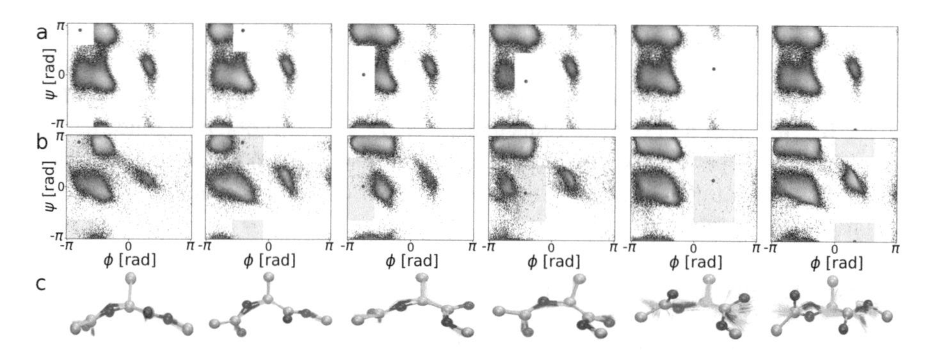

Fig. 16.7 Reproduced from [131]: DeepGenMSMs can generate physically realistic structures in areas that were not included in the training data. (**a**) Distribution of training data. (**b**) Generated stationary distribution. (**c**) Representative "real" molecular configuration (from MD simulation) in each of the metastable states (sticks and balls), and the 100 closest configurations generated by the deep generative MSM (lines)

benchmarking dataset is less straightforward as in supervised learning. Commonly used datasets for the evaluation of long-time MD models are the fast folding protein trajectories produced by D. E. Shaw research on the Anton supercomputer [52], which can be obtained from them on request. We provide datasets for small peptides via the Python package mdshare (https://markovmodel.github.io/mdshare/).

16.9 Conclusions

The present chapter attempts to establish a better connection of the MD and ML research areas by coining the SAME (Sampling, Analysis, Model, Experiment) MD problems as ML tasks and pointing out existing solutions for them. Besides the open problems mentioned *en passant* in this chapter, I would like to highlight two outstanding challenges:

Firstly, both ML research on the approximation of quantum-chemical energies and forces [5–7, 13, 14, 91, 99, 100] (see other chapters in this book), and the analysis of long-time molecular thermodynamics and kinetics described in the present chapter have made impressive strides in the last decade. However, these two areas have not yet been combined. The time is ripe to do so—ML learned force fields can be used to simulate ensembles of MD trajectories, and the methods described in this chapter can be used to learn models of their stationary and long-time behavior. This combination, that we may loosely call "Quantum Kinetics," will likely teach us much new physics, as traditional ab initio MD that attempts to compute direct approximations of the Schrödinger equation in every MD time step is prohibitively expensive, and this territory is consequently uncharted.

Secondly, most of the methods described herein are fundamentally descriptive for a given molecular system but do not make predictions for other molecular systems or even simulation conditions. The role of held-out datasets is merely to select hyper-parameters and therefore to balance over- and underfitting. As generating vast MD datasets is extremely expensive even when classical force fields are used, it would be a fundamental progress if kinetic models could be learned that are transferable across chemical space. While much research on transferable models has been done in the ML prediction of quantum-mechanical energies, this problem is much harder for intensive properties such as rates or metastable states. A possible starting point might be the treatment of metastable kinetics as the combination of local switches, as proposed recently [73].

Acknowledgments I am grateful to Brooke E. Husic and Klaus-Robert Müller for valuable comments on this chapter. Funding is acknowledged from the European Commission (ERC CoG 772230 "ScaleCell"), Deutsche Forschungsgemeinschaft (SFB 1114/A04) and the MATH+ excellence cluster (Projects AA1-8, EF1-2).

References

1. A. Altis, P.H. Nguyen, R. Hegger, G. Stock, Dihedral angle principal component analysis of molecular dynamics simulations. J. Chem. Phys. **126**, 244111 (2007)
2. S. Bacallado, J.D. Chodera, V.S. Pande, Bayesian comparison of Markov models of molecular dynamics with detailed balance constraint. J. Chem. Phys. **131**, 045106 (2009)
3. C. Bartels, Analyzing biased Monte Carlo and molecular dynamics simulations. Chem. Phys. Lett. **331**, 446–454 (2000)
4. C. Bartels, M. Karplus, Multidimensional adaptive umbrella sampling: application to main chain and side chain peptide conformations. J. Comput. Chem. **18**, 1450–1462 (1997)
5. A.P. Bartók, R. Kondor, G. Csányi, On representing chemical environments. Phys. Rev. B **87**, 184115 (2013)
6. J. Behler, M. Parrinello, Generalized neural-network representation of high-dimensional potential-energy surfaces. Phys. Rev. Lett. **98**, 146401 (2007)
7. T. Bereau, R.A. DiStasio Jr, A. Tkatchenko, O.A. Von Lilienfeld, Non-covalent interactions across organic and biological subsets of chemical space: physics-based potentials parametrized from machine learning. J. Chem. Phys. **148**, 241706 (2018)
8. F. Bießmann, F.C. Meinecke, A. Gretton, A. Rauch, G. Rainer, N.K. Logothetis, K.-R. Müller, Temporal kernel CCA and its application in multimodal neuronal data analysis. Mach. Learn. **79**, 5–27 (2010)
9. G.R. Bowman, K.A. Beauchamp, G. Boxer, V.S. Pande, Progress and challenges in the automated construction of Markov state models for full protein systems. J. Chem. Phys. **131**, 124101 (2009)
10. G.R. Bowman, D.L. Ensign, V.S. Pande, Enhanced modeling via network theory: adaptive sampling of Markov state models. J. Chem. Theory Comput. **6**(3), 787–794 (2010)
11. G.R. Bowman, V.S. Pande, F. Noé (eds.), *An Introduction to Markov State Models and Their Application to Long Timescale Molecular Simulation.* Advances in Experimental Medicine and Biology, vol. 797 (Springer, Heidelberg, 2014)
12. N.V. Buchete, G. Hummer, Coarse master equations for peptide folding dynamics. J. Phys. Chem. B **112**, 6057–6069 (2008)
13. S. Chmiela, A. Tkatchenko, H.E. Sauceda, I. Poltavsky, K.T. Schütt, K.-R. Müller, Machine learning of accurate energy-conserving molecular force fields. Sci. Adv. **3**, e1603015 (2017)
14. S. Chmiela, H.E. Sauceda, K.-R. Müller, A. Tkatchenko, Towards exact molecular dynamics simulations with machine-learned force fields. Nat. Commun. **9**, 3887 (2018)
15. J.D. Chodera, F. Noé, Probability distributions of molecular observables computed from Markov models. II: Uncertainties in observables and their time-evolution. J. Chem. Phys. **133**, 105102 (2010)

16. J.D. Chodera, K.A. Dill, N. Singhal, V.S. Pande, W.C. Swope, J.W. Pitera, Automatic discovery of metastable states for the construction of Markov models of macromolecular conformational dynamics. J. Chem. Phys. **126**, 155101 (2007)
17. R.R. Coifman, S. Lafon, A.B. Lee, M. Maggioni, B. Nadler, F. Warner, S.W. Zucker, Geometric diffusions as a tool for harmonic analysis and structure definition of data: Diffusion maps. Proc. Natl. Acad. Sci. U. S. A. **102**, 7426–7431 (2005)
18. P. Das, M. Moll, H. Stamati, L.E. Kavraki, C. Clementi, Low-dimensional, free-energy landscapes of protein-folding reactions by nonlinear dimensionality reduction. Proc. Natl. Acad. Sci. U. S. A. **103**, 9885–9890 (2008)
19. P. Deuflhard, M. Weber, Robust Perron cluster analysis in conformation dynamics, in *Linear Algebra Appl.*, ed. by M. Dellnitz, S. Kirkland, M. Neumann, C. Schütte, vol. 398C (Elsevier, New York, 2005), pp. 161–184
20. P.D. Dixit, K.A. Dill, Caliber corrected Markov modeling (C2M2): correcting equilibrium Markov models. J. Chem. Theory Comput. **14**, 1111–1119 (2018)
21. S. Doerr, G. De Fabritiis, On-the-fly learning and sampling of ligand binding by high-throughput molecular simulations. J. Chem. Theory Comput. **10**, 2064–2069 (2014)
22. S. Doerr, M.J. Harvey, F. Noé, G. De Fabritiis, HTMD: high-throughput molecular dynamics for molecular discovery. J. Chem. Theory Comput. **12**, 1845–1852 (2016)
23. K. Fackeldey, M. Weber, Genpcca – Markov state models for non-equilibrium steady states. WIAS Rep. **29**, 70–80 (2017)
24. A.M. Ferrenberg, R.H. Swendsen, Optimized Monte Carlo data analysis. Phys. Rev. Lett. **63**, 1195–1198 (1989)
25. C.W. Fox, S.J. Roberts, A tutorial on variational Bayesian inference. Artif. Intell. Rev. **38**, 85–95 (2012)
26. H. Fukunishi, O. Watanabe, S. Takada, On the Hamiltonian replica exchange method for efficient sampling of biomolecular systems: application to protein structure prediction. J. Chem. Phys. **116**, 9058 (2002)
27. E. Gallicchio, M. Andrec, A.K. Felts, R.M. Levy, Temperature weighted histogram analysis method, replica exchange, and transition paths. J. Phys. Chem. B **109**, 6722–6731 (2005)
28. S. Gerber, I. Horenko, Toward a direct and scalable identification of reduced models for categorical processes. Proc. Natl. Acad. Sci. U. S. A. **114**, 4863–4868 (2017)
29. I. Goodfellow, J. Pouget-Abadie, M. Mirza, B. Xu, D. Warde-Farley, S. Ozair, A. Courville, J. Bengio, Generative adversarial networks, in *NIPS'14 Proceedings of the 27th International Conference on Neural Information Processing Systems*, vol. 2 (MIT Press, Cambridge, 2014), pp. 2672–2680
30. H. Grubmüller, Predicting slow structural transitions in macromolecular systems: conformational flooding. Phys. Rev. E **52**, 2893 (1995)
31. U.H.E. Hansmann, Parallel tempering algorithm for conformational studies of biological molecules. Chem. Phys. Lett. **281**(1–3), 140–150 (1997)
32. S. Harmeling, A. Ziehe, M. Kawanabe, K.-R. Müller, Kernel-based nonlinear blind source separation. Neural Comput. **15**, 1089–1124 (2003)
33. M.P. Harrigan, V.S. Pande, Landmark kernel tICA for conformational dynamics (2017). bioRxiv, 123752
34. C.X. Hernández, H.K. Wayment-Steele, M.M. Sultan, B.E. Husic, V.S. Pande, Variational encoding of complex dynamics. Phys. Rev. E **97**, 062412 (2018)
35. A. Hyvärinen, J. Karhunen, E. Oja, *Independent Component Analysis* (Wiley, New York, 2001)
36. N.S. Hinrichs, V.S. Pande, Calculation of the distribution of eigenvalues and eigenvectors in Markovian state models for molecular dynamics. J. Chem. Phys. **126**, 244101 (2007)
37. G. Hummer, J. Köfinger, Bayesian ensemble refinement by replica simulations and reweighting. J. Chem. Phys. **143**, 243150 (2015)
38. G. Hummer, A. Szabo, Optimal dimensionality reduction of multistate kinetic and Markov-state models. J. Phys. Chem. B **119**, 9029–9037 (2015)

39. B.E. Husic, V.S. Pande, Ward clustering improves cross-validated Markov state models of protein folding. J. Chem. Theory Comp. **13**, 963–967 (2017)
40. B.E. Husic, V.S. Pande, Markov state models: from an art to a science. J. Am. Chem. Soc. **140**, 2386–2396 (2018)
41. B.E. Husic, R.T. McGibbon, M.M. Sultan, V.S. Pande, Optimized parameter selection reveals trends in Markov state models for protein folding. J. Chem. Phys. **145**, 194103 (2016)
42. A. Jain, G. Stock, Identifying metastable states of folding proteins. J. Chem. Theory Comput. **8**, 3810–3819 (2012)
43. W. Kabsch, A solution for the best rotation to relate two sets of vectors. Acta Cryst. **A32**, 922–923 (1976)
44. B.G. Keller, X. Daura, W.F. van Gunsteren, Comparing geometric and kinetic cluster algorithms for molecular simulation data. J. Chem. Phys. **132**, 074110 (2010)
45. D.P. Kingma, M. Welling, Auto-encoding variational Bayes, in *Proceedings of the 2nd International Conference on Learning Representations (ICLR)* (2014). arXiv:1312.6114
46. P. Koltai, G. Ciccotti, Ch. Schütte, On metastability and Markov state models for non-stationary molecular dynamics. J. Chem. Phys. **145**, 174103 (2016)
47. P. Koltai, H. Wu, F. Noé, C. Schütte, Optimal data-driven estimation of generalized Markov state models for non-equilibrium dynamics. Computation **6**, 22 (2018)
48. S. Kube, M. Weber, A coarse graining method for the identification of transition rates between molecular conformations. J. Chem. Phys. **126**, 024103 (2007)
49. A. Laio, M. Parrinello, Escaping free energy minima. Proc. Natl. Acad. Sci. U. S. A. **99**, 12562–12566 (2002)
50. O. Ledoit, M. Wolf, Honey, I shrunk the sample covariance matrix. J. Portfolio Manag. **30**, 110–119 (2004)
51. Q. Li, F. Dietrich, E.M. Bollt, I.G. Kevrekidis, Extended dynamic mode decomposition with dictionary learning: a data-driven adaptive spectral decomposition of the Koopman operator. Chaos **27**, 103111 (2017)
52. K. Lindorff-Larsen, S. Piana, R.O. Dror, D.E. Shaw, How fast-folding proteins fold. Science **334**, 517–520 (2011)
53. F. Litzinger, L. Boninsegna, H. Wu, F. Nüske, R. Patel, R. Baraniuk, F. Noé, C. Clementi, Rapid calculation of molecular kinetics using compressed sensing. J. Chem. Theory Comput. **24**, 2771–2783 (2018)
54. B. Lusch, S.L. Brunton J.N. Kutz, Deep learning for universal linear embeddings of nonlinear dynamics (2017). arXiv:1712.09707
55. A. Mardt, L. Pasquali, H. Wu, F. Noé, VAMPnets: deep learning of molecular kinetics. Nat. Commun. **9**, 5 (2018)
56. R.T. McGibbon, V.S. Pande, Variational cross-validation of slow dynamical modes in molecular kinetics. J. Chem. Phys. **142**, 124105 (2015)
57. P. Metzner, F. Noé, C. Schütte, Estimation of transition matrix distributions by Monte Carlo sampling. Phys. Rev. E **80**, 021106 (2009)
58. P. Metzner, C. Schütte, E. Vanden-Eijnden, Transition path theory for Markov jump processes. Multiscale Model. Simul. **7**, 1192–1219 (2009)
59. A.S.J.S. Mey, H. Wu, F. Noé, xTRAM: Estimating equilibrium expectations from time-correlated simulation data at multiple thermodynamic states. Phys. Rev. X **4**, 041018 (2014)
60. I. Mezić, Spectral properties of dynamical systems, model reduction and decompositions. Nonlinear Dyn. **41**, 309–325 (2005)
61. L. Molgedey, H.G. Schuster, Separation of a mixture of independent signals using time delayed correlations. Phys. Rev. Lett. **72**, 3634–3637 (1994)
62. Y. Naritomi, S. Fuchigami, Slow dynamics in protein fluctuations revealed by time-structure based independent component analysis: the case of domain motions. J. Chem. Phys. **134**(6), 065101 (2011)
63. F. Noé, Probability distributions of molecular observables computed from Markov Models. J. Chem. Phys. **128**, 244103 (2008)

64. F. Noé, C. Clementi, Kinetic distance and kinetic maps from molecular dynamics simulation. J. Chem. Theory Comput. **11**, 5002–5011 (2015)
65. F. Noé, C. Clementi, Collective variables for the study of long-time kinetics from molecular trajectories: theory and methods. Curr. Opin. Struct. Biol. **43**, 141–147 (2017)
66. F. Noé, F. Nüske, A variational approach to modeling slow processes in stochastic dynamical systems. Multiscale Model. Simul. **11**, 635–655 (2013)
67. F. Noé, I. Horenko, C. Schütte, J.C. Smith, Hierarchical analysis of conformational dynamics in biomolecules: transition networks of metastable states. J. Chem. Phys. **126**, 155102 (2007)
68. F. Noé, C. Schütte, E. Vanden-Eijnden, L. Reich, T.R. Weikl, Constructing the full ensemble of folding pathways from short off-equilibrium simulations. Proc. Natl. Acad. Sci. U. S. A. **106**, 19011–19016 (2009)
69. F. Noé, S. Doose, I. Daidone, M. Löllmann, J.D. Chodera, M. Sauer, J.C. Smith, Dynamical fingerprints for probing individual relaxation processes in biomolecular dynamics with simulations and kinetic experiments. Proc. Natl. Acad. Sci. U. S. A. **108**, 4822–4827 (2011)
70. F. Noé, H. Wu, J.-H. Prinz, N. Plattner, Projected and hidden Markov models for calculating kinetics and metastable states of complex molecules. J. Chem. Phys. **139**, 184114 (2013)
71. F. Noé, S. Olsson, J. Köhler, H. Wu, Boltzmann generators – sampling equilibrium states of many-body systems with deep learning (2019). arXiv:1812.01729
72. F. Nüske, B.G. Keller, G. Pérez-Hernández, A.S.J.S. Mey, F. Noé, Variational approach to molecular kinetics. J. Chem. Theory Comput. **10**, 1739–1752 (2014)
73. S. Olsson, F. Noé, Dynamic graphical models of molecular kinetics. Proc. Natl. Acad. Sci. U. S. A. **116**, 15001–15006 (2019)
74. S. Olsson, H. Wu, F. Paul, C. Clementi, F. Noé, Combining experimental and simulation data of molecular processes via augmented Markov models. Proc. Natl. Acad. Sci. U. S. A. **114**, 8265–8270 (2017)
75. S. Orioli, P. Faccioli, Dimensional reduction of Markov state models from renormalization group theory. J. Chem. Phys. **145**, 124120 (2016)
76. S.E. Otto, C.W. Rowley, Linearly-recurrent autoencoder networks for learning dynamics (2017). arXiv:1712.01378
77. F. Paul, C. Wehmeyer, E.T. Abualrous, H. Wu, M.D. Crabtree, J. Schöneberg, J. Clarke, C. Freund, T.R. Weikl, F. Noé, Protein-ligand kinetics on the seconds timescale from atomistic simulations. Nat. Commun. **8**, 1095 (2017)
78. F. Paul, H. Wu, M. Vossel, B.L. de Groot, F. Noé, Identification of kinetic order parameters for non-equilibrium dynamics. J. Chem. Phys. **150**, 164120 (2019)
79. G. Perez-Hernandez, F. Noé, Hierarchical time-lagged independent component analysis: computing slow modes and reaction coordinates for large molecular systems. J. Chem. Theory Comput. **12**, 6118–6129 (2016)
80. G. Perez-Hernandez, F. Paul, T. Giorgino, G. De Fabritiis, F. Noé, Identification of slow molecular order parameters for Markov model construction. J. Chem. Phys. **139**, 015102 (2013)
81. N. Plattner, S. Doerr, G. De Fabritiis, F. Noé, Protein-protein association and binding mechanism resolved in atomic detail. Nat. Chem. **9**, 1005–1011 (2017)
82. J. Preto, C. Clementi, Fast recovery of free energy landscapes via diffusion-map-directed molecular dynamics. Phys. Chem. Chem. Phys. **16**, 19181–19191 (2014)
83. J.-H. Prinz, J.D. Chodera, V.S. Pande, W.C. Swope, J.C. Smith, F. Noé, Optimal use of data in parallel tempering simulations for the construction of discrete-state Markov models of biomolecular dynamics. J. Chem. Phys. **134**, 244108 (2011)
84. J.-H. Prinz, H. Wu, M. Sarich, B.G. Keller, M. Senne, M. Held, J.D. Chodera, C. Schütte, F. Noé, Markov models of molecular kinetics: generation and validation. J. Chem. Phys. **134**, 174105 (2011)
85. R. Ramakrishnan, P.O. Dral, M. Rupp, O.A. von Lilienfeld, Quantum chemistry structures and properties of 134 kilo molecules. Sci. Data **1**, 140022 (2014)
86. J.M.L. Ribeiro, P. Bravo, Y. Wang, P. Tiwary, Reweighted autoencoded variational Bayes for enhanced sampling (RAVE). J. Chem. Phys. **149**, 072301 (2018)

87. S. Röblitz, M. Weber, Fuzzy spectral clustering by PCCA+: application to Markov state models and data classification. Adv. Data Anal. Classif. **7**, 147–179 (2013)
88. M.A. Rohrdanz, W. Zheng, M. Maggioni, C. Clementi, Determination of reaction coordinates via locally scaled diffusion map. J. Chem. Phys. **134**, 124116 (2011)
89. E. Rosta, G. Hummer, Free energies from dynamic weighted histogram analysis using unbiased Markov state model. J. Chem. Theory Comput. **11**, 276–285 (2015)
90. C.W. Rowley, I. Mezić, S. Bagheri, P. Schlatter, D.S. Henningson, Spectral analysis of nonlinear flows. J. Fluid Mech. **641**, 115 (2009)
91. M. Rupp, A. Tkatchenko, K.-R. Müller, O.A. Von Lilienfeld, Fast and accurate modeling of molecular atomization energies with machine learning. Phys. Rev. Lett. **108**, 058301 (2012)
92. M. Sarich, F. Noé, C. Schütte, On the approximation quality of Markov state models. Multiscale Model. Simul. **8**, 1154–1177 (2010)
93. J. Schäfer, K. Strimmer, A shrinkage approach to large-scale covariance matrix estimation and implications for functional genomics, in *Statistical Applications in Genetics and Molecular Biology*, vol. 4 (Walter de Gruyter GmbH & Co. KG, Berlin, 2005), pp. 2194–6302
94. M.K. Scherer, B. Trendelkamp-Schroer, F. Paul, G. Perez-Hernandez, M. Hoffmann, N. Plattner, C. Wehmeyer, J.-H. Prinz, F. Noé, PyEMMA 2: a software package for estimation, validation and analysis of Markov models. J. Chem. Theory Comput. **11**, 5525–5542 (2015)
95. M.K. Scherer, B.E. Husic, M. Hoffmann, F. Paul, H. Wu, F. Noé, Variational selection of features for molecular kinetics. J. Chem. Phys. **150**, 194108 (2019)
96. P.J. Schmid, Dynamic mode decomposition of numerical and experimental data. J. Fluid Mech. **656**, 5–28 (2010)
97. P.J. Schmid, J. Sesterhenn, Dynamic mode decomposition of numerical and experimental data, in *61st Annual Meeting of the APS Division of Fluid Dynamics* (American Physical Society, Philadelphia, 2008)
98. B. Schölkopf, A. Smola, K.-R. Müller, Nonlinear component analysis as a kernel eigenvalue problem. Neural Comput. **10**, 1299–1319 (1998)
99. K.T. Schütt, F. Arbabzadah, S. Chmiela, K.R. Müller, A. Tkatchenko, Quantum-chemical insights from deep tensor neural networks. Nat. Commun. **8**, 13890 (2017)
100. K.T. Schütt, H.E. Sauceda, P.J. Kindermans, A. Tkatchenko, K.R. Müller, SchNet – a deep learning architecture for molecules and materials. J. Chem. Phys. **148**(24), 241722 (2018)
101. C. Schütte, A. Fischer, W. Huisinga, P. Deuflhard, A direct approach to conformational dynamics based on hybrid Monte Carlo. J. Comput. Phys. **151**, 146–168 (1999)
102. C.R. Schwantes, V.S. Pande, Improvements in Markov state model construction reveal many non-native interactions in the folding of NTL9. J. Chem. Theory Comput. **9**, 2000–2009 (2013)
103. C.R. Schwantes, V.S. Pande, Modeling molecular kinetics with tICA and the kernel trick. J. Chem. Theory Comput. **11**, 600–608 (2015)
104. D.E. Shaw, J.P. Grossman, J.A. Bank, B. Batson, J.A. Butts, J.C. Chao, M.M. Deneroff, R.O. Dror, A. Even, C.H. Fenton, A. Forte, J. Gagliardo, G. Gill, B. Greskamp, C.R. Ho, D.J. Ierardi, L. Iserovich, J.S. Kuskin, R.H. Larson, T. Layman, L.-S. Lee, A.K. Lerer, C. Li, D. Killebrew, K.M. Mackenzie, S. Yeuk-Hai Mok, M.A. Moraes, R. Mueller, L.J. Nociolo, J.L. Peticolas, Anton 2: raising the bar for performance and programmability in a special-purpose molecular dynamics supercomputer, in *SC '14: Proceedings of the International Conference for High Performance Computing, Networking, Storage and Analysis* (IEEE, Piscataway, 2014)
105. F.K. Sheong, D.-A. Silva, L. Meng, Y. Zhao, X. Huang, Automatic state partitioning for multibody systems (APM): an efficient algorithm for constructing Markov state models to elucidate conformational dynamics of multibody systems. J. Chem. Theory Comput. **11**, 17–27 (2015)
106. M.R. Shirts, J.D. Chodera, Statistically optimal analysis of samples from multiple equilibrium states. J. Chem. Phys. **129**, 124105 (2008)
107. N. Singhal, V.S. Pande, Error analysis and efficient sampling in Markovian state models for molecular dynamics. J. Chem. Phys. **123**, 204909 (2005)

108. W.C. Swope, J.W. Pitera, F. Suits, Describing protein folding kinetics by molecular dynamics simulations: 1. Theory. J. Phys. Chem. B **108**, 6571–6581 (2004)
109. W.C. Swope, J.W. Pitera, F. Suits, M. Pitman, M. Eleftheriou, Describing protein folding kinetics by molecular dynamics simulations: 2. Example applications to alanine dipeptide and beta-hairpin peptide. J. Phys. Chem. B **108**, 6582–6594 (2004)
110. G. Székely, M. Rizzo, Testing for equal distributions in high dimension. InterStat **5**, 1249–1272 (2004)
111. J.B. Tenenbaum, V. de Silva, J.C. Langford, A global geometric framework for nonlinear dimensionality reduction. Science **290**, 2319–2323 (2000)
112. D.L. Theobald, Rapid calculation of RMSDs using a quaternion-based characteristic polynomial. Acta Cryst. **A61**, 478–480 (2005)
113. G.M. Torrie, J.P. Valleau, Nonphysical sampling distributions in Monte Carlo free-energy estimation: umbrella sampling. J. Comput. Phys. **23**, 187–199 (1977)
114. B. Trendelkamp-Schroer, F. Noé, Efficient Bayesian estimation of Markov model transition matrices with given stationary distribution. J. Phys. Chem. **138**, 164113 (2013)
115. B. Trendelkamp-Schroer, F. Noé, Efficient estimation of rare-event kinetics. Phys. Rev. X (2015). Preprint. arXiv:1409.6439
116. B. Trendelkamp-Schroer, H. Wu, F. Paul, F. Noé, Estimation and uncertainty of reversible Markov models. J. Chem. Phys. **143**, 174101 (2015)
117. J.H. Tu, C.W. Rowley, D.M. Luchtenburg, S.L. Brunton, J.N. Kutz, On dynamic mode decomposition: theory and applications. J. Comput. Dyn. **1**(2), 391–421 (2014)
118. O. Valsson, M. Parrinello, Variational approach to enhanced sampling and free energy calculations. Phys. Rev. Lett. **113**, 090601 (2014)
119. W. Wang, R. Gómez-Bombarelli, Variational coarse-graining for molecular dynamics (2018). arXiv:1812.02706
120. J. Wang, C. Wehmeyer, F. Noé, C. Clementi, Machine learning of coarse-grained molecular dynamics force fields. ACS Cent. Sci. **5**, 755–767 (2019)
121. C. Wehmeyer, F. Noé, Time-lagged autoencoders: deep learning of slow collective variables for molecular kinetics. J. Chem. Phys. **148**, 241703 (2018)
122. C. Wehmeyer, M.K. Scherer, T. Hempel, B.E. Husic, S. Olsson, F. Noé, Introduction to Markov state modeling with the PyEMMA software. LiveCoMS **1**, 5965 (2018)
123. M.O. Williams, C.W. Rowley, I.G. Kevrekidis, A kernel-based approach to data-driven Koopman spectral analysis (2014). arXiv:1411.2260
124. M.O. Williams, I.G. Kevrekidis, C.W. Rowley, A data-driven approximation of the Koopman operator: extending dynamic mode decomposition. J. Nonlinear Sci. **25**, 1307–1346 (2015)
125. W. Wojtas-Niziurski, Y. Meng, B. Roux, S. Bernèche, Self-learning adaptive umbrella sampling method for the determination of free energy landscapes in multiple dimensions. J. Chem. Theory Comput. **9**, 1885–1895 (2013)
126. H. Wu, F. Noé, Optimal estimation of free energies and stationary densities from multiple biased simulations. Multiscale Model. Simul. **12**, 25–54 (2014)
127. H. Wu, F. Noé, Variational approach for learning Markov processes from time series data (2017). arXiv:1707.04659
128. H. Wu, A.S.J.S. Mey, E. Rosta, F. Noé, Statistically optimal analysis of state-discretized trajectory data from multiple thermodynamic states. J. Chem. Phys. **141**, 214106 (2014)
129. H. Wu, F. Paul, C. Wehmeyer, F. Noé, Multiensemble Markov models of molecular thermodynamics and kinetics. Proc. Natl. Acad. Sci. U. S. A. **113**, E3221–E3230 (2016)
130. H. Wu, F. Nüske, F. Paul, S. Klus, P. Koltai, F. Noé, Variational Koopman models: slow collective variables and molecular kinetics from short off-equilibrium simulations. J. Chem. Phys. **146**, 154104 (2017)
131. H. Wu, A. Mardt, L. Pasquali, F. Noé, Deep generative Markov state models, in *NIPS* (2018). Preprint. arXiv:1805.07601
132. Y. Yao, J. Sun, X. Huang, G.R. Bowman, G. Singh, M. Lesnick, L.J. Guibas, V.S. Pande, G. Carlsson, Topological methods for exploring low-density states in biomolecular folding pathways. J. Chem. Phys. **130**, 144115 (2009)

133. Y. Yao, R.Z. Cui, G.R. Bowman, D.-A. Silva, J. Sun, X. Huang, Hierarchical Nyström methods for constructing Markov state models for conformational dynamics. J. Chem. Phys. **138**, 174106 (2013)
134. L. Zhang, J. Han, H. Wang, R. Car, W. E, DeePCG: constructing coarse-grained models via deep neural networks. J. Chem. Phys. **149**, 034101 (2018)
135. A. Ziehe, K.-R. Müller, TDSEP – an efficient algorithm for blind separation using time structure, in *ICANN 98* (Springer Science and Business Media, New York, 1998), pp. 675–680
136. A. Ziehe, P. Laskov, G. Nolte, K.-R. Müller, A fast algorithm for joint diagonalization with non-orthogonal transformations and its application to blind source separation. J. Mach. Learn. Res. **5**, 777–800 (2004)
137. M.I. Zimmerman, G.R. Bowman, Fast conformational searches by balancing exploration/exploitation trade-offs. J. Chem. Theory Comput. **11**, 5747–5757 (2015)

Part V
Discovery and Design

Discovery and Design: Preface

The discovery and design of novel molecules and materials with desired chemical properties is crucial for a wide range of technologies, ranging from the development of new drugs to improving the efficiency of batteries and solar cells [1]. High-throughput computational methods which combine electronic structure calculations with data analysis methods have proven to be valuable tools to tackle this challenge [2]. Yet, the computational cost of quantum-chemical calculations constitutes a serious bottleneck to the design of novel systems. The final part of this book shows that machine learning approaches can not only be used to accelerate quantum-chemical computations, but also to effectively guide the search of chemical compound space and even directly generate promising candidate systems.

Armiento [3] starts off with a review of combining high-throughput ab initio calculations with machine learning predictions of chemical properties for materials design. It covers the whole design pipeline from building a dataset of reference calculations to the training of machine learning models with focus on bulk materials. Connecting to this, Chandrasekaran et al. [4] present an application of the introduced principles to soft materials for the example of polymers in Chap. 18. Beyond that, an online platform for polymer design is introduced that provides an easily accessible interface to the machine learning models.

Since the reference calculations used for training machine learning models populate only a fraction of the vast chemical space, it is important to detect when one strays too far from the known domain and the model becomes unreliable. This topic has already been described in Chap. 15 [5] covering uncertainty estimation for machine learning force fields. However, when searching for novel materials, we are actively looking simultaneously for *uncharted territory* and promising candidates. In Chap. 19, Hou and Tsuda [6] give an introduction to Bayesian optimization—a machine learning-based, global optimization method that tackles this issue. The chapter reviews several applications of this technique to materials science, such as discovery of functional materials and crystal structure prediction. Next, Chap. 20 [7] discusses two alternative recommender systems for materials discovery: a descriptor-free approach based on tensor decomposition and a knowledge-based approach.

Part III showed how deep neural networks can learn representations of molecules and materials to obtain accurate property predictions and to accelerate atomistic simulation. Beyond that, generative neural networks can be used to solve the inverse problem and, thus, are a promising tool for chemical design [8–10]. Given the desired properties, one could in principle sample suitable candidate structures from these models in the form of SMILES strings, molecular graphs, or even the full equilibrium geometries. Chapter 21 by Schwalbe-Koda and Gómez-Bombarelli [11] concludes in this part with an extensive review of such deep generative models covering the predominant architectures—variational auto-encoders, generative adversarial networks, and auto-regressive models. In this way, this last part of the book also aims to bridge the gap between machine learning for quantum simulations and ML approaches in materials and chemo-informatics.

Berlin, Germany — Kristof T. Schütt
Berlin, Germany — Stefan Chmiela
Basel, Switzerland — O. Anatole von Lilienfeld
Luxembourg, Luxembourg — Alexandre Tkatchenko
Kashiwa, Japan — Koji Tsuda
Berlin, Germany — Klaus-Robert Müller
September 2019

References

1. D.P. Tabor, L.M. Roch, S.K. Saikin, C. Kreisbeck, D. Sheberla, J.H. Montoya, S. Dwaraknath, M. Aykol, C. Ortiz, H. Tribukait, et al., Nat. Rev. Mater. **3**(5), 5 (2018)
2. E.O. Pyzer-Knapp, C. Suh, R. Gómez-Bombarelli, J. Aguilera-Iparraguirre, A. Aspuru-Guzik, Annu. Rev. Mater. Res. **45**, 195 (2015)
3. R. Armiento, in *Machine Learning for Quantum Simulations of Molecules and Materials*, ed. by K.T. Schütt, S. Chmiela, A. von Lilienfeld, A. Tkatchenko, K. Tsuda, K.-R. Müller. Lecture Notes Physics (Springer, Berlin, 2019)
4. A. Chandrasekaran, C. Kim, R. Ramprasad, in *Machine Learning for Quantum Simulations of Molecules and Materials*, ed. by K.T. Schütt, S. Chmiela, A. von Lilienfeld, A. Tkatchenko, K. Tsuda, K.-R. Müller. Lecture Notes Physics (Springer, Berlin, 2019)
5. A. Shapeev, K. Gubaev, E. Tsymbalov, E. Podryabinkin, in *Machine Learning for Quantum Simulations of Molecules and Materials*, ed. by K.T. Schütt, S. Chmiela, A. von Lilienfeld, A. Tkatchenko, K. Tsuda, K.-R. Müller. Lecture Notes Physics (Springer, Berlin, 2019)
6. Z. Hou, K. Tsuda, in *Machine Learning for Quantum Simulations of Molecules and Materials*, ed. by K.T. Schütt, S. Chmiela, A. von Lilienfeld, A. Tkatchenko, K. Tsuda, K.-R. Müller. Lecture Notes Physics (Springer, Berlin, 2019)
7. A. Seko, H. Hayashi, H. Kashima, I. Tanaka, in *Machine Learning for Quantum Simulations of Molecules and Materials*, ed. by K.T. Schütt, S. Chmiela, A. von Lilienfeld, A. Tkatchenko, K. Tsuda, K.-R. Müller. Lecture Notes Physics (Springer, Berlin, 2019)
8. Q. Liu, M. Allamanis, M. Brockschmidt, A. Gaunt, *Advances in Neural Information Processing Systems* (2018), pp. 7795–7804
9. J. You, B. Liu, Z. Ying, V. Pande, J. Leskovec, in *Advances in Neural Information Processing Systems* (2018), pp. 6410–6421

10. N.W. Gebauer, M. Gastegger, K.T. Schütt, *Advances in Neural Information Processing Systems 33* (in press). Preprint arXiv:1906.00957
11. D. Schwalbe-Koda, R. Gómez-Bombarelli, in *Machine Learning for Quantum Simulations of Molecules and Materials*, ed. by K.T. Schütt, S. Chmiela, A. von Lilienfeld, A. Tkatchenko, K. Tsuda, K.-R. Müller. Lecture Notes Physics (Springer, Berlin, 2019)

Database-Driven High-Throughput Calculations and Machine Learning Models for Materials Design

17

Rickard Armiento

Abstract

This chapter reviews past and ongoing efforts in using high-throughput ab-initio calculations in combination with machine learning models for materials design. The primary focus is on bulk materials, i.e., materials with fixed, ordered, crystal structures, although the methods naturally extend into more complicated configurations. Efficient and robust computational methods, computational power, and reliable methods for automated database-driven high-throughput computation are combined to produce high-quality data sets. This data can be used to train machine learning models for predicting the stability of bulk materials and their properties. The underlying computational methods and the tools for automated calculations are discussed in some detail. Various machine learning models and, in particular, descriptors for general use in materials design are also covered.

17.1 Background

Design of new materials with desired properties is a crucial step in making many innovative technologies viable. The aim is to find materials that fulfill requirements on efficiency, cost, environmental impact, length of life, safety, and other properties. During the past decades, we have seen major progress in theoretical materials science due to the combination of improved computational methods and a massive increase in available computational power. It is now standard practice to obtain insights into the physics of materials by using supercomputers to find numerical solutions to the basic equations of quantum mechanics. When using the appropriate

R. Armiento (✉)
Department of Physics, Chemistry and Biology, Linköping University, Linköping, Sweden
e-mail: rickard.armiento@liu.se

K. T. Schütt et al. (eds.), *Machine Learning Meets Quantum Physics*, Lecture Notes in Physics 968, https://doi.org/10.1007/978-3-030-40245-7_17

level of theory, these calculations can be robust enough to run in unsupervised high-throughput. Hence, materials design can be done via automated theoretical screening of candidate materials and substances, picking out those with desired properties. Early examples of this methodology include works in the fields of catalysts [1], battery materials [2, 3], detector materials for ionizing radiation [4], superconductivity [5], thermoelectricity [6, 7], piezoelectrics [8, 9], transparent conducting oxides [10], and two-dimensional materials [11]. There is a wealth of further examples in the literature, see, e.g., the reviews in Ref. [12–15].

Early adoption of high-throughput methodology for materials design has invoked the ambition that it may be possible to computationally predict the existence and basic properties of essentially *every single material*, i.e., any composition that, in principle, can be synthesized as a reasonably long-lived "stable" compound (in the context of an environment.) This ambition has been referred to as the *materials genome project* [13, 16, 17], which in 2011 was endorsed as a White House initiative [18]. The idea is that access to materials genome data with sufficient coverage would greatly accelerate materials design. It would be possible to perform queries against this data to pick out compositions that have some sought combination of desired properties for a specific application at, essentially, no additional computational cost [12, 13].

A large number of databases of materials-genome-type are now available, many of them open and free for access over the Internet. Some notable examples include: the *Electronic Structure Project* (http://gurka.physics.uu.se/esp/; 2002), the *Automatic FLOW repository* (aflowlib.org; 2011), *the Materials Project* (materialsproject.org; 2011), the *Open Materials Database* (openmaterialsdb.se; 2013), the *Open Quantum Materials Database* (oqmd.org; 2013), the *Theoretical Crystallographic Open Database* (www.crystallography.net/tcod; 2013), the *Novel Materials Discovery Repository* (nomad-repository.eu; 2014), the *High Performance Computing Center Materials Database—NREL MatDb* (materials.nrel.gov; 2015), and the *Materials Cloud* (materialscloud.org; 2017).

To use machine learning models for, e.g., molecular dynamics simulations of systems with up to a few chemical species has become increasingly popular (i.e., to accelerate simulations of the movement of some types of atoms in a material.) To train more general models with data from materials genome-type databases opens a way forward towards the vision of a complete coverage of materials and their properties. This chapter reviews the use of high-throughput techniques and tools to produce training data for these models and recent developments in the area of models with the aim of a general description of atomistic systems (i.e., molecules and materials.) This development is, at its core, the adoption of an informatics perspective to materials science and design, which has been referred to as *materials informatics* [19, 20].

It has been posed as a hypothesis that the progress of general AI methods will eventually reach "the singularity," a moment in time when self-improving AI methods set off a runaway technological development that fundamentally changes society.[1] One can, in a similar way, formulate the hypothesis that the development of increasingly sophisticated machine learning models for atomistic systems will reach a singularity of its own, i.e., a point in time of fundamental change in our theoretical description of physical matter. This change would happen when fully trained, general, machine learning models appear that are capable of predictions at the same accuracy as physics-based quantum mechanical simulations but at negligible computational effort. The result would turn the present materials genome-type databases obsolete and enable true inverse design of molecules and materials with desired properties across the full chemical space at near zero computational expense. Such a development would bring far-reaching changes across the natural sciences.

In conclusion, advancing the present state of materials design with machine learning models requires progress in three key areas: (1) progress in the theory and methods used in physics-based calculations that can be used to improve the quality of training data. This requires developing methods with improved accuracy without sacrificing the low computational demand and the high level of generality that are necessary for the methods to be useful for high-throughput calculations; (2) further improved methods and tools for running automated calculations in high throughput. While there are many software packages and solutions available today for running calculations in high-throughput, major work of both practical and theoretical nature remains to turn methods that were developed and tested only on a few systems into automated workflows capable of running unsupervised at large scale without human interference; and (3) further improved machine learning models for general atomistic systems.

17.2 Computational Methods

Kohn–Sham density functional theory [23, 24] (KS-DFT) is the standard theoretical framework for high-throughput computation in present materials property databases. There is a range of software implementations for performing the numerical solution of the basic equations of DFT. A few prominent examples include the *Vienna Ab-initio Simulation Package—VASP* (vasp.at), *ABINIT* (www.abinit.org), *Wien2K* (susi.theochem.tuwien.ac.at), and *Quantum ESPRESSO* (www.quantum-espresso.org). Of primary concern for these software packages is the numerical convergence towards the exact solution with respect to the approximations used. Most approximations are fairly straightforward to systematically improve towards a converged result, which has led to a number of standard practices for setting convergence parameters that are typically documented in relation to the respective database. See, e.g., Ref. [17] for the practices used in the Materials Project database.

[1]The term was recently popularized by a 2006 book by Kurzweil [21], but its use goes back to a 1958 account by Stanislaw Ulam of a discussion with John von Neumann that references a point in time of fundamental change due to runaway technological development [22].

One aspect of numerical convergence that frequently is in focus when discussing the accuracy of KS-DFT calculations in the context of chemistry-oriented calculations is the basis set used to represent the single particle wave functions (also known as the *KS orbitals*.) While more or less all basis sets can be systematically extended towards numerical convergence, this can be impractical for some choices. Nevertheless, in the context of materials design of bulk materials, we are mostly concerned with fully periodic crystals where the most common choice is a plane-wave basis set where systematic convergence is more straightforward.

In contrast to the numerical approximations that can be, at least in principle, systematically refined to arbitrary accuracy towards the solution of the KS-DFT equations, there is one aspect of the calculations where this is not possible. This is the choice of exchange-correlation density functional. This choice is crucial for the description of the physics of the system and, by extension, which properties are available in the output. The kind of systems and properties for which one can obtain reliable data is of key importance in the present context of using high-throughput computation to produce reliable training data for machine learning models. Hence, we will in the following review the important aspects of this choice in detail.

The level of theory that so far has been the standard for high-throughput computation in first-principles materials property databases is the semi-local, "second-rung" [25] level, which uses exchange-correlation functionals on the generalized gradient approximation (GGA) form. The most commonly used functional in the context of high-throughput calculations for materials databases is the one by Perdew, Burke, and Ernzerhof [26] (PBE) with the $+U$ correction [27]. This level of theory strikes a desirable balance between computational speed and accuracy while maintaining a high level of transferability. Nevertheless, the most popular GGA-type functionals, including PBE, have known shortcomings in their description of the electronic structure. The primary issues include: (1) a tendency to give energetics that in geometrical relaxations lead to a systematic over- or underestimation of bond lengths (the local density approximation, LDA, overbinds, whereas PBE underbinds); (2) an insufficient description of the physics of weak dispersion forces/van der Waals bonding; and (3) a systematically overdelocalized description of the KS orbitals that leads to inaccuracies in a number of properties that are derived from the orbitals. These three issues will be discussed in some more detail in the subsections below.

17.2.1 Overdelocalized Orbitals

The fundamental issue of overdelocalized KS orbitals is related to various aspects of the self-interaction error present at the semi-local exchange-correlation functional level of theory. A simplified picture is that the self-interaction introduces a repulsive electrostatic interaction of an electron with itself, leading to a delocalization that becomes more severe the more localized the correct representation of the orbital was supposed to be, i.e., the effect more severely impacts the more localized d-, and even more so, the f orbitals, compared to the less localized s and p states.

The result is a number of deficiencies in predicted materials properties. Examples of problematic properties include redox reaction energies [28, 29], the polarizability of extended systems [30, 31], and the silicon interstitial formation energy [32, 33].

In addition to these examples, issues are also seen in a number of properties calculated from the single-particle orbitals from the KS-DFT framework, where they are used as approximations of the "true quasi-electron orbitals" of the many-electron system (to the extent that such can be defined). However, from a fundamental perspective, the discussion of the accuracy of such properties is delicate because the DFT orbitals and the quasi-electron orbitals are not the same thing, even in theory for the exact exchange-correlation functional. Hence, one cannot a priori assume that an improved functional increases the agreement with the experimental values of, e.g., optical properties calculated from the KS band structure. Nevertheless, if one compares common GGA functionals to higher order methods that are still within the framework of KS-DFT (e.g., exact DFT exchange) one finds a qualitative difference in the orbital physics. This difference translates to that when materials properties which are directly associated with the electronic structure are calculated using higher-order theory, the results come out qualitatively closer to experiments than those calculated using standard GGAs. One can, therefore, take the position that it is a worthwhile improvement over standard semi-local functionals if improved functionals can make the orbitals to more closely mimic the orbital features given by higher order methods. This motivation is independent of the justification, or lack thereof, of using KS states to approximate quasi-particle bands for calculating materials properties. For an expanded discussion on this delicate topic, see, e.g., Ref. [34].

There are a range of well-known methods to address the description of localized states in semi-local DFT, (i) an explicit orbital-dependent correction that removes the surplus electrostatic term (sic correction) [35–37]; (ii) exact exchange DFT [38]; (iii) interpolating the DFT functional with Hartree–Fock exchange energy (hybrid functionals) [39–41]; (iv) use of the many-body Green's function for a more precise description of the localized quasi-particle orbitals (GW) [42]; (v) the DFT+U correction that adds an effective Hubbard-like term to the Hamiltonian to make selected localized orbitals energetically preferable [27]; and (vi) various attempts to modify the KS potential directly to make it reproduce essential features of exact exchange [31, 43–48]. All these methods, except for the last two (v, vi), require a vastly increased computational expense. Hybrid density functional methods (iii) are increasingly adopted for resolving these issues when the extra computational cost is acceptable. However, at a cost of roughly 50 times of that of standard GGAs, they are very inconvenient, or even completely unsuitable for, e.g., larger systems and high-throughput-type calculations.

Of the two less computationally expensive methods, DFT+U (v) is widely adopted as, arguably, the standard way of dealing with the issue of overdelocalized orbitals in high-throughput calculations and materials genome-type databases. However, DFT+U is not a highly transferable method; it requires attention in the assignment of site-specific "U-values." In setting the value of U, one selects how strongly a given localized orbital on a specific site prefers full occupation over partial occupation. In low throughput calculations, it is common to somewhat thoroughly investigate a system to arrive at a value of U that reasonably reproduces the expected physics of the system, but this is clearly not an option for high-throughput calculations. There are schemes to obtain sets of values that work well for systems with some specific type of physics, e.g., for typical oxides. However, in systems of mixed chemistries and intermixed types of bonding physics, the non-universality of U values becomes a serious problem. Energies obtained for different systems using different U-values for the same species cannot easily be mixed. Furthermore, since U values are usually only assigned to specific orbital projections on a pre-selected set of transition metal species, they cannot help with overdelocalized states of different origin, e.g., for defect states that are not atomic-orbital-like.

The second computationally less expensive method in the list above is (v) the approach to model the exchange-correlation potential directly to make it reproduce essential "non-local" features of exact exchange, instead of obtaining it as a functional derivative of an energy functional. Such potentials are known as *model potentials,* and have in some cases been quite successful [31, 43–48]. Some recent interest has been generated by the model potential of Becke and Johnson (BJ) [45], which was observed to mimic some of the crucial features of exact exchange for atoms. With various adjustments and extensions, it improves the polarization of hydrogen chains [31, 47], gives closer correspondence to experimental band gaps [48], and, to some extent, gives other improved properties [49, 50]. These model potentials seem promising for future adoption in high-throughput calculations to access properties that would otherwise be problematic because of orbital delocalization.

However, there are some fundamental issues with the general approach of model potentials. Since they directly model the exchange-correlation potential, the corresponding energy functionals are not merely unknown, they usually do not *exist* [46, 51, 52], and this deficiency cannot easily be corrected [53]. Since the KS equations are derived from a variational treatment of an energy equation, the use of such potentials has to be regarded on a weak formal basis, and are, strictly, outside the framework of KS-DFT. One cannot calculate any energy-derived properties from model potentials, e.g., one cannot do a geometry optimization that is consistent with the potential. Hence, if one starts from, e.g., theoretically generated structure candidates, one would have to use another method first to pre-relax the structure.

A closely related promising direction of functional development is the Armiento–Kümmel exchange functional (AK13) [54] (co-authored by the author of this chapter.) This is *a normal GGA exchange energy functional* that mimics the behavior of the BJ potential while avoiding the fundamental issues with model potentials. Similar to the modified BJ-based model potentials, the AK13 exchange energy functional gives qualitatively different orbitals from common GGA functionals. The results are a KS potential with improved atomic shell structure [54], improved ionization potentials from the highest eigenvalue [54] (but see the discussion in Ref. [55]), overall a KS band structure that better match that of higher order methods, including enlarged band gaps, and improved optical properties [34, 54, 56, 57].

As mentioned, the AK13 functional avoids the problem of undefined energies and energetics in model potentials. However, their values are not as accurate as those of commonly used GGAs and mostly insufficient. In addition, other issues appear from the AK13 construction that prevent its broader indiscriminate application [55,58,59]. We are hopeful that further research into modifications of the expression can overcome the difficulties while still retaining the favorable exchange potential features.

17.2.2 Under- and Overestimated Lattice Constants

On the issue of systematic under- and overestimation of lattice constants, this has mostly been resolved in functional development beyond PBE. The Armiento–Mattsson 2005 functional (AM05) [60, 61] is a semi-local functional with the same computational difficulty as PBE, but which gives roughly half the error for lattice constants. The comprehensive testing of Haas et al. finds for the lattice constants of 60 solids that the mean absolute error is 0.053 Å for PBE and 0.033 Å for AM05 [62–64]. Later functionals developed by Wu–Cohen in 2006 [65–67], SOGGA by Zaho, and Truhlar in 2008 [68], and PBEsol by Perdew et al. in 2009 [69–72] report similar improvements [63, 64, 70]. Further progress has been made by Perdew and coworkers on the meta-GGA level of theory, where, in addition to the electron density and its derivatives, a functional may also depend on the local value of the kinetic energy density of the KS particles. While meta-GGAs are technically more complex expressions than GGAs, implementations can be made that do not significantly increase the computational cost. The 2015 *Strongly Constrained and Appropriately Normed Semilocal Density Functional* (SCAN) meta-GGA [73] reportedly performs well for a wide range of properties for both solids and molecules, including lattice constants [74, 75]. However, some issues have recently been reported in the description of systems with itinerant magnetism [76].

17.2.3 Weak Dispersion Forces

On the topic of the description of van der Waals/London dispersion forces/weak interactions by semi-local DFT functionals, there exist a range of post-correction schemes of the energy to handle such interactions that can be deployed without any significant additional computational cost, see, e.g., Refs. [77–83]. Furthermore, there is a series of successful exchange-correlation functionals known as the vdW-DF from a collaboration between Chalmers University and Rutgers University [84–86] which allow a self-consistent treatment of these interactions. These functionals are not semi-local, but still fairly computationally inexpensive compared to, e.g., hybrid functionals. Furthermore, it has been shown that information about weak interactions can be extracted from local values of the kinetic energy density which are available to meta-GGAs [73, 87], at least to a level where the region around the equilibrium in van der Waals bonds can be described. This development has been incorporated in the SCAN functional [73].

17.3 Materials Properties

One of the central questions with the materials genome effort is what basic properties are within reach to be collected and included in these databases. This is determined by a combination of what can be described by the level of theory used for the computations (as carefully reviewed in the previous section), and what methods are available as automated workflows. The starting point, crucial for building any materials genome-type resource, is the crystal structures and corresponding formation energies. The importance of the formation energies is due to their use in creating composition phase diagrams to estimate the zero temperature thermodynamic stability of a material. The composition phase diagram gives the ground state crystal structure of a material at zero temperature as a function of composition. It is constructed by determining the convex hull of the predicted formation energies of all competing crystal structures in a chemical subspace [16, 88, 89]. A compound with a formation energy on the convex hull is stable, whereas a compound that ends up above the hull is unstable. The distance to the hull can be used as a rough estimate of the degree to which a material is unstable (i.e., how unlikely it is to be observed, and if observed, how quickly it would deteriorate into a combination of lower energy structures.) Crystal structures with a small hull distance (very roughly up to $\sim$50 meV) may still be regarded as candidates for materials that in practice may be stable since such an "error margin" can account for meta-stability, stability at limited elevated temperatures, and the computational inaccuracy of the methods.

Several works have investigated the accuracy of DFT calculations of formation energies. The standard deviation of formation energies calculated with PBE+U to experiments for the formation of ternary oxides from binary oxides was found to be 0.024 eV/atom; meaning 90% of the errors are within 0.047 eV/atom, which corresponds to a mean absolute error of approximately 0.02 eV/atom [90]. Kirklin et al. determined a mean absolute error of PBE formation energies of systems over all chemistries to be 0.136 eV/atom, but with energy corrections that are often used in high-throughput databases to some of the elemental phase energies, this lowers to 0.081 eV/atom [91]. However, the same paper notes that for 75 intermetallic structures they found experimental results from more than one source, giving an estimate for the mean absolute error in the experiments of 0.082 eV/atom. (Note that the latter estimate may be affected by selection bias, i.e., there may be a larger probability of finding multiple experimental values if the results are uncertain.)

Presently the set of materials properties beyond stability and formation energies available for large data sets is somewhat limited. There is an ongoing competition between the online materials genome-type databases to grow the data they provide both in terms of included structures and materials properties. There is a wealth of methods in the literature that could potentially be used to produce data for many different properties. However, to turn these methods into a form where they can run reliably in high-throughput is non-trivial. Among the available databases, the Materials Project is quite comprehensive in terms of properties. In addition to structural information and formation energies, they have over the years added the KS-DFT band structure (in some cases corrected using the GW approximation [42, 92]), elastic tensors [93], piezoelectric tensors [94], dielectric properties [95], phonon spectra [96], synthesis descriptors [97], and X-ray absorption spectra [98].

17.4 Database-Driven High-Throughput Calculations

A basic flowchart for materials design using database-driven high-throughput calculations is shown in Fig. 17.1. There are many software packages with partially overlapping aims for helping with the steps in the flowchart. Some recognized open source examples are the *atomic simulation environment—ASE* (wiki.fysik.dtu.dk/ase), *pymatgen, custodian, and fireworks* (pymatgen.org, see also the information at materialsproject.org/infrastructure; connected to the Materials Project), *aflow* (materials.duke.edu/AFLOW; connected to the AFLOW repository), *AiiDA* (aiida.net; connected to materials cloud), *qmpy* (connected to the open quantum materials database). The author is involved in the development of the open source *high-throughput toolkit—httk* (httk.org) framework, which we use extensively for high-throughput computation in our own research, and which provides the backend for the open materials database. This toolkit provides functionality for preparing and running unsupervised workflows of calculations (electronic structure, mostly targeted towards the software package VASP), analyzing the results, and storing

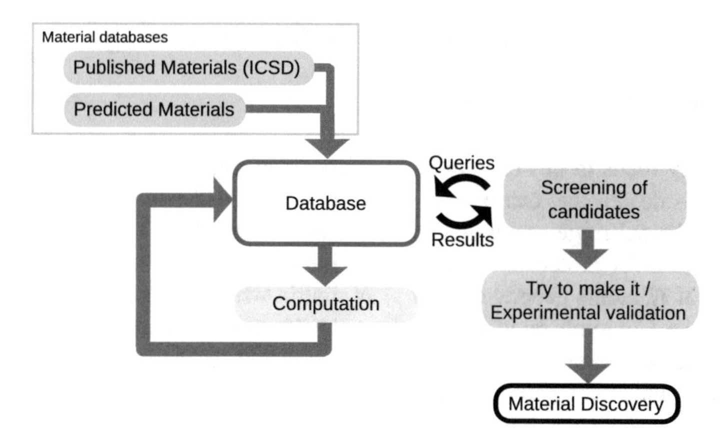

Fig. 17.1 A schematic flowchart representation of database-driven high-throughput materials design, largely inspired by the setup used in the Materials Project [16]. The steps on the right-hand sides represent the use of the database to find materials with desirable properties. In the context of machine learning models, the materials and materials properties in the database can be used for training and validation

them in a global and/or in a personalized database. The basic functionalities of these software packages are quite similar; in the following, we discuss the functionality of *httk*.

The primary focus of *httk* is for running automated calculations with as little human intervention as possible. This is crucial when working with large data sets, but can also be convenient when working with smaller projects. The toolkit consists of a software library developed in Python and a set of script programs that enable the interaction with supercomputers. The primary strengths of this framework compared to common alternatives are (1) the Python library provides a very integrated object-relational mapper, where classes in object-oriented Python are used to introduce abstractions that remove much of the difficulty in setting up a personal database of SQL type in which one can store, search, retrieve, and analyze results; (2) *httk* consistently allows the use of exact rational numbers in place of the more commonly used floating-point numbers. The exact rational numbers allow processing of crystal structures, application of transforms, etc., without the usual loss of precision. Hence, *httk* can deterministically produce an internal representation of structures read from a source file (e.g., on the cif file format), which is not the case in most other frameworks due to their use of floating-point numbers means the precise end result is influenced by the computer architecture.

The *httk* framework is distributed in several ways, including the PyPI service. Hence, it can easily be installed by issuing: `pip install httk` on a system with a modern distribution of Python. There is a set of tutorial steps and a large number of examples available to show how the framework can be utilized in the various steps of database-driven high-throughput as shown in Fig. 17.1. These are available via the project website (httk.openmaterialsdb.se).

17.5 Machine Learning Models for Materials Design

17.5.1 Models for Molecules

The primary focus in this chapter is on a type of machine learning models for use in materials design that can be said to begin with a 2012 paper by Rupp et al. on the use of kernel ridge regression for small molecules [99]. They define a matrix representation for molecules named the "Coulomb matrix." In this representation a system of N atoms generates an $N \times N$ matrix where the off-diagonal elements (i,j) are the Coulomb repulsion between the ith and jth bare atomic cores, and the diagonal elements are based on a polynomial fit to energies of free atoms to represent a static energy contribution pertaining to the ith atom,

$$C_{ij} = \begin{cases} 0.5 Z_i^{2.4} & \text{if } i = j \\ Z_i Z_j / (\|\mathbf{r}_i - \mathbf{r}_j\|_2) & \text{if } i \neq j \end{cases} \qquad (17.1)$$

One may note that the Coulomb interaction between the bare atomic cores is not a good indicator of the physics of the bonds in a system. However, the representation does not aim to push the machine learning model into a specific physics-based description, but just to constitute a well-formed way to represent the structural information (i.e., the positions of the atoms) so that the machine is free to learn the physics from the data. This model was trained on small organic molecules (with up to 7 atoms from the elements C, O, N, and S, and with the valencies satisfied by hydrogen atoms; this data set is named *qm7*.) It was shown in the original paper that the machine can be trained to predict atomization energies of molecules not in the training set down to a mean absolute error of 10 kcal/mol at a training set size of 7k. In units more common for materials, this model reaches 20 meV/atom at a training set of 3000 molecules from qm7 [100].

17.5.2 General Models for Periodic Systems

In a 2015 work Faber, Lindmaa, von Lilienfeld, and Armiento (the author of the present chapter) extended the Coulomb matrix construct into a suitable form for periodic crystals [100]. This extension is non-trivial, since there exist more than one way to choose a unit cell in a periodic system, and therefore representations based on the Coulomb matrix easily become non-unique. As pointed out in that paper, the aim when seeking a representation for atomistic systems is to find one that is (1) *complete*: incorporates all features of the structural information that are relevant for the underlying problem, but at the same time; (2) *compact*: avoids representation of features irrelevant for the underlying problem (e.g., static rotations); (3) *descriptive*: structural similarity should give representations that are close; and (4) *simple*: low computational effort to compute, and conceptually easy to understand.

The end result of Ref. [100] was three alternative Coulomb matrix inspired representations applicable to periodic crystals. The first one was based on replacing the bare Coulomb interactions in the off-diagonal matrix elements with the corresponding expression for fully periodic systems, i.e., the sum of the total Coulomb interaction energy per unit cell between the infinite periodic lattices of the bare cores of repetitions of two separate atoms in the unit cell. These expressions are evaluated via Ewald sums [101]. The issue with this expression is that it is somewhat computationally expensive and non-trivial to evaluate correctly. The second generalization of the Coulomb matrix was to duplicate the unit cell a number of times and then use the same expression as for the non-periodic Coulomb matrix, however, with a screened Coulomb interaction (i.e., where the interaction decays exponentially to give a finite reach.) This is very similar to just using the short range term in the Ewald sum. To get an even simpler descriptor, a third expression was invented. It was shown how the Ewald sum can be replaced by an expression that mimics the basic shape and periodicity of the Ewald expression, but which still remains on a simple closed form that is easy to evaluate. This expression was named the “sine” or “sinusoidal” descriptor, because of how it reproduced the periodicity over the unit cell via a sine function.

The three alternative extensions of the Coulomb matrix to periodic systems were tested on a data set that is now known as FLAA (from the authors' initials). It consists of structures with up to 25 atoms that were randomly selected out of the Materials Project database. In these structures most atomic species occur, in proportions roughly similar to their occurrence in structures published in the literature and extracted into the inorganic crystal structure database (ICSD) [102, 103] which is the main source of crystal structures for the Materials Project. The conclusion of the 2015 paper [100] was that all three alternative extensions of the Coulomb matrix to periodic systems performed approximately equal. The sine descriptor did slightly better than the others, with a 370 meV/atom mean absolute error for predicting formation energies when trained on 3k structures from the FLAA data set.

Two main conclusions follow from the above results. Firstly, the performance of kernel ridge regression-based machines for atomistic systems does not appear to be particularly sensitive to the exact details of how the generalized Coulomb matrix descriptors are constructed, as long as they reasonably well adhere to the aims for a good representation listed above. Secondly, at first glance it may appear as if the performance of the models for molecules far outperforms the corresponding ones for periodic crystals (20 meV/atom vs. 0.370 meV/atom). However, the sizes of the chemical space for the two cases are not comparable, and arguably the one used for crystals in Ref. [100] is far larger.

17.5.3 Crystal-Structure Specific Models

To demonstrate that these types of models are capable of reaching a level of accuracy directly useful for applications if one restricts the chemical space, the same authors investigated in 2016 a machine learning model operating on such a smaller space [104]. This work considered all substitutions of main group elements into four sites of one specific quaternary crystal structure, the elpasolite. This structure was selected because it is the quaternary crystal most frequently occurring with different substitutions in the ICSD database, indicating that this structure can accommodate many types of bonds and thus to be rewarding to characterize fully. High-throughput DFT calculations using the *httk* framework were used to produce data for ca 10k substitutions of elements into the elpasolite crystal structure out a total of two million possibilities. Furthermore, a subset of 12 main group elements was selected to give a reduced chemical space of 12k possible substitutions, which were run exhaustively.

A substitution into a fixed crystal structure can be uniquely specified by giving which chemical species are at which atomic site in the structure. Hence, the 2016 paper used a very straightforward representation of, essentially, a 2×4 matrix that specified the row and column in the periodic table of the atom species at each of the four sites in the elpasolite structure. This leaves out the precise structural information of the system from the descriptor, i.e., the bond lengths between the atoms. The 2×4 matrix descriptor should be understood to technically refer to the system relaxed while confined to the elpasolite crystal structure.

A kernel ridge regression machine learning model was trained using this descriptor on formation energies for structures in the elpasolite data set, and it was shown that (1) by training on a sufficiently large subset of the exhaustive 12k data set, the model can reach essentially any level of accuracy for predictions of structures outside the training set, at least below <10 meV/atom which is significantly less than the errors in the DFT data. (See the discussions of accuracy of DFT formation energies in Sect. 17.3.) This shows that the performance of this machine learning model is merely a question of having a large enough training set; (2) when training on data in the larger chemical space of two million possible substitutions of main group elements into the elpasolite structure, it was sufficient to train on about 10k structures to reach roughly the accuracy of the DFT calculations, 100 meV/atom. This result means that the machine learning model was capable of producing DFT-quality formation energies with a net $\times 200$ speedup, including all the time used to produce the training data. The resulting two million formation energies are illustrated in Fig. 17.2 reproduced from the original paper.

Furthermore, the 2016 paper also demonstrated a practical use of the large set of predicted formation energies. Phase diagrams were created for most of the elpasolite systems by using information about competing compositions from the Materials Project using the pymatgen Python library (some systems were outright dismissed on grounds of containing rare-gas elements). From these phase diagrams a number of candidates for thermodynamically stable materials were obtained by identifying compositions with a predicted formation energy on the convex hull. These candidates were validated by DFT calculations and 90 systems were confirmed to be thermodynamically stable within this level of theory. However, the compounds that passed validation only constituted a small fraction of the candidates. As explained in the paper, the reason is that the process of identifying structures on the convex hull is a screening for systems with the lowest formation energies, which are outliers in the full data set. The interpolative nature of machine learning models leads to them being significantly less accurate in predicting properties of outlier systems. Nevertheless, even with this limitation, the scheme far reduced the number of DFT calculations needed to identify thermodynamically stable elpasolite systems compared to just obtaining all formation energies from DFT calculations. The net result was a $\times 11$ speedup, including the full time spent both on the training set and the calculations used to validate the materials picked out as candidates for stability.

Hence, the crystal-structure-specific machine was demonstrated to be very successful for generating large amounts of formation energy data which is useful for greatly accelerating predictions of stable compounds in a considered crystal structure. The predictions allow extending the available data in materials genome-type databases. The structures identified as stable in the work discussed above are now available (with some singular exceptions) via the Materials Project and, e.g., enters the predictions of convex hulls for user-generated phase diagrams via their online service, thus contributing to the accuracy of those predictions.

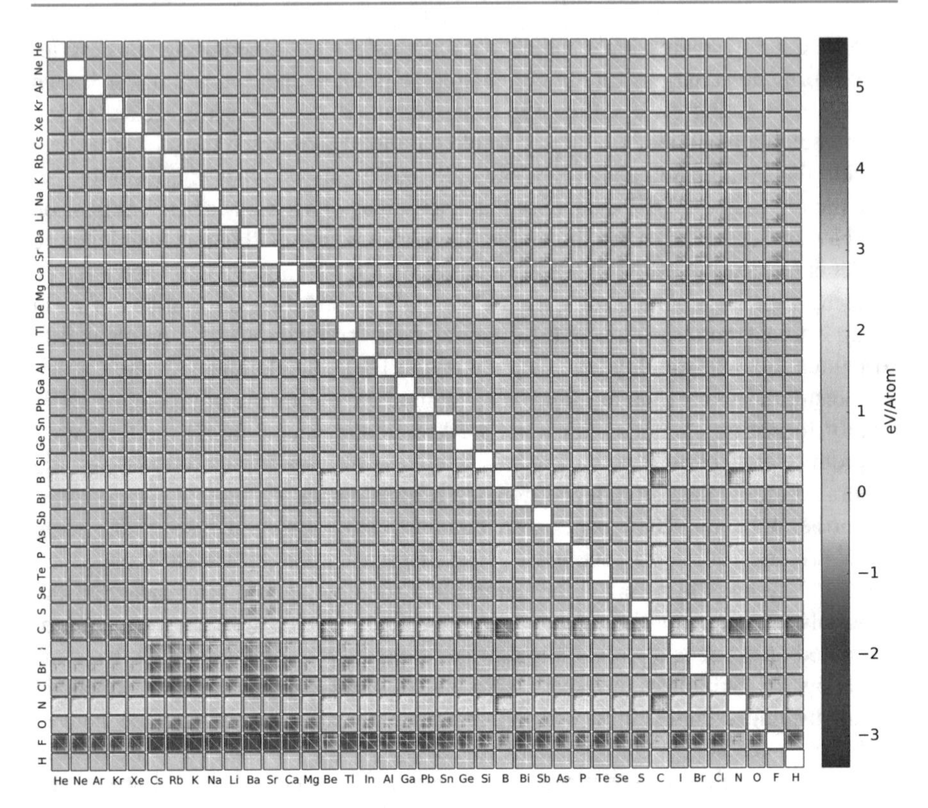

Fig. 17.2 Color matrix of the two million elpasolite energies predicted with the crystal-structure specific machine learning model of Faber et al. [104]. The x- and y-axes specify which atomic species sits on two of the four sites in the crystal structure. At those coordinates one finds a miniature diagram over the species at the remaining two sites. Every pixel in the miniature diagram shows a formation energy of the corresponding composition of four atomic species. The figure is reproduced from the original paper and is licensed under the Creative Commons Attribution 3.0 License

17.5.4 Models for Predicting Composition Phase Diagrams, Crystal Structures

The success of the crystal-structure-specific machine notwithstanding, it does not directly answer the most typical materials design problem. It is, arguably, more common to seek the stable crystal structures that can be formed from a given set of chemical species, rather than all the stable chemical compositions that share the same crystal structure. This is, in essence, the crystal structure prediction problem.

In 2016, Tholander, Andersson, Armiento, Tasnádi, and Alling [105] (TAATA) produced a data set by high-throughput calculations using the *httk* framework. The aim was to seek stable crystal structures in the ternary chemical systems Ti-Zn-N, Zr-Zn-N, and Hf-Zn-N for possible use in piezoelectrics. This high-throughput data set is a good real-world test case to evaluate the possible acceleration of the generation of phase diagrams for identifying stable structures using machine learning models.

The author of this chapter and coworkers have since then engaged in a project of trying out new machine learning models on this problem and to develop new ones for it; the progress on this was recently reported in, e.g., Ref. [106]. At the present stage, it appears the original Coulomb matrix-based descriptors from Ref. [100] perform similar on this data set as for the original FLAA data set, which is encouraging in establishing the generality of these models. However, the resulting accuracy is not sufficient to be useful for accelerating the production of the phase diagrams. Compared to the FLAA set, the TAATA data set has much fewer atomic species, but at the same time is comprised of structures over a very wide range of formation energies. The origin of the structures in the FLAA set is the Materials Project which, as explained above, are based on structures from the ICSD database. The ICSD primarily indexes materials seen in nature which means most are thermodynamically stable and have comparably low formation energies. This restriction lowers the dimensionality of the chemical space of FLAA relative to that of TAATA.

Other recent machine learning models perform better; e.g., in Ref. [106] it was found that a descriptor by Ward et al. that encodes structural information using a Voronoi tessellation reaches a mean absolute error of 0.28 eV/atom for 10k structures from the TAATA data set [107]. While errors on this level are not small enough to replace the need for DFT calculations with model predictions, one may still be able to use predictions to identify and remove competing structures that are highly unstable and therefore would not influence the phase diagram, thus reducing the number of DFT calculations necessary, giving an overall reduction in the effort of producing the phase diagram. The field moves rapidly forward, and some other interesting recent developments are found in Refs. [108–111].

17.6 Conclusions and Outlook

This chapter has reviewed several aspects of producing training data by database-driven high-throughput calculations, and the use of this data to train machine learning models with the aim of accelerating materials design. All these aspects are making rapid and encouraging progress. The research-front machine learning methods are now on the edge of producing results that are accurate and reliable enough to accelerate theoretical prediction of thermodynamic stability via the creation of convex hulls; i.e., the crystal prediction problem which arguably is the most important first step for materials design of bulk materials with desired properties. Further progress towards this goal, and for predicting other properties, is continuously being made. Looking forward, two crucial points can be raised: (1) further development of general machine learning models for atomistic systems with improved accuracy and a reduced need for training data is needed; but how far can that development go before it hits a fundamental wall where not enough information about the underlying physics is present in the data?; (2) the rapid development of machine learning models will drive a need for more accurate training data. Will the progress of physics-based computational methods be able to keep up with this need of methods with improved accuracy but low enough computational effort to be useful in high-throughput?; or will the lack of a sufficient amount of high quality training data become a major bottleneck for further progress? Future research needs to target both these areas.

Acknowledgments The author thanks Anatole von Lilienfeld and Felix Faber for many insightful discussions on topics in the overlap of machine learning and materials design. Joel Davidsson is acknowledged for help with supervising the master's thesis discussed in the text as Ref. [106]. The author acknowledges support from the Swedish e-Science Centre (SeRC), Swedish Research Council (VR) Grants No. 2016-04810, and the Centre in Nano science and Nanotechnology (CeNano) at Linköping University. Some of the discussed computations were performed on resources provided by the Swedish National Infrastructure for Computing (SNIC) at the National Supercomputer Centre (NSC) at Linköping University.

References

1. J. Greeley, T.F. Jaramillo, J. Bonde, I. Chorkendorff, J.K. Nørskov, Nat. Mater. **5**(11), 909 (2006)
2. K. Kang, Y.S. Meng, J. Bréger, C.P. Grey, G. Ceder, Science **311**(5763), 977 (2006)
3. S. Kirklin, B. Meredig, C. Wolverton, Adv. Energy Mater. **3**(2), 252 (2013)
4. C. Ortiz, O. Eriksson, M. Klintenberg, Comput. Mater. Sci. **44**(4), 1042 (2009)
5. M. Klintenberg, O. Eriksson, Comput. Mater. Sci. **67**, 282 (2013)
6. G.K.H. Madsen, J. Am. Chem. Soc. **128**(37), 12140 (2006)
7. S. Wang, Z. Wang, W. Setyawan, N. Mingo, S. Curtarolo, Phys. Rev. X **1**(2), 021012 (2011)
8. R. Armiento, B. Kozinsky, M. Fornari, G. Ceder, Phys. Rev. B **84**(1) (2011)
9. R. Armiento, B. Kozinsky, G. Hautier, M. Fornari, G. Ceder, Phys. Rev. B **89**(13), 134103 (2014)
10. G. Hautier, A. Miglio, G. Ceder, G.M. Rignanese, X. Gonze, Nat. Commun. **4**, 2292 (2013)
11. S. Lebègue, T. Björkman, M. Klintenberg, R.M. Nieminen, O. Eriksson, Phys. Rev. X **3**(3), 031002 (2013)
12. S. Curtarolo, G.L.W. Hart, M.B. Nardelli, N. Mingo, S. Sanvito, O. Levy, Nat. Mater. **12**(3), 191 (2013)
13. G. Ceder, K.A. Persson, Sci. Amer. **309**(6), 36 (2013)
14. K. Alberi, M.B. Nardelli, A. Zakutayev, L. Mitas, S. Curtarolo, A. Jain, M. Fornari, N. Marzari, I. Takeuchi, M.L. Green, M. Kanatzidis, M.F. Toney, S. Butenko, B. Meredig, S. Lany, U. Kattner, A. Davydov, E.S. Toberer, V. Stevanovic, A. Walsh, N.G. Park, A. Aspuru-Guzik, D.P. Tabor, J. Nelson, J. Murphy, A. Setlur, J. Gregoire, H. Li, R. Xiao, A. Ludwig, L.W. Martin, A.M. Rappe, S.-H. Wei, J. Perkins, J. Phys. D: Appl. Phys. **52**(1), 013001 (2019)
15. F. Oba, Y. Kumagai, Appl. Phys. Express **11**(6), 060101 (2018)
16. A. Jain, G. Hautier, C.J. Moore, S. Ping Ong, C.C. Fischer, T. Mueller, K.A. Persson, G. Ceder, Comput. Mater. Sci. **50**(8), 2295 (2011)
17. A. Jain, S.P. Ong, G. Hautier, W. Chen, W.D. Richards, S. Dacek, S. Cholia, D. Gunter, D. Skinner, G. Ceder, K.A. Persson, APL Mater. **1**(1), 011002 (2013)
18. Executive Office of the President National Science and Technology Council, Washington. Materials Genome Initiative for Global Competitiveness (2011). https://www.mgi.gov/sites/default/files/documents/materials_genome_initiative-final.pdf; https://www.mgi.gov/
19. K. Rajan, Mater. Today **8**(10), 38 (2005)

20. J.R. Rodgers, D. Cebon, MRS Bull. **31**(12), 975 (2006)
21. R. Kurzweil, *The Singularity Is Near: When Humans Transcend Biology* (Penguin Books, New York, 2006)
22. S. Ulam, Bull. Amer. Math. Soc. **64**(3), 1 (1958)
23. P. Hohenberg, W. Kohn, Phys. Rev. **136**(3B), B864 (1964)
24. W. Kohn, L.J. Sham, Phys. Rev. **140**(4A), A1133 (1965)
25. J.P. Perdew, K. Schmidt, in *AIP Conference Proceedings*, vol. 577 (AIP, College Park, 2001), pp. 1–20
26. J.P. Perdew, K. Burke, M. Ernzerhof, Phys. Rev. Lett. **77**(18), 3865 (1996)
27. V.I. Anisimov, F. Aryasetiawan, A.I. Lichtenstein, J. Phys. Condens. Matter **9**(4), 767 (1997)
28. F. Zhou, M. Cococcioni, C.A. Marianetti, D. Morgan, G. Ceder, Phys. Rev. B **70**(23), 235121 (2004)
29. V.L. Chevrier, S.P. Ong, R. Armiento, M.K.Y. Chan, G. Ceder, Phys. Rev. B **82**(7), 075122 (2010)
30. S. Kümmel, L. Kronik, J.P. Perdew, Phys. Rev. Lett. **93**(21), 213002 (2004)
31. R. Armiento, S. Kümmel, T. Körzdörfer, Phys. Rev. B **77**(16), 165106 (2008)
32. A.E. Mattsson, R.R. Wixom, R. Armiento, Phys. Rev. B **77**(15), 155211 (2008)
33. P. Rinke, A. Janotti, M. Scheffler, C.G. Van de Walle, Phys. Rev. Lett. **102**(2), 026402 (2009)
34. V. Vlček, G. Steinle-Neumann, L. Leppert, R. Armiento, S. Kümmel, Phys. Rev. B **91**(3), 035107 (2015)
35. J.P. Perdew, Chem. Phys. Lett. **64**(1), 127 (1979)
36. J.P. Perdew, A. Zunger, Phys. Rev. B **23**(10), 5048 (1981)
37. R.O. Jones, O. Gunnarsson, Rev. Mod. Phys. **61**(3), 689 (1989)
38. M. Städele, M. Moukara, J.A. Majewski, P. Vogl, A. Görling, Phys. Rev. B **59**(15), 10031 (1999)
39. A.D. Becke, J. Chem. Phys. **98**(7), 5648 (1993)
40. J. Heyd, G.E. Scuseria, M. Ernzerhof, J. Chem. Phys. **118**(18), 8207 (2003)
41. J. Heyd, G.E. Scuseria, M. Ernzerhof, J. Chem. Phys. **124**(21), 219906 (2006)
42. L. Hedin, Phys. Rev. **139**(3A), A796 (1965)
43. R. van Leeuwen, E.J. Baerends, Phys. Rev. A **49**(4), 2421 (1994)
44. O. Gritsenko, R. van Leeuwen, E. van Lenthe, E.J. Baerends, Phys. Rev. A **51**(3), 1944 (1995)
45. A.D. Becke, E.R. Johnson, J. Chem. Phys. **124**(22), 221101 (2006)
46. N. Umezawa, Phys. Rev. A **74**(3), 032505 (2006)
47. E. Räsänen, S. Pittalis, C.R. Proetto, J. Chem. Phys. **132**(4), 044112 (2010)
48. F. Tran, P. Blaha, Phys. Rev. Lett. **102**(22), 226401 (2009)
49. M.J.T. Oliveira, E. Räsänen, S. Pittalis, M.A.L. Marques, J. Chem. Theory Comput. **6**(12), 3664 (2010)
50. D.J. Singh, Phys. Rev. B **82**(20), 205102 (2010)
51. R. van Leeuwen, E.J. Baerends, Phys. Rev. A **51**(1), 170 (1995)
52. A.P. Gaiduk, V.N. Staroverov, Phys. Rev. A **83**(1), 012509 (2011)
53. A. Karolewski, R. Armiento, S. Kümmel, J. Chem. Theory Comput. **5**(4), 712 (2009)
54. R. Armiento, S. Kümmel, Phys. Rev. Lett. **111**(3), 036402 (2013)
55. T. Aschebrock, R. Armiento, S. Kümmel, Phys. Rev. B **96**(7), 075140 (2017)
56. T.F.T. Cerqueira, M.J.T. Oliveira, M.A.L. Marques, J. Chem. Theory Comput. **10**(12), 5625 (2014)
57. F. Tran, P. Blaha, M. Betzinger, S. Blügel, Phys. Rev. B **91**(16), 165121 (2015)
58. A. Lindmaa, R. Armiento, Phys. Rev. B **94**(15), 155143 (2016)
59. T. Aschebrock, R. Armiento, S. Kümmel, Phys. Rev. B **95**(24), 245118 (2017)
60. R. Armiento, A.E. Mattsson, Phys. Rev. B **72**(8), 085108 (2005)
61. A.E. Mattsson, R. Armiento, Phys. Rev. B **79**(15), 155101 (2009)
62. A.E. Mattsson, R. Armiento, J. Paier, G. Kresse, J.M. Wills, T.R. Mattsson, J. Chem. Phys. **128**(8), 084714 (2008)
63. P. Haas, F. Tran, P. Blaha, Phys. Rev. B **79**(8), 085104 (2009)
64. P. Haas, F. Tran, P. Blaha, Phys. Rev. B **79**(20), 209902 (2009)

65. Z. Wu, R.E. Cohen, Phys. Rev. B **73**(23), 235116 (2006)
66. Y. Zhao, D.G. Truhlar, Phys. Rev. B **78**(19), 197101 (2008)
67. Z. Wu, R.E. Cohen, Phys. Rev. B **78**(19), 197102 (2008)
68. Y. Zhao, D.G. Truhlar, J. Chem. Phys. **128**(18), 184109 (2008)
69. J.P. Perdew, A. Ruzsinszky, G.I. Csonka, O.A. Vydrov, G.E. Scuseria, L.A. Constantin, X. Zhou, K. Burke, Phys. Rev. Lett. **100**(13), 136406 (2008)
70. A.E. Mattsson, R. Armiento, T.R. Mattsson, Phys. Rev. Lett. **101**(23), 239701 (2008)
71. J.P. Perdew, A. Ruzsinszky, G.I. Csonka, O.A. Vydrov, G.E. Scuseria, L.A. Constantin, X. Zhou, K. Burke, Phys. Rev. Lett. **101**(23), 239702 (2008)
72. J.P. Perdew, A. Ruzsinszky, G.I. Csonka, O.A. Vydrov, G.E. Scuseria, L.A. Constantin, X. Zhou, K. Burke, Phys. Rev. Lett. **102**(3), 039902 (2009)
73. J. Sun, A. Ruzsinszky, J.P. Perdew, Phys. Rev. Lett. **115**(3), 036402 (2015)
74. J. Sun, R.C. Remsing, Y. Zhang, Z. Sun, A. Ruzsinszky, H. Peng, Z. Yang, A. Paul, U. Waghmare, X. Wu, M.L. Klein, J.P. Perdew, Nat. Chem. **8**(9), 831 (2016). https://doi.org/10.1038/nchem.2535. https://www.nature.com/articles/nchem.2535
75. Y. Zhang, D.A. Kitchaev, J. Yang, T. Chen, S.T. Dacek, R.A. Sarmiento-Pérez, M.A.L. Marques, H. Peng, G. Ceder, J.P. Perdew, J. Sun, npj Comput. Mater. **4**(1), 9 (2018). https://doi.org/10.1038/s41524-018-0065-z. https://www.nature.com/articles/s41524-018-0065-z
76. M. Ekholm, D. Gambino, H.J.M. Jönsson, F. Tasnádi, B. Alling, I.A. Abrikosov, Phys. Rev. B **98**(9), 094413 (2018). https://doi.org/10.1103/PhysRevB.98.094413. https://link.aps.org/doi/10.1103/PhysRevB.98.094413
77. S. Grimme, J. Comput. Chem. **27**(15), 1787 (2006)
78. S. Grimme, J. Antony, S. Ehrlich, H. Krieg, J. Chem. Phys. **132**(15), 154104 (2010)
79. A. Tkatchenko, M. Scheffler, Phys. Rev. Lett. **102**(7), 073005 (2009)
80. A. Tkatchenko, R.A. DiStasio, R. Car, M. Scheffler, Phys. Rev. Lett. **108**(23), 236402 (2012)
81. A. Ambrosetti, A.M. Reilly, R.A. DiStasio, A. Tkatchenko, J. Chem. Phys. **140**(18), 18A508 (2014)
82. S.N. Steinmann, C. Corminboeuf, J. Chem. Theory Comput. **7**(11), 3567 (2011)
83. S.N. Steinmann, C. Corminboeuf, J. Chem. Phys. **134**(4), 044117 (2011)
84. M. Dion, H. Rydberg, E. Schröder, D.C. Langreth, B.I. Lundqvist, Phys. Rev. Lett. **92**(24), 246401 (2004)
85. K. Lee, E.D. Murray, L. Kong, B.I. Lundqvist, D.C. Langreth, Phys. Rev. B **82**(8), 081101 (2010)
86. K. Berland, P. Hyldgaard, Phys. Rev. B **89**(3), 035412 (2014)
87. J. Sun, B. Xiao, Y. Fang, R. Haunschild, P. Hao, A. Ruzsinszky, G.I. Csonka, G.E. Scuseria, J.P. Perdew, Phys. Rev. Lett. **111**(10), 106401 (2013)
88. A.R. Akbarzadeh, V. Ozoliņš, C. Wolverton, Adv. Mater. **19**(20), 3233 (2007)
89. S.P. Ong, L. Wang, B. Kang, G. Ceder, Chem. Mater. **20**(5), 1798 (2008)
90. G. Hautier, S.P. Ong, A. Jain, C.J. Moore, G. Ceder, Phys. Rev. B **85**(15), 155208 (2012)
91. S. Kirklin, J.E. Saal, B. Meredig, A. Thompson, J.W. Doak, M. Aykol, S. Rühl, C.M. Wolverton, npj Comput. Mater. **1**, 15010 (2015)
92. I.E. Castelli, F. Hüser, M. Pandey, H. Li, K.S. Thygesen, B. Seger, A. Jain, K.A. Persson, G. Ceder, K.W. Jacobsen, Adv. Energy Mater. **5**(2), 1400915 (2015)
93. M. de Jong, W. Chen, T. Angsten, A. Jain, R. Notestine, A. Gamst, M. Sluiter, C. Krishna Ande, S. van der Zwaag, J.J. Plata, C. Toher, S. Curtarolo, G. Ceder, K.A. Persson, M. Asta, Sci. Data **2**, 150009 (2015)
94. M. de Jong, W. Chen, H. Geerlings, M. Asta, K.A. Persson, Sci. Data **2**, 150053 (2015)
95. I. Petousis, D. Mrdjenovich, E. Ballouz, M. Liu, D. Winston, W. Chen, T. Graf, T.D. Schladt, K.A. Persson, F.B. Prinz, Sci. Data **4**, 160134 (2017)
96. G. Petretto, S. Dwaraknath, H.P.C. Miranda, D. Winston, M. Giantomassi, M.J. van Setten, X. Gonze, K.A. Persson, G. Hautier, G.M. Rignanese, Sci. Data **5**, 180065 (2018)
97. E. Kim, K. Huang, A. Saunders, A. McCallum, G. Ceder, E. Olivetti, Chem. Mater. **29**(21), 9436 (2017)

98. K. Mathew, C. Zheng, D. Winston, C. Chen, A. Dozier, J.J. Rehr, S.P. Ong, K.A. Persson, Sci. Data **5**, 180151 (2018)
99. M. Rupp, A. Tkatchenko, K.R. Müller, O.A. von Lilienfeld, Phys. Rev. Lett. **108**(5), 058301 (2012)
100. F. Faber, A. Lindmaa, O.A.V. Lilienfeld, R. Armiento, Int. J. Quantum Chem. **115**(16), 1094 (2015)
101. P.P. Ewald, Ann. Phys. **369**(3), 253 (1921)
102. G. Bergerhoff, R. Hundt, R. Sievers, I.D. Brown, J. Chem. Inf. Comput. Sci. **23**(2), 66 (1983)
103. A. Belsky, M. Hellenbrandt, V.L. Karen, P. Luksch, Acta Cryst. B **58**(3–1), 364 (2002)
104. F.A. Faber, A. Lindmaa, O.A.v. Lilienfeld, R. Armiento, Phys. Rev. Lett. **117**(13), 135502 (2016)
105. C. Tholander, C.B.A. Andersson, R. Armiento, F. Tasnádi, B. Alling, J. Appl. Phys. **120**(22), 225102 (2016)
106. C. Bratu, Machine Learning of Crystal Formation Energies with Novel Structural Descriptors. Master's Thesis, Linköping University, Sweden, 2017. http://urn.kb.se/resolve?urn=urn:nbn:se:liu:diva-143203
107. L. Ward, R. Liu, A. Krishna, V.I. Hegde, A. Agrawal, A. Choudhary, C. Wolverton, Phys. Rev. B **96**(2), 024104 (2017)
108. F.A. Faber, A.S. Christensen, B. Huang, O.A. von Lilienfeld, J. Chem. Phys. **148**(24), 241717 (2018)
109. H. Huo, M. Rupp (2017). arXiv:1704.06439
110. K.T. Schütt, H.E. Sauceda, P.J. Kindermans, A. Tkatchenko, K.R. Müller, J. Chem. Phys. **148**(24), 241722 (2018)
111. W. Ye, C. Chen, Z. Wang, I.H. Chu, S.P. Ong, Nat. Commun. **9**(1), 3800 (2018)

Polymer Genome: A Polymer Informatics Platform to Accelerate Polymer Discovery

18

Anand Chandrasekaran, Chiho Kim, and Rampi Ramprasad

Abstract

The Materials Genome Initiative has brought about a paradigm shift in the design and discovery of novel materials. In a growing number of applications, the materials innovation cycle has been greatly accelerated as a result of insights provided by data-driven materials informatics platforms. High-throughput computational methodologies, data descriptors, and machine learning are playing an increasingly invaluable role in research development portfolios across both academia and industry. Polymers, especially, have long suffered from a lack of data on electronic, mechanical, and dielectric properties across large chemical spaces, causing a stagnation in the set of suitable candidates for various applications. The nascent field of polymer informatics seeks to provide tools and pathways for accelerated polymer property prediction (and materials design) via surrogate machine learning models built on reliable past data. With this goal in mind, we have carefully accumulated a dataset of organic polymers whose properties were obtained either computationally (bandgap, dielectric constant, refractive index, and atomization energy) or experimentally (glass transition temperature, solubility parameter, and density). A fingerprinting scheme that captures atomistic to morphological structural features was developed to numerically represent the polymers. Machine learning models were then trained by mapping the polymer fingerprints (or features) to their respective properties. Once developed, these models can rapidly predict properties of new polymers (within the same chemical class as the parent dataset) and can also provide uncertainties underlying the predictions. Since different properties depend on different length-scale features, the prediction models were built on an optimized

---

A. Chandrasekaran · C. Kim · R. Ramprasad (✉)
School of Materials Science and Engineering, Georgia Institute of Technology, Atlanta, GA, USA
e-mail: chiho.kim@gatech.edu; rampi.ramprasad@mse.gatech.edu

K. T. Schütt et al. (eds.), *Machine Learning Meets Quantum Physics*, Lecture Notes in Physics 968, https://doi.org/10.1007/978-3-030-40245-7_18

set of features for each individual property. Furthermore, these models are incorporated in a user friendly online platform named Polymer Genome (www.polymergenome.org). Systematic and progressive expansion of both chemical and property spaces are planned to extend the applicability of Polymer Genome to a wide range of technological domains.

18.1 Introduction: Applications of Machine Learning in Materials Science

The past few years have been witness to a surge in the application of data-driven techniques to a broad spectrum of research and development fields. The discipline of machine learning [1], responsible for bringing such techniques to light, has seen multiple breakthroughs over the past two decades. One of the factors responsible for such rapid advancements in the field is the development of novel algorithms and quantitative approaches capable of learning any arbitrary mapping between a given input and the corresponding output. The increased availability of vast amounts of data and the reduction in the cost of fast computational resources are other reasons that have abetted in the preeminence of the field of machine learning.

The materials science and chemistry communities have greatly benefited from machine learning approaches over the past few years. In these communities, there have been many large-scale efforts to curate accurate and reliable databases of materials properties (both computational [2–4] and experimental). Large-scale programs such as the Materials Genome Initiative [5] (in the USA), NOMAD [6] (in Europe), and MARVEL [7] (in Switzerland) have contributed to the development of novel database infrastructures tailored to materials science challenges and have also resulted in high-throughput frameworks capable of leveraging the power of modern high-performance computing facilities [8, 9].

In materials science, the increasing availability of large amounts of data (both computational and experimental) has led to the prominent field of materials informatics [10–24]. The overarching goal of the field materials informatics is to accelerate the development of novel materials for specific applications. To this end, the materials science community has used machine learning to accelerate various stages of the materials discovery pipeline. For example, a variety of machine learning force-fields [25–28] have been developed to provide rapid predictions of energies and forces with quantum-mechanical accuracy. Other approaches involve the utilization of machine learning approaches to bypass the Kohn–Sham equations to directly obtain important electronic properties such as the charge density [29].

This chapter, however, focuses on the ability of surrogate models to directly predict higher length-scale properties of materials. As shown in Fig. 18.1, the measurement of materials properties has traditionally involved computationally expensive quantum-mechanical simulations or perhaps the utilization of a time-consuming or laborious experimental technique. A novel paradigm has emerged in recent years wherein the properties of materials can be directly and rapidly obtained using predictive frameworks employing machine learning methodologies.

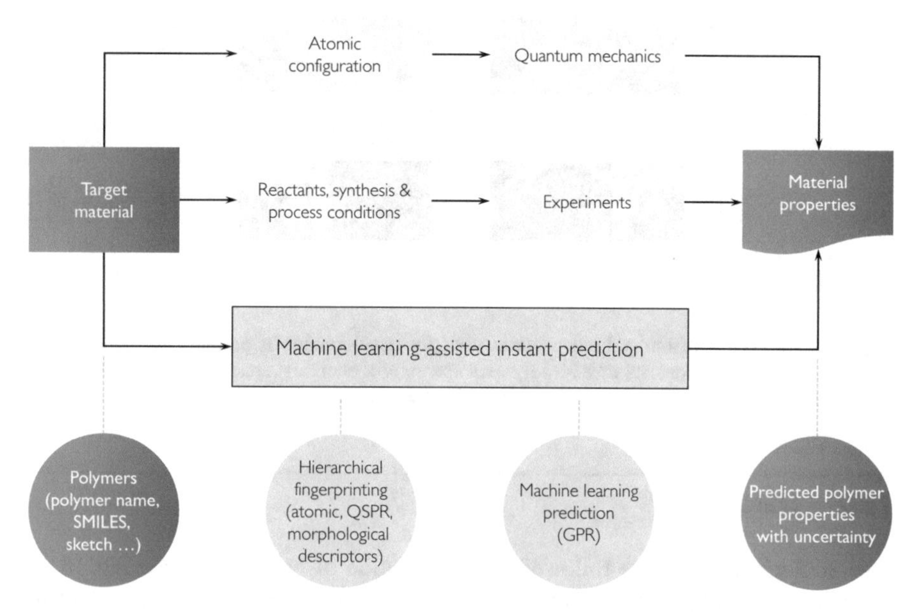

Fig. 18.1 The top two workflows indicate how the physical properties of materials can be obtained using traditional computational or experimental pipelines. Recent efforts, such as the *Polymer Genome* paradigm, seeks to accelerate the prediction of materials properties using machine learning approaches

A specific example where such a paradigm has been of great utility is the nascent field of polymer informatics. Polymers form an important (and challenging) materials class and they are pervasive with applications ranging from daily products, e.g., plastic packaging and containers, to state-of-the-art technological components, e.g., high-energy density capacitors, electrolytes for Li-ion batteries, polymer light-emitting diodes, and photovoltaic materials. Their chemical and morphological spaces are immensely vast and complex [30], leading to fundamental obstacles in polymer discovery. Some recent successes in rationally designing polymer dielectrics via experiment-computation synergies [10, 11, 19, 23, 31–38] indicate that there may be opportunities for machine learning and informatics approaches in this challenging research and development area.

We have created an informatics platform capable of predicting a variety of important polymer properties on-demand. This platform utilizes surrogate (or machine learning) models, which link key features of polymers to properties, trained on high-throughput DFT calculations and experimental data from literature and existing databases. The main elements of the polymer property prediction pipeline are summarized in the lowermost pipeline of Fig. 18.1.

In the following sections, we explain in detail the various stages of abovementioned pipeline [39], starting from the curation of the dataset all the way up to the machine learning algorithms that we have employed.

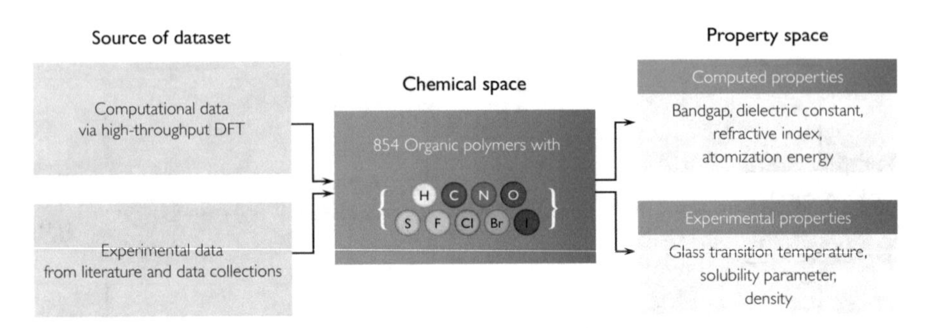

Fig. 18.2 Overview of our polymer dataset used for development of property prediction models [39]. The dataset consists of 854 polymers spanning a chemical space of nine elements and comprises properties obtained using computations as well as experiments

18.2 Dataset

Two strategic tracks were followed for the creation of our dataset (see Fig. 18.2): (1) via high-throughput computation using density functional theory (DFT) as presented earlier [31,40,41] and (2) by utilizing experimentally measured properties from literature and data collections [42, 43]. The overall dataset includes 854 polymers made up of a subset of the following species: H, C, N, O, S, F, Cl, Br, and I. Seven different properties were included in the present study. The bandgap, dielectric constant, refractive index, and atomization energy were determined using DFT computations whereas the T_g, solubility parameter, and density were obtained from experimental measurements.

All the computational data were generated through a series of studies related to advanced polymer dielectrics [31, 40, 41]. The computational dataset includes polymers containing the following building blocks, CH_2, CO, CS, NH, C_6H_4, C_4H_2S, CF_2, CHF, and O [19, 22, 40, 41, 44]. Repeat units contained 4–8 building blocks, and 3D structure prediction algorithms were used to determine their structure [31,40,41]. The building blocks considered in the dataset are found in common polymeric materials including polyethylene (PE), polyesters, and polyureas, and could theoretically produce an enormous variety of different polymers. The bandgap was computed using the hybrid Heyd–Scuseria–Ernzerhof (HSE06) electronic exchange-correlation functional [45]. Dielectric constant and refractive index (the square root of the electronic part of the dielectric constant) were computed using density functional perturbation theory (DFPT) [46]. The atomization energy was computed for all the polymers following previous work [33–36, 41, 44, 47–52]. The DFT computed properties and associated 3D structures are available from Khazana [53](khazana.gatech.edu).

The T_g, solubility parameter, and density data were obtained from the existing databases of experimental measurements [42, 43]. T_g, which is an indication of the transition point between the glassy and supercooled liquid phases in an amorphous polymer, is important in many polymer applications because the structural characteristics (and, consequently, other properties) of the polymer changes dramatically at this point. The solubility parameter of a polymer is typically used to determine a suitable solvent to use during polymer synthesis. In this particular study we consider the Hildebrand solubility parameter.

We have determined the chemical formula and the associated topological structure from the name of polymers listed in the literature. The dataset contains a total of 854 organic polymers composed of 9 frequently found atomic species, i.e., C, H, O, N, S, F, Cl, Br, and I with properties listed in the right side panel of Fig. 18.2. Figure 18.3 shows a summary of the property space for the polymer dataset, including the range of property values, distribution, standard deviation, and the number of polymers associated with each property.

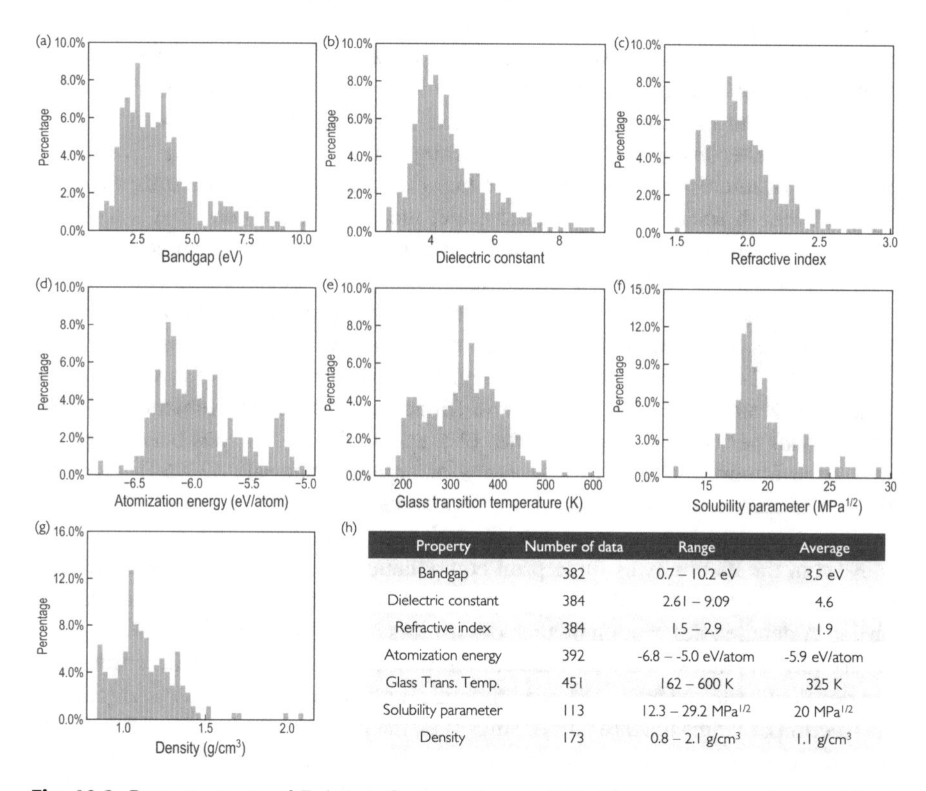

Property	Number of data	Range	Average
Bandgap	382	0.7 – 10.2 eV	3.5 eV
Dielectric constant	384	2.61 – 9.09	4.6
Refractive index	384	1.5 – 2.9	1.9
Atomization energy	392	-6.8 – -5.0 eV/atom	-5.9 eV/atom
Glass Trans. Temp.	451	162 – 600 K	325 K
Solubility parameter	113	12.3 – 29.2 MPa$^{1/2}$	20 MPa$^{1/2}$
Density	173	0.8 – 2.1 g/cm^3	1.1 g/cm^3

Fig. 18.3 Property space of Polymer Genome dataset [39]. The seven properties considered in this study were the (**a**) bandgap, (**b**) dielectric constant, (**c**) refractive index, (**d**) atomization energy, (**e**) T_g, (**f**) solubility parameter, and (**g**) density. The histograms represent the distribution of each individual property. The solid line depicts the mean of the distribution whereas the distance between the solid line and dashed line represents the standard deviation. (**h**) Table detailing the number of data-points, range, and mean of each individual property considered

18.3 Hierarchical Fingerprinting

Fingerprinting is a crucial step of our data-driven property prediction pipeline. In this step, the geometric and chemical information of the polymers is converted to a numerical representation. This numerical representation, more often than not, is a vector of fixed number of dimensions that can be provided as an input to any given machine learning algorithm. The different dimensions of this vector would represent different characteristics of the polymer repeat unit. Such numerical descriptors of organic molecules have been utilized extensively in the past in the form quantitative structure-property relationship (QSPR) or quantitative structure-activity relationship (QSAR) models. In the current work, we go beyond existing QSPR/QSAR descriptors in order to systematically capture different length-scale features that are specific to polymeric materials built up of very long polymer chains. In essence, given a particular repeat unit, we assume that the polymer chain constructed from that repeat unit is infinitely long and therefore the descriptors that we construct must take into account this "one-dimensional" periodicity.

To comprehensively capture the key features that may control the diversity of properties of interest, we consider three hierarchical levels of descriptors spanning different length scales. At the atomic-scale, the number of times that a fixed set of atomic fragments (or motifs) occur are counted [54]. An example of such a fragment is `O1-C3-C4`, made up of three contiguous atoms, namely, a one-fold coordinated oxygen, a three-fold coordinated carbon, and a four-fold coordinated carbon, in this order. For a given polymer repeat unit, we count the number of times the `O1-C3-C4` fragment occurs and then proceed to normalize this value by the number of atoms in polymer repeat unit (to account for the abovementioned one-dimensional periodicity). Such a series of predefined "triplets" has been shown to be a good fingerprint for a diverse range of organic materials [23, 54]. A vector of such triplets form the fingerprint components at the lowest hierarchy. For the polymer class under study, there are 108 such components.

Next in the hierarchy of fingerprint components are larger length-scale descriptors of the quantitative structure-property relationship (QSPR) type mentioned earlier. A detailed description of such descriptors can be found in the RDKit Python library [55–57] that was used for the current work. Examples of such descriptors are van der Waals surface area [58], the topological polar surface area (TPSA) [59,60], the fraction of atoms that are part of rings (i.e., the number of atoms associated with rings divided by the total number of atoms in the formula unit), and the fraction of rotatable bonds. TPSA is the sum of surfaces of polar atoms in the molecule and we observed this descriptor to be strongly correlated to the solubility. Descriptors such as the fraction of ring atoms and fraction of rotatable bonds strongly influenced properties such as T_g and density. Such descriptors, 99 in total, form the next set of components of our overall fingerprint vector.

The highest length-scale fingerprint components we considered may be classified as "morphological descriptors." These include features such as the shortest topological distance between rings, fraction of atoms that are part of side chains, and the length of the largest side-chain. Properties such as T_g strongly depend on such features which influence the way the chains are packed in the polymer. For instance, if two rings are very close, the stiffness of the polymer backbone is much higher than if the rings were separated by a larger topological distance. Both the number and the length of the side chains strongly influence the amount of free volume in the polymeric material and therefore directly influence T_g. The larger the free volume, the lower the T_g. We include 22 such morphological descriptors in our overall fingerprint.

Figure 18.4a shows the hierarchy of polymer fingerprints, including atomic level, QSPR and morphological descriptors. The overall fingerprint of a polymer is constructed by concatenating the three classes of fingerprint components. In total, this leads to a fingerprint with 229 components. Since certain descriptors are more relevant for certain properties, in the next section, we outline a methodology to discard irrelevant descriptors for every target property. Moreover, during performance assessment, we use different combinations of the three fingerprint hierarchies. For

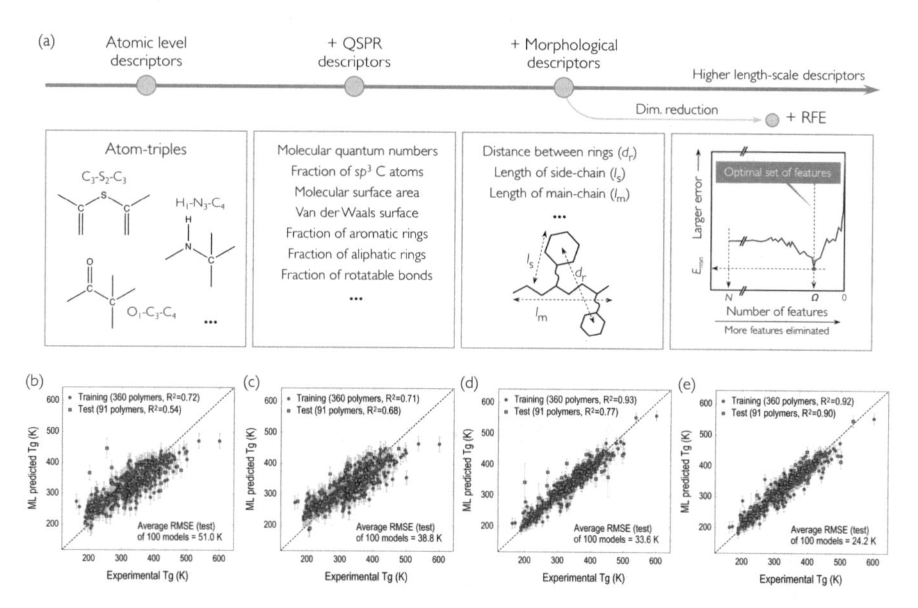

Fig. 18.4 Hierarchy of descriptors used to fingerprint the polymers, and an example demonstration for the systematic improvement of model performance depending on the type of fingerprint considered. (**a**) Classification of descriptors according to the physical scale and chemical characteristics are shown with representative examples. Dimension of the fingerprint in each level can be reduced by a recursive feature elimination (RFE) process. In the "+RFE" panel, N, Ω, and E_{min} are total number of features in fingerprint, optimal number of features determined by RFE, and minimum error of prediction model, respectively. Plots at the bottom panel show the performance of machine learning prediction models for glass transition temperature (T_g) with (**b**) only atomic level descriptors, (**c**) atomic level and QSPR descriptors, and (**d**) entire fingerprint components including morphological descriptors. (**e**) Shows how the optimal subset selected by RFE improves the prediction model for T_g [39]

clarity of the ensuing discussion, we introduce some nomenclature. The atom triples fingerprint, QSPR descriptors, and morphological descriptors are denoted by "A," "Q," and "M," respectively. Therefore, "AQ" implies a combination of just the atom triples and QSPR descriptors.

In order to visualize the chemical diversity of polymers considered here, we have performed principal component analysis (PCA) of the complete fingerprint vector. PCA identifies orthogonal linear combinations of the original fingerprint components that provide the highest variance; the first few principal components account for much of the variability in the data [13]. Figure 18.5 displays the dataset with the horizontal and vertical axes chosen as the first two principal components, PC_1 and PC_2. Molecular models of some common polymers are shown explicitly, and symbol color, symbol size, and symbol type are used to represent the fraction of sp^3 bonded C atoms, fraction of rings, and TPSA of polymers, respectively. As an example from the figure, PE is composed of only sp^3 bonded C without any rings in the chain, while poly(1,4-phenylene sulfide) contains no sp^3 bonded C atoms, and more than 90% of its atoms are part of rings. As a result, these two polymers are situated far from each other in 2D principal component space.

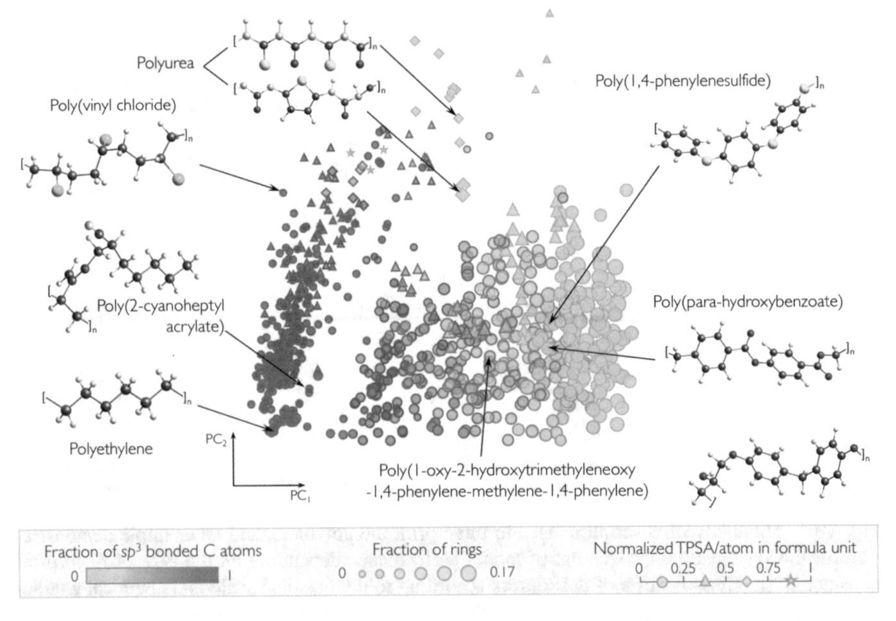

Fig. 18.5 Graphical summary of chemical space of polymers considered. 854 chemically unique organic polymers generated by structure prediction method (minima-hopping [61]) and experimental sources [42, 43] distributed in 2D principal component space. Two leading components, PC_1 and PC_2, are produced by principal component analysis, and assigned to axes of the plot. Fraction of sp^3 bonded C atoms, fraction of rings, and normalized TPSA per atoms in a formula unit are used for color code, size, and symbol of each polymer. A few representative structures with various number of aromatic and/or aliphatic rings and their position on the map are shown [39]

18.4 Surrogate (Machine Learning) Model Development

18.4.1 Recursive Feature Elimination

As alluded to earlier, our general fingerprint is rather high in dimensionality, and not all of the components may be relevant for describing a particular property. In fact, irrelevant features often lead to a poor prediction capability. On the practical side, large fingerprint dimensionality also implies longer training times. There is thus a need to determine the optimal subset of the complete fingerprint necessary for the prediction of a particular property (i.e., different properties may require different subsets of the fingerprint vector). Rather than manually deciding which fingerprint components to use, one may utilize a wide variety of dimensionality reduction techniques to automatically select a set of features that best represent a particular property. In the current work, we utilize the recursive feature elimination (RFE) algorithm to sequentially eliminate the least important features for a given property [62]. First, linear regression is performed using the complete fingerprint vector via support vector regression. Through this process, each of the features are weighted by certain coefficients and are then ranked based on the square of these coefficients [62]. The feature with the lowest rank is subsequently eliminated and the iteration is repeated to remove the next least-important-feature. As shown in right-most panel of Fig. 18.4, the optimal number of features for a given property can be obtained by plotting the cross-validated root mean square error (RMSE) as a function of the number of descriptors. The final set of features is passed forward to the non-linear machine learning algorithm described next in Sect. 18.4.2. These features can also be used to obtain an intuitive understanding of how certain key fingerprint components influence particular materials properties.

18.4.2 Gaussian Process Regression

In our past work [12, 19, 31], we have successfully utilized kernel ridge regression (KRR) [63] to learn the non-linear relationship between a polymer's fingerprint and its properties. However, in this work we utilize Gaussian process regression (GPR) because of two key benefits. Firstly, GPR learns a generative, probabilistic model of the target property and thus provides meaningful uncertainties/confidence intervals for the prediction. Secondly, the optimization of the model hyperparameters is relatively faster in GPR because one may perform gradient-ascent on the marginal likelihood function as opposed to the cross-validated grid-search which is required for KRR. We use a radial basis function (RBF) kernel defined as

$$k(\mathbf{x}_i, \mathbf{x}_j) = \sigma^2 \exp\left\{\left[\frac{-\left(\mathbf{x}_i - \mathbf{x}_j\right)^2}{2l^2}\right]\right\} + \sigma_n^2 \delta(\mathbf{x}_i, \mathbf{x}_j), \tag{18.1}$$

where σ, l, and σ_n are hyperparameters to be determined during the training process (in the machine learning parlance, these hyperparameters are referred to as signal variance, length-scale parameter, and noise level parameter, respectively). $\mathbf{x}_i$ and $\mathbf{x}_j$ are the fingerprint vectors for two polymers i and j. ($\mathbf{x}_i$ is an m dimensional vector with components $x_i^1, x_i^2, x_i^3, \ldots, x_i^m$, determined and optimized by the RFE step described above). Performance of the model was evaluated based on the root mean square error (RMSE) and the coefficient of determination (R^2). 80% of the data was used for training and the remaining 20% was set aside as a test set.

18.5 Model Performance Validation

The final machine learning models for each of the properties under consideration here were constructed using the entire polymer dataset for each property. To avoid overfitting the data, and to ensure that the models are generalizable, we employed five-fold cross-validation, wherein the dataset is divided into 5 different subsets and one subset was used for testing while remaining sets were employed for training. Table 18.1 summarizes the best fingerprint, dimension of fingerprint vector, and performance based on RMSE for the entire dataset. As shown in the table, the best machine learning model for the atomization energy can be constructed using just the atom triples and QSPR descriptors (i.e., "AQ") whereas most of the other properties necessitate the inclusion of morphological descriptors (i.e., 'AQM"). In Fig. 18.6, we demonstrate the sensitivity of the bandgap and dielectric constant models to the size of the training set. We see a convergence in the train and test errors as the training set size increases. Therefore, the accuracy of the ML models may be systematically improved as more polymer property values are added to the dataset.

Parity plots in Fig. 18.7 are shown to compare experimental or DFT computed properties with respect to machine learning predicted values with percentage relative error distribution. Several error metrics, such as RMSE, mean absolute error (MAE),

Table 18.1 Summary of fingerprint used for development of machine learning prediction model, and the performance of prediction for each property [39]

Property	Best fingerprint	Dimension of fingerprint	RMSE
Bandgap	AQM + RFE	88	0.30 eV
Dielectric constant	AQ + RFE	35	0.48
Refractive index	AQM + RFE	19	0.08
Atomization energy	AQ	207	0.01 eV/atom
Glass transition temperature	AQM + RFE	69	18 K
Solubility parameter	AQM + RFE	24	0.56 $\text{MPa}^{1/2}$
Density	AQ + RFE	9	0.05 g/cm^3

Best fingerprint is selected based on average RMSE of test set for 100 models. (*A* Atomic level descriptors; *Q* QSPR descriptors; *M* Morphological descriptors; *+RFE* subject to the RFE process)

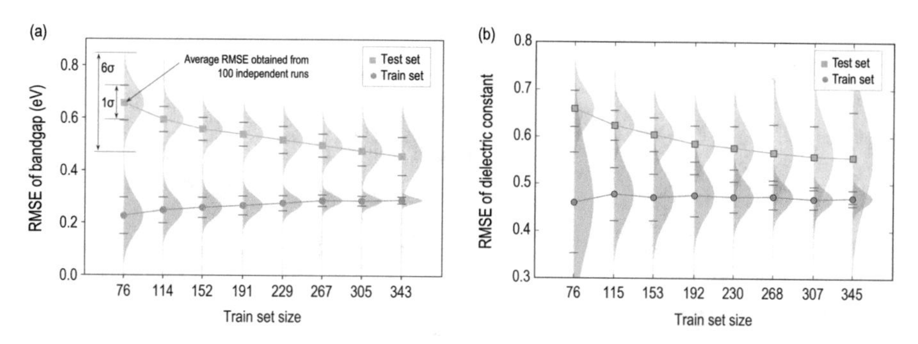

Fig. 18.6 Learning curves constructed from the RMSE of the machine learning models for (**a**) bandgap and (**b**) dielectric constant. For each model, data was obtained from 100 independent runs with different selection of train and test set

mean absolute relative error (MARE), and $1 - R^2$ were considered to evaluate the performance of these models, and shown together in Fig. 18.7h.

As mentioned earlier, the utilization of GPR provides meaningful uncertainties associated with each prediction. Moreover, the noise parameter of the GPR kernel gives insights into the overall errors and uncertainties associated with the prediction of that particular property for a given dataset. These uncertainties could arise as a result of variation in measurement techniques (in the case of T_g, for example) or it may even arise as a result of limitations of our representation technique. For example, we are providing estimates of the bandgap through purely the SMILES string rather than the 3D crystal structure of the polymer. Therefore, the representation technique itself results in partial loss of information and this underlying uncertainty can be estimated statistically using the GPR noise parameter.

18.6 Polymer Genome Online Platform

For easy access and use of the prediction models developed here, an online platform called Polymer Genome has been created. This platform is available at www.polymergenome.org [64]. The Polymer Genome application was developed using Python and standard web languages such as Hypertext Preprocessor (php) and Hypertext Markup Language (HTML). As user input, the repeat unit of a polymer or its SMILES string may be used (following a prescribed format described in the Appendix). One may also use an integrated drawing tool to sketch the repeat unit of the polymer.

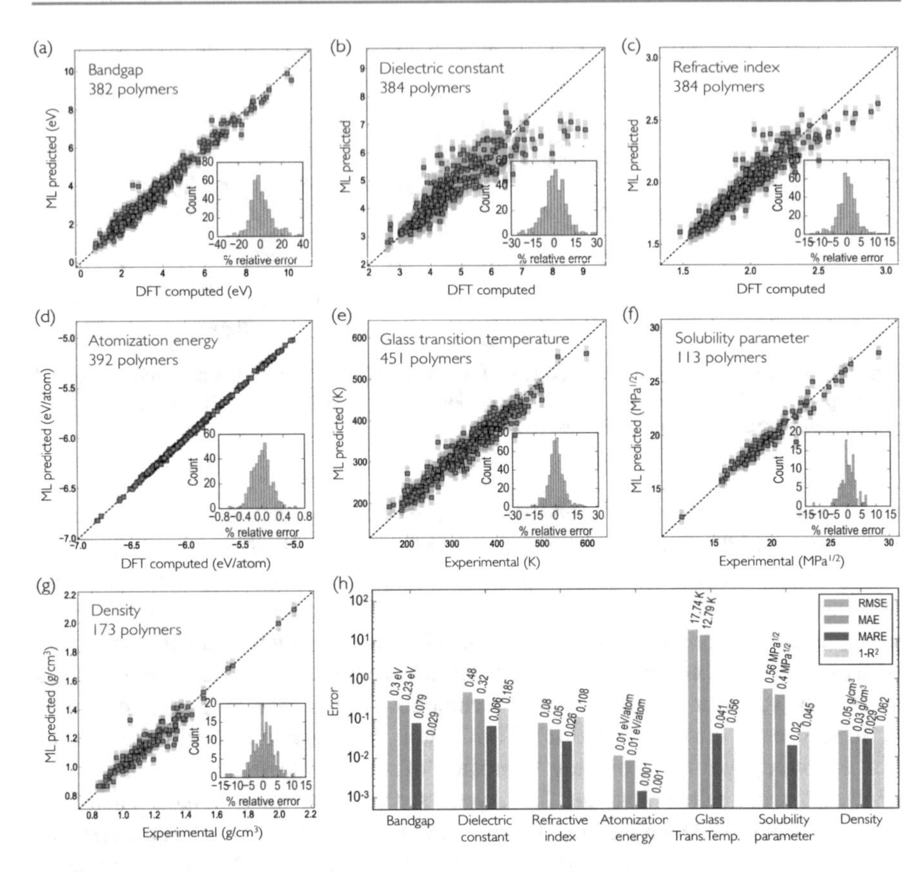

Fig. 18.7 The performance of the cross-validated machine learning models developed by GPR with combination of RBF and white noise kernels [39]. Comparison of DFT computed (**a**) bandgap, (**b**) dielectric constant, (**c**) refractive index, (**d**) atomization energy, experimental (**e**) T_g, (**f**) Hildebrand solubility parameter, and (**g**) density for the predicted values are shown with inset of distribution of % relative error, $(y - Y)/Y \times 100$ where Y is DFT computed or experimental value, and y is machine learning predicted value. The error bars in the parity plots represent uncertainties (standard deviations) obtained using GPR. Other error metrics including RMSE, mean absolute error (MAE), mean absolute relative error (MARE), and $1 - R^2$ are summarized in (**h**)

Once the user input is delivered to Polymer Genome by the user, property predictions (with uncertainty) are made, and the results are shown in an organized table. The names of polymers (if there are more than one meeting the search criteria) with SMILES and repeat unit are provided with customizable collection of properties. Upon selection of any polymer from this list, comprehensive information is reported. This one-page report provides the name and class of the polymer, 3D visualization of the structure with atomic coordinates (if such is available), and properties determined using our machine learning models. A typical user output of Polymer Genome is captured in Fig. 18.8.

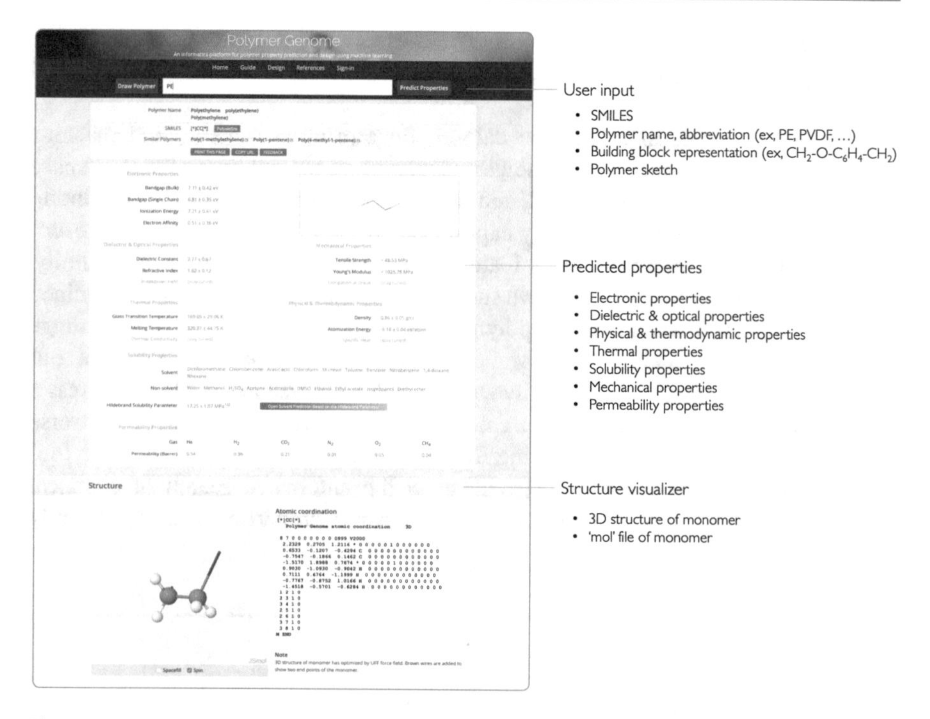

Fig. 18.8 Overview of Polymer Genome online platform available at www.polymergenome.org. Keyword PE is used as an example user input to show resulting Polymer details page [39]

18.7 Conclusions and Outlook

The Materials Genome Initiative and similar other initiatives around the world have provided the impetus for data-centric informatics approaches in several subfields of materials research. Such informatics approaches seek to provide tools and pathways for accelerated property prediction (and materials design) via surrogate models built on reliable past data. Here, we have presented a polymer informatics platform capable of predicting a variety of important polymer properties on-demand. This platform utilizes surrogate (or machine learning) models that link key features of polymers (i.e., their "fingerprint") to properties. The models are trained on high-throughput DFT calculations (of the bandgap, dielectric constant, refractive index, and atomization energy) and experimental data from polymer data handbooks (on the glass transition temperature, solubility parameter, and density). Certain properties, like the atomization energy, depend mainly on the atomic constituents and short-range bonding, whereas other properties, such as the glass transition temperature, are strongly influenced by morphological characteristics like the chain-stiffness and branching. Our polymer fingerprinting scheme is thus necessarily hierarchical and captures features at multiple length scales ranging from atomic connectivity to the size and density of side chains. The property prediction models are incorporated in a user friendly online platform named Polymer Genome (www.polymergenome.org), which utilizes a custom Python-based machine learning and polymer querying framework.

Polymer Genome, including the dataset, fingerprinting scheme, and machine learning models, remains in early stages. Coverage of the polymer chemical space needs to be progressively increased, and further developments on the fingerprinting scheme are necessary to adequately capture conformational (e.g., *cis* versus *trans*, tacticity, etc.) and morphological features (e.g., copolymerization, crystallinity, etc.). Systematic pathways to achieve such expansion are presently being examined to extend the applicability of the polymer informatics paradigm to a wide range of technological domains. Moreover, looking to the future, the ability of our informatics platform to automatically suggest polymers that are likely to possess a given set of properties would be of tremendous value within the context of "inverse design" [65]. Approaches involving Bayesian active learning techniques [66] and variational autoencoders [67] will allow the automated search of chemical and morphological space for materials with desired properties at a significantly accelerated pace.

References

1. M.I. Jordan, T.M. Mitchell, Science **349**(6245), 255 (2015)
2. M. Rupp, A. Tkatchenko, K.R. Müller, O.A. von Lilienfeld, Phys. Rev. Lett. **108**, 058301 (2012)
3. L. Ruddigkeit, R. van Deursen, L.C. Blum, J.L. Reymond, J. Chem. Inf. Model. **52**(11), 2864 (2012). https://doi.org/10.1021/ci300415d. PMID: 23088335
4. R. Ramakrishnan, P.O. Dral, M. Rupp, O.A. Von Lilienfeld, Sci. Data **1**, 140022 (2014)
5. Materials Genome Initiative. https://www.mgi.gov/
6. The Novel Materials Discovery (nomad) Laboratory. https://nomad-coe.eu/
7. National Center for Competence in Research - Marvel. nccr-marvel.ch
8. G. Pizzi, A. Cepellotti, R. Sabatini, N. Marzari, B. Kozinsky, Comput. Mater. Sci. **111**, 218 (2016)
9. K. Mathew, J.H. Montoya, A. Faghaninia, S. Dwarakanath, M. Aykol, H. Tang, I.h. Chu, T. Smidt, B. Bocklund, M. Horton, et al., Comput. Mater. Sci. **139**, 140 (2017)
10. A. Mannodi-Kanakkithodi, A. Chandrasekaran, C. Kim, T.D. Huan, G. Pilania, V. Botu, R. Ramprasad, Mater. Today **21**, 785–796 (2017). https://doi.org/10.1016/j.mattod.2017.11.021
11. R. Ramprasad, R. Batra, G. Pilania, A. Mannodi-Kanakkithodi, C. Kim, npj Comput. Mater. **3**, 54 (2017). https://doi.org/10.1038/s41524-017-0056-5
12. A. Mannodi-Kanakkithodi, G. Pilania, R. Ramprasad, Comput. Mater. Sci. **125**, 123 (2016). https://doi.org/10.1016/j.commatsci.2016.08.039
13. T. Mueller, A.G. Kusne, R. Ramprasad, *Machine Learning in Materials Science: Recent Progress and Emerging Applications*, vol. 29 (Wiley, Hoboken, 2016), pp. 186–273
14. G. Hautier, C.C. Fischer, A. Jain, T. Mueller, G. Ceder, Chem. Mater. **22**(12), 3762 (2010)
15. A.O. Oliynyk, E. Antono, T.D. Sparks, L. Ghadbeigi, M.W. Gaultois, B. Meredig, A. Mar, Chem. Mater. **28**(20), 7324 (2016)
16. P. Pankajakshan, S. Sanyal, O.E. de Noord, I. Bhattacharya, A. Bhattacharyya, U. Waghmare, Chem. Mater. **29**(10), 4190 (2017). https://doi.org/10.1021/acs.chemmater.6b04229
17. C. Kim, G. Pilania, R. Ramprasad, Chem. Mater. **28**(5), 1304 (2016). https://doi.org/10.1021/acs.chemmater.5b04109

18. A. Jain, Y. Shin, K.A. Persson, Nat. Rev. Mater. **1**, 15004 (2016). https://doi.org/10.1038/natrevmats.2015.4
19. A. Mannodi-Kanakkithodi, G. Pilania, T.D. Huan, T. Lookman, R. Ramprasad, Sci. Rep. **6**, 20952 (2016). https://doi.org/10.1038/srep20952
20. L. Ghadbeigi, J.K. Harada, B.R. Lettiere, T.D. Sparks, Energy Environ. Sci. **8**, 1640 (2015). https://doi.org/10.1039/C5EE00685F
21. J. Hattrick-Simpers, C. Wen, J. Lauterbach, Catal. Lett. **145**(1), 290 (2015). https://doi.org/10.1007/s10562-014-1442-y
22. J. Hill, A. Mannodi-Kanakkithodi, R. Ramprasad, B. Meredig, *Materials Data Infrastructure and Materials Informatics* (Springer International Publishing, Cham, 2018), pp. 193–225. https://doi.org/10.1007/978-3-319-68280-8_9
23. A. Mannodi-Kanakkithodi, T.D. Huan, R. Ramprasad, Chem. Mater. **29**(21), 9001 (2017). https://doi.org/10.1021/acs.chemmater.7b02027
24. C. Kim, T.D. Huan, S. Krishnan, R. Ramprasad, Sci. Data **4**, 170057 (2017). https://doi.org/10.1038/sdata.2017.57
25. J. Behler, M. Parrinello, Phys. Rev. Lett. **98**(14), 146401 (2007)
26. V. Botu, R. Batra, J. Chapman, R. Ramprasad, J. Phys. Chem. C **121**(1), 511 (2017). https://doi.org/10.1021/acs.jpcc.6b10908
27. T.D. Huan, R. Batra, J. Chapman, S. Krishnan, L. Chen, R. Ramprasad, npj Comput. Mater. **3**(1), 37 (2017). https://doi.org/10.1038/s41524-017-0042-y
28. V. Botu, J. Chapman, R. Ramprasad, Comput. Mater. Sci. **129**, 332 (2017). https://doi.org/10.1016/j.commatsci.2016.12.007
29. F. Brockherde, L. Vogt, L. Li, M.E. Tuckerman, K. Burke, K.R. Müller, Nat. Commun. **8**(1), 872 (2017)
30. L. Chen, T.D. Huan, R. Ramprasad, Sci. Rep. **7**(1), 6128 (2017)
31. A. Mannodi-Kanakkithodi, G.M. Treich, T.D. Huan, R. Ma, M. Tefferi, Y. Cao, G.A. Sotzing, R. Ramprasad, Adv. Mater. **28**(30), 6277 (2016). https://doi.org/10.1002/adma.201600377
32. G.M. Treich, M. Tefferi, S. Nasreen, A. Mannodi-Kanakkithodi, Z. Li, R. Ramprasad, G.A. Sotzing, Y. Cao, IEEE Trans. Dielectr. Electr. Insul. **24**(2), 732 (2017). https://doi.org/10.1109/TDEI.2017.006329
33. A.F. Baldwin, T.D. Huan, R. Ma, A. Mannodi-Kanakkithodi, M. Tefferi, N. Katz, Y. Cao, R. Ramprasad, G.A. Sotzing, Macromolecules **48**, 2422 (2015)
34. Q. Zhu, V. Sharma, A.R. Oganov, R. Ramprasad, J. Chem. Phys. **141**(15), 154102 (2014). https://doi.org/10.1063/1.4897337
35. R. Lorenzini, W. Kline, C. Wang, R. Ramprasad, G. Sotzing, Polymer **54**(14), 3529 (2013). https://doi.org/10.1016/j.polymer.2013.05.003
36. A.F. Baldwin, R. Ma, T.D. Huan, Y. Cao, R. Ramprasad, G.A. Sotzing, Macromol. Rapid Commun. **35**, 2082 (2014)
37. A. Mannodi-Kanakkithodi, G. Pilania, R. Ramprasad, T. Lookman, J.E. Gubernatis, Comput. Mater. Sci. **125**, 92 (2016). https://doi.org/10.1016/j.commatsci.2016.08.018
38. T.D. Huan, S. Boggs, G. Teyssedre, C. Laurent, M. Cakmak, S. Kumar, R. Ramprasad, Prog. Mater. Sci. **83**, 236 (2016). https://doi.org/10.1016/j.pmatsci.2016.05.001
39. C. Kim, A. Chandrasekaran, T.D. Huan, D. Das, R. Ramprasad, J. Phys. Chem. C **122**(31), 17575 (2018). https://doi.org/10.1021/acs.jpcc.8b02913
40. T.D. Huan, A. Mannodi-Kanakkithodi, C. Kim, V. Sharma, G. Pilania, R. Ramprasad, Sci. Data **3**, 160012 (2016). https://doi.org/10.1038/sdata.2016.12
41. V. Sharma, C.C. Wang, R.G. Lorenzini, R. Ma, Q. Zhu, D.W. Sinkovits, G. Pilania, A.R. Oganov, S. Kumar, G.A. Sotzing, S.A. Boggs, R. Ramprasad, Nat. Commun. **5**, 4845 (2014)
42. J. Bicerano, *Prediction of Polymer Properties* (Dekker, New York, 2002)
43. A.F.M. Barton, *Handbook of Solubility Parameters and Other Cohesion Parameters* (CRC Press, Florida, 1983)
44. C.C. Wang, G. Pilania, S.A. Boggs, S. Kumar, C. Breneman, R. Ramprasad, Polymer **55**, 979 (2014)
45. J. Heyd, G.E. Scuseria, M. Ernzerhof, J. Chem. Phys. **118**(18), 8207 (2003)

46. S. Baroni, S. de Gironcoli, A. Dal Corso, P. Giannozzi, Rev. Mod. Phys. **73**(2), 515 (2001). https://doi.org/10.1103/RevModPhys.73.515
47. T.D. Huan, M. Amsler, V.N. Tuoc, A. Willand, S. Goedecker, Phys. Rev. B **86**, 224110 (2012)
48. H. Sharma, V. Sharma, T.D. Huan, Phys. Chem. Chem. Phys. **17**, 18146 (2015)
49. T.D. Huan, V. Sharma, G.A. Rossetti, R. Ramprasad, Phys. Rev. B **90**, 064111 (2014)
50. T.D. Huan, M. Amsler, R. Sabatini, V.N. Tuoc, N.B. Le, L.M. Woods, N. Marzari, S. Goedecker, Phys. Rev. B **88**, 024108 (2013)
51. A.F. Baldwin, R. Ma, A. Mannodi-Kanakkithodi, T.D. Huan, C. Wang, J.E. Marszalek, M. Cakmak, Y. Cao, R. Ramprasad, G.A. Sotzing, Adv. Matter. **27**, 346 (2015)
52. R. Ma, V. Sharma, A.F. Baldwin, M. Tefferi, I. Offenbach, M. Cakmak, R. Weiss, Y. Cao, R. Ramprasad, G.A. Sotzing, J. Mater. Chem. A **3**, 14845 (2015). https://doi.org/10.1039/C5TA01252J
53. Khazana, a Computational Materials Knowledgebase. https://khazana.gatech.edu
54. T.D. Huan, A. Mannodi-Kanakkithodi, R. Ramprasad, Phys. Rev. B **92**(014106), 14106 (2015). https://doi.org/10.1103/PhysRevB.92.014106
55. C. Nantasenamat, C. Isarankura-Na-Ayudhya, T. Naenna, V. Prachayasittikul, EXCLI J. **8**, 74 (2009)
56. C. Nantasenamat, C. Isarankura-Na-Ayudhya, V. Prachayasittikul, Expert Opin. Drug Discov. **5**(7), 633 (2010). https://doi.org/10.1517/17460441.2010.492827. PMID: 22823204
57. Rdkit, Open Source Toolkit for Cheminformatics. http://www.rdkit.org/
58. P. Labute, J. Mol. Graph. Model. **18**(4), 464 (2000). https://doi.org/10.1016/S1093-3263(00)00068-1
59. P. Ertl, B. Rohde, P. Selzer, J. Med. Chem. **43**(20), 3714 (2000). https://doi.org/10.1021/jm000942e. PMID: 11020286
60. S. Prasanna, R. Doerksen, Curr. Med. Chem. **16**, 21 (2009)
61. M. Sicher, S. Mohr, S. Goedecker, J. Chem. Phys. **134**(4), 044106 (2011)
62. I. Guyon, J. Weston, S. Barnhill, V. Vapnik, Mach. Learn. **46**(1), 389 (2002). https://doi.org/10.1023/A:1012487302797
63. K. Vu, J.C. Snyder, L. Li, M. Rupp, B.F. Chen, T. Khelif, K.R. Müller, K. Burke, Int. J. Quantum Chem. **115**(16), 1115 (2015). https://doi.org/10.1002/qua.24939
64. Polymer Genome. http://www.polymergenome.org
65. B. Sanchez-Lengeling, A. Aspuru-Guzik, Science **361**(6400), 360 (2018)
66. D.A. Cohn, Z. Ghahramani, M.I. Jordan, J. Artif. Intell. Res **4**, 129 (1996)
67. H. Dai, Y. Tian, B. Dai, S. Skiena, L. Song (2018, preprint). arXiv:1802.08786

Bayesian Optimization in Materials Science 19

Zhufeng Hou and Koji Tsuda

Abstract
Bayesian optimization (BO) algorithm is a global optimization approach, and it has been recently gained growing attention in materials science field for the search and design of new functional materials. Herein, we briefly give an overview of recent applications of BO algorithm in the determination of physical parameters of physics model, the design of experimental synthesis conditions, the discovery of functional materials with targeted properties, and the global optimization of atomic structures. The basic methodologies of BO in these applications are also addressed.

19.1 Introduction

The materials design and discovery mostly involve the choice of atomic elements, chemical compositions, structure, or processing condition of a material to meet a design criteria. Mathematically, the design or discovery of a material with a targeted property is often formulated as an optimization problem of a black-box function [1]. The traditional trial-and-error approach for discovering a new functional material is laborious, time-consuming, and costly. Its success also depends on the intuition and prior knowledge of researcher. Thanks to the rapid advancement of computing power and of predictive power of first-principles calculations, the high-throughput

Z. Hou
Research and Services Division of Materials Data and Integrated System, National Institute for Materials Science, Tsukuba, Ibaraki, Japan

K. Tsuda (✉)
Graduate School of Frontier Sciences, The University of Tokyo, Kashiwa, Japan
e-mail: tsuda@k.u-tokyo.ac.jp

K. T. Schütt et al. (eds.), *Machine Learning Meets Quantum Physics*, Lecture Notes in Physics 968, https://doi.org/10.1007/978-3-030-40245-7_19

computational method becomes an alternative approach for materials design [2]. In both approaches, the exhaustive search is mostly prohibited because of the limitation in assigned budget. The more valuable experiment is desired to try in a higher priority.

Bayesian optimization (BO), aka kriging, is a well-established technique for the black-box optimization [1, 3, 4]. Bayesian prediction model, most commonly Gaussian processes (GP) [5], is used to predict the black-box function, where the uncertainty of the predicted function is also evaluated as predictive variance. Based on the predicted values and variances, next experiments are suggested. In the early stage, the BO was popular in engineering [6]. Recently, the BO has gained momentum in machine-learning research due to the success of hyperparameter tuning in deep learning algorithms [7,8]. The application of BO in materials science started from 2015 [9]. We are aware of one chapter [10] and one monograph [11] that have discussed the basic methodologies of BO for materials design. Nowadays, the BO has been gained more attention for its applications in either the experimental condition design [12, 13] or the computational design of functional materials [14]. Herein, we will give an overview of the recent applications of BO in several subjects of materials science.

19.2 Bayesian Optimization

The BO proceeds by iteratively developing a global statistical model of objective function $f(\mathbf{x})$. We start with a prior distribution $P(f)$ over the objective function and a likelihood function $P(\mathcal{D}|f)$ that describes the data generation process. Then, at each iteration, a posterior distribution $P(f|\mathcal{D})$ is computed by conditioning on the previous evaluations of the objective function, namely $P(f|\mathcal{D}) \propto P(\mathcal{D}|f)P(f)$, treating them as observations in a Bayesian nonlinear regression with Gaussian process prior. An acquisition function is then used to map beliefs about the objective function to a measure of how promising each location in the input space is, if it was to be evaluated next. The goal is then to find the input that maximizes the acquisition function, and select it for function evaluation. Once the function evaluation is complete, the new observation is added to the dataset and we begin the next iteration. The termination condition is often a maximum number of function evaluations. This procedure is illustrated in Algorithm 5.

Algorithm 5: General algorithm of Bayesian optimization

Input: search space $\mathcal{X}$, model $\mathcal{M}$, initial design $\mathcal{D}$, acquisition function α, objective f

1: **repeat**
2: Fit the model $\mathcal{M}$ to the data $\mathcal{D}$
3: Maximize the acquisition function: $\hat{\mathbf{x}} = \text{argmax}_{\mathbf{x} \in \mathcal{X}} \alpha(\mathbf{x}, \mathcal{M})$
4: Evaluate the function: $\hat{y} = f(\hat{\mathbf{x}})$
5: Add the new data to the dataset: $\mathcal{D} = \mathcal{D} \cup \{(\hat{\mathbf{x}}, \hat{y})\}$
6: **until** termination condition is met

Output: the optimal candidate $\mathbf{x}^* = \text{argmax}_{\mathbf{x} \in \mathcal{X}} \mathbb{E}_{\mathcal{M}}[f(\mathbf{x})]$

In the recent years, several open-source packages of BO were designed especially for the materials science. We list them as below:

- COMBO (COMmon Bayesian Optimization Library): https://github.com/tsudalab/combo;
- BASC (Bayesian Active Site Calculator): https://gitlab.com/caml/basc;
- Phoenics (Probabilistic Harvard Optimizer Exploring Non-Intuitive Complex Surfaces): https://github.com/aspuru-guzik-group/phoenics;
- Matpredict: https://gitlab.com/tammal/matpredict;
- MOE (Metric Optimization Engine): https://github.com/Yelp/MOE.

Herein, we briefly present the technical features of BO in the COMBO code [15], which was developed by Ueno et al. in Tsuda's research group. Both the most commonly used GP [5] and the Bayesian linear regression (BLR) were implemented for the prediction model. The GP in the COMBO code was constructed using the squared-exponential kernel function, while the BLR was constructed using the random feature map [16] to approximate the kernel function. Since the details of GP have been well explained in a book by Rasmussen and Williams [5], herein we will skip them and give a bit more about the implementation of BLR in the COMBO code. Let us now assume that we have observed a set of data $\mathcal{D} = \{(y_i, \mathbf{x}_i), i = 1, \ldots, n, \mathbf{x}_i \in \mathcal{R}^d\}$. The BLR model for this dataset can be specified as follows:

$$\mathbf{y} = \mathbf{w}^\top \phi(\mathbf{x}) + \epsilon, \tag{19.1}$$

where $\phi : \mathcal{R}^d \to \mathcal{R}^\ell$ is a feature map, $\mathbf{w} \in \mathcal{R}^\ell$ is a weight vector, and ϵ is the noise subject to $\mathcal{N}(0, \sigma^2)$. The feature map is defined so that the inner product $\phi(\mathbf{x})^\top \phi(\mathbf{x}')$ corresponds to the Gaussian kernel:

$$\phi(\mathbf{x})^\top \phi(\mathbf{x}') = \exp\left(-\frac{\|\mathbf{x} - \mathbf{x}'\|^2}{\eta^2}\right). \tag{19.2}$$

Using ℓ random samples $\{\boldsymbol{\omega}_i, b_i\}_{i=1}^{\ell}$, the feature map is defined as below:

$$\phi(\mathbf{x}) = \left(z_{\omega_1, b_1}\left(\frac{\mathbf{x}}{\eta}\right), \ldots, z_{\omega_\ell, b_\ell}\left(\frac{\mathbf{x}}{\eta}\right) \right)^\top, \tag{19.3}$$

where $z_{\omega,b}(\mathbf{x}) = \sqrt{2}\cos(\boldsymbol{\omega}^\top + b)$, in which $\boldsymbol{\omega}$ is drawn from the Fourier transform $p(\boldsymbol{\omega})$ of kernel and b is drawn uniformly from $[0, 2\pi]$. A $\ell \times n$ matrix Φ can be constructed by choosing its ith column to be $\phi(\mathbf{x}_i)$. The posterior distribution of $\mathbf{w}$ at given data $\mathcal{D}$ is described as below:

$$\mathbf{w}|D \sim \mathcal{N}(\boldsymbol{\mu}, \Sigma), \tag{19.4}$$

where

$$\boldsymbol{\mu} = \left(\Phi\Phi^\top + \sigma^2 \mathrm{I}\right)^{-1} \Phi \mathbf{y}, \tag{19.5}$$

$$\Sigma = \sigma^2 \left(\Phi\Phi^\top + \sigma^2 \mathrm{I}\right)^{-1}. \tag{19.6}$$

The predicted value for candidate point $\mathbf{x}_i$ is given by $\mathbf{w}^\top \phi(\mathbf{x}_i)$. The hyperparameters σ and η in the kernel functions can be tuned automatically by maximizing the type-II likelihood [5]. Three types of acquisition functions, namely maximum probability of improvement (MPI) [1], maximum expected of improvement (MEI) [17], and Thompson sampling (TS) [18], have been implemented in the COMBO code. The combination of TS and BLR was designed to improve the optimization efficiency for the large-scale problems. To use BO in the COMBO code, the user shall feed a search space as input and prepare the evaluation of objective function via calling the simulator or feeding the experimental measurement results as input. Depending on the evaluation approach of objective function, the materials design and discovery using BO in the COMBO code can be performed in an automatic or interactive manner. The flowchart of BO in the COMBO code is illustrated in Fig. 19.1.

19.3 Application of Bayesian Optimization in Materials Science

19.3.1 Determine the Parameters in a Physics Model

In materials science the understanding of material properties at a microscopic or macroscopic scale in many cases can be rationalized using some (semi-)empirical physics models, such as tight-binding (TB) model, model Hamiltonian, classical molecular dynamics, and so on. Such physics models may contain lots of ad hoc parameters, whose values can be determined by fitting the model to a reference of available experimental data or of pre-generated high-level (e.g., first-principles)

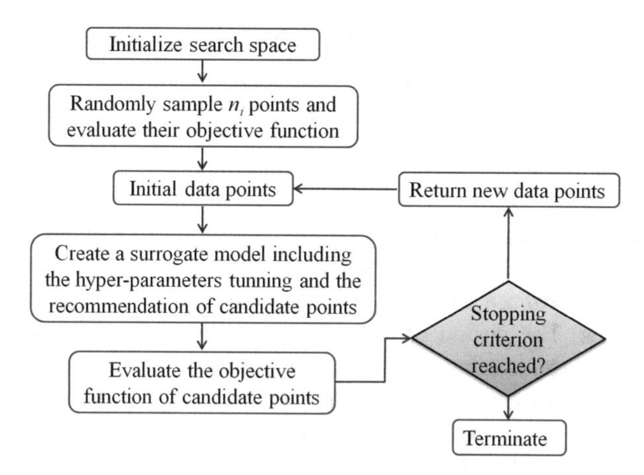

Fig. 19.1 Flowchart of BO in the COMBO code. The evaluation of objective function may be experimental measurement or theoretical simulations. The candidate points for next evaluation of objective function can be recommended by maximum expected of improvement, which is a widely used acquisition function, or Thompson sampling for a large-scale problem. The stopping criterion may be the allowed maximum iteration steps or the predefined convergence threshold of objective function

simulation data [19, 20], although some of which can be rigorously derived in an ab initio manner. In the former case, the fitting problem of a set of parameters $\boldsymbol{\theta} = \{\theta_1, \theta_2, \ldots, \theta_m\}$ in a physics model can be formulated by minimizing a loss function as defined below to search for optimal parameters $\boldsymbol{\theta}^*$:

$$L(\boldsymbol{\theta}) = \sum_{\alpha=1}^{n} |f_\alpha(\boldsymbol{\theta}) - f_\alpha|^2, \tag{19.7}$$

where $f_\alpha(\boldsymbol{\theta})$ is the outputted data of the considered physical properties from the employed physics model at a given set of parameters $\boldsymbol{\theta}$, while f_α is the reference data of the considered physical properties that may be collected from experimental measurement or more accurate ab initio simulation. We note that the derivatives of loss function (Eq. 19.7) with respect to $\boldsymbol{\theta}$ may be numerically inaccessible, particularly in the case of reference data obtained from an experimental measurement. Therefore, a derivative-free optimization approach is highly desired for the determination of optimal parameters $\boldsymbol{\theta}^*$ in Eq. 19.7.

Since BO is derivative free, it has become a successful tool for the hyper-parameter optimization of machine-learning algorithms, such as support vector machines or deep neural networks [8]. The BO has also gained attention for fitting the parameters in a physics model. Recently, Tamura and Hukushima have demonstrated the BO for estimating spin–spin interactions in the classic Ising model and in the quantum Heisenberg model under magnetic field H [21, 22]. The BO showed much better efficiency than the random search method, steepest descent method, and Monte Carlo method in the estimation of three or five parameters in these model Hamiltonians [22].

19.3.2 Discovery of New Functional Materials

The discovery of new materials, or innovative use of existing materials, is essential to make progress in many areas such as electronics, information technologies, automotive and aerospace transportation, biomedicine, energy storage as well as nanotechnologies. The design of materials with the optimal properties for each individual application is a long-standing topic in materials research. The approach from an integration of experimental, computational, and data sciences increasingly shows promising powerfulness in accelerating the materials discovery and design.

The BO, which is also known as a machine-learning framework to optimize expensive black-box functions, can be easily integrated not only with experimentation to effectively optimize developmental stage processes for synthesis of novel materials [13,23–26], but also with first-principles simulations for high-throughput virtual screening of large-scale space of candidate compounds [9, 27]. In the former case, the materials composition and operating parameters in synthesis are of most interest, and they can form a continuous and large search space. Generally, both material property of interest and process-related variables are used as inputs, the machine-learning algorithm in BO recommends a next experimental setup or specification, the experimental synthesis and measurement for the recommended specification are performed, and the return of experimental data values to the algorithm is proceeded in an iterative cycle until the target goal of material property is achieved. The whole procedure is schematically illustrated in Fig. 19.2. Very recently, we applied such an approach to assist the synthesis of off-stoichiometric samples of $Al_2Fe_3Si_3$ compound and were able to find its optimal composition ratio for a significant enhancement of thermoelectric performance within several iterations [28].

Thanks to rapid advances in computational power and techniques over recent decades, many material properties at atomic scale can now be predicted reliably from first-principles calculations [29,30]. Recently, high-throughput first-principles computation has been recognized as an efficient tool to accelerate materials discovery [2]. However, because of the computational cost, a straightforward high-throughput screening with first-principles techniques is usually limited to hundreds or thousands of stoichiometric materials, which are a small fraction of the overall phase space of inorganic compounds [31]. The integration of BO with high-throughput first-principles calculations enables a possibility of virtual screening to reduce computational cost and to cover the overall phase space. In this approach, the search space may be chosen from the experimental crystal structure database, such as Inorganic Crystal Structure Database (ICSD) [32], Pauling File [33], Atomwork [34], and Crystallography Open Database (COD) [35], or be generated for prototype structures by enumerating the ways in which the constituent elements of periodic table can be combined. In this context, the search space is usually

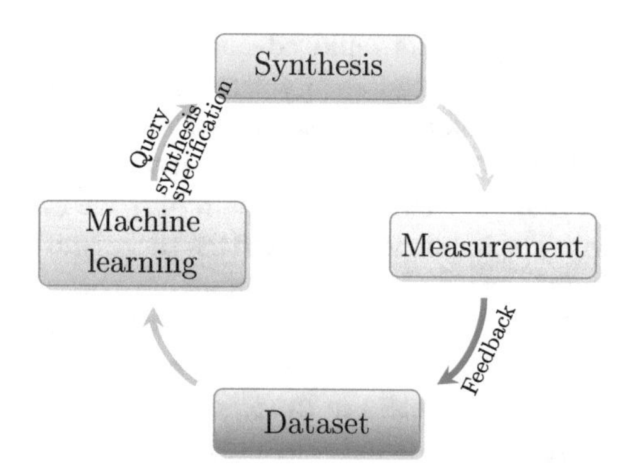

Fig. 19.2 Framework of machine-learning assisted synthesis of functional materials with targeted properties. The initial dataset comprises the material property for already explored specifications, which may be pre-generated or even collected from literature. The machine-learning algorithm recommends a next experimental specification for synthesis according to a selection policy. Next, synthesis of the recommended specification and measurements of its properties are performed. The measured data is appended to the dataset as a feedback, and the next iteration is repeated until the target goal of material property is achieved

represented by material descriptors and thus is discrete. The features of constituent elements, the crystalline volume, or the mass density of every candidate compound can be used in the material descriptors, which form a vector [36–38]. The features of constituent elements may include the information of atoms in periodic table, such as the mass, radius, electron affinity, ionization energy, and so on [36–39], and they are usually called elemental descriptors. The choice of materials descriptors can affect the practice efficiency of BO in the virtual screening of materials. This has been pointed out by Seko et al. [9] in the application of BO for seeking low-thermal-conductivity compounds.

19.3.3 Global Optimization of Atomic Structure

One of the most fundamental properties of a material is its atomic structures, such as cluster, crystal, surface, and interface structures. Knowledge of a material's atomic structure is a prerequisite to many computational materials studies. Once the structure is known, a large number of physical properties can be predicted using first-principles calculations. Computationally finding the most stable (lowest in energy or free energy) structure of a large assembly of atoms is a very difficult problem because of two aspects [40]. One is that the number of minima in the potential energy surface (PES) of a large system increases exponentially with the number of atoms [40–42]. The other one is that finding the global minimum energy structure with certainty presumably involves visiting every local minimum and consequently the computational cost also increases exponentially with the number of atoms [40]. Despite these difficulties, steady progress has been made over the last two decades owing to the rapid advances in computing power and the improved efficiency in ab initio total energy calculation methods. So far, a variety of approaches have been developed for the atomic structure prediction, including the genetic algorithm (GA) [43], the evolutionary algorithm (EA) [42], the particle swarm optimization (PSO) [44], the random search [40], the minima hopping (MH) method [45], and so on. The implementation and the successful applications of these approaches for the global search of atomic structure have been reviewed recently in three books [46–48]. Compared with the aforementioned approaches for the global search of atomic structure, the BO is a new comer and shows competing features, namely the free of ad hoc parameters in the algorithm itself and the capability of predicting the total energies of unvisited structures based on the visited PES. Herein, we discussed the applications of BO in the global optimization of atomic structure for three typical systems, namely the interface structure, the adsorption structure of a molecule on solid surface, and the crystal structure.

19.3.3.1 Optimization of Interface Structure

A solid interface is defined as a small number of atomic layers that separate two solids in intimate contact with one another, where the properties differ significantly from those of the bulk material it separates [49]. Grain boundaries (GBs) are a special type of interface that form when grains of the same phase but different crystallographic orientations abut [50]. The simulation of an interface structure using the supercell model requires a larger number of atoms than those of bulk structures. Because of the increased computational cost, it is much more difficult to perform straightforward structure search using accurate quantum mechanics methods. Much less work on the global prediction of interface structures has been reported [51–53].

Very recently, Kiyohara and his coworkers [54] have taken the case of simplified coincidence-site lattice (CSL) GB of a metal as an example and employed the BO algorithm with the empirical potential method to determine its stable atomic structure. For such a simplified CSL GB, each side of the interface could be approximated as a rigid body and thus the corresponding atomic structure is characterized by three parameters that arise from three-dimensional translations in the used supercell, namely the in-plane relative displacement (denoted as Δx and Δy) of the two rigid atomic layers along the interface plane (assumed in the xy plane) and the interlay distance (denoted as Δz) along the normal direction of interface plane. Here, we shall mention that this treatment is also applicable in the simulation of vertical heterostructure of two-dimensional (2D) materials [55], since the structure reconstruction rarely occurs therein. So that, the search of stable atomic structure of simplified CSL GBs is formulated as the minimization of GB energy in a three-dimensional space. For a given structure (namely Δx, Δy, and Δz) of GB, the GB energy can be evaluated from a static lattice calculation using the empirical potential method. Kiyohara et al. showed that the stable interface structure among ten hundreds possible structures of GB can be determined by only several tens to a hundred calculations using the BO algorithm [54].

19.3.3.2 Adsorption Structure of Molecule on Solid Surface

Molecule adsorption on solid surfaces plays a key role in many surface chemical processes, including heterogeneous catalysis, gas sensing, and building nanoelectronic device from a bottom-up approach. The process of a molecule binding to a surface (called adsorption) involves searching for energetically favorable binding sites. That is to say, for a given solid surface and a given molecule, which adsorption structure does correspond to the globally minimized potential energy?

By taking several assumptions to simplify the degrees of freedom, i.e., no dissociation for molecule adsorbate, negligible deformation of molecule caused by the adsorption, and no molecular-adsorption-induced surface reconstruction, which might be reasonable particularly in the gas sensing and in the self-organization of organic molecules on solid surface, the molecule adsorbate can be treated as a rigid body. So that the search of adsorption structure could be formulated as the minimization of a low-dimensional objective function, that is, the energy minimization with respect to six parameters. These six parameters include the location of the center of molecule along the surface plane, which can be characterized by two parameters, x and y; the distance of the center of molecule with respect to the topmost surface atoms, which is denoted as z; and the orientation of molecule adsorbate in space, which can be fully specified by three parameters, φ, θ, and ψ, commonly known as the Euler angles [56–58]. By taking into account the symmetry, e.g., the in-plane translation symmetry of solid surface due to the periodic boundary along the surface plane and the molecule symmetry, one may rationally choose the constraint ranges for the parameters x, y, φ, θ, and ψ. The constraint range for the parameter z can be chosen according to the typical bonding range that covers from a strong covalent bonding to a weak Van der Waals one. Very recently, Carr et al. [56, 57] and Todorović et al. [58] have independently demonstrated the BO algorithm in treating the energy minimization of such a six-dimensional potential surface. The results in both of their studies showed that BO can efficiently accelerate the search of adsorption structure of a molecule on solid surface. To find the minimum potential energy, the number of density functional theory evaluations called in BO was much smaller than that in the other two used algorithms, namely the differential evolution and the constrained minima hopping algorithms [56, 57].

For the self-organization of medium-sized organic molecules on solid surface, not only the molecule adsorption but also the molecule arrangement are needed to be considered. Taking the adsorption of two molecules on solid surface as an example for simplicity, the distance between the centers of two adsorbed molecule shall be taken into account, which increases the dimensionality of the potential energy surface. However, it does not affect the efficiency of BO algorithm in the determination of molecule arrangement on solid surface, as demonstrated by Packwood and Hitosugi [59]. Their results showed that BO can optimize the arrangement of two medium-sized aromatic adsorbates on a copper (111) surface within tens of density functional theory energy evaluations.

19.3.3.3 Crystal Structure Prediction

Crystal structure is a unique arrangement of atoms in a crystalline material, which is represented by the coordinates of atoms and the lattice vectors. The PES of a crystal structure is exceedingly multidimensional, namely the number of degree of freedom is $3N + 3$ [42], where N is the number of atoms in the unit cell. The crystal structure prediction might be much more complex, as compared with the aforementioned optimizations of a CSL GB structure and a molecule adsorption structure. The central problem in crystal structure prediction is to find the globally stable structure for a given chemical composition from first-principles [60, 61].

Based on the BO algorithm, as implemented in the COMBO code [15], Yamashita et al. [62] have developed the approach of BO for crystal structure prediction. It involves the following steps.

The search space in the BO for crystal structure prediction was predefined by a large number of initial structures that were randomly generated. The random generation of an initial structure was performed in three steps. First, the space group of a structure was randomly selected. And then the unique lattice parameters of a structure with the selected space group were randomly taken from their predefined constraint ranges. Finally, a combination of the Wyckoff positions corresponding to the selected space group is randomly selected so that the number of atoms in the cell meets the predefined chemical composition. Such a scheme for randomly generating initial structures has also been employed in other approaches [40, 42, 44] for crystal structure prediction. The number of initial structures required to find the globally stable structure depends on the system size.

In the BO approach for crystal structure prediction, the descriptors of a crystal structures are required to describe similarity of crystal structures and to establish the correlation between energy and structure. The general aspect of structure descriptors has been discussed by Bartók et al. [63] The choice of structure descriptors is not unique. One of them is the fingerprint function of Oganov and Valle [64], as adopted by Yamashita et al. [62] in the BO approach for crystal structure prediction. For the structure of substitution alloy, the structure descriptor can be properly constructed using the bits of 0 or 1 for the host or substitution sites. Such a choice of structure descriptors has been used in the applications of BO for the design of SiGe nanostructure [14] and for the determination of the stable structures of boron-doped graphene [65].

In crystal structure prediction, the local structure optimization was usually performed using gradient-based methods for every calculated structure, namely the total energy, the residual forces on atoms, and the residual stress on cell were relaxed until their predefined convergence thresholds were reached. If all generated structures were relaxed to reach their respective local minima, it will be very time-consuming. In the BO approach for crystal structure prediction, one first randomly selected ten or twenty structures from the all generated crystal structures and then optimized their structures using the first-principles calculations. The structure descriptors of the optimized structures were calculated. The obtained total energies and structure descriptors of the optimized structures were used as a training dataset in the BO, and then the next candidate structures to be optimized were recommended by the BO according to an acquisition function. Such a procedure of local structure optimization and structure selection was repeated in the BO until the most stable structure was found.

Yamashita et al. [62] has taken the NaCl and Y_2Co_{17} as the test cases to check the efficiency of BO approach for crystal structure prediction. The most stable structures of NaCl and Y_2Co_{17} at ambient are known to be the rocksalt and Th_2Zn_{17} structures, respectively. The BO approach can reduce the average number of trials by 31 and 39% to find their most stable structures among the 800 and 1000 initial structures, respectively, leading to a much less computational cost. This is because the BO enables us to balance the trade-off between the exploration and exploitation of search space in crystal structure prediction.

19.4 Conclusions

The Bayesian optimization, which is a machine-learning-based global optimization method, has been gained attention in materials science. It has been shown to accelerate the fitting of low-dimensional physical parameters and the search of global stable crystal structure in a high-dimensional potential energy surface. The integration of the high-throughput first-principles calculations with the Bayesian optimization provides a more efficient way for materials design and discovery.

In materials design and discovery, multiple targeted properties are often encountered. For searching new functional materials, not only the thermodynamic stability but also the figure of merits are vital objectives, some of which may compete. To solve such a problem, the optimal candidates could be discovered by using a multi-stage screening procedure [66, 67], by converting the multiple objectives into a single objective, or via the Pareto front [68–74]. The multi-objective Bayesian optimization is called to treat more complex problems in materials science. The extension of COMBO code for such a purpose will be released in its coming development version.

Acknowledgments This work was supported by the "Materials research by Information Integration" Initiative (MI^2I) project of the Support Program for Starting Up Innovation Hub from Japan Science and Technology Agency (JST).

References

1. D.R. Jones, M. Schonlau, W.J. Welch, J. Glob. Optim. **13**(4), 455 (1998)
2. S. Curtarolo, G.L.W. Hart, M.B. Nardelli, N. Mingo, S. Sanvito, O. Levy, Nat. Mater. **12**, 191 (2013). https://doi.org/10.1038/nmat3568
3. H.J. Kushner, J. Basic. Eng. **86**(1), 97 (1964). https://doi.org/10.1115/1.3653121
4. J. Mockus, *Bayesian Approach to Global Optimization: Theory and Applications* (Kluwer Academic, Dordrecht, 1989). https://doi.org/10.1007/978-94-009-0909-0

5. C.E. Rasmussen, C.K.I. Williams (eds.), *Gaussian Processes for Machine Learning* (MIT Press, Cambridge, 2006)
6. A.J. Booker, J.E. Dennis, P.D. Frank, D.B. Serafini, V. Torczon, M.W. Trosset, Struct. Optim. **17**(1), 1 (1999). https://doi.org/10.1007/BF01197708
7. J. Snoek, H. Larochelle, R.P. Adams, *Proceedings of the 25th International Conference on Neural Information Processing Systems - Volume 2*, NIPS'12 (Curran Associates Inc., Red Hook, 2012), pp. 2951–2959
8. A. Klein, S. Falkner, S. Bartels, P. Hennig, F. Hutter, Electron. J. Statist. **11**(2), 4945 (2017). https://doi.org/10.1214/17-EJS1335SI
9. A. Seko, A. Togo, H. Hayashi, K. Tsuda, L. Chaput, I. Tanaka, Phys. Rev. Lett. **115**, 205901 (2015). https://doi.org/10.1103/PhysRevLett.115.205901
10. P.I. Frazier, J. Wang, in *Information Science for Materials Discovery and Design*, ed. by T. Lookman, F.J. Alexander, K. Rajan (Springer International Publishing, Cham, 2016), pp. 45–75. https://doi.org/10.1007/978-3-319-23871-5_3
11. D. Packwood, *Bayesian Optimization for Materials Science* (Springer, Singapore, 2017). https://doi.org/10.1007/978-981-10-6781-5
12. P.B. Wigley, P.J. Everitt, A. van den Hengel, J.W. Bastian, M.A. Sooriyabandara, G.D. McDonald, K.S. Hardman, C.D. Quinlivan, P. Manju, C.C.N. Kuhn, I.R. Petersen, A.N. Luiten, J.J. Hope, N.P. Robins, M.R. Hush, Sci. Rep. **6**, 25890 (2016). https://doi.org/10.1038/srep25890
13. C. Li, D. Rubín de Celis Leal, S. Rana, S. Gupta, A. Sutti, S. Greenhill, T. Slezak, M. Height, S. Venkatesh, Sci. Rep. **7**(1), 5683 (2017). https://doi.org/10.1038/s41598-017-05723-0
14. S. Ju, T. Shiga, L. Feng, Z. Hou, K. Tsuda, J. Shiomi, Phys. Rev. X **7**, 021024 (2017). https://doi.org/10.1103/PhysRevX.7.021024
15. T. Ueno, T.D. Rhone, Z. Hou, T. Mizoguchi, K. Tsuda, Mat. Discov. **4**, 18 (2016). https://doi.org/10.1016/j.md.2016.04.001
16. A. Rahimi, B. Recht, in *Advances in Neural Information Processing Systems 20*, ed. by J.C. Platt, D. Koller, Y. Singer, S.T. Roweis (Curran Associates, Inc., Red Hook, 2008), pp. 1177–1184
17. J. Močkus, in *Optimization Techniques IFIP Technical Conference Novosibirsk, July 1–7, 1974*, ed. by G.I. Marchuk (Springer, Berlin, 1975), pp. 400–404. https://doi.org/10.1007/3-540-07165-2_55
18. O. Chapelle, L. Li, in *Advances in Neural Information Processing Systems 24*, ed. by J. Shawe-Taylor, R.S. Zemel, P.L. Bartlett, F. Pereira, K.Q. Weinberger (Curran Associates, Inc., Red Hook, 2011), pp. 2249–2257
19. G.L.W. Hart, V. Blum, M.J. Walorski, A. Zunger, Nat. Mater. **4**(5), 391 (2005). https://doi.org/10.1038/nmat1374
20. R.A. DiStasio, E. Marcotte, R. Car, F.H. Stillinger, S. Torquato, Phys. Rev. B **88**, 134104 (2013). https://doi.org/10.1103/PhysRevB.88.134104
21. R. Tamura, K. Hukushima, Phys. Rev. B **95**, 064407 (2017). https://doi.org/10.1103/PhysRevB.95.064407
22. R. Tamura, K. Hukushima, PLoS One **13**(3), 1 (2018). https://doi.org/10.1371/journal.pone.0193785
23. D. Xue, P.V. Balachandran, J. Hogden, J. Theiler, D. Xue, T. Lookman, Nat. Comm. **7**, 11241 (2016). https://doi.org/10.1038/ncomms11241
24. D. Xue, P.V. Balachandran, R. Yuan, T. Hu, X. Qian, E.R. Dougherty, T. Lookman, Proc. Natl. Acad. Sci. USA **113**(47), 13301 (2016). https://doi.org/10.1073/pnas.1607412113
25. J. Gao, Y. Liu, Y. Wang, X. Hu, W. Yan, X. Ke, L. Zhong, Y. He, X. Ren, J. Phys. Chem. C **121**(24), 13106 (2017). https://doi.org/10.1021/acs.jpcc.7b04636
26. P.V. Balachandran, B. Kowalski, A. Sehirlioglu, T. Lookman, Nat. Comm. **9**(1), 1668 (2018). https://doi.org/10.1038/s41467-018-03821-9
27. R. Jalem, K. Kanamori, I. Takeuchi, M. Nakayama, H. Yamasaki, T. Saito, Sci. Rep. **8**(1), 5845 (2018). https://doi.org/10.1038/s41598-018-23852-y

28. Z. Hou, Y. Takagiwa, Y. Shinohara, Y. Xu, K. Tsuda, ACS Appl. Mater. Interfaces **11**(12), 11545 (2019). https://doi.org/10.1021/acsami.9b02381
29. E.A. Carter, Science **321**(5890), 800 (2008). https://doi.org/10.1126/science.1158009
30. C.K. Skylaris, Science **351**(6280), 1394 (2016). https://doi.org/10.1126/science.aaf3412
31. D. Davies, K. Butler, A. Jackson, A. Morris, J. Frost, J. Skelton, A. Walsh, Chem **1**(4), 617 (2016). https://doi.org/10.1016/j.chempr.2016.09.010
32. Royal Society of Chemistry. CDS: National Chemical Database Service. http://icsd.cds.rsc.org.
33. P. Villars, M. Berndt, K. Brandenburg, K. Cenzual, J. Daams, F. Hulliger, T. Massalski, H. Okamoto, K. Osaki, A. Prince, H. Putz, S. Iwata, J. Alloys. Compd. **367**(1), 293 (2004). https://doi.org/10.1016/j.jallcom.2003.08.058. http://paulingfile.com/
34. Y. Xu, M. Yamazaki, P. Villars, Jap. J. Appl. Phys. **50**(11S), 11RH02 (2011). https://doi.org/10.1143/JJAP.50.11RH02. https://atomwork-adv.nims.go.jp/
35. S. Gražulis, D. Chateigner, R.T. Downs, A.F.T. Yokochi, M. Quirós, L. Lutterotti, E. Manakova, J. Butkus, P. Moeck, A. Le Bail, J. Appl. Crystallogr. **42**(4), 726 (2009). https://doi.org/10.1107/S0021889809016690. http://www.crystallography.net/cod/
36. L.M. Ghiringhelli, J. Vybiral, S.V. Levchenko, C. Draxl, M. Scheffler, Phys. Rev. Lett. **114**, 105503 (2015). https://doi.org/10.1103/PhysRevLett.114.105503
37. R. Jalem, M. Nakayama, Y. Noda, T. Le, I. Takeuchi, Y. Tateyama, H. Yamazaki, Sci. Tech. Adv. Mater. **19**(1), 231 (2018). https://doi.org/10.1080/14686996.2018.1439253
38. L. Ward, A. Dunn, A. Faghaninia, N.E. Zimmermann, S. Bajaj, Q. Wang, J. Montoya, J. Chen, K. Bystrom, M. Dylla, K. Chard, M. Asta, K.A. Persson, G.J. Snyder, I. Foster, A. Jain, Comp. Mater. Sci. **152**, 60 (2018). https://doi.org/10.1016/j.commatsci.2018.05.018
39. A. Seko, T. Maekawa, K. Tsuda, I. Tanaka, Phys. Rev. B **89**, 054303 (2014). https://doi.org/10.1103/PhysRevB.89.054303
40. C.J. Pickard, R.J. Needs, J. Phys.: Condens. Matter **23**(5), 053201 (2011). https://doi.org/10.1088/0953-8984/23/5/053201
41. F.H. Stillinger, Phys. Rev. E **59**, 48 (1999). https://doi.org/10.1103/PhysRevE.59.48
42. A.R. Oganov, C.W. Glass, J. Chem. Phys. **124**(24), 244704 (2006). https://doi.org/10.1063/1.2210932
43. D.M. Deaven, K.M. Ho, Phys. Rev. Lett. **75**, 288 (1995). https://doi.org/10.1103/PhysRevLett.75.288
44. Y. Wang, J. Lv, L. Zhu, Y. Ma, Phys. Rev. B **82**, 094116 (2010). https://doi.org/10.1103/PhysRevB.82.094116
45. M. Amsler, S. Goedecker, J. Chem. Phys. **133**(22), 224104 (2010). https://doi.org/10.1063/1.3512900
46. A.R. Oganov (ed.), *Modern Methods of Crystal Structure Prediction* (Wiley, Weinheim, 2010). https://doi.org/10.1002/9783527632831
47. C.V. Ciobanu, C. Wang, K. Ho, *Atomic Structure Prediction of Nanostructures, Clusters and Surfaces* (Wiley, Weinheim, 2013). https://doi.org/10.1002/9783527655021
48. Ş. Atahan-Evrenk, A. Aspuru-Guzik (eds.), *Prediction and Calculation of Crystal Structures: Methods and Applications* (Springer International Publishing, Switzerland, 2014). https://doi.org/10.1007/978-3-319-05774-3
49. L. H, *Solid Surfaces, Interfaces and Thin Films* (Springer, Berlin, 2010). https://doi.org/10.1007/978-3-642-13592-7_1
50. W.C. Carter, Nat. Mater. **9**, 383–385 (2010). https://doi.org/10.1038/nmat2754
51. A.L.S. Chua, N.A. Benedek, L. Chen, M.W. Finnis, A.P. Sutton, Nat. Mater. **9**, 418–422 (2010). https://doi.org/10.1038/nmat2712
52. X. Zhao, Q. Shu, M.C. Nguyen, Y. Wang, M. Ji, H. Xiang, K.M. Ho, X. Gong, C.Z. Wang, J. Phys. Chem. C **118**(18), 9524 (2014). https://doi.org/10.1021/jp5010852
53. G. Schusteritsch, C.J. Pickard, Phys. Rev. B **90**, 035424 (2014). https://doi.org/10.1103/PhysRevB.90.035424
54. S. Kiyohara, H. Oda, K. Tsuda, T. Mizoguchi, Jpn. J. Appl. Phys. **55**(4), 045502 (2016). https://doi.org/10.7567/JJAP.55.045502

55. B.V. Lotsch, Annu. Rev. Mater. Res. **45**(1), 85 (2015). https://doi.org/10.1146/annurev-matsci-070214-020934
56. S.F. Carr, R. Garnett, C.S. Lo, J. Chem. Phys. **145**(15), 154106 (2016). https://doi.org/10.1063/1.4964671
57. S. Carr, R. Garnett, C. Lo, in *Proceedings of The 33rd International Conference on Machine Learning Research*, vol. 48, ed. by M.F. Balcan, K.Q. Weinberger (PMLR, New York, 2016), pp. 898–907
58. M. Todorović, M.U. Gutmann, J. Corander, P. Rinke, npj Comput. Mater. **5**(1), 35 (2019). https://doi.org/10.1038/s41524-019-0175-2
59. D.M. Packwood, T. Hitosugi, Appl. Phys. Express **10**(6), 065502 (2017). https://doi.org/10.7567/APEX.10.065502
60. J. Maddox, Nature **335**, 201 (1988). https://doi.org/10.1038/335201a0
61. S.M. Woodley, R. Catlow, Nat. Mater. **7**, 937 (2008). https://doi.org/10.1038/nmat2321
62. T. Yamashita, N. Sato, H. Kino, T. Miyake, K. Tsuda, T. Oguchi, Phys. Rev. Materials **2**, 013803 (2018). https://doi.org/10.1103/PhysRevMaterials.2.013803
63. A.P. Bartók, R. Kondor, G. Csányi, Phys. Rev. B **87**, 184115 (2013). https://doi.org/10.1103/PhysRevB.87.184115
64. A.R. Oganov, M. Valle, J. Chem. Phys. **130**(10), 104504 (2009). https://doi.org/10.1063/1.3079326
65. T. M. Dieb, Z. Hou, K. Tsuda, J. Chem. Phys. **148**(24), 241716 (2018). https://doi.org/10.1063/1.5018065
66. D. Davies, K.T. Butler, J.M. Skelton, C. Xie, A.R. Oganov, A. Walsh, Chem. Sci. **9**, 1022 (2018). https://doi.org/10.1039/C7SC03961A
67. R. Matsumoto, Z. Hou, H. Hara, S. Adachi, H. Takeya, T. Irifune, K. Terakura, Y. Takano, Appl. Phys. Express **11**(9), 093101 (2018). https://doi.org/10.7567/apex.11.093101
68. A.M. Gopakumar, P.V. Balachandran, D. Xue, J.E. Gubernatis, T. Lookman, Sci. Rep. **8**(1), 3738 (2018). https://doi.org/10.1038/s41598-018-21936-3
69. M. Núñez-Valdez, Z. Allahyari, T. Fan, A.R. Oganov, Comput. Phys. Commun. **222**, 152 (2018). https://doi.org/10.1016/j.cpc.2017.10.001
70. P. Singh, I. Couckuyt, K. Elsayed, D. Deschrijver, T. Dhaene, J. Optimiz. Theory App. **175**(1), 172 (2017). https://doi.org/10.1007/s10957-017-1114-3
71. I. Couckuyt, D. Deschrijver, T. Dhaene, J. Global Optim. **60**(3), 575 (2014). https://doi.org/10.1007/s10898-013-0118-2
72. A. Solomou, G. Zhao, S. Boluki, J.K. Joy, X. Qian, I. Karaman, R. Arróyave, D.C. Lagoudas, Mater. Des. **160**, 810 (2018). https://doi.org/10.1016/j.matdes.2018.10.014
73. M.T.M. Emmerich, A.H. Deutz, J.W. Klinkenberg, in *2011 IEEE Congress of Evolutionary Computation (CEC)* (2011), pp. 2147–2154. https://doi.org/10.1109/CEC.2011.5949880
74. A. Talapatra, S. Boluki, P. Honarmandi, A. Solomou, G. Zhao, S.F. Ghoreishi, A. Molkeri, D. Allaire, A. Srivastava, X. Qian, E.R. Dougherty, D.C. Lagoudas, R. Arróyave, Front. Mater. **6**, 82 (2019). https://doi.org/10.3389/fmats.2019.00082

Recommender Systems for Materials Discovery

20

Atsuto Seko, Hiroyuki Hayashi, Hisashi Kashima, and Isao Tanaka

Abstract

Chemically relevant compositions (CRCs) and atomic arrangements of inorganic compounds have been collected as inorganic crystal structure databases. Machine learning is a unique approach to search for currently unknown CRCs from a vast number of candidates. Firstly, we show matrix- and tensor-based recommender system approaches to predict currently unknown CRCs from database entries of CRCs. Secondly, we demonstrate classification approaches using compositional similarity defined by descriptors obtained from a set of well-known elemental representations. They indicate that the recommender system has great potential to accelerate the discovery of new compounds.

20.1 Introduction

A reliable and quantitative prediction of chemically relevant compositions (CRCs) where stable crystals are formed is highly demanded because the synthesis of new inorganic compounds is an important target of physicists, chemists, and material scientists. Density functional theory (DFT) calculations have played a central role in such predictions. However, exhaustive DFT calculations without prior knowledge of the crystal structures are expensive even for a given composition. New compounds have also been discovered by inspecting the similarity between chemical elements and their compositions. Similarity has been measured using heuristic quantities

A. Seko (✉) · H. Hayashi · I. Tanaka
Department of Materials Science and Engineering, Kyoto University, Kyoto, Japan
e-mail: seko@cms.mtl.kyoto-u.ac.jp

H. Kashima
Department of Intelligence Science and Technology, Kyoto University, Kyoto, Japan

K. T. Schütt et al. (eds.), *Machine Learning Meets Quantum Physics*,
Lecture Notes in Physics 968, https://doi.org/10.1007/978-3-030-40245-7_20

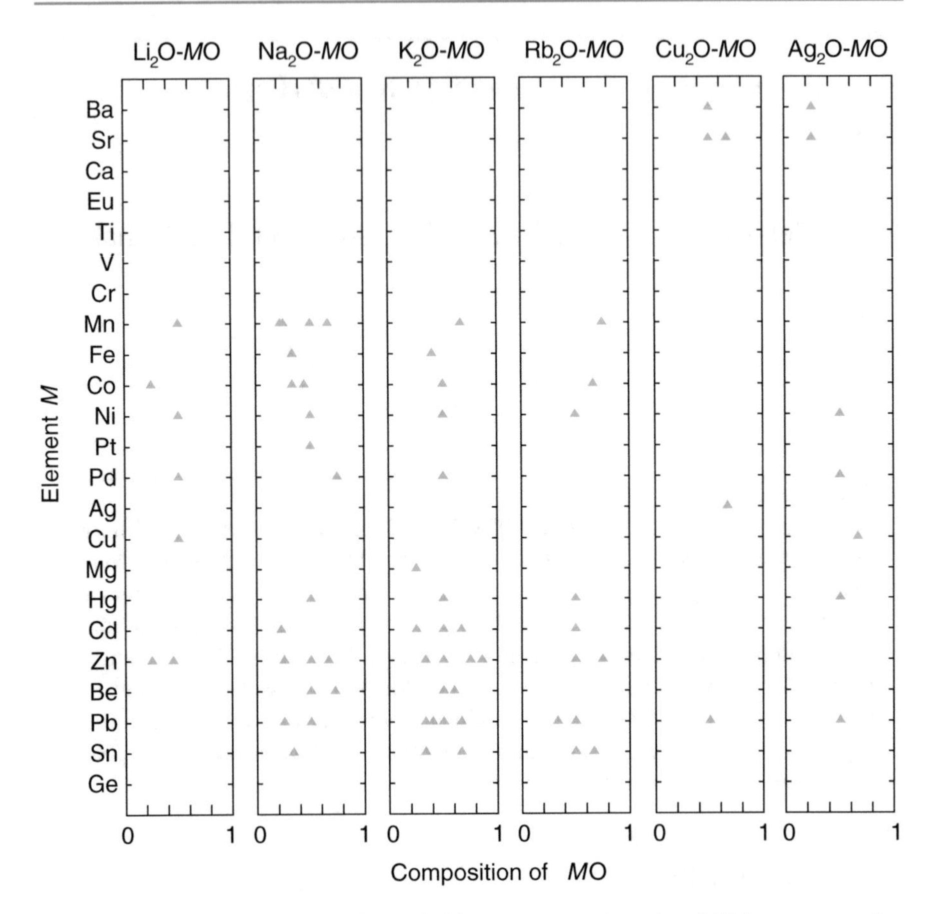

Fig. 20.1 ICSD entries in A_2O-MO pseudo-binary systems, where A and M denote monovalent and divalent cations, respectively. Closed triangles indicate the compositions of the ICSD entries. Divalent cations are indicated in the order of Mendeleev number conceived by D.G. Pettifor [1] (reproduced from [2], with the permission of AIP Publishing)

such as the proximity in the periodic table, electronegativity, ionicity, and ionic radius. These quantities are derived from either simplified theoretical considerations or chemists' intuition. Based on the similarity between given compositions and entries of experimental databases for existing crystals, currently unknown CRCs may be discovered. As an example, consider searching for CRCs in Li_2O-MO (M: divalent cation) pseudo-binary systems from known CRCs. Figure 20.1 shows the compositions in both Li_2O-MO and A_2O-MO (A: monovalent cation) systems where the Inorganic Crystal Structure Database (ICSD) [3] entries exist. Known CRCs are widely scattered and depend on both elements M and A, suggesting that cationic similarity may identify many currently unknown CRCs. However, a quantitative figure of merit such as phase stability cannot be given for numerous compositions without using machine learning (ML)-based methods.

Recent developments and the popularization of ML have facilitated the prediction of currently unknown CRCs. For example, a ML model with respect to only the composition, which was estimated from a DFT database of the formation energies, predicts currently unknown CRCs (e.g., [4–6]). Another ML-based approach to search for currently unknown CRCs uses inorganic crystal structure databases such as ICSD. This approach estimates the CRC probability for a candidate composition using a given compositional similarity. A procedure to estimate the CRC probability uses the compositional similarity on the basis of the entries in the database itself [7, 8].

These approaches based on inorganic crystal structure databases can be referred to as "recommender systems" for the discovery of currently unknown CRCs. Recommender systems [9, 10] developed in the ML community have become increasingly popular in a variety of scientific and non-scientific areas. In the field of commerce, recommender systems suggest items to a user. Such a recommender system predicts the rating or preference for an item from an existing dataset comprised of the users' history such as items purchased and numerical ratings given to items to generate a recommendation for the user.

Simple matrix- and tensor-based recommender system approaches can be adopted to discover currently unknown CRCs. These approaches show a robust performance of recommendations for a wide variety of datasets in the ML community [11–13]. Here we introduce recommender systems to discover currently unknown CRCs using matrix- and tensor-based approaches, which are descriptor-free approaches [14]. A different procedure adopts the compositional descriptors defined by a set of well-known elemental representations (for example, Ref. [15]) as prior knowledge [2, 16]. We also show ML approach using compositional similarity defined by descriptors obtained from a set of well-known elemental representations [2]. We demonstrate the potential possibility of descriptors for predicting currently unknown CRCs. This method corresponds to a kind of knowledge-based recommender system that utilizes prior knowledge about compositions. A major advantage of knowledge-based recommender systems is avoiding the so-called cold-start problems. In the present case, the cold-start problem is that a meaningful figure of merit cannot be estimated for a given composition due to the lack of related known CRCs. As a result, few compositions are recommended. This may occur in applications to multicomponent systems such as a pseudo-ternary system where few known CRCs exist, which are the most interesting application of ML-based methods. Therefore, the use of a knowledge-based method should contribute significantly to the recommendation of multicomponent CRCs.

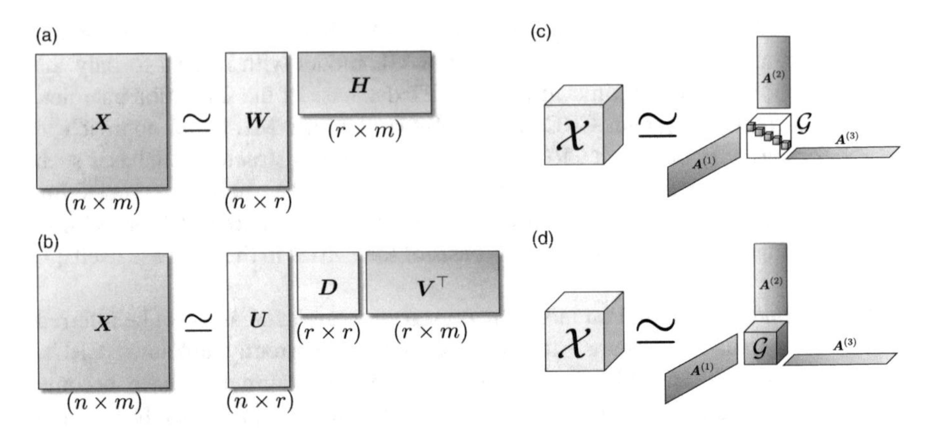

Fig. 20.2 Schematic illustration of the approximation of the rating matrix by (**a**) NMF and (**b**) SVD and the rating tensor by (**c**) CP and (**d**) Tucker decomposition. Dimensions of factorized matrices are also shown

20.2 Matrix- and Tensor-Based Recommender System

In this section, we introduce recommender systems to discover currently unknown CRCs using matrix- and tensor-based approaches, which are descriptor-free approaches.[1]

20.2.1 Matrix and Tensor Factorization

A historical dataset used to build a recommender system is simply described as a form of an incomplete matrix or tensor with missing values [13], called a "rating matrix" or a "rating tensor." The underlying assumption of many recommender system approaches is that the complete rating matrix or tensor has a low-rank structure. Using this assumption, the ratings of missing elements are predicted as an approximated rating matrix or tensor with a given reduced rank. They are hereafter called "predicted rating." Compositions are recommended on the basis of the values of the predicted ratings.

The nonnegative matrix factorization (NMF) factorizes nonnegative rating matrix $\boldsymbol{X}$ into two nonnegative matrices $\boldsymbol{W}$ and $\boldsymbol{H}$ with given rank r. $\boldsymbol{W}$ and $\boldsymbol{H}$ do not have negative elements. Figure 20.2a illustrates the behavior of NMF. The original (n, m)-dimensional matrix $\boldsymbol{X}$ is approximated as

$$\boldsymbol{X} \simeq \boldsymbol{W}\boldsymbol{H} \tag{20.1}$$

[1]This section is reproduced from [14].

using (n, r)-dimensional matrix $\boldsymbol{W}$ and (r, m)-dimensional matrix $\boldsymbol{H}$ with a given rank r. Each of the r dimensions corresponds to a distinctive rating trend. The distinctive rating trends enable predicting similar ratings for similar elements.

Another matrix factorization approach is the singular value decomposition (SVD). Figure 20.2b illustrates the behavior of SVD. SVD factorizes original rating matrix $\boldsymbol{X}$ into three matrices. The rating matrix is approximated as

$$\boldsymbol{X} \simeq \boldsymbol{U}\boldsymbol{D}\boldsymbol{V}^{\top} \tag{20.2}$$

using (n, r)-dimensional matrix $\boldsymbol{U}$, r-dimensional diagonal matrix $\boldsymbol{D}$, and (m, r)-dimensional matrix $\boldsymbol{V}$. The diagonal matrix contains only r largest singular values. Hence, each of the r dimensions corresponds to a distinctive rating trend as well as the NMF.

Matrix factorization approaches require that the original data is transformed into a matrix form when the rating depends on three or more factors. On the other hand, tensor factorization approaches can include information of all factors naturally as a form of the rating tensor. A tensor factorization technique is the tensor rank decomposition or canonical polyadic (CP) decomposition [17]. Figure 20.2c illustrates the behavior of CP decomposition. CP decomposition factorizes rating tensor $\mathcal{X}$ into a set of matrices and a super diagonal core tensor. Third-order rating tensor $\mathcal{X}$ is approximated using three matrices $\boldsymbol{A}^{(1)}$, $\boldsymbol{A}^{(2)}$, and $\boldsymbol{A}^{(3)}$ as

$$\mathcal{X} \simeq \mathcal{G} \times_1 \boldsymbol{A}^{(1)} \times_2 \boldsymbol{A}^{(2)} \times_3 \boldsymbol{A}^{(3)} \tag{20.3}$$

where $\mathcal{G}$ denotes the core tensor.

A more flexible tensor factorization is the Tucker decomposition [18], which is also known as higher-order SVD (HOSVD) [19]. The Tucker decomposition is an extension of SVD and decomposes rating tensor $\mathcal{X}$ into a set of matrices and a small dense core tensor. In the Tucker decomposition, $\boldsymbol{A}^{(1)}$, $\boldsymbol{A}^{(2)}$, and $\boldsymbol{A}^{(3)}$ contain the orthonormal vector called mode-1, mode-2, and mode-3 singular vectors, respectively. Figure 20.2d illustrates the behavior of Tucker decomposition. Third-order tensor $\mathcal{X}$ can also be approximated as Eq. (20.3), where $\mathcal{G}$ is also the core tensor, but unlike in CP decomposition, the core tensor is not a super diagonal tensor. In Tucker decomposition, the rank value can be independently given for each mode of the rating tensor. Distributed packages such as SCIKIT-TENSOR [20] are available for tensor factorization methods.

20.2.2 Datasets

We employ three inorganic crystal structure databases of the ICSD, Powder Diffraction File (PDF) by the International Centre for Diffraction Data (ICDD) [21], and SpringerMaterials (SpMat) [22]. Only ICSD entries, including entries reported to show a partial occupancy behavior, are regarded as known CRCs. Both ICDD

Table 20.1 Numbers of ICSD, ICDD, and SpMat entries

	Ternary	Quaternary	Quinary
ICSD	9313	7742	1321
ICDD	2369 (9278)	2647 (7864)	639 (1326)
SpMat	2708 (10,461)	3066 (8141)	1169 (1893)
ICDD+SpMat	4134 (12,573)	4961 (11,307)	1616 (2562)
Candidates	7,405,200	1,188,038,460	23,104,706,560

Numbers of candidate compositions are also shown. ICSD entries are regarded as known CRCs. In the rows of ICDD and SpMat databases, the numbers of entries that are not included in the ICSD are shown. Numbers in the parentheses indicate the numbers of all entries

and SpMat entries, which are not included in the ICSD, are used to validate the recommender systems.

To build a recommender system for ternary CRCs, only a set of ternary known CRCs is employed. Candidate ternary compositions of $A_aB_bX_x$ are generated, where elements A and B correspond to a cation (66 possible) and element X corresponds to an anion (10 possible). The possible cations and anions correspond to elements included in the dataset of known CRCs. One hundred seventy (170) integer sets (a, b, x) that satisfy the condition of $\max(a, b, x) \leq 8$ and are included in the dataset of known CRCs are also considered. Therefore, there are $66^2 \times 10 \times 170 = 7{,}405{,}200$ ternary compositions. Of these, 9313 CRCs are known. The combined database of ICDD and SpMat has 4134 ternary compositions that are not included in ICSD. This corresponds to 0.056% of the candidate compositions. Table 20.1 summarizes the numbers of database entries and candidate compositions.

In addition to ternary compositions, candidate quaternary $A_aB_bC_cX_x$ compositions are generated, where elements A, B, and C denote a cation and element X denotes an anion. Integer sets (a, b, c, x) satisfy $\max(a, b, c, x) \leq 20$. The dataset of known quaternary CRCs includes 53 cations, 10 anions, and 798 integer sets. Therefore the number of quaternary compositions is $53^3 \times 10 \times 798 = 1{,}188{,}038{,}460$, which includes 7742 known CRCs. The combined database of ICDD and SpMat includes 4961 quaternary compositions (4.2×10^{-4}% of the candidate compositions) that are not the entries of ICSD.

Candidate quinary $A_aB_bC_cD_dX_x$ compositions are also generated, where elements A, B, C, and D denote cations and element X denotes an anion. Integer sets (a, b, c, d, x) satisfy $\max(a, b, c, d, x) \leq 20$. The dataset of known quinary CRCs includes 52 cations, 10 anions, 316 integer sets. Therefore, the number of quinary compositions is $52^4 \times 10 \times 316 = 23{,}104{,}706{,}560$, where only 1321 known CRCs are included. The combined database of ICDD and SpMat contains 1616 quinary compositions (7.0×10^{-6}% of the candidate compositions) that are not ICSD entries.

20.2.3 Rating Matrix and Tensor Representations

The performance of a recommender system strongly depends on the representation of the composition dataset as a rating matrix or tensor. Experts' knowledge is required to introduce a good representation. In particular, a composition dataset must be transformed into only two sets of features, which corresponds to users and items in a user-item rating matrix. The ratings of missing elements are approximately predicted on the basis of feature similarity given by the representation.

We introduce three kinds of matrix representations for the ternary composition dataset. A composition is decomposed into two feature sets in the following three ways:

1. $\{A\}$ and $\{B, X, (a, b, x)\}$
2. $\{A, X\}$ and $\{B, (a, b, x)\}$
3. $\{A, B\}$ and $\{X, (a, b, x)\}$

Each of the feature sets corresponds to the row or column of the rating matrix. For example, the first matrix representation is schematically illustrated in Fig. 20.3a. In the first representation, the row corresponds to cation A. The column corresponds to the combination of cation B, anion X, and integer set (a, b, x). Each composition is expressed by the collection of features included in the row and column. Only the values of the rating matrix elements corresponding to known CRCs are set to unity. The other compositions are classified as either currently unknown CRCs or nonexistent compositions, and their values are set to zero in the original rating matrix.

Figure 20.3b illustrates the tensor representation used in this study. A mode of the rating tensor means each cation type A, cation type B, anion type X, and integer set $\{a, b, x\}$ for ternary compositions $A_aB_bX_x$. Therefore, the rating tensor for

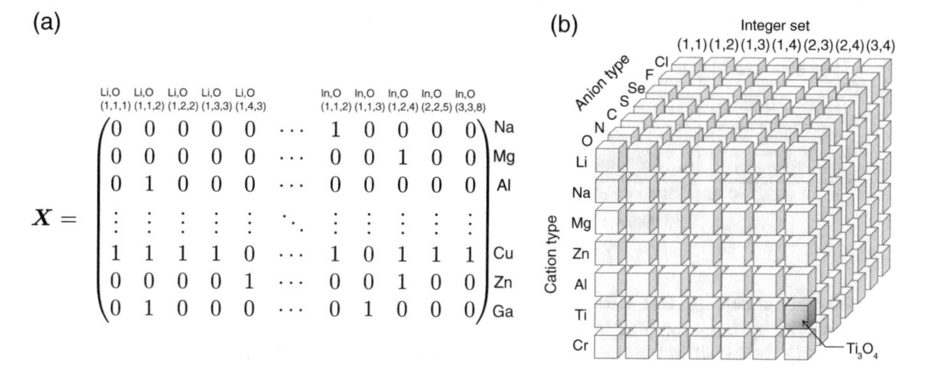

Fig. 20.3 Schematic illustration of (**a**) a matrix representation and (**b**) a tensor representation for a composition dataset. As a simple example, a tensor representation for binary compositions is illustrated. In a tensor element highlighted by the orange block, modes for cation type, anion type, and integer set indicate Ti, O, and (3, 4), respectively. Hence, this tensor element represents the composition of Ti_3O_4

ternary compositions is a (66, 66, 10, 170)-dimensional tensor. Similar to the matrix representations, the composition is expressed by the collections of features included in all modes, corresponding to an element of the rating tensor. Only the values of the rating tensor elements corresponding to known CRCs are set to unity. The values of tensor elements corresponding to the other compositions are assumed to be zero in the original rating tensor. By representing quaternary and quinary compositions in tensor forms as well as ternary compositions, the numbers of modes of the rating tensor correspond to five and six for quaternary and quinary compositions, respectively.

20.2.4 Discovery Performance of Unknown CRCs

The performance of matrix factorization based on the recommender systems to discover currently unknown ternary CRCs is evaluated. Only 4134 of 7,405,200 compositions (0.056%) are used as the test dataset of known CRCs (test CRCs). The test CRCs correspond to ICDD and SpMat entries that are not included in ICSD. The low percentage obviously confirms that discovering ICDD and SpMat entries by random sampling is not efficient.

Figure 20.4 shows the rank dependence of the number of test CRCs found in the top 3000 compositions of NMF and SVD recommender systems. The rank dependences of NMF and SVD are qualitatively similar but weak. They also exhibit similar matrix representation dependences. The performance of discovering test

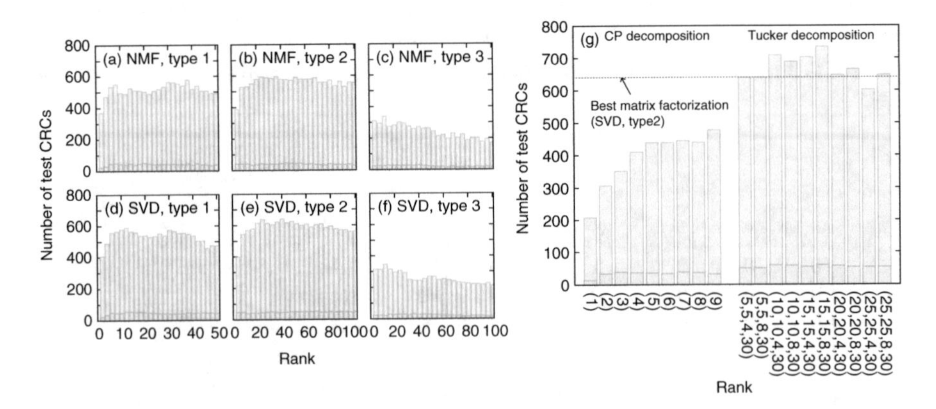

Fig. 20.4 Numbers of test CRCs included in the top 100 and 3000 compositions of NMF recommender systems for (**a**) type-1, (**b**) type-2, and (**c**) type-3 matrix representations. Numbers of test CRCs included in the top 100 and 3000 compositions of SVD recommender systems for (**d**) type-1, (**e**) type-2, and (**f**) type-3 matrix representations. Horizontal axis indicates a given rank for NMF and SVD. Orange and blue bars denote the top 100 and 300 compositions, respectively. (**g**) Rank dependence of the number of test CRCs included in the top 100 and 3000 compositions of the CP and Tucker decomposition recommender systems. Here only the performance of Tucker decomposition with a rank value of 30 for the composition integer set is shown

CRCs of type-1 and type-2 matrix representations is much better than that of type-3 matrix representations. This implies that it is important to consider the cation and anion combination as a feature when building a rule to distinguish CRCs and non-existent compositions.

When adopting type-1 and type-2 matrix representations, NMF recommender systems with type-1 and type-2 matrix representations show the best performance at $r = 38$ and $r = 24$, respectively, indicating that the maximum number of test CRCs are included in the top 3000 compositions of the recommender system. For type-1 and type-2 representations, 576 and 594 test CRCs (19.2 and 19.8%) are found in the top 3000 compositions, respectively. These percentages are much larger than the number of test CRCs discovered by random sampling (0.056%). A much larger discovery rate of test CRCs is seen in the sampling of top 100 compositions. The numbers of test CRCs are 44 and 41 (44.0 and 41.0%), respectively. SVD recommender systems with type-1 and type-2 matrix representations show the best performance at $r = 12$ and $r = 40$, respectively. The top 3000 compositions include 590 and 640 test CRCs (19.7 and 21.3%), respectively. Additionally, 53 and 45 of the top 100 compositions (53.0 and 45.0%) correspond to test CRCs. These results indicate that SVD recommender systems perform slightly better than NMF recommender systems.

Figure 20.4g shows the number of test CRCs included in the top 3000 ternary compositions of the ranking by CP decomposition. The horizontal axis indicates a given rank for CP decomposition. Since only a single value of rank is given in CP decomposition, the maximum value of rank is identical to the number of anion types included in the ternary composition dataset. The CP recommender system with $r = 9$ shows the best performance. That is, 478 of the top 3000 compositions (15.9%) correspond to test CRCs. In the top 100 compositions, 32 test CRCs (32.0%) are found. However, CP decompositions detect fewer test CRCs than the best matrix decomposition-based recommender system. This is attributed to the fact that only a single value of rank is given in CP decomposition.

Figure 20.4g shows the number of test CRCs included in the top 3000 compositions of the ranking by Tucker decomposition. The candidate values of ranks for the cation mode, anion mode, and integer set mode are given by arithmetic sequences of $\{5, 10, 15, 20, 25\}$, $\{4, 8\}$, and $\{10, 20, 30, \ldots, 100\}$, respectively. The optimal set of ranks is obtained by a grid search using these candidates. This means that the performance of discovering test CRCs is examined for 100 combinations of rank values. The Tucker decomposition recommender system with ranks of (15, 15, 8, 30) shows the best performance; 735 test CRCs (24.5%) are found in the top 3000 compositions. In the top 100 compositions, 59 CRCs (59.0%) are found in the test dataset. The Tucker recommender system detects more test CRCs than the best matrix-based recommender system. This means that Tucker decompositions allow a hidden low-rank data-structure included in the composition dataset to be extracted more flexibly than matrix factorizations and CP decompositions.

The Tucker decomposition should be the best approach to discover currently unknown CRCs. Therefore, the performance to discover currently unknown quaternary and quinary CRCs using the Tucker decomposition is examined. The performances for quaternary and quinary systems are particularly important features required for recommender systems because the numbers of candidate compositions and currently unknown CRCs should be exponentially huge. As shown in Table 20.1, only 7742 and 4961 of 1.2×10^9 (6.5×10^{-4} and 4.2×10^{-4}%) quaternary compositions are ICSD and ICDD+SpMat entries, respectively. Even for quinary compositions, only 1321 and 1616 of 2.3×10^{10} (5.7×10^{-6} and 7.0×10^{-6}%) compositions correspond to ICSD and ICDD+SpMat entries, respectively.

Figure 20.5a and b shows the performance to determine test ternary, quaternary, and quinary CRCs of the Tucker recommender system. The discovery rate gradually decreases as the number of samples all for ternary, quaternary, and quinary compositions increases. In the top 20 compositions, 16 and 15 compositions (80 and 75%) correspond to test CRCs for ternary and quaternary compositions, respectively. The discovery rate of quaternary test CRCs is higher than expected. The rate is close to that of ternary test CRCs. For quaternary compositions, 52 and 524 CRCs are found in the top 100 and 3000 compositions (52.0 and 17.5%), respectively. The discovery rates for the top 20, 100, and 3000 compositions are approximately 180,000, 120,000, and 40,000 times larger than the random sampling discovery rate, respectively. Even for quinary compositions, seven test CRCs (35%) are found in the top 20 compositions. In addition, 14 and 82 CRCs are found in the top 100 and 3000 compositions (14.0 and 2.7%), respectively. Although the discovery rate of quinary test CRCs is lower than those of ternary and quaternary test CRCs, it is much larger than the random sampling. In particular, the discovery rate for the top 100 compositions is astonishingly large despite the fact that few known quinary CRCs are included in the quinary composition dataset.

Figure 20.5c and d shows the relationship between the predicted rating and the discovery rate of test CRCs. Figure 20.5c shows the histogram of the discovery rate of the test CRCs obtained with a bin width of predicted rating of 0.1. Figure 20.5d replots the relationship between the median of the bin and discovery rate from the histogram of the discovery rate. For ternary, quaternary, and quinary compositions, the discovery rate of test CRCs is almost proportional to the median of the bin for predicted rating, and slightly smaller than the predicted rating. This is strong evidence that the predicted rating can be regarded as a figure of merit to identify currently unknown CRCs. This may also indicate that the predicted rating is almost identical to the discovery rate of currently unknown CRCs because test CRCs can be regarded as a part of currently unknown CRCs. In addition, quinary compositions with a predicted rating exceeding 0.7 are not observed. Consequently, the discovery rate of test CRCs is smaller than that of the ternary and quaternary test CRCs. This is mainly ascribed to the lack of quinary known CRCs. The number of quinary known CRCs is inadequate to predict currently unknown quinary CRCs.

The discovery of new ternary compounds is also demonstrated using a combination of the Tucker recommender system and DFT calculations. A Tucker decomposition recommender system is constructed using all of the available entries of the ternary compositions included in ICSD, ICDD, and SpMat as known CRCs. The rank values are the same values as those optimized above. The existence

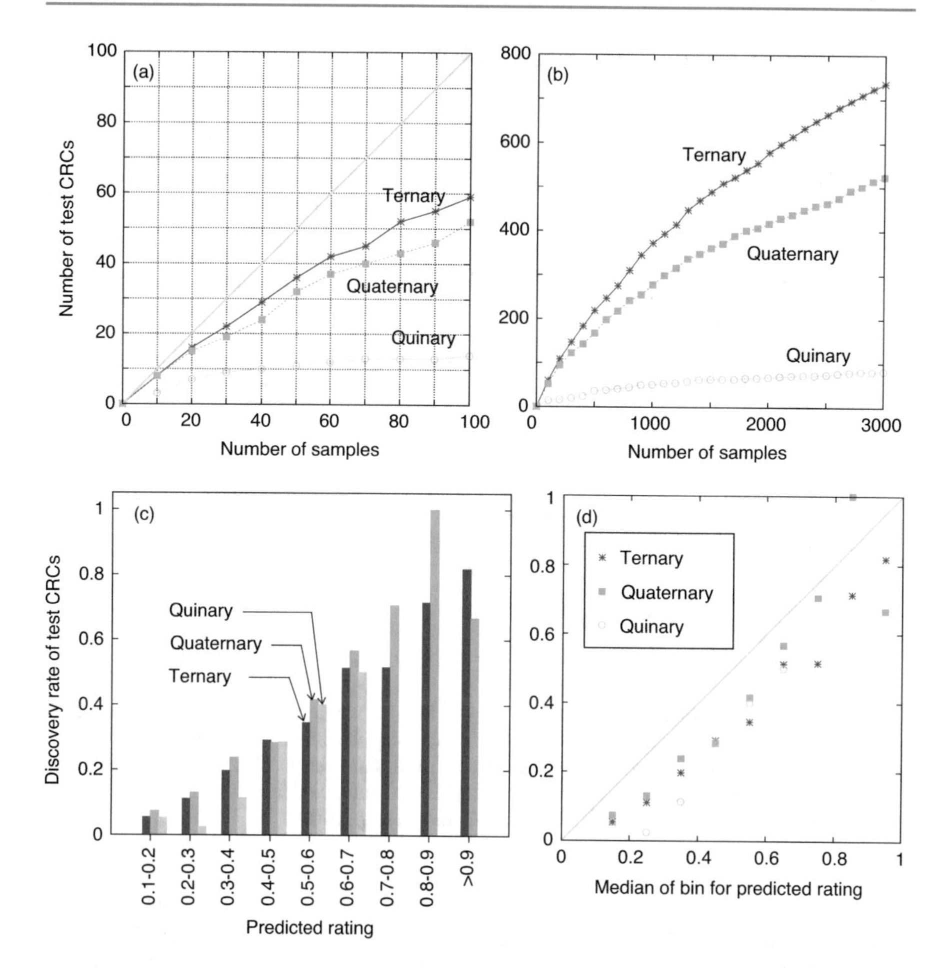

Fig. 20.5 Dependence of the number of test quaternary and quinary CRCs on the number of samples by the Tucker decomposition recommender system up to (**a**) 100 and (**b**) 3000 samples. Diagonal line in (**a**) indicates the performance of the ideal sampling (i.e., the number of test CRCs is identical to the number of samples). (**c**) Dependence of the discovery rate of test CRCs on the predicted rating for ternary, quaternary, and quinary compositions. (**d**) Relationship between the predicted rating and the discovery rate of the test CRCs. Horizontal axis indicates the median of bin for the predicted rating

of stable compounds at compositions with high predicted ratings is examined by evaluating the phase stability of the corresponding pseudo-binary systems using DFT calculations. DFT calculations were performed only for pseudo-binary systems containing the top 27 compositions composed only of elements in Groups 1, 2, 3, 4, 5, 6, 12, 13, 14, 15, 16, and 17 in the periodic table. For each pseudo-binary system, DFT calculations were performed for all possible prototype structures included in the ICSD. Totally, the number of DFT calculations was 13,274.

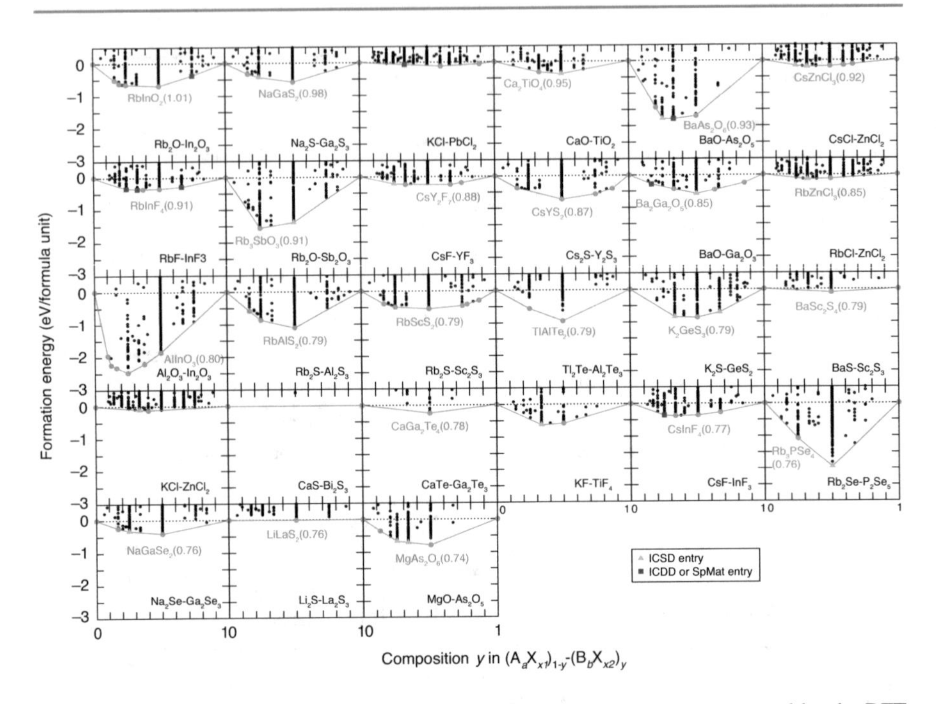

Fig. 20.6 Stable compounds on the convex hull of the formation energy computed by the DFT calculations for the 27 pseudo-binary systems containing the top 27 compositions with high predicted ratings. Compositions with high predicted ratings are shown along with their predicted ratings as the values in parentheses. Closed triangles and squares denote the stable compounds corresponding to ICSD entries and ICDD+SpMat entries, respectively

All DFT calculations were performed using the plane-wave basis projector augmented wave (PAW) method [23, 24] within the Perdew–Burke–Ernzerhof exchange-correlation functional [25] as implemented in the VASP code [26,27]. The cutoff energy was set to 400 eV. The total energy converged to less than 10^{-3} meV. The atomic positions and lattice constants were optimized until the residual forces became less than 10^{-2} eV/Å.

Figure 20.6 shows the stable compounds or CRCs obtained by the DFT calculations for the 27 pseudo-binary systems. The predicted ratings of the top 27 compositions range from 0.74 to 1.01. Among the 27 compositions, most (23 compositions) are located on the convex hull, meaning that they are CRCs according to the DFT calculations. In other words, 85% of the recommended compositions can be regarded as new CRCs. We emphasize that they are not included in the databases of ICSD, ICDD, and SpMat. Such compounds can be discovered with a high probability.

The results of the DFT calculations also demonstrate that the predicted rating can be regarded as the discovery rate of new CRCs as described above. In addition to the top 27 compositions, some other compositions can be CRCs depending on the predicted rating. In compositions with a predicted rating of 0.2–0.4, only two of the 14 compositions (14%) correspond to CRCs. On the other hand, four of the five compositions (80%), 13 of the 16 compositions (81%), and 11 of the 12 compositions (92%) are CRCs with predicted ratings of 0.4–0.6, 0.6–0.8, and 0.8–1.0, respectively. In other words, compounds with a predicted rating exceeding 0.4 have a high probability of being a CRC.

The remaining 15% of recommended compositions have a high predicted rating, but do not correspond to the CRCs in the DFT calculations. Plausible explanations include the following: (1) A stable or metastable compound does not truly exist at the composition. (2) A metastable compound can be found at the composition. (3) A stable compound with a structure other than the prototype structures is observed.

In addition, stable compounds can be found at the other compositions with a high predicted rating. The compositions are classified into two types. The first corresponds to compositions omitted in the candidate dataset, which do not satisfy the condition of $\max(a, b, x) \leq 8$. This type may be predicted with a high rating if a larger maximum value of the integer is considered. The other corresponds to compositions with a low predicted rating. They may be the true DFT stable compound when a low rating is predicted due to the lack of known CRCs similar to the corresponding composition. On the other hand, they may be artificial stable compounds attributed to the use of only the prototype structures for the DFT calculations.

20.3 Compositional Descriptor-Based Approach

We will show a different procedure adopting the compositional descriptors defined by a set of well-known elemental representations as prior knowledge.[2] The potential possibility of descriptors for predicting currently unknown CRCs will be shown.

20.3.1 Classification

A figure of merit for CRC is estimated on the basis of a ML two-class classification, where responses have two distinct values of $y = 1$ and 0. Since a supervised or semi-supervised classification approach requires a dataset with observations for both responses $y = 1$ and 0 (i.e., datasets for CRCs as well as non-existent compositions), we initially labeled compositions based on the criterion of whether the composition exists. Entries in a crystal structure database are regarded as $y = 1$ because they are known to exist. On the other hand, it is not as simple to judge if a compound is non-existent at a given composition because the absence of a compound at a specific composition in the database does not necessarily mean that the compound does not exist. There are two reasons for a lack of entry: (1) A stable compound does not actually exist for a given composition. (2) The composition has not been

[2]This section is reproduced from [2], with the permission of AIP Publishing.

well investigated. It should be emphasized that inorganic compound databases are biased to common metals, their intermetallics, and oxides. Few experiments have been devoted to other compositions. In the present study, we simply assume that all candidate compositions are $y = 0$ in the training process. Then the predicted response, $\hat{y}$, is regarded as a figure of merit for CRC. Therefore, a candidate composition with high figure of merit is expected to be a CRC. Here three kinds of classifiers: logistic regression [28, 29], gradient boosting [30], and random forest classifiers [29, 31] are adopted.

20.3.2 Descriptors

Herein, the compositional similarity is defined by a set of 165 descriptors composed of means, standard deviations, and covariances of established elemental representations [15], which is also similar to descriptors used in the literature [16]. This set of descriptors can cover a wide range of compositions. Structural representations were not used because crystal structures with candidate compositions are unknown. We adopted twenty-two elemental representations: (1) atomic number, (2) atomic mass, (3) period and (4) group in the periodic table, (5) first ionization energy, (6) second ionization energy, (7) electron affinity, (8) Pauling electronegativity, (9) Allen electronegativity, (10) van der Waals radius, (11) covalent radius, (12) atomic radius, (13) pseudopotential radius for the s orbital, (14) pseudopotential radius for the p orbital, (15) melting point, (16) boiling point, (17) density, (18) molar volume, (19) heat of fusion, (20) heat of vaporization, (21) thermal conductivity, and (22) specific heat.

20.3.3 Datasets

The training dataset is composed of entries in the ICSD (known CRCs) and candidate compositions. The known CRCs correspond to compounds up to septenary compositions. Compounds reported to show a partial occupancy behavior are excluded. Thus, the number of the known CRCs is 33,367. The candidate compositions are used as the training data and the prediction data to find unknown CRCs. Although the present method is applicable to any kind of compound, we restrict the results to ionic compounds with normal cation/anion charge states. Candidates of pseudo-binary compositions $A_aB_bX_x$ are generated by considering combinations of $\{A, B, X, a, b, x\}$. We consider 930,142,080 chemical compositions expressed by integers satisfying the condition of $\max(a, b, x) \leq 15$. Here, all charge states are adopted whenever Shannon's ionic radii are reported. Compositions that do not satisfy the charge neutrality condition of $n_A a + n_B b + n_X x = 0$ are removed where n_A, n_B, and n_X denote the valences for elements A, B, and X, respectively. Finally, known CRCs are removed from the set of candidate compositions. Thereafter, 1,294,591 pseudo-binary compositions remain, which are used as the candidate composition data. Additionally, pseudo-ternary oxides $(AO_{x_A})_a(BO_{x_B})_b(CO_{x_C})_c$, nitrides $(AN_{x_A})_a(BN_{x_B})_b(CN_{x_C})_c$, and sulfides $(AS_{x_A})_a(BS_{x_B})_b(CS_{x_C})_c$ are also considered to be candidate composition data. Only a smaller number of elements and their charge states are adopted for pseudo-ternary compounds. In all, there are 3,846,928 pseudo-ternary compositions.

20.3.4 Discovery Performance of Unknown CRCs

The power of discovering currently unknown CRCs is the most important feature of the classification model. We measured the efficiency for finding compositions of entries included in another inorganic compound database, PDF by the ICDD, from pseudo-binary and pseudo-ternary oxide candidate compositions of the ICSD. Only 3731 (0.3%) of the 1,294,591 pseudo-binary and 842 (0.04%) of the 1,933,994 pseudo-ternary oxide candidate compositions in the ICSD are included in the ICDD, respectively. Therefore, it is obvious that discovering ICDD entries by random sampling is not effective.

Figure 20.7a shows the efficiency of the three classifiers for discovering ICDD entries from the pseudo-binary compositions. Sampling of the candidate composi-

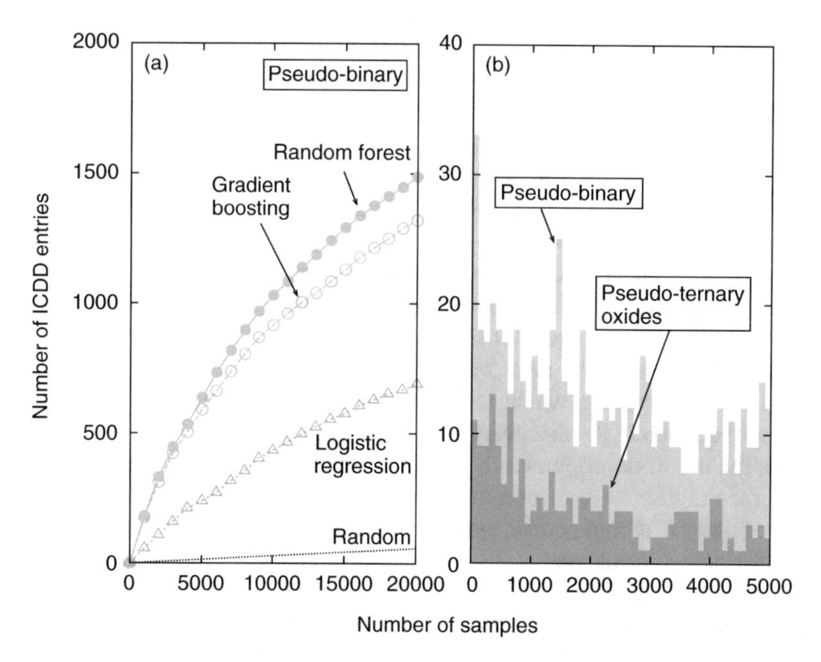

Fig. 20.7 (**a**) Number of ICDD entries found in "no entries" pseudo-binary compositions sampled by classification models and random sampling. (**b**) Distributions of the increment of the number of ICDD entries as "no entries" pseudo-binary and pseudo-ternary oxide samples increase in the random forest model (reproduced from [2], with the permission of AIP Publishing)

tions is performed in decreasing order of the predicted y. Figure 20.7b shows the increment of the number of ICDD entries that can be discovered by the random forest model for pseudo-binary and pseudo-ternary oxide compositions. When sampling 100 and 1000 candidate pseudo-binary compositions according to the ranking by the random forest model, which is the best among the three models, 33 and 180 compositions are found in the ICDD, respectively. Even for pseudo-ternary oxide compositions, 11 and 85 compositions are found in the ICDD. These discovery rates are approximately 250 and 190 times higher than that of random sampling (0.04%), respectively. Figure 20.7b also indicates that the increment of the number of ICDD entries tends to decrease as the predicted y decreases. This provides evidence that the predicted y can be regarded as a figure of merit for exploring currently unknown CRCs.

The discovery rates of recommender systems are much higher than that of random sampling, demonstrating that the use of descriptors as a prior knowledge for compositional similarity is helpful for the discovery of currently unknown CRCs that are not present in the training database. The most meaningful application of such descriptor-based recommender systems for currently unknown CRCs can be demonstrated in multicomponent systems, where only a small number of CRCs are known and a huge number of compositions can be candidates of CRCs. Therefore, the knowledge-based method of the present study should contribute significantly to the prediction of multicomponent currently unknown CRCs.

20.4 Conclusion

In this chapter, we show ML approaches to discover currently unknown CRCs, including matrix- and tensor-based recommender system and knowledge-based recommender system approaches, only from existing inorganic crystal structure databases. The present study indicates that the performance of the recommender systems depends on not only the given rank corresponding to a low-rank structure hidden in the rating matrix and tensor but also on the representation of rating matrix and tensor. The discovery rates of recommender systems are much higher than that of random sampling, which indicates that the recommender system has great potential to accelerate the discovery of new compounds.

Also, the use of descriptors as a prior knowledge for compositional similarity is helpful for the discovery of currently unknown CRCs. A major advantage of the knowledge-based recommender systems should be avoiding cold-start problems. The cold-start problems may occur in applications to multicomponent systems such as a pseudo-ternary system where few known CRCs exist, which are the most interesting application of recommender systems. Although matrix- and tensor-based recommender systems may work well in a general case, the use of a knowledge-based method should contribute significantly to the prediction of multicomponent CRCs.

References

1. D. Pettifor, J. Phys. C Solid State Phys. **19**(3), 285 (1986)
2. A. Seko, H. Hayashi, I. Tanaka, J. Chem. Phys. **148**(24), 241719 (2018). https://doi.org/10.1063/1.5016210
3. G. Bergerhoff, I.D. Brown, in *Crystallographic Databases*, ed. by F.H. Allen et al. (International Union of Crystallography, Chester, 1987)
4. B. Meredig, A. Agrawal, S. Kirklin, J.E. Saal, J.W. Doak, A. Thompson, K. Zhang, A. Choudhary, C. Wolverton, Phys. Rev. B **89**, 094104 (2014). https://doi.org/10.1103/PhysRevB.89.094104
5. F.A. Faber, A. Lindmaa, O.A. Von Lilienfeld, R. Armiento, Phys. Rev. Lett. **117**(13), 135502 (2016)
6. J. Schmidt, J. Shi, P. Borlido, L. Chen, S. Botti, M.A. Marques, Chem. Mater. **29**(12), 5090 (2017)
7. G. Hautier, C.C. Fischer, A. Jain, T. Mueller, G. Ceder, Chem. Mater. **22**(12), 3762 (2010)
8. G. Hautier, C. Fischer, V. Ehrlacher, A. Jain, G. Ceder, Inorg. Chem. **50**(2), 656 (2010)
9. P. Resnick, H.R. Varian, Commun. ACM **40**(3), 56 (1997)
10. C.C. Aggarwal, *Recommender Systems* (Springer, Berlin, 2016)
11. B. Sarwar, G. Karypis, J. Konstan, J. Riedl, Application of dimensionality reduction in recommender system-a case study. Technical Report, Minnesota Univ Minneapolis Dept of Computer Science (2000)
12. E. Frolov, I. Oseledets, Wiley Interdiscip. Rev. Data Min. Knowl. Discov. **7**(3), e1201 (2017)
13. P. Symeonidis, A. Zioupos, *Matrix and Tensor Factorization Techniques for Recommender Systems* (Springer, Berlin, 2016)
14. A. Seko, H. Hayashi, H. Kashima, I. Tanaka, Phys. Rev. Materials **2**, 013805 (2018). https://doi.org/10.1103/PhysRevMaterials.2.013805
15. A. Seko, H. Hayashi, K. Nakayama, A. Takahashi, I. Tanaka, Phys. Rev. B **95**, 144110 (2017). https://doi.org/10.1103/PhysRevB.95.144110
16. L. Ward, A. Agrawal, A. Choudhary, C. Wolverton, npj Comput. Mater. **2**, 16028 (2016)
17. F.L. Hitchcock, Stud. Appl. Math. **6**(1–4), 164 (1927)
18. L.R. Tucker, Psychometrika **31**(3), 279 (1966)
19. L. De Lathauwer, B. De Moor, J. Vandewalle, SIAM J. Matrix Anal. Appl. **21**(4), 1253 (2000)
20. M. Nickel. SCIKIT-TENSOR. Available Online (2013). https://pypi.org/project/scikit-tensor/
21. ICDD, PDF-4/Organics 2011 (Database), ed. by S. Kabekkodu, International Centre for Diffraction Data, Newtown Square (2010)
22. SpringerMaterials. http://materials.springer.com
23. P.E. Blöchl, Phys. Rev. B **50**(24), 17953 (1994)
24. G. Kresse, D. Joubert, Phys. Rev. B **59**(3), 1758 (1999)
25. J.P. Perdew, K. Burke, M. Ernzerhof, Phys. Rev. Lett. **77**(18), 3865 (1996)
26. G. Kresse, J. Hafner, Phys. Rev. B **47**(1), 558 (1993)
27. G. Kresse, J. Furthmüller, Phys. Rev. B **54**(16), 11169 (1996)
28. D.R. Cox, The Regression Analysis of Binary Sequences. J. Roy. Stat. Soc. Ser. B **20**, 215–242 (1958)
29. T. Hastie, R. Tibshirani, J. Friedman, *The Elements of Statistical Learning*, 2nd edn. (Springer, New York, 2009)
30. L. Breiman, Arcing the edge. Technical Report 486, Statistics Department, University of California at Berkeley (1997)
31. T.K. Ho, in *Proceedings of the Third International Conference on Document Analysis and Recognition*, vol. 1 (IEEE, Piscataway, 1995), pp. 278–282

Generative Models for Automatic Chemical Design

21

Daniel Schwalbe-Koda and Rafael Gómez-Bombarelli

Abstract

Materials discovery is decisive for tackling urgent challenges related to energy, the environment, health care, and many others. In chemistry, conventional methodologies for innovation usually rely on expensive and incremental strategies to optimize properties from molecular structures. On the other hand, inverse approaches map properties to structures, thus expediting the design of novel useful compounds. In this chapter, we examine the way in which current deep generative models are addressing the inverse chemical discovery paradigm. We begin by revisiting early inverse design algorithms. Then, we introduce generative models for molecular systems and categorize them according to their architecture and molecular representation. Using this classification, we review the evolution and performance of important molecular generation schemes reported in the literature. Finally, we conclude highlighting the prospects and challenges of generative models as cutting edge tools in materials discovery.

21.1 Introduction

Innovation in materials is the key driver for many recent technological advances. From clean energy [1] to the aerospace industry [2] or drug discovery [3], research in chemical and materials science is constantly pushed forward to develop compounds and formulae with novel applications, lower cost, and better performance. Conventional methods for the discovery of new materials start from a well-defined

D. Schwalbe-Koda · R. Gómez-Bombarelli (✉)
Department of Materials Science and Engineering, Massachusetts Institute of Technology, Cambridge, MA, USA
e-mail: dskoda@mit.edu; rafagb@mit.edu

K. T. Schütt et al. (eds.), *Machine Learning Meets Quantum Physics*, Lecture Notes in Physics 968, https://doi.org/10.1007/978-3-030-40245-7_21

set of substances from which properties of interest are derived. Then, intensive research on the relationship between structures and properties is performed. The gained insights from this procedure lead to incremental improvements in the compounds and the cycle is restarted with a new search space to be explored. This trial-and-error approach to innovation often leads to costly and incremental steps towards the development of new technologies and in occasion relies on serendipity for leap progress. Materials development may require billions of dollars in investments [4] and up to 20 years to be deployed to the market [1,4].

Despite the challenges associated with such direct approaches, they have not prevented data-driven discovery of materials from happening. High-throughput materials screening [5–14] and data mining [15–20] have been responsible for several breakthroughs in the last two decades [21,22], leading to the establishment of the Materials Genome Initiative [23] and multiple collaborative projects around the world built around databases and analysis pipelines [24–27]. Automated, scalable approaches leverage from datasets in the thousands to millions of simulations to offer a cornucopia of insights on materials composition, structure, and synthesis.

Developing materials with the inverse perspective departs from these traditional methods. Instead of exhaustively deriving properties from structures, the performance parameters are chosen beforehand and unknown materials satisfying these requirements are inferred. Hence, innovation in this setting is achieved by reverting the mapping between structures and their properties. Unfortunately, this approach is even harder than the conventional one. Inverting a given Hamiltonian is not a well-defined problem, and the absence of a systematic exploratory methodology may result in delays, or outright failure, of the discovery cycle of materials [28]. Furthermore, another major obstacle to the design of arbitrary compounds is the dimensionality of the missing data for known and unknown compounds [29]. As an example, the breadth of accessible drug-like molecules can be on the order of 10^{60} [30, 31], rendering manual searches or enumerations through the chemical space an intractable problem. In addition, molecules and crystal structures are discrete objects, which hinders automated optimization, and computer-generated candidates must follow a series of hard (valence rules, thermal stability) and soft (synthetic accessibility, cost, safety) constraints that may be difficult to state in explicit form. As the inverse chemical design holds great promise for economic, environmental, and societal progress, one can ask how to rationalize the exploration of unknown substances and accelerate the discovery of new materials.

21.1.1 Early Inverse Design Strategies for Materials

The inverse chemical design is usually posed as an optimization problem in which molecular properties are extremized with respect to given parameters [32]. This concept splits the inverse design problem into two parts: (i) efficiently sampling materials from an enormous configuration space and (ii) searching for global maxima in their properties [33] corresponding to minima in their potential energy surface [34, 35]. Early approaches towards the inverse materials design used chemical intuition to address (i), narrowing down and navigating the space of structures under investigation with probabilistic methods [33, 36–40]. Nevertheless, even constrained spaces can be too large to be exhaustively enumerated. Especially in the absence of an efficient exploratory policy, this discovery process demands considerable computational resources and time. Several different strategies are required to simultaneously navigate the chemical space and evaluate the properties of the materials under investigation.

Monte Carlo methods resort to statistical sampling to avoid enumerating a space of interest. When combined with simulated annealing [41], for example, they become adequate to locate extrema within property spaces. In physics, reverse Monte Carlo methods have long been developed to determine structural information from experimental data [42–44]. However, the popularization of similar methods to *de novo* design of materials is more recent. Wolverton et al. [40] employed such methods to aid the design of alloys and avoid expensive enumeration of compositions and Franceschetti and Zunger [45] improved the idea to design $Al_xGa_{1-x}As$ and $Ga_xIn_{1-x}P$ superlattices with a tailored band gap. They started with configurations sampled using Monte Carlo, relaxed the atomic positions using valence-force-field methods, and calculated their band gap by fast diagonalization of pseudopotential Hamiltonians. Through this practical process, they predicted superlattices with optimal band gaps after analyzing less than 10^4 compounds among 10^{14} structures [45].

Other popular techniques for multidimensional optimization that also involve a stochastic component are genetic algorithms (GAs) [46]. Based on evolution principles, GAs refine specific parameters of a population that improve a targeted property. In materials design, GAs have been vastly employed in the inverse design of small molecules [47, 48], polymers [49, 50], drugs [51, 52], biomolecules [53, 54], catalysts [55], alloys [56, 57], semiconductors [58–60], and photovoltaic materials [61]. Furthermore, evolution-inspired approaches have been used as a general modeling tool to predict stable structures [62–67] and Hamiltonian parameters [68, 69]. Many more applications of GAs in materials design are still being demonstrated after decades of its inception [31, 70–73].

Monte Carlo and evolutionary algorithms are interpretable and often produce powerful implementations. The combination of sampling and optimization is a great improvement over random searches or full enumeration of a chemical space. Nonetheless, they still correspond to discrete optimization techniques in a combinatorial space and require individual evaluation of their properties at every step. This discrete form hinders chemical interpolations and the definition of property gradients during optimization processes, thus retaining a flavor of "trial-and-error" in the computational design of materials, rather than an invertible structure–property mapping. One of the first attempts to use a continuous representation on the molecular design was performed by Kuhn and Beratan [33]. The authors varied coefficients in linear combination of atomic orbitals while keeping the energy eigenvalues fixed to optimize linear chains of atoms. Later, von Lilienfeld et al. [74] generalized the discrete nature of atoms by approximating atomic numbers by continuous functions and defining property gradients with respect to this "alchemical potential." They used this theory to design ligands for proteins [74] and tune electronic properties of derivatives of benzene [75]. A similar strategy was proposed by Wang et al. [76] around the same time. Instead of atomic numbers, a linear combination of atomic potentials was used as a basis for optimizations in property landscapes. Following the bijectiveness between potential and electronic density in the Hohenberg–Kohn theory [77], nuclei–electrons interaction potentials were employed as quasi-invertible representations of molecules. Potentials resulting from optimizations with property gradients can be later interpolated or approximated by a discrete molecular structure whose atomic coordinates give rise to a similar potential. Over the years, the approach was further refined within the tight-binding framework [78, 79] and gradient-directed Monte Carlo method [80, 81], its applicability demonstrated in the design of molecules with improved hyperpolarizability [76, 78, 80] and acidity [82].

Despite these promising approaches, many challenges in inverse chemical design remain unsolved. Monte Carlo and genetic algorithms share the complexity of discrete optimization methods over graphs, particularly exacerbated by the rugged property surfaces. They rely on stochastic steps that struggle to capture the interrelated hard and soft constraints of chemical design: converting a single into a double bond may produce a formally valid, but impractical and unacceptable molecule depending on chemical context. On the other hand, a compromise between validity and diversity of the chemical space is difficult to achieve with continuous representations. Lastly, finding optimal points in the 3D potential energy surface that produce a desired output is still not the same as molecular optimization, since the generated "atom cloud" may not be a local minimum, stable enough in operating conditions, or synthetically attainable. An ideal inverse chemical design tool would offer the best of the two worlds: an efficient way to sample valid and acceptable regions of the chemical space; a fast method to calculate properties from the given structures; a differentiable representation for a wide spectrum of materials; and the capacity to optimize them using property gradients. Furthermore, it should operate on the manifold of synthetically accessible, stable compounds. This is where modern machine learning (ML) algorithms come into play.

21.1.2 Deep Learning and Generative Models

Deep learning (DL) is emerging as a promising tool to address the inverse design of many different applications. Particularly through generative models, algorithms in DL push forward how machines understand real data. Roughly speaking, the role of a generative model is to capture the underlying rules of a data distribution. Given a collection of (training) data points $\{X_i\}$ in a space $\mathcal{X}$, a model is trained to match the data distribution P_X by means of a generative process P_G in such a way that the generated data $Y \sim P_G$ resembles the real data $X \sim P_X$. Earlier generative models such as Boltzmann Machines [83, 84], Restricted Boltzmann Machines [85], Deep Belief Networks [86], or Deep Boltzmann Machines [87] were the first to tackle the problem of learning probability distributions based on training examples. Their lack of flexibility, tractability, and generalizing ability, however, rendered them obsolete in favor of more modern ones [88].

Current generative models have been successful in learning and generating novel data from different types of real-world examples. Deep neural networks trained on image datasets are able to produce realistic-looking house interiors, animals, buildings, objects, and human faces [89, 90], as well as embed pictures with artistic style [91] or enhance it with super-resolution [92]. Other examples include convincing text [93, 94], music [95], voices [96], and videos [97] synthesized by such networks. Most interesting is the creation of novel data conditioned on latent features, which allows tuning models with vector and arithmetic operations in a property space [98, 99]. The adaptable architectures of these models also enable straightforward training procedures based on backpropagation [100]. Within the DL framework, a proper loss function drives gradients so that the generative model, typically parameterized by a neural network, learns to minimize the distance between the two distributions.

Among the popular architectures for generating data from deep neural networks, the Variational Auto-Encoder (VAE) [101] is a particularly robust architecture. It couples inference and generation by mapping data to a manifold conditioned to implicit data descriptors. To do so, the model is trained to learn the identity function while constrained by a dimensional bottleneck called latent space (see Fig. 21.1a). In this scheme, data is first encoded to a probability distribution $Q_\phi(\mathbf{z}|X)$ matching a given prior distribution $P_z(\mathbf{z})$, where $\mathbf{z}$ is called latent vector. Then, a sample from the latent space is reconstructed with the generative algorithm $P_\theta(X|\mathbf{z})$. In the VAE [101], outcomes of both processes are parameterized by ϕ and θ to maximize a lower bound for the log-likelihood of the output with respect to the input data distribution. The VAE objective is, therefore,

$$\mathcal{L}(\theta, \phi) = -D_{KL}\left(Q_\phi(\mathbf{z}|X)||P_z(\mathbf{z})\right) + \mathbb{E}_{z\sim Q_\phi}\left[\log P_\theta(X|\mathbf{z})\right]. \quad (21.1)$$

The encoder is regularized with a divergence term D_{KL}, while the decoder is penalized by a reconstruction error $\log P_\theta(X|\mathbf{z})$, usually in the form of mean-squared or cross entropy losses. This maximization can then be performed by stochastic gradient ascent.

The probabilistic nature of VAE manifolds approximately accounts for many complex interactions between data points. Although functional in many cases, the modeled data distribution does not always converge to real data distributions [102]. Furthermore, Kullback–Leibler or Jensen–Shannon divergences cannot be analytically computed for an arbitrary prior, and most works are restricted to Gaussian distributions. Avoiding high-variance methods to determine this regularizing term is also an important concern. Recently, this limitation was simplified by employing the Wasserstein distance as a penalty for the encoder regularization [102, 103]. As a result, richer latent representations are computed more efficiently within Wasserstein Auto-Encoders, resulting in disentanglement, latent shaping, and improved reconstruction [102–104].

Another approach to generative models are the Generative Adversarial Networks (GANs) [90]. Recognized by their sharp reconstructions, GANs are constructed by making two neural networks compete against each other until a Nash equilibrium is found. One of the networks is a deterministic generative model. It applies a non-linear set of transformations to a prior probability distribution P_z in order to match the real data distribution P_X. Interestingly, the generator (or actor) only receives the prior distribution as input and has no contact with the real data whatsoever. It can only be trained through a second network, called discriminator or critic. The latter tries to distinguish real data $X \sim P_X$ from fake data $Y = G(\mathbf{z}) \sim P_G$, as depicted in Fig. 21.1b. The objective of the critic is to perfectly distinguish between P_X and P_G, thus maximizing the prediction accuracy. On the other hand, the generator tries to fool the discriminator by creating data points that look like real data points, minimizing the prediction accuracy of the critic. Consequently, the complete GAN objective is written as [90]

$$\min_G \max_D V(D, G) = \mathbb{E}_{X \sim P_X}\left[\log D(X)\right] + \mathbb{E}_{\mathbf{z} \sim P_z}\left[\log\left(1 - D(G(\mathbf{z}))\right)\right]. \quad (21.2)$$

Despite the impressive results from GANs, their training process is highly unstable. The min–max problem requires a well-balanced training from both networks to ensure non-vanishing gradients and convergence to a successful model. Furthermore, GANs do not reward diversity of generated samples and the system is prone to mode collapse. There is no reason why the generated distribution P_G should have the same support of the original data P_X, and the actor produces only a handful of different examples which are realistic enough. This does not happen for the VAE, since the log-likelihood term gives an infinite loss for a generated data distribution with a disjoint support with respect to the original data distribution. Several different architectures have been proposed to address these issues among GANs [102, 105–116]. Although many of them may be equivalent to a certain extent [117], steady progress is being made in this area, especially through more complex ways of approximating data distributions, such as with f-divergence [112] or optimal transport [102, 114, 115].

Other models such as the auto-regressive PixelRNN [118] and PixelCNN [119, 120] have also been successful as generators of images [118–120], video [121], text [122], and sound [96]. Differently from VAE and GANs, these models approximate the data distribution by a tractable factorization P_X. For example, in an $n \times n$ image, the generative model $P(X)$ is written as [118]

$$P(X) = \prod_{i=1}^{n^2} P\left(x_i | x_1, \ldots, x_{i-1}\right), \quad (21.3)$$

where each x_i is a pixel generated by the model (see Fig. 21.1c). These models with explicit distributions yield samples with very good negative log-likelihood and diversity [118]. The model evaluation is also straightforward, given the explicit

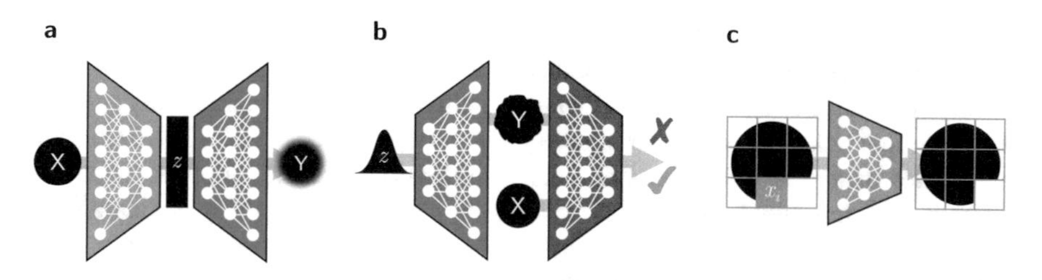

Fig. 21.1 Schematic diagrams for three popular generative models: (**a**) VAE, (**b**) GAN, and (**c**) auto-regressive

computation of $P(X)$. As a drawback, however, these models rely on the sequential generation of data, which is a slow process. A diagram of the architectures of the three generative models here discussed is seen in Fig. 21.1.

21.1.3 Generative Models Meet Chemical Design

Apart from their numerous aforementioned applications, generative models are also attracting attention in chemistry and materials science. DL is being employed not only for the prediction and identification of properties of molecules, but also to generate new chemical compounds [100]. In the context of inverse design, generative models provide benefits such as generating complex samples from simple probability distributions; providing meaningful latent representations, over which optimizations can be performed; and the ability to perform inference when coupled to supervised models. Therefore, unifying generative models with chemical design is a promising venue to accelerate innovation in chemistry and related fields.

To go beyond the limitations of traditional inverse design strategies, an ideal way to discover new materials should satisfy some requisites [123]. To be a completely hands-free model, the model should be data-driven, thus avoiding fixed libraries and expensive labeling. It is also desirable that it outputs as many potential molecules as possible under a subset of interest, which means that the model needs a powerful generator coupled with a continuous representation for molecules. Furthermore, such a representation should be interpretable, allowing a correct description of structure–property relationships within molecules. If, additionally, the model is differentiable, it would be possible to optimize certain properties using gradient techniques and, later, look for molecules satisfying such constraints.

The development of such a tool is currently a priority for ML models in chemistry and for the inverse chemical design. It relies primarily on two decisions: which model to use and how to represent a molecule in a computer-friendly way. Following our brief introduction to the early inverse design strategies and main generative models in the literature, we describe which molecular representations are possible.

In quantum mechanics, a molecular system is represented by a wave function that is a solution of the Schrödinger equation for that particular molecule. To derive most properties of interest, the spatial wave function is enough. Computing such a representation, however, is equivalent to solving an (approximate) version of the Schrödinger equation itself. Many methods for theoretical chemistry, such as Hartree–Fock [124, 125] or Density Functional Theory [77, 126], represent molecules using wave functions or electronic densities and obtain other properties from it. Solving quantum chemical calculations is computationally demanding in many cases, though. The idea with many ML methods is not only to avoid these calculations, but also to make a generalizable model that highlights different aspects of chemical intuition. Therefore, we should look for other representations for chemical structures.

Thousands of different descriptors are available for chemical prediction methods [127]. Several relevant features for ML have demonstrated their capabilities for predicting properties of molecules, such as fingerprints [128], bag-of-bonds [129], Coulomb matrices [130], deep tensor neural networks train on the distance matrix [131], many-body tensor representation [132], SMILES strings [133], and graphs [134–136]. Not all representations are invertible for human interpretation, however. To teach a generative model how to create a molecule, it may suffice for it to produce a fingerprint, for example. However, how can one map any possible fingerprint to a molecule is an extra step of complexity equivalent to the generation of libraries. This is undesirable in a practical generative model. In this chapter, we focus on two easily interpretable representations, SMILES strings and molecular graphs, and how generative models perform with these representations. Examples of these two forms of writing a molecule are shown in Fig. 21.2.

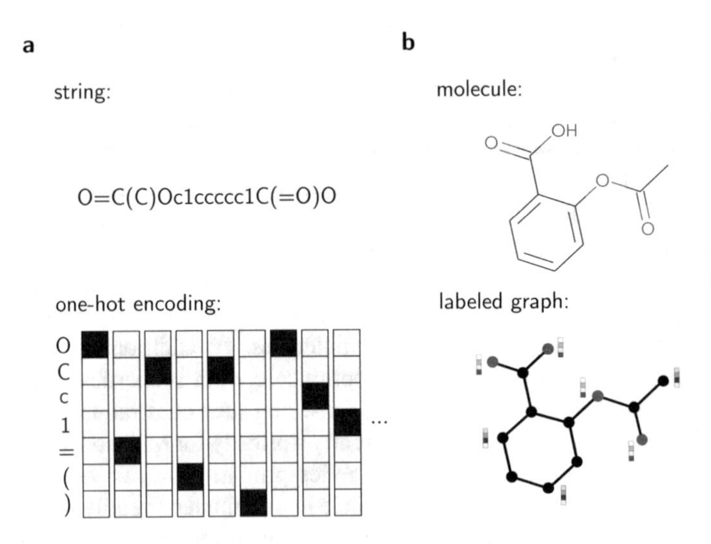

Fig. 21.2 Two popular ways of representing a molecule using: (**a**) SMILES strings converted to one-hot encoding or (**b**) a graph derived from the Lewis structure

21.2 Chemical Generative Models

21.2.1 SMILES Representation

SMILES (Simplified Molecular Input Line Entry System) strings have been widely adopted as representation for molecules [133]. Through graph-to-text mapping algorithms, it determines atoms by atomic number and aromaticity and can capture branching, cycles, ionization, etc. The same molecule can be represented by multiple SMILES strings, and thus a canonical representation is typically chosen, although some works leverage non-canonical strings as a data augmentation and regularization strategy. Although SMILES are inferior to the more modern InChI (International Chemical Identifier) representation in their ability to address key challenges in representing molecules as strings such as tautomerism, mesomerism, and some forms of isomerism, SMILES follow a much simpler syntax that has proven easier to learn for ML models.

Since SMILES rely on a sequence-based representation, natural language processing (NLP) algorithms in deep learning can be naturally extended to them. This allows the transferability of several architectures from the NLP community to interpret the chemical world. Mostly, these systems make use of recurrent neural networks (RNNs) to condition the generation of the next character on the previous ones, creating arbitrarily long sequences character by character [88]. The order of the sequence is very relevant to generate a valid molecule, and observation of such restrictions can be typically incorporated in RNNs with long short-term memory cells (LSTM) [137], gated recurrent units (GRUs) [138], or stack-augmented memory [139].

A simple form of generating molecules using only RNN architectures is to extensively train them with valid SMILES from a database of molecules. This requires post-processing analyses, as it resembles traditional library generation. As a proof of concept, Ikebata et al. [140] used SMILES strings to design small organic molecules by employing Bayesian sampling with sequential Monte Carlo. Ertl et al. [141] instead generated molecules using LSTM cells and later employed them in a virtual screening for properties.

Generating libraries, however, is not enough for the automatic discovery of chemical compounds. Asking an RNN-based model to simply create SMILES strings does not improve on the rational exploration of the chemical space. In general, the design of new molecules is also oriented towards certain properties, like solubility, toxicity, and drug-likeness [123], which are not necessarily incorporated in the training process of RNNs. In order to skew the generation of molecules and better investigate a subset of the chemical space, Segler et al. [142] used transfer learning to first train the RNN on a whole dataset of molecules and later fine-tune the model towards the generation of molecules with physico-chemical properties of interest. This two-part approach allows the model to first learn the grammar inherent to SMILES to then create new molecules based only on the most interesting ones. In line with this depth-search, Gupta et al. [143] demonstrated the application of transfer learning to grow molecules from fragments. This technique is particularly useful for drug discovery [3, 144], in which the search of the chemical space usually begins from a known substructure with certain desired functionalities.

Recently, the usage of reinforcement learning (RL) to generate molecules with certain properties became popular among generative models. Since the representation of a molecule using SMILES requires the generator to output a sequence of characters, each decision can be considered as an action. The successful completion of a valid SMILES string is associated with a reward, for example, and undesired features in the sequence are penalized. Jaques et al. [145] used RL to impose a structure on sequence generation, avoiding repeating patterns not only in SMILES strings but also in text and music. By penalizing large rings, short sequences of characters, and long, monotonous carbon chains, they were able to increase the number of valid molecules their model produced. Olivecrona et al. [146] demonstrated the usage of augmented episodic likelihood and traditional policy gradient methods to tune the generation of molecules from an RNN. Their method achieved 94% of validity on generating molecules sampled from a prior distribution. It was also taught to avoid functional groups containing sulfur and to generate structures similar to a given structure or with certain target activities. Similarly, Popova et al. [139] designed molecules for drugs using a stack-augmented RNN. It demonstrated improved capacity to capture the grammar of SMILES while using RL to tune their synthetic accessibility, solubility, inhibition, and other properties.

As the degree of abstraction grows in the molecule design, more complex generative models are proposed to explore the chemical space. VAEs, for example, can include a direct mapping between structures and properties and vice versa. Its joint training with an encoder and a decoder is capable of approximating very complex data distributions using a real-valued and compressed representation, which is essential for improving the search for chemical compounds. Since the latent space is meaningful, the generator learns to associate patterns in the latent space with properties of the real data. After both the encoding and the decoding networks are jointly trained, the generative model can be decoupled from the inference step and latent variables then become the field for exploration. Therefore, VAEs map the original chemical space to a continuous, differentiable space conveying all the information about the original molecules, over which optimization can be performed. Additionally, conditional generation of molecules based on properties is made possible without hand-made constraints in SMILES, semi-supervised methods can be used to tune the model with relevant properties. This approach is closer to the model of an ideal, automatic, chemical generative model as discussed earlier.

Constructed over RNNs as both encoder and decoder, Gómez-Bombarelli et al. [123] trained a VAE on prediction and reconstruction tasks for molecules extracted from the QM9 and ZINC datasets. The latent space allowed not only sampling of molecules but also interpolations, reconstruction, and optimization using a Gaussian process predictor trained on the latent space (Fig. 21.3). Kang and Cho [147] used partial annotation on molecules to train a semi-supervised VAE to decrease the error for property prediction and to generate molecules conditioned on targets. It can also be enhanced in combination with other dimensionality reduction algorithms

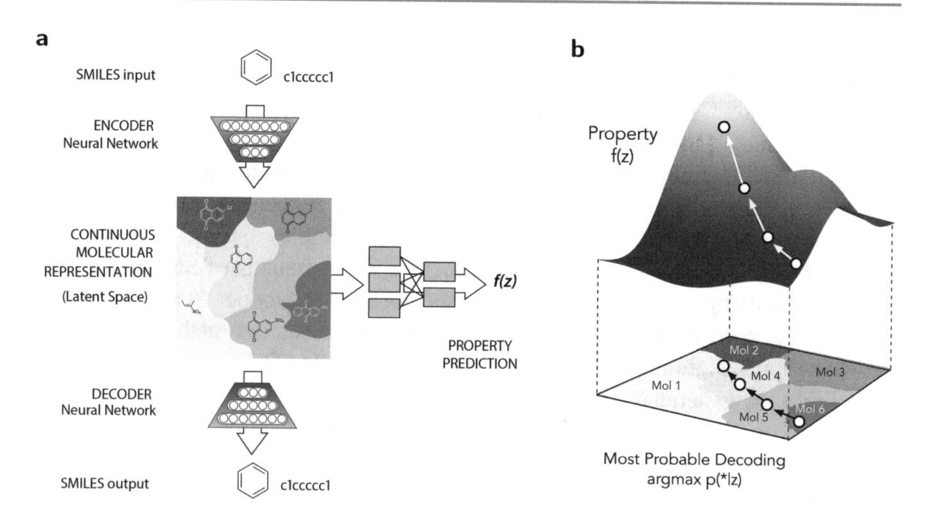

Fig. 21.3 Variational Auto-Encoder for chemical design. The architecture in (**a**) allows for property optimization in the latent space, as depicted in (**b**). Figure reproduced from [123]

[148]. Within the chemical world, VAEs based on sequences also show promise for investigating proteins [149], learning chemical interactions between molecules [150], designing organic light-emitting diodes [151], and generating ligands [152, 153].

In the field of molecule generation, GANs usually appear associated with RL. To fine-tune the generation of long SMILES strings, Guimaraes et al. [154] employed a Wasserstein GAN [102] with a stochastic policy that increased the diversity, optimized the properties, and maintained the drug-likeness of the generated samples. Sanchez-Lengeling et al. [155] and Putin et al. [156] further improved upon this work to bias the distribution of generated molecules towards a goal. In addition, Mendez-Lucio et al. [157] used a GAN to generate molecules conditioned on gene expression signatures, which is particularly useful to create active compounds towards a certain target. Similarly to what is done with molecules, Killoran et al. [158] employed a GAN to create realistic samples of DNA sequences from a small subset of configurations. The model was also tuned to design DNA chains adapted to protein binding and look for motifs representing functional roles. Adversarial training was also employed in the discovery of drugs for using molecular fingerprints as opposed to a reversible representation [159–161] and SMILES [162]. However, avoiding the unstable training and mode collapse while generating molecules is still a hindrance for the usage of GANs in chemical design.

Although SMILES have proved to be a reliable representation for molecule generation, their sequential nature imposes some constraints to the architectures being learned. Forcing an RNN to implicitly learn their linguistic rules poses additional difficulties to the model under training. Additionally, decoding a sequence of generated characters into a valid molecule is especially difficult. In [123], the rate of success when decoding molecules depended on the proximity of the latent point to the valid molecule and could be as low as 4% for random points on the latent space. Although RL is as an alternative to reward the generation of valid molecules [145, 154, 155], other architecture changes can also circumvent this difficulty. Techniques to generate valid sequences imported from NLP studies include: using revision to improve the outcome of sequences [163]; adding a validator to the decoder to generate more valid samples [164]; introducing a grammar within the VAE to teach the model the fundamentals of SMILES strings [165]; using compiler theory to constrain the decoder to produce syntactically and semantically correct data [166]; and using machine translation methods to convert between representations of sequences and/or grammar [167].

Validity of generated sequences, however, is not the only thing that makes working with SMILES difficult. The sequential representation cannot represent similarity between molecules within edit distances [168] and a single molecule may have several different SMILES strings [169, 170]. The trade-off between processing this representation with text-based algorithms and discarding its chemical intuition calls for other approaches in the study and design of molecules.

21.2.2 Molecular Graphs

An intuitive way of representing molecules is by means of its Lewis structure, computationally translated as a molecular graph. Given a graph $\mathcal{G} = (\mathcal{V}, \mathcal{E})$, the atoms are represented as nodes $v_i \in \mathcal{V}$ and chemical bonds as edges $(v_i, v_j) \in \mathcal{E}$. Then, nodes and edges are decorated with labels indicating the atom type, bond type, and so on. Many times, hydrogen atoms are treated implicitly for simplicity, since their presence can be inferred from traditional chemistry rules.

Molecular graphs and DL were first jointly used for property prediction. Molecules were treated as undirected cyclic graphs and further processed using RNNs [171]. Using graph convolutional networks [172], Duvenaud et al. [135] demonstrated the usage of machine-learned fingerprints to achieve better prediction of properties on neural networks. This approach started with a molecular graph and led to fixed-size fingerprints after several graph convolutions and graph pooling layers. Kearnes et al. [134] and Coley et al. [173] also evaluated the flexibility and promise of learned fingerprints from graph structures, especially because models could learn how to associate its chemical structure to their properties. Later, Gilmer et al. [136] unified graph convolutions as message-passing neural networks for quantum chemistry predictions, achieving DFT accuracy within their predictions of quantum properties, interpreting molecular 3D geometries as graphs with distance-labeled edges. Many more studies have explored the representative power of graphs within prediction tasks [174, 175]. These frameworks paved the way for using graph-based representations of molecules, especially because of their proximity with chemistry and geometrical interpretation.

The generation of graphs is, however, non-trivial, especially because of the challenges imposed by graph isomorphism. As in SMILES strings, one way to generate molecular graphs is by sequentially adding nodes and edges to the graph. The sequential nature of decisions over graphs has already been implemented using an RNN [176] for arbitrary graphs. Specifically for a small subset of graphs corresponding to valid molecules, Li et al. [177] used a decoder policy to improve the outcomes of the model. The conditional generation of graphs allowed for molecules to be created with improved drug-likeness, synthetic accessibility, as well as allowed scaffold-based generations from a template (Fig. 21.4a). Similar procedure was adopted by Li et al. [177], in which a graph-generating decision process using RNNs was proposed for molecules. These node-by-node generations rely on the ordering of nodes in the molecular graph and thus suffer with random permutations of the nodes.

In the VAE world, several methods have been proposed to deal with the problem of directly generating graphs from a latent code [178–182]. However, when working with reconstructions, the problem of graph isomorphism cannot be addressed without expensive calculations [179]. Furthermore, graph reconstructions suffer from validity and accuracy [179], except when these constraints are enforced in the graph generation process [181–183]. Currently, one of the most successful approaches to translate molecular graphs into a meaningful latent code while avoiding node-by-node generation is the Junction Tree Variational Auto-Encoder (JT-VAE) [168]. In this framework, the molecular graph is first decomposed into a vocabulary of subpieces extracted from the training set, which include rings, functional groups, and atoms (see Fig. 21.4b). Then, the model is trained to encode the full graph and the tree structure resulting from the decomposition into two latent spaces. A two-part reconstruction process is necessary to recover the original molecule from the two vector representations. Remarkably, the JT-VAE achieves 100% of validity when generating small molecules, as well as 100% of novelty when sampling the latent code from a prior. Moreover, a meaningful latent space is also seen for this method, which is essential for optimization and the automatic design of molecules. The authors later improve over the JT-VAE with graph-to-graph translation and auto-regressive methods towards molecular optimization tasks [184, 185].

Other auto-regressive approaches combining VAE and sequential graph generation have been proposed to generate and optimize molecules. Assouel et al. [186] introduced a decoding strategy to output arbitrarily large molecules based on their graph representation. The model, named DEFactor, is end-to-end differentiable, dispenses retraining during the optimization procedure, and achieved high reconstruction accuracy (>80%) even for molecules with about 25 heavy atoms. Despite the restrictions on node permutations, DEFactor allows the direct optimization of the graph conditioned to properties of interest. This and other similar models also allow the generation of molecules based on given scaffolds [187].

Auto-regressive methods for molecules have also been reported with the use of RL. Zhou et al. [188] created a Markov decision process to produce molecules with targeted properties through multi-objective RL. Similarly to what is done with graphs, this strategy adds bonds and atoms sequentially. However, as the actions are restricted to chemically valid ones, the model scores 100% of validity in the

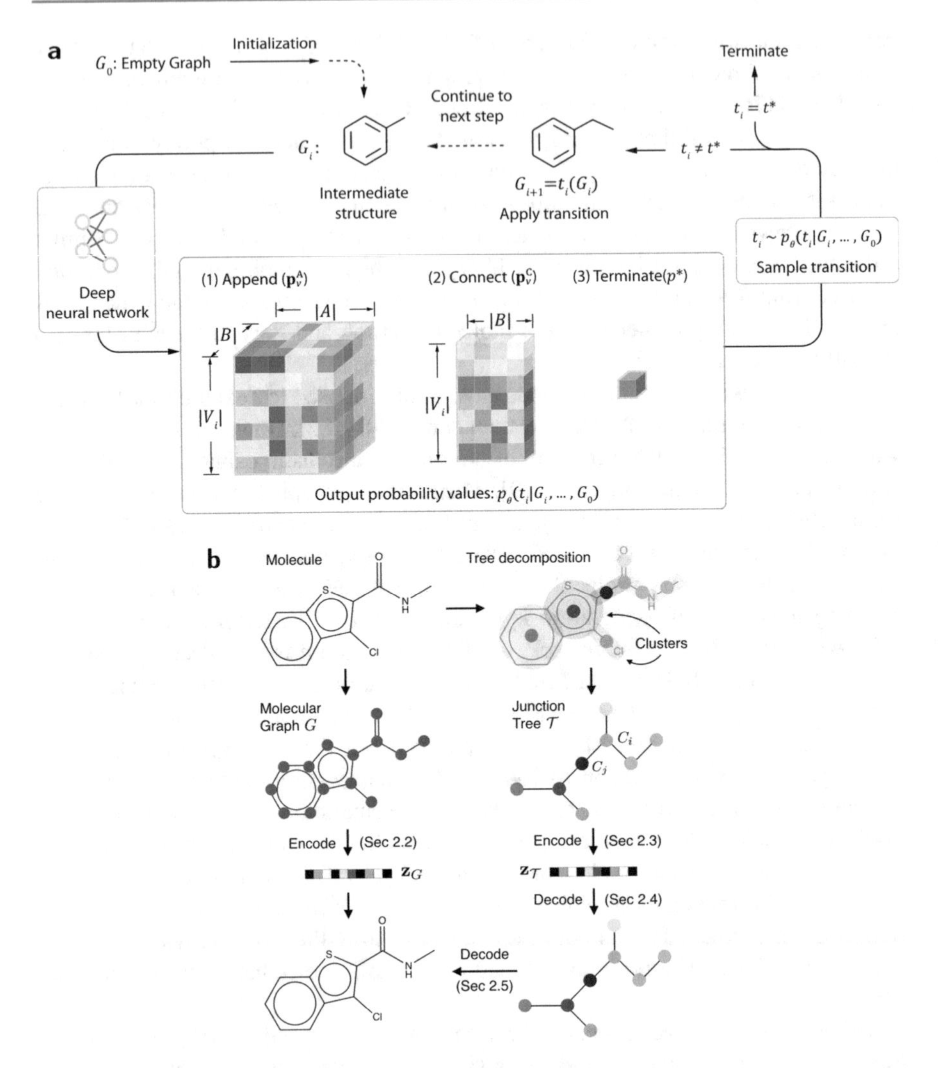

Fig. 21.4 Generative models for molecules using graphs. (**a**) Decision process for sequential generation of molecules from Li et al. [177]. (**b**) Junction Tree VAE for molecular graphs [168]. Figures reproduced from [168, 177] with the permission of the authors

generated compounds. The optimization process forgoes pre-training and allows flexibility in the choice of the importances for each objectives. As a follow-up to this work, the same group reports the usage of this generation scheme as a decoder in a RL-enhanced VAE for molecules [189].

In line with the usage of sequences of actions to create graphs, several groups have been working on different ways to represent and generate graphs through sequences. One approach is to split a graph in permutation-invariant N-gram path sets [190], in analogy with NLP with atoms as words and molecules as sentences.

This representation performs competitively with message-passing neural networks in classification and regression tasks. The combination of strings and graph methods is also seen in the work of Krenn et al. [191], which developed a sequence representation for general-purpose graphs. Their scheme shows high robustness against mutations in sequences and outperforms other representations (including SMILES strings) in terms of diversity, validity, and reconstruction accuracy when employed in sequence-based VAEs.

The adversarial generation of graphs is still very incipient, and few models of GANs with graphs have been demonstrated [192–194]. De Cao and Kipf [195] demonstrated MolGAN, a GAN trained with RL for generating molecular graphs, but their system is too prone to mode collapse. The output structure can be made discrete by differentiable processes such as Gumbel-softmax [196, 197], but balancing the adversarial training with molecular constraints requires more study. Pölsterl and Wachinger [198] build on MolGAN by adding an adversarial training to avoid calculating the reconstruction loss and extending the graph isomorphism network [199] to multigraphs. Further improvements include the approach from Maziarka et al. [200], which relies on the latent space of a pretrained JT-VAE to produce and optimize molecules, and the work of Fan and Huang [201], which aims to generate labeled graphs.

While the combination of DL with graph theory and molecular design seems promising, large room for improvement is available in the field of graph generation. Outputting an arbitrary graph is still an open problem and scalability to larger graphs is still an issue [136]. Computing graph isomorphism is a class-NP problem, and the measure of similarity between two graphs usually resorts to expensive kernels or edit distances [202], as are other problems with reconstruction, ordering, and so on [203]. In some cases, a distance metric can be defined for such data structures [204, 205] or a set of networks can be trained to recognize similarity patterns within graphs [206]. Furthermore, adding attention to graphs could also help in classification tasks [207] or in the extraction of structure–property relationships [208], and specifying grammar rules for graph reconstruction may lead to improved results in molecular validity and stereochemistry [209].

21.3 Challenges and Outlook for Generative Models

The use of deep generative models is a powerful approach for teaching computers to observe and understand the real world. Far from being just a big-data crunching tool, DL algorithms can provide insights that augment human creativity [122]. Completely evaluating a generative model is difficult [210], since we lack an expression for the statistical distribution being learned. Nevertheless, by approximating real-life data with an appropriate representation, we are embedding intuition in the machine's understanding. In a sense, this is what we do, as human beings, when formulating theoretical concepts on chemistry, physics, and many other fields of study. Furthering our limited ability to probe the inner workings of deep neural networks will allow us to transform learned embeddings into logical rules.

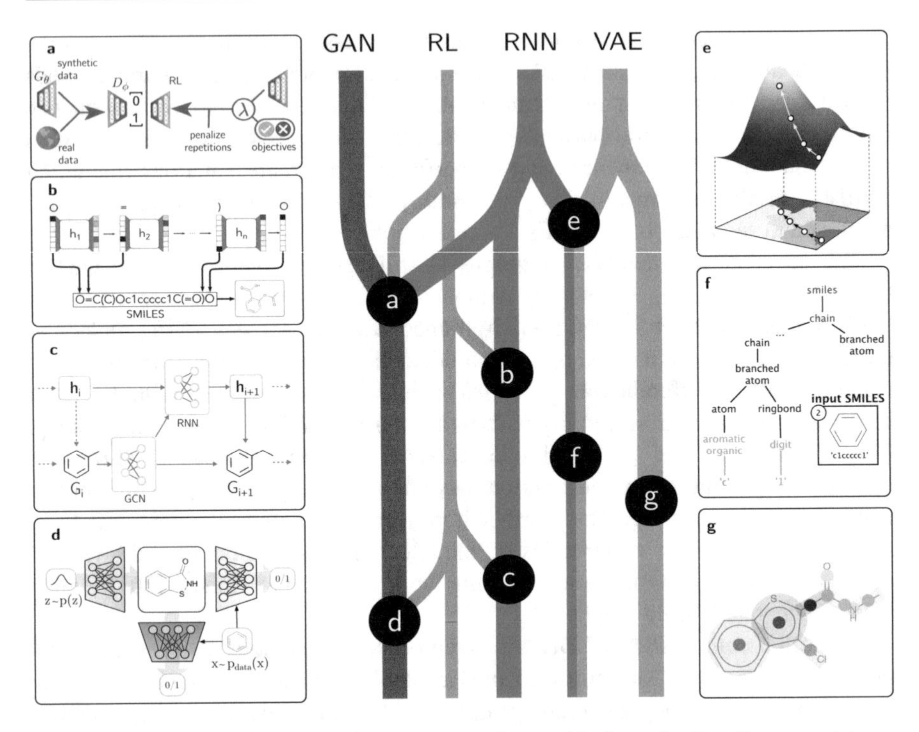

Fig. 21.5 Summary and timeline of current generative models for molecules. Newer models are located in the bottom of the diagram. Figures reproduced from [123, 154, 165, 168, 195, 203]. (**a**) SMILES generators with GANs papers. (**b**) SMILES generators with RNNs papers. (**c**) Graph generators with RNNs papers. (**d**) Graph generators with GANs papers. (**e**) SMILES VAE papers. (**f**) NLP + SMILES VAE papers. (**g**) Graph VAE papers

In the field of chemical design, generative models are still in their infancy (see timeline summary in Fig. 21.5). While many achievements have been reported for such models, all of them share many challenges before a "closed loop" approach can be effectively implemented. Some of the trials are still inherent to all generative models: the generalization capability of a model, its power to make inferences on the real world, and the capacity to bring novelty to it. In the chemical space, originality can be translated as the breadth and quality of possible molecules that the model can generate. To push forward the development of new technologies, we want our generative models to explore further regions of the chemical space in search of new solutions to current problems and extrapolate the training set, avoiding mode collapses or naïve interpolations. At the same time, we want it to capture rules inherent to the synthetically accessible space. Finally, we want to critically evaluate the performance of such models. Several benchmarks are being developed to assess the evolution of chemical generative models, providing quantitative comparisons beyond the mere prediction of solubility or drug-likeness [211–214].

The ease of navigation throughout the chemical space alone is not enough to determine a good model, however. Tailoring the generation of valid molecules for certain applications such as drug design [142] is also an important task. It reflects how well a generative model focuses on the structure–property relationships for certain applications. This interpretation leads to even more powerful understandings of chemistry and is closely tied to Gaussian processes [123], Bayesian optimization [215], and virtual screening.

In the generation process, outputting an arbitrary molecule is still an open problem and is closely conditioned to the representation. While SMILES have been demonstrated useful to represent molecules, graphs are able to convey real chemical features in it, which is useful for learning properties from structures. However, three-dimensional atomic coordinates should be considered for decoding as well. Recent works are going well beyond the connectivity of a molecule to provide equilibrium geometries of molecules using generative models [216–220]. This is crucial to bypass expensive sampling of low-energy configurations from the potential energy surface of molecules. We should expect advances not only on decoding and generating graphs from latent codes, but also in invertible molecular representations in terms of sequences, connectivity, and spatial arrangement.

Finally, as the field of generative models advances, we should expect even more exciting models to design molecules. The normalizing-flow based Boltzmann Generator [217] and GraphNVP [221] are examples of models based on more recent strategies. Furthermore, the use of generative models to understand molecules in an unsupervised way advances along with the inverse design, from coarse-graining [222, 223] and synthesizability of small molecules [224, 225] to genetic variation in complex biomolecules [226].

In summary, generative models hold promise to revolutionize the chemical design. Not only they allow optimizations or learn directly from data, but also bypass the necessity of a human supervising the generation of materials. Facing the challenges among these models is essential for accelerating the discovery cycle of new materials and, perhaps, improvement of the human understanding of the nature.

Acknowledgments D.S.-K. acknowledges the MIT Nicole and Ingo Wender Fellowship and the MIT Robert Rose Presidential Fellowship for financial support. R.G.-B. thanks MIT DMSE and Toyota Faculty Chair for support.

References

1. D.P. Tabor, L.M. Roch, S.K. Saikin, C. Kreisbeck, D. Sheberla, J.H. Montoya, S. Dwaraknath, M. Aykol, C. Ortiz, H. Tribukait, C. Amador-Bedolla, C.J. Brabec, B. Maruyama, K.A. Persson, A. Aspuru-Guzik, Nat. Rev. Mater. **3**(5), 5 (2018)
2. R.F. Gibson, Compos. Struct. **92**(12), 2793 (2010)
3. H. Chen, O. Engkvist, Y. Wang, M. Olivecrona, T. Blaschke, Drug Discov. Today **23**(6), 1241 (2018)
4. J. A. DiMasi, H. G. Grabowski, R. W. Hansen, J. Health Econ. **47**, 20 (2016)
5. B. K. Shoichet, Nature **432**(7019), 862 (2004)
6. J. Greeley, T.F. Jaramillo, J. Bonde, I. Chorkendorff, J.K. Nørskov, Nat. Mater. **5**(11), 909 (2006)
7. S.V. Alapati, J.K. Johnson, D.S. Sholl, J. Phys. Chem. B **110**(17), 8769 (2006)

8. W. Setyawan, R.M. Gaume, S. Lam, R.S. Feigelson, S. Curtarolo, ACS Comb. Sci. **13**(4), 382 (2011)
9. S. Subramaniam, M. Mehrotra, D. Gupta, Bioinformation **3**(1), 14 (2008)
10. R. Armiento, B. Kozinsky, M. Fornari, G. Ceder, Phys. Rev. B **84**(1) (2011)
11. A. Jain, G. Hautier, C.J. Moore, S.P. Ong, C.C. Fischer, T. Mueller, K.A. Persson, G. Ceder, Comput. Mater. Sci. **50**(8), 2295 (2011)
12. S. Curtarolo, G.L.W. Hart, M.B. Nardelli, N. Mingo, S. Sanvito, O. Levy, Nat. Mater. **12**(3), 191 (2013)
13. E.O. Pyzer-Knapp, C. Suh, R. Gómez-Bombarelli, J. Aguilera-Iparraguirre, A.A.A. Aspuru-Guzik, R. Gomez-Bombarelli, J. Aguilera-Iparraguirre, A.A.A. Aspuru-Guzik, D.R. Clarke, Annu. Rev. Mater. Res. **45**(1), 195 (2015)
14. R. Gómez-Bombarelli, J. Aguilera-Iparraguirre, T.D. Hirzel, D. Duvenaud, D. Maclaurin, M.A. Blood-Forsythe, H.S. Chae, M. Einzinger, D.-G. Ha, T. Wu, G. Markopoulos, S. Jeon, H. Kang, H. Miyazaki, M. Numata, S. Kim, W. Huang, S.I. Hong, M. Baldo, R.P. Adams, A. Aspuru-Guzik, Nat. Mater. **15**(10), 1120 (2016)
15. D. Morgan, G. Ceder, S. Curtarolo, Meas. Sci. Technol. **16**(1), 296 (2004)
16. C. Ortiz, O. Eriksson, M. Klintenberg, Comput. Mater. Sci. **44**(4), 1042 (2009)
17. L. Yu, A. Zunger, Phys. Rev. Lett. **108**(6) (2012)
18. K. Yang, W. Setyawan, S. Wang, M.B. Nardelli, S. Curtarolo, Nat. Mater. **11**(7), 614 (2012)
19. L.-C. Lin, A.H. Berger, R.L. Martin, J. Kim, J.A. Swisher, K. Jariwala, C.H. Rycroft, A.S. Bhown, M.W. Deem, M. Haranczyk, B. Smit, Nat. Mater. **11**(7), 633 (2012)
20. N. Mounet, M. Gibertini, P. Schwaller, D. Campi, A. Merkys, A. Marrazzo, T. Sohier, I.E. Castelli, A. Cepellotti, G. Pizzi, et al., Nat. Nanotechnol. **13**(3), 246 (2018)
21. R. Potyrailo, K. Rajan, K. Stoewe, I. Takeuchi, B. Chisholm, H. Lam, ACS Comb. Sci. **13**(6), 579 (2011)
22. A. Jain, Y. Shin, K.A. Persson, Nat. Rev. Mater. **1**(1) (2016)
23. National Science and Technology Council (US), *Materials Genome Initiative for Global Competitiveness* (Executive Office of the President, National Science and Technology Council, Washington, 2011)
24. S. Curtarolo, W. Setyawan, S. Wang, J. Xue, K. Yang, R.H. Taylor, L.J. Nelson, G.L. Hart, S. Sanvito, M. Buongiorno-Nardelli, N. Mingo, O. Levy, Comput. Mater. Sci. **58**, 227 (2012)
25. C.E. Calderon, J.J. Plata, C. Toher, C. Oses, O. Levy, M. Fornari, A. Natan, M.J. Mehl, G. Hart, M.B. Nardelli, S. Curtarolo, Comput. Mater. Sci. **108**, 233 (2015)
26. A. Jain, S.P. Ong, G. Hautier, W. Chen, W.D. Richards, S. Dacek, S. Cholia, D. Gunter, D. Skinner, G. Ceder, K.A. Persson, APL Mater. **1**(1), 011002 (2013)
27. J.E. Saal, S. Kirklin, M. Aykol, B. Meredig, C. Wolverton, JOM **65**(11), 1501 (2013)
28. B. Sanchez-Lengeling, A. Aspuru-Guzik, Science **361**(6400), 360 (2018)
29. A. Zunger, Nat. Rev. Chem. **2**(4), 0121 (2018)
30. P.G. Polishchuk, T.I. Madzhidov, A. Varnek, J. Comput. Aided Mol. Des. **27**(8), 675 (2013)
31. A.M. Virshup, J. Contreras-García, P. Wipf, W. Yang, D.N. Beratan, J. Am. Chem. Soc. **135**(19), 7296 (2013)
32. K.G. Joback, Designing Molecules Possessing Desired Physical Property Values. Ph.D. Thesis, Massachusetts Institute of Technology, 1989
33. C. Kuhn, D.N. Beratan, J. Phys. Chem. **100**(25), 10595 (1996)
34. D.J. Wales, H.A. Scheraga, Science **285**(5432), 1368 (1999)
35. J. Schön, M. Jansen, Z. Kristallogr. Cryst. Mater. **216**(6) (2001)
36. R. Gani, E. Brignole, Fluid Phase Equilib. **13**, 331 (1983)
37. S.R. Marder, D.N. Beratan, L.T. Cheng, Science **252**(5002), 103 (1991)
38. P.M. Holmblad, J.H. Larsen, I. Chorkendorff, L.P. Nielsen, F. Besenbacher, I. Stensgaard, E. Lægsgaard, P. Kratzer, B. Hammer, J.K. Nøskov, Catal. Lett. **40**(3–4), 131 (1996)
39. O. Sigmund, S. Torquato, J. Mech. Phys. Solids **45**(6), 1037 (1997)
40. C. Wolverton, A. Zunger, B. Schönfeld, Solid State Commun. **101**(7), 519 (1997)
41. N. Metropolis, A.W. Rosenbluth, M.N. Rosenbluth, A.H. Teller, E. Teller, J. Chem. Phys. **21**(6), 1087 (1953)

42. R. Kaplow, T.A. Rowe, B.L. Averbach, Phys. Rev. **168**(3), 1068 (1968)
43. V. Gerold, J. Kern, Acta Metall. **35**(2), 393 (1987)
44. R.L. McGreevy, L. Pusztai, Mol. Simul. **1**(6), 359 (1988)
45. A. Franceschetti, A. Zunger, Nature **402**(6757), 60 (1999)
46. J.H. Holland, *Adaptation in Natural and Artificial Systems* (MIT Press, Cambridge, 1992)
47. R. Judson, E. Jaeger, A. Treasurywala, M. Peterson, J. Comput. Chem. **14**(11), 1407 (1993)
48. R.C. Glen, A.W.R. Payne, J. Comput. Aided Mol. Des. **9**(2), 181 (1995)
49. V. Venkatasubramanian, K. Chan, J. Caruthers, Comput. Chem. Eng. **18**(9), 833 (1994)
50. V. Venkatasubramanian, K. Chan, J.M. Caruthers, J. Chem. Inf. Model. **35**(2), 188 (1995)
51. A.L. Parrill, Drug Discov. Today **1**(12), 514 (1996)
52. G. Schneider, M.-L. Lee, M. Stahl, P. Schneider, J. Comput. Aided Mol. Des. **14**(5), 487 (2000)
53. D.B. Gordon, S.L. Mayo, Structure **7**(9), 1089 (1999)
54. M.T. Reetz, Proc. Natl. Acad. Sci. **101**(16), 5716 (2004)
55. D. Wolf, O. Buyevskaya, M. Baerns, Appl. Catal. A **200**(1–2), 63 (2000)
56. G.H. Jóhannesson, T. Bligaard, A.V. Ruban, H.L. Skriver, K.W. Jacobsen, J.K. Nørskov, Phys. Rev. Lett. **88**(25) (2002)
57. S.V. Dudiy, A. Zunger, Phys. Rev. Lett. **97**(4) (2006)
58. P. Piquini, P.A. Graf, A. Zunger, Phys. Rev. Lett. **100**(18) (2008)
59. M. d'Avezac, J.-W. Luo, T. Chanier, A. Zunger, Phys. Rev. Lett. **108**(2) (2012)
60. L. Zhang, J.-W. Luo, A. Saraiva, B. Koiller, A. Zunger, Nat. Commun. **4**(1) (2013)
61. L. Yu, R.S. Kokenyesi, D.A. Keszler, A. Zunger, Adv. Energy Mater. **3**(1), 43 (2012)
62. T. Brodmeier, E. Pretsch, J. Comput. Chem. **15**(6), 588 (1994)
63. S.M. Woodley, P.D. Battle, J.D. Gale, C.R.A. Catlow, Phys. Chem. Chem. Phys. **1**(10), 2535 (1999)
64. C.W. Glass, A.R. Oganov, N. Hansen, Comput. Phys. Commun. **175**(11–12), 713 (2006)
65. A.R. Oganov, C.W. Glass, J. Chem. Phys. **124**(24), 244704 (2006)
66. N.S. Froemming, G. Henkelman, J. Chem. Phys. **131**(23), 234103 (2009)
67. L.B. Vilhelmsen, B. Hammer, J. Chem. Phys. **141**(4), 044711 (2014)
68. G.L.W. Hart, V. Blum, M.J. Walorski, A. Zunger, Nat. Mater. **4**(5), 391 (2005)
69. V. Blum, G.L.W. Hart, M.J. Walorski, A. Zunger, Phys. Rev. B **72**(16) (2005)
70. C. Rupakheti, A. Virshup, W. Yang, D.N. Beratan, J. Chem. Inf. Model. **55**(3), 529 (2015)
71. J.L. Reymond, Acc. Chem. Res. **48**(3), 722 (2015)
72. T.C. Le, D.A. Winkler, Chem. Rev. **116**(10), 6107 (2016)
73. P.C. Jennings, S. Lysgaard, J.S. Hummelshøj, T. Vegge, T. Bligaard, npj Comput. Mater. **5**(1) (2019)
74. O.A. von Lilienfeld, R.D. Lins, U. Rothlisberger, Phys. Rev. Lett. **95**(15) (2005)
75. V. Marcon, O.A. von Lilienfeld, D. Andrienko, J. Chem. Phys. **127**(6), 064305 (2007)
76. M. Wang, X. Hu, D.N. Beratan, W. Yang, J. Am. Chem. Soc. **128**(10), 3228 (2006)
77. P. Hohenberg, W. Kohn, Phys. Rev. **136**(3B), B864 (1964)
78. D. Xiao, W. Yang, D.N. Beratan, J. Chem. Phys. **129**(4), 044106 (2008)
79. D. Balamurugan, W. Yang, D.N. Beratan, J. Chem. Phys. **129**(17), 174105 (2008)
80. S. Keinan, X. Hu, D.N. Beratan, W. Yang, J. Phys. Chem. A **111**(1), 176 (2007)
81. X. Hu, D.N. Beratan, W. Yang, J. Chem. Phys. **129**(6), 064102 (2008)
82. F.D. Vleeschouwer, W. Yang, D.N. Beratan, P. Geerlings, F.D. Proft, Phys. Chem. Chem. Phys. **14**(46), 16002 (2012)
83. G.E. Hinton, T.J. Sejnowski, in *Parallel Distributed Processing: Explorations in the Microstructure of Cognition, Vol. 1*, ed. by D.E. Rumelhart, J.L. McClelland, C. PDP Research Group (MIT Press, Cambridge, 1986), pp. 282–317
84. G.E. Hinton, T.J. Sejnowski, in *Proceedings of the IEEE Conference on Computer Vision and Pattern Recognition* (1983)
85. P. Smolensky, in *Parallel Distributed Processing: Explorations in the Microstructure of Cognition, Vol. 1*, ed. by D.E. Rumelhart, J.L. McClelland, C. PDP Research Group (MIT Press, Cambridge, 1986), pp. 194–281

86. G.E. Hinton, S. Osindero, Y.-W. Teh, Neural Comput. **18**(7), 1527 (2006)
87. R. Salakhutdinov, G. Hinton, in *Proceedings of the Twelth International Conference on Artificial Intelligence and Statistics*. PMLR, vol. 5, 2009, pp. 448–455
88. I. Goodfellow, Y. Bengio, A. Courville, *Deep Learning* (MIT Press, Cambridge, 2016)
89. T. Karras, T. Aila, S. Laine, J. Lehtinen (2017). arXiv:1710.10196
90. I.J. Goodfellow, J. Pouget-Abadie, M. Mirza, B. Xu, D. Warde-Farley, S. Ozair, A. Courville, Y. Bengio (2014). arXiv:1406.2661
91. L.A. Gatys, A.S. Ecker, M. Bethge (2015). arXiv:1508.06576
92. C. Ledig, L. Theis, F. Huszar, J. Caballero, A. Cunningham, A. Acosta, A. Aitken, A. Tejani, J. Totz, Z. Wang, W. Shi (2016). arXiv:1609.04802
93. S.R. Bowman, L. Vilnis, O. Vinyals, A.M. Dai, R. Jozefowicz, S. Bengio, G. Brain (2015), pp. 1–15. arXiv:1511.06349
94. K. Xu, J. Ba, R. Kiros, K. Cho, A. Courville, R. Salakhutdinov, R. Zemel, Y. Bengio (2015). arXiv:1502.03044
95. S. Mehri, K. Kumar, I. Gulrajani, R. Kumar, S. Jain, J. Sotelo, A. Courville, Y. Bengio (2016). arXiv:1612.07837
96. A. van den Oord, S. Dieleman, H. Zen, K. Simonyan, O. Vinyals, A. Graves, N. Kalchbrenner, A. Senior, K. Kavukcuoglu (2016). arXiv:1609.03499
97. C. Vondrick, H. Pirsiavash, A. Torralba (2016). arXiv:1609.02612
98. A. Radford, L. Metz, S. Chintala (2015). arXiv:1511.06434
99. J. Engel, M. Hoffman, A. Roberts (2017). arXiv:1711.05772
100. Y. LeCun, Y. Bengio, G. Hinton, Nature **521**(7553), 436 (2015)
101. D.P. Kingma, M. Welling (2013). arXiv:1312.6114
102. M. Arjovsky, S. Chintala, L. Bottou (2017). arXiv:1701.07875
103. I. Tolstikhin, O. Bousquet, S. Gelly, B. Schölkopf, B. Schoelkopf (2017). arXiv:1711.01558
104. P.K. Rubenstein, B. Schoelkopf, I. Tolstikhin, B. Schölkopf, I. Tolstikhin (2018). arXiv:1802.03761
105. M. Mirza, S. Osindero (2014). arXiv:1411.1784
106. X. Chen, Y. Duan, R. Houthooft, J. Schulman, I. Sutskever, P. Abbeel (2016). arXiv:1606.03657
107. T. Che, Y. Li, A.P. Jacob, Y. Bengio, W. Li (2016). arXiv:1612.02136
108. A. Odena, C. Olah, J. Shlens (2016). arXiv:1610.09585
109. X. Mao, Q. Li, H. Xie, R. Y. K. Lau, Z. Wang, S.P. Smolley (2016). arXiv:1611.04076
110. R.D. Hjelm, A.P. Jacob, T. Che, A. Trischler, K. Cho, Y. Bengio (2017). arXiv:1702.08431
111. J. Zhao, M. Mathieu, Y. LeCun (2016). arXiv:1609.03126
112. S. Nowozin, B. Cseke, R. Tomioka (2016). arXiv:1606.00709
113. J. Donahue, P. Krähenbühl, T. Darrell (2016). arXiv:1605.09782
114. D. Berthelot, T. Schumm, L. Metz (2017). arXiv:1703.10717
115. I. Gulrajani, F. Ahmed, M. Arjovsky, V. Dumoulin, A. Courville (2017). arXiv:1704.00028
116. Z. Yi, H. Zhang, P. Tan, M. Gong (2017). arXiv:1704.02510
117. M. Lucic, K. Kurach, M. Michalski, S. Gelly, O. Bousquet (2017). arXiv:1711.10337
118. A. van den Oord, N. Kalchbrenner, K. Kavukcuoglu (2016). arXiv:1601.06759
119. A. van den Oord, N. Kalchbrenner, O. Vinyals, L. Espeholt, A. Graves, K. Kavukcuoglu (2016). arXiv:1606.05328
120. T. Salimans, A. Karpathy, X. Chen, D.P. Kingma (2017). arXiv:1701.05517
121. N. Kalchbrenner, A. van den Oord, K. Simonyan, I. Danihelka, O. Vinyals, A. Graves, K. Kavukcuoglu (2016). arXiv:1610.00527
122. N. Kalchbrenner, L. Espeholt, K. Simonyan, A. van den Oord, A. Graves, K. Kavukcuoglu (2016). arXiv:1610.10099
123. R. Gómez-Bombarelli, J.N. Wei, D. Duvenaud, J.M. Hernández-Lobato, B. Sánchez-Lengeling, D. Sheberla, J. Aguilera-Iparraguirre, T.D. Hirzel, R.P. Adams, A. Aspuru-Guzik, ACS Cent. Sci. **4**(2), 268 (2018)
124. D.R. Hartree, Math. Proc. Cambridge Philos. Soc. **24**(01), 89 (1928)
125. V. Fock, Z. Phys. A At. Nucl. **61**(1–2), 126 (1930)

126. W. Kohn, L.J. Sham, Phys. Rev. **140**(4A), A1133 (1965)
127. R. Todeschini, V. Consonni, *Handbook of Molecular Descriptors*. Methods and Principles in Medicinal Chemistry (Wiley-VCH, Weinheim, 2000)
128. D. Rogers, M. Hahn, J. Chem. Inf. Model. **50**(5), 742 (2010)
129. K. Hansen, F. Biegler, R. Ramakrishnan, W. Pronobis, O.A. von Lilienfeld, K.-R. Müller, A. Tkatchenko, J. Phys. Chem. Lett. **6**(12), 2326 (2015)
130. M. Rupp, A. Tkatchenko, K.-R. Müller, O.A. von Lilienfeld, Phys. Rev. Lett. **108**(5), 058301 (2012)
131. K.T. Schütt, F. Arbabzadah, S. Chmiela, K.R. Müller, A. Tkatchenko, Nat. Commun. **8**, 13890 (2017)
132. H. Huo, M. Rupp (2017). arXiv:1704.06439
133. D. Weininger, J. Chem. Inf. Model. **28**(1), 31 (1988)
134. S. Kearnes, K. McCloskey, M. Berndl, V. Pande, P. Riley, J. Comput. Aided Mol. Des. **30**(8), 595 (2016)
135. D.K. Duvenaud, D. Maclaurin, J. Aguilera-Iparraguirre, R. Gómez-Bombarelli, T. Hirzel, A. Aspuru-Guzik, R.P. Adams, *Advances in Neural Information Processing Systems* (2015), pp. 2215–2223
136. J. Gilmer, S.S. Schoenholz, P.F. Riley, O. Vinyals, G.E. Dahl (2017). arXiv:1704.01212
137. S. Hochreiter, J. Schmidhuber, Neural Comput. **9**(8), 1735 (1997)
138. J. Chung, C. Gulcehre, K. Cho, Y. Bengio (2014). arXiv:1412.3555
139. M. Popova, O. Isayev, A. Tropsha, Sci. Adv. **4**(7), eaap7885 (2018)
140. H. Ikebata, K. Hongo, T. Isomura, R. Maezono, R. Yoshida, J. Comput. Aided Mol. Des. **31**(4), 379 (2017)
141. P. Ertl, R. Lewis, E. Martin, V. Polyakov (2017). arXiv:1712.07449
142. M.H.S. Segler, T. Kogej, C. Tyrchan, M.P. Waller, ACS Cent. Sci. **4**(1), 120 (2018)
143. A. Gupta, A.T. Müller, B.J.H. Huisman, J.A. Fuchs, P. Schneider, G. Schneider, Mol. Inf. **37**(1–2), 1700111 (2017)
144. T. Ching, D.S. Himmelstein, B.K. Beaulieu-Jones, A.A. Kalinin, B.T. Do, G.P. Way, E. Ferrero, P.-M. Agapow, M. Zietz, M.M. Hoffman, W. Xie, G.L. Rosen, B.J. Lengerich, J. Israeli, J. Lanchantin, S. Woloszynek, A.E. Carpenter, A. Shrikumar, J. Xu, E.M. Cofer, C.A. Lavender, S.C. Turaga, A.M. Alexandari, Z. Lu, D.J. Harris, D. DeCaprio, Y. Qi, A. Kundaje, Y. Peng, L.K. Wiley, M.H.S. Segler, S.M. Boca, S.J. Swamidass, A. Huang, A. Gitter, C.S. Greene, J. R. Soc. Interface **15**(141), 20170387 (2018)
145. N. Jaques, S. Gu, D. Bahdanau, J.M. Hernández-Lobato, R.E. Turner, D. Eck, in *Proceedings of the 34th International Conference on Machine Learning Research*, vol. 70, ed. by D. Precup, Y.W. Teh (PMLR, International Convention Centre, Sydney, 2017), pp. 1645–1654
146. M. Olivecrona, T. Blaschke, O. Engkvist, H. Chen, J. Cheminf. **9**(1), 48 (2017)
147. S. Kang, K. Cho, J. Chem. Inf. Model. **59**(1), 43 (2018)
148. B. Sattarov, I.I. Baskin, D. Horvath, G. Marcou, E.J. Bjerrum, A. Varnek, J. Chem. Inf. Model. **59**(3), 1182 (2019)
149. S. Sinai, E. Kelsic, G.M. Church, M.A. Nowak (2017), pp. 1–6. arXiv:1712.03346
150. S. Kwon, S. Yoon, in *Proceedings of the 8th ACM International Conference on Bioinformatics, Computational Biology, and Health Informatics - ACM-BCB '17* (ACM Press, New York, 2017), pp. 203–212
151. K. Kim, S. Kang, J. Yoo, Y. Kwon, Y. Nam, D. Lee, I. Kim, Y.-S. Choi, Y. Jung, S. Kim, W.-J. Son, J. Son, H.S. Lee, S. Kim, J. Shin, S. Hwang, npj Comput. Mater. **4**(1) (2018)
152. V. Mallet, C.G. Oliver, N. Moitessier, J. Waldispuhl (2019). arXiv:1905.12033
153. J. Lim, S. Ryu, J.W. Kim, W.Y. Kim, J. Cheminf. **10**(1) (2018)
154. G.L. Guimaraes, B. Sanchez-Lengeling, C. Outeiral, P.L.C. Farias, A. Aspuru-Guzik, C. Outeiral, P.L.C. Farias, A. Aspuru-Guzik (2017). arXiv:1705.10843
155. B. Sanchez-Lengeling, C. Outeiral, G.L.L. Guimaraes, A.A. Aspuru-Guzik (2017), pp. 1–18. chemRxiv:5309668
156. E. Putin, A. Asadulaev, Y. Ivanenkov, V. Aladinskiy, B. Sanchez-Lengeling, A. Aspuru-Guzik, A. Zhavoronkov, J. Chem. Inf. Model. **58**(6), 1194 (2018)

157. O. Mendez-Lucio, B. Baillif, D.-A. Clevert, D. Rouquié, J. Wichard (2018). chemrXiv:7294388
158. N. Killoran, L.J. Lee, A. Delong, D. Duvenaud, B.J. Frey (2017). arXiv:1712.06148
159. A. Kadurin, S. Nikolenko, K. Khrabrov, A. Aliper, A. Zhavoronkov, Mol. Pharm. **14**(9), 3098 (2017)
160. A. Kadurin, A. Aliper, A. Kazennov, P. Mamoshina, Q. Vanhaelen, K. Khrabrov, A. Zhavoronkov, Oncotarget **8**(7), 10883 (2017)
161. T. Blaschke, M. Olivecrona, O. Engkvist, J. Bajorath, H. Chen, Mol. Inf. **37**(1–2), 1700123 (2018)
162. D. Polykovskiy, A. Zhebrak, D. Vetrov, Y. Ivanenkov, V. Aladinskiy, P. Mamoshina, M. Bozdaganyan, A. Aliper, A. Zhavoronkov, A. Kadurin, Mol. Pharm. **15**(10), 4398 (2018)
163. J. Mueller, D. Gifford, T. Jaakkola, in *Proceedings of the 34th International Conference on Machine Learning Research*, vol. 70, ed. by D. Precup, Y.W. Teh (PMLR, International Convention Centre, Sydney, 2017), pp. 2536–2544
164. J.P. Janet, L. Chan, H.J. Kulik, J. Phys. Chem. Lett. **9**(5), 1064 (2018)
165. M.J. Kusner, B. Paige, J.M. Hernández-Lobato (2017). arXiv:1703.01925
166. H. Dai, Y. Tian, B. Dai, S. Skiena, L. Song (2018). arXiv:1802.08786
167. R. Winter, F. Montanari, F. Noé, D.-A. Clevert, Chem. Sci. **10**(6), 1692 (2019)
168. W. Jin, R. Barzilay, T. Jaakkola (2018). arXiv:1802.04364
169. E.J. Bjerrum (2017). arXiv:1703.07076
170. Z. Alperstein, A. Cherkasov, J.T. Rolfe (2019). arXiv:1905.13343
171. A. Lusci, G. Pollastri, P. Baldi, J. Chem. Inf. Model. **53**(7), 1563 (2013)
172. J. Bruna, W. Zaremba, A. Szlam, Y. LeCun (2013). arXiv:1312.6203
173. C.W. Coley, R. Barzilay, W.H. Green, T.S. Jaakkola, K.F. Jensen, J. Chem. Inf. Model. **57**(8), 1757 (2017)
174. P. Hop, B. Allgood, J. Yu, Mol. Pharmaceutics **15**(10), 4371 (2018)
175. K. Yang, K. Swanson, W. Jin, C. Coley, P. Eiden, H. Gao, A. Guzman-Perez, T. Hopper, B. Kelley, M. Mathea, A. Palmer, V. Settels, T. Jaakkola, K. Jensen, R. Barzilay (2019). arXiv:1904.01561
176. J. You, R. Ying, X. Ren, W.L. Hamilton, J. Leskovec (2018). arXiv:1802.08773
177. Y. Li, L. Zhang, Z. Liu (2018). arXiv:1801.07299
178. T.N. Kipf, M. Welling (2016). arXiv:1611.07308
179. M. Simonovsky, N. Komodakis (2018). arXiv:1802.03480
180. A. Grover, A. Zweig, S. Ermon (2018). arXiv:1803.10459
181. B. Samanta, A. De, N. Ganguly, M. Gomez-Rodriguez (2018). arXiv:1802.05283
182. Q. Liu, M. Allamanis, M. Brockschmidt, A.L. Gaunt (2018). arXiv:1805.09076
183. T. Ma, J. Chen, C. Xiao (2018). arXiv:1809.02630
184. W. Jin, K. Yang, R. Barzilay, T. Jaakkola, *International Conference on Learning Representations* (2019)
185. W. Jin, R. Barzilay, T.S. Jaakkola (2019). chemrXiv:8266745
186. R. Assouel, M. Ahmed, M.H. Segler, A. Saffari, Y. Bengio (2018). arXiv:1811.09766
187. J. Lim, S.-Y. Hwang, S. Kim, S. Moon, W.Y. Kim (2019). arXiv:1905.13639
188. Z. Zhou, S. Kearnes, L. Li, R.N. Zare, P. Riley (2018). arXiv:1810.08678
189. S. Kearnes, L. Li, P. Riley (2019). arXiv:1904.08915
190. S. Liu, T. Chandereng, Y. Liang (2018). arXiv:1806.09206
191. M. Krenn, F. Häse, A. Nigam, P. Friederich, A. Aspuru-Guzik (2019). arXiv:1905.13741
192. X. Guo, L. Wu, L. Zhao (2018). arXiv:1805.09980
193. A. Bojchevski, O. Shchur, D. Zügner, S. Günnemann (2018). arXiv:1803.00816
194. Y. Xiong, Y. Zhang, H. Fu, W. Wang, Y. Zhu, P.S. Yu, *Database Systems for Advanced Applications* (Springer International Publishing, Cham, 2019), pp. 536–552
195. N. De Cao, T. Kipf (2018). arXiv:1805.11973
196. E. Jang, S. Gu, B. Poole (2016). arXiv:1611.01144
197. M.J. Kusner, J.M. Hernández-Lobato (2016). arXiv:1611.04051
198. S. Pölsterl, C. Wachinger (2019). arXiv:1905.10310

199. K. Xu, W. Hu, J. Leskovec, S. Jegelka (2018). arXiv:1810.00826
200. Ł. Maziarka, A. Pocha, J. Kaczmarczyk, K. Rataj, M. Warchoł (2019). arXiv:1902.02119
201. S. Fan, B. Huang (2019). arXiv:1906.03220
202. M. Neuhaus, H. Bunke, *Bridging the Gap Between Graph Edit Distance and Kernel Machines* (World Scientific Publishing, River Edge, 2007)
203. Y. Li, O. Vinyals, C. Dyer, R. Pascanu, P. Battaglia (2018). arXiv:1803.03324
204. T.A. Schieber, L. Carpi, A. Díaz-Guilera, P.M. Pardalos, C. Masoller, M.G. Ravetti, Nat. Commun. **8**, 13928 (2017)
205. H. Choi, H. Lee, Y. Shen, Y. Shi (2018). arXiv:1807.00252
206. S.I. Ktena, S. Parisot, E. Ferrante, M. Rajchl, M. Lee, B. Glocker, D. Rueckert (2017). arXiv:1703.02161
207. K. Do, T. Tran, T. Nguyen, S. Venkatesh (2018). arXiv:1804.00293
208. S. Ryu, J. Lim, W.Y. Kim (2018). arXiv:1805.10988
209. H. Kajino (2018). arXiv:1809.02745
210. L. Theis, A. van den Oord, M. Bethge (2015). arXiv:1511.01844
211. K. Preuer, P. Renz, T. Unterthiner, S. Hochreiter, G. Klambauer, J. Chem. Inf. Model. **58**(9), 1736 (2018)
212. D. Polykovskiy, A. Zhebrak, B. Sanchez-Lengeling, S. Golovanov, O. Tatanov, S. Belyaev, R. Kurbanov, A. Artamonov, V. Aladinskiy, M. Veselov, A. Kadurin, S. Nikolenko, A. Aspuru-Guzik, A. Zhavoronkov (2018). arXiv:1811.12823
213. Z. Wu, B. Ramsundar, E.N. Feinberg, J. Gomes, C. Geniesse, A.S. Pappu, K. Leswing, V. Pande, Chem. Sci. **9**(2), 513 (2018)
214. N. Brown, M. Fiscato, M.H. Segler, A.C. Vaucher, J. Chem. Inf. Model. **59**(3), 1096 (2019)
215. F. Häse, L.M. Roch, C. Kreisbeck, A. Aspuru-Guzik (2018). arXiv:1801.01469
216. N.W.A. Gebauer, M. Gastegger, K.T. Schütt (2018). arXiv:1810.11347
217. F. Noé, H. Wu (2018). arXiv:1812.01729
218. N.W.A. Gebauer, M. Gastegger, K.T. Schütt (2019). arXiv:1906.00957
219. M.S. Jørgensen, H.L. Mortensen, S.A. Meldgaard, E.L. Kolsbjerg, T.L. Jacobsen, K.H. Sørensen, B. Hammer (2019). arXiv:1902.10501
220. E. Mansimov, O. Mahmood, S. Kang, K. Cho (2019). arXiv:1904.00314
221. K. Madhawa, K. Ishiguro, K. Nakago, M. Abe (2019). arXiv:1905.11600
222. J. Wang, S. Olsson, C. Wehmeyer, A. Perez, N.E. Charron, G. de Fabritiis, F. Noe, C. Clementi (2018). arXiv:1812.01736
223. W. Wang, R. Gómez-Bombarelli (2018). arXiv:1812.02706
224. J. Bradshaw, M.J. Kusner, B. Paige, M.H.S. Segler, J.M. Hernández-Lobato, *International Conference on Learning Representations* (2019)
225. J. Bradshaw, B. Paige, M.J. Kusner, M.H.S. Segler, J.M. Hernández-Lobato (2019). arXiv:1906.05221
226. A.J. Riesselman, J.B. Ingraham, D.S. Marks, Nat. Methods **15**(10), 816 (2018)

Printed in the United States
By Bookmasters